WOODY PLANTS
of the
NORTH
CENTRAL
PLAINS

WOODY PLANTS
of the
NORTH
CENTRAL
PLAINS

by H. A. Stephens

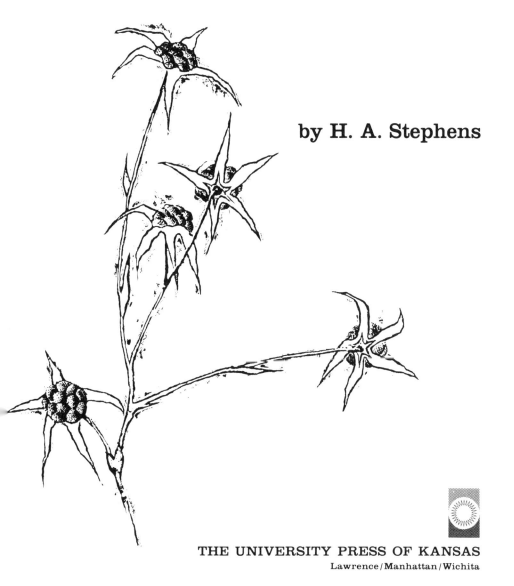

THE UNIVERSITY PRESS OF KANSAS
Lawrence/Manhattan/Wichita

© Copyright 1973 by the University Press of Kansas
Standard Book Number 7006-0107-4
Library of Congress Catalog Card Number 72-97834
Printed in the United States of America

Designed by Fritz Reiber

ACKNOWLEDGMENTS

I am especially indebted to Ronald L. McGregor, a professor of botany at the University of Kansas and curator of the herbarium, for his help and advice and for arranging parts of my regular research so I could be in a certain area during the period of flowering or fruiting of a species. Also, I wish to extend my appreciation to the following curators and professors for permitting me to work in their herbaria: O. A. Stevens and William T. Barker, North Dakota State University; C. A. Taylor, South Dakota State University; Theodore Van Bruggen, University of South Dakota; S. V. Froiland, Augustana College; John F. Davidson, University of Nebraska; T. M. Barkley and Loren C. Anderson, Kansas State University; Howard Reynolds, Fort Hays Kansas State College; Theodore M. Sperry, Kansas State College of Pittsburg; James S. Wilson, Kansas State Teachers College; George J. Goodman, University of Oklahoma; and U. T. Waterfall, Oklahoma State University.

In addition to these, I am grateful to the many laymen and students who have helped in one way or another. I especially wish to thank Ralph Brooks, a student who traveled with me during five summers and several winter trips and helped locate the plants or even to dig them from under the snow so as to get the plants in winter condition.

H. A. (Steve) Stephens

CONTENTS

INTRODUCTION

The purpose of this book is to bring together the present information and to add new information about the woody and the more common suffrutescent plants of the Central Plains. As is well known, authors do not always agree on the correct name for a plant. Therefore, the name used here appears to the author as the best name in light of recent publications and from his own field and laboratory experiences. Some previously separated species are here combined because the material examined does not show sufficient consistency of characters to maintain separation. This inconsistency may be due to the transitional nature of this area, for the book specifically covers the plants of North Dakota, South Dakota, Nebraska, and Kansas, and the term "our area" in the text refers to these states. In order to gain an overall picture of the plants of the area and of the area itself, collections were also made in the surrounding states.

Since the travel was done by camper, laboratory equipment could be carried into the field and, except where noted, all of the descriptions and drawings were made by the author from fresh material in the field. This meant traveling to all parts of the area during all seasons of the year for a period of about eight years.

The drawings on any one page are not in proportion to each other but the measurements are given in the text to provide the comparison of size. In some of the drawings parts of the structure have been removed or omitted in order to show the main organ more clearly. For example, part of a flower may have been removed to show the inner structure, most of the hairs of a pappus omitted so as to show the seed, or the pine needles omitted at the end of a twig in order to bring out the strobile.

The dot maps are by counties although the county lines do not show. Each dot represents at least one specimen seen and identified by the author either in the field or in a herbarium. They are used only to give a general idea of distribution; in no way do they indicate the density of a plant population. The same is true of the range. In general, the range is given as a perimeter of the region in which the species is known, but it may surround large areas where the plant does not exist.

The Prairies and Plains Province is a comparatively open area between the Eastern Deciduous Forest and the Rocky Mountains. As the name implies, it is a rolling or nearly flat area with grass as the dominant cover. This cover varies from east to west; and the taller grasses, *Andropogon* and *Sorghastrum*, occupy a strip along the eastern edge. Along the western edge are the short grasses, *Buchloe* and *Bouteloua*. In between, and without definite border lines, is land with a mixture of the two grass types that is spoken of as the mixed-grass prairie. Scattered throughout the whole region are numerous herbaceous plants with more or less showy flowers. Although in much smaller numbers, the woody plants are well dispersed; the more obvious ones are the trees that border the streams and cover many of the larger hills. It is this latter group that is of concern here, specifically in the states of North and South Dakota, Nebraska, and Kansas.

As might be expected, the main drainage systems extend from the west to the east, the streams originating directly in the mountains or their foothills, passing through the prairie and into the Eastern Deciduous Forest Province. However, in eastern South Dakota the streams run almost directly south, originating in the glaciated prairies, and eventually joining the Missouri River and flowing toward the southeast. In eastern and north central North Dakota the general course is north

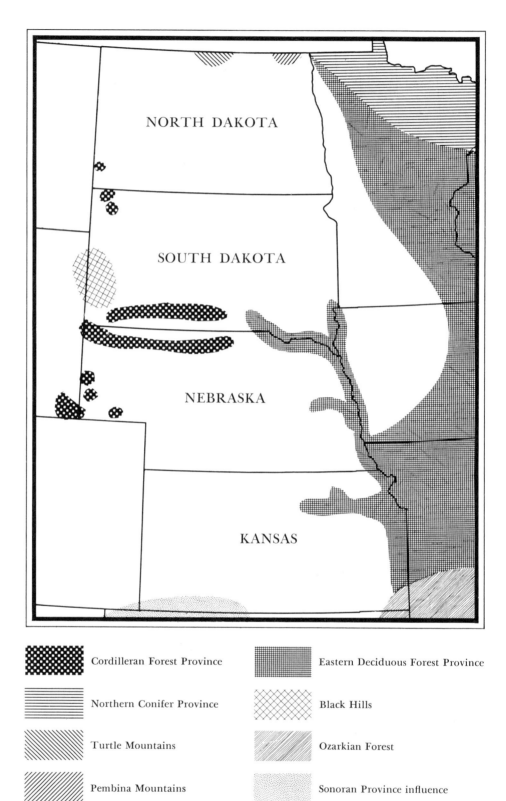

Cordilleran Forest Province	Eastern Deciduous Forest Province
Northern Conifer Province	Black Hills
Turtle Mountains	Ozarkian Forest
Pembina Mountains	Sonoran Province influence

Extensions of Other Floristic Provinces into the Prairie Province

States and Counties of the North Central Plains

into Canada, thus entering the Northern Conifer Province. Since the trees of a prairie area normally follow the course of the streams, this gives an easy access into the prairie for both the woody and the herbaceous plants from three outside vegetation provinces.

Needless to say, there are many other factors which make this access difficult. The general climate and the amount and reliability of rainfall are two of the principal barriers to plant migration. One could hardly expect the mountainous plants of the upper Arkansas River to migrate into western Kansas where the rainfall and humidity are low and the summer temperatures high. Nor would one expect to see the plants of the mountains of western Montana and Wyoming travel along the Missouri River and establish a lush growth in the badlands of North and South Dakota.

The same might be true of the plants entering from the desert areas of the southwest. Nevertheless, some of the plants from all of the bordering vegetative provinces are tolerant enough to establish themselves in the plains. These established plants usually appear in finger-like patterns extending along the streams for some distance out into the prairie. Examples of this could well be *Pinus ponderosa* in western Nebraska and South Dakota; *Opuntia imbricata* and *Prosopis glandulosa* in southwestern Kansas; and *Quercus macrocarpa* throughout the Missouri River valley of North and South Dakota.

The migration to the west is facilitated by the close proximity of the Eastern Deciduous Forest, a division of which, the Ozarkian, covers the southeastern corner of Kansas. This area is characterized by such woody plants as *Quercus, Carya, Fraxinus, Cornus florida, Viburnum rufidulum,* and *Vaccinium arboreum.* Beneath these are *Trillium sessile* and *Polystichum acrostichoides.*

Quercus and *Carya* extend north along the eastern edge of Kansas and into the corner of Nebraska where all of them but *Quercus macrocarpa* disappear. About midway along the east edge of Kansas, *Acer saccharum, Tilia americana, Fraxinus pennsylvanica,* and *Ulmus americana* appear in numbers. The understory consists of shrubs such as *Aesculus glabra, Cornus drummondii,* and *Ribes missouriense* and an abundance of herbaceous plants, including *Asarum, Arisaema, Hydrophyllum,* and *Polygonatum.* All of these extend westward along the Missouri River and its tributaries well into the Prairie Province. *Quercus macrocarpa* and *Fraxinus pennsylvanica* are the principal trees with sufficient tolerance to exist in abundance far to the northwest. Additional fingers of the Eastern Deciduous Forest protrude into the prairie all along the eastern edge of our area.

The extension of the southwestern flora into our area is not nearly as distinct. Here the abundance of a migrating species is much lower and the plants appear as individuals or in small, broken colonies scattered through the prairie. In no place can *Opuntia imbricata* or *Prosopis glandulosa* be called abundant or even common. How they became established is a matter of speculation. Some stories indicate they were brought in from Texas during the days of the big cattle and sheep drives. On the other hand, it will be noted that the principal stand of *Opuntia* is on and along the flood plain of the Cimarron River, which enters Kansas after flowing through New Mexico, Oklahoma, and Colorado.

The flora of the Rocky Mountain part of the Cordilleran Forest has a little more difficulty penetrating and existing in the plains area. The rather sharp line of reduced rainfall and the change of soil type on the east side of the Rockies are major factors inhibiting a migration. However, certain species of woody plants have bridged the gap and are well established. *Pinus flexilis* has done this in a small stand in southwestern Nebraska and a much more extensive area in southwestern North Dakota. In both places it occupies, in open stands, the tops of the low hills. *Pinus ponderosa* has penetrated in a most interesting pattern that does not show on a dot distribution map by counties. It is found in small, isolated, hilltop stands across the westernmost part of the Nebraska panhandle. In the northern part of Nebraska it extends eastward on the hilltops above the Niobrara River to at least halfway across the state. In southern South Dakota again it extends eastward in a narrow finger nearly to the Missouri River, and apparently is not connected with the strip in Nebraska. It appears again as isolated and scattered groups in the northwest corner of South Dakota. Beneath all these stands will be found the prairie grasses typical of the area, as well as the western species of *Ribes* and *Calochortus.*

It may be noted that the Black Hills area, where *Pinus ponderosa* is abundant, was omitted from the above statements. Geologically these hills belong to the Rocky Mountain system. From the vegetative point of view, they are a delightful mixture of the Cordilleran Forest, Northern Conifer Forest, Eastern Deciduous Forest, and the Prairie Provinces. By taking only a few steps one can find plants, both woody and herbaceous, typical of all four areas. This is exactly what could be expected when the geography and climate are considered. The Black Hills have a latitude of about 44° while the nearest portion of the Northern Conifer Forest in Minnesota is only 48° and at 5,000 feet less elevation than the highest peak of the Black Hills.

The Black Hills are completely surrounded by prairie, above which they rise as much as 4,000 vertical feet. Comparatively, that is well over half of the vertical distance between Manitou Springs and the top of Pikes Peak. The soils of the Black Hills vary from solid granite, massive limestone ledges, sandstone, and red gypsum soil to the rich, alluvial soils along the permanent streams. The Belle Fourche and Cheyenne rivers meander through the plains and almost completely encircle the Black Hills. The average annual rainfall in the Black Hills is around 28 inches, while that of the surrounding prairie is only about 16 inches. During the winter the snowfall may be as much as 100 inches, compared to the 25 inches on the plains. The annual temperature range in the Black Hills covers about 140° and a range as high as 164° has been recorded. With such wide variations of moisture and temperature in an area only 120 miles long and 50 miles wide, it is no wonder that the vegetation in the Black Hills is quite different from that on the plains.

In the Black Hills area, typical plants from the Rocky Mountain region are *Pinus ponderosa, Pinus contorta,* a few *Pinus flexilis,* and an abundance of *Populus tremuloides.* Beneath these are several species of *Ribes, Berberis, Ceanothus, Lupinus, Dodecatheon,* and a few *Saxifraga.* The plants typical of the Northern Conifer Forest are *Betula papyrifera, Populus balsamifera, Picea glauca, Shepherdia canadensis, Betula glandulosa, Arctostaphylos uva-ursi,* and several *Orchidaceae* and *Pyrolaceae.* The Eastern Deciduous Forest is represented by *Quercus macrocarpa, Ostrya virginiana, Ulmus americana,* and *Fraxinus pennsylvanica.* The Prairie plants in greatest abundance are the *Andropogon, Bouteloua, Astragalus, Aster,* and *Solidago.*

East of the Black Hills is a large area of badlands where the vegetation is scarce, and although the soil is different, another such area exists in the western part of North Dakota. It is on the eroded sides of these areas that one finds *Sarcobatus, Atriplex, Eurotia, Chrysothamnus,* and *Suaeda,* some of which are from the Great Basin Province. Directly above these on the tops of the tablelands is shortgrass prairie where *Bouteloua gracilis* is abundant.

Along the northern border of North Dakota are the Turtle Mountains and the Pembina Mountains; both are low, wooded hills with many lakes and bogs. Here the flora is mainly Northern Conifer Forest, or of the type found in the northern portion of the Eastern Deciduous Forest. Such plants as *Populus tremuloides, Populus balsamifera, Alnus incana, Salix* ssp., *Cornus canadensis,* and *Petasites* are common.

Through the central portion of all four states is an area of agricultural importance, the eastern part under cultivation and the western part for grazing or under cultivation. Portions of northeast South Dakota and central North Dakota are dotted with large or small glacial lakes; and a great deal of the surrounding land is used for grazing and, increasingly, for recreational purposes.

Although the area covered in this book is predominately prairie, woody plants cover a great deal of it, far more than a great many people have been led to believe. It can easily be seen that the bordering floristic provinces have a heavy influence on the vegetation of the prairie. Many of the species within the region are at the extreme edge of their range. This is true from all four directions. Since the characters of transitional plants may vary to some extent from what they are in the center of the range, the description given herein may not always coincide with the description given elsewhere.

KEY TO GROUPS

If the page number is not given, the plant is not discussed in the text.

GROUP I

Leaves needle-like or scale-like

GROUP II

Leaves broad, simple, opposite or whorled

GROUP III

Leaves simple, alternate; trees

GROUP IV

Simple, alternate; shrubs with entire leaves; vines with entire, toothed or lobed leaves

GROUP V

Shrubs with simple, alternate, toothed leaves

GROUP VI

Leaves broad, compound, opposite

GROUP VII

Leaves broad, alternate, 3-foliolate

GROUP VIII

Leaves broad, compound, alternate, more than 3 leaflets

QUERCUS

CRATAEGUS

POPULUS

SALIX

RIBES

PRUNUS

RUBUS

ROSA

DRAWINGS AND DESCRIPTIONS
OF THE PLANTS

PINACEAE
Picea glauca (Moench) Voss

Pinus glauca Moench; *Picea canadensis* var. *glauca* Sudw.; *Abies canadensis* Mill.; *Picea canadensis* B.S.P.; *Picea laxa* Sarg.; *Picea alba* Ait.; *Picea albertiana* Brown

White spruce, Canadian spruce, Black Hills spruce

A conical tree to 20 m high, with central trunk, rather densely branched.

LEAVES: Simple, distributed around the stem with more on the upper side; needles rigid, 4-angled, usually less angled on old stems, 8-18 mm long, 1-1.5 mm wide, curved, sharply pointed, often with a resinous tip, yellow-green to blue-green, glaucous in lines of minute white dots between the angles; margin entire; constricted to a brown, petiole-like base 0.5 mm long; no stipules.

STROBILES: Early June; monoecious. Male strobiles from buds on wood of previous year. Stalk 5 mm long surrounded by brown bud scales of the same length; strobile 10-15 mm long, 5-7 mm wide, cylindric, tan to pale red; scales peltate 1.5-2 mm across, irregularly circular with the longer side toward the strobile tip; the stalk 1.5-2 mm long, 2 pollen sacs attached along the lower side of the stalk. Female strobile terminal on growth of the previous year; strobile sessile, ellipsoid, 1 cm long, 5 mm wide, red-brown with brown, papery bud scales around the base; cone scales somewhat circular, concave, 3 mm across; a small, obovate scale 2 mm long on the basal side of the main scale; an ovule on each side of a central ridge on the apical (concave) side of the scale; ovary pinkish, nearly circular, 0.5 mm across, wing 0.75 mm long, terminal, extending down along the outer side of the ovule; pore at the base of the ovule.

CONES: August-September. Cones brown, ascending or drooping, 3-4 cm long, 2-2.5 cm wide when open, mature in one season, scales rounded or emarginate, margin thin. Seeds dark brown with lighter wing, obliquely ovoid, 2.5-3 mm long, 1.8-1.9 mm wide, 1.4-1.6 mm thick; wing obovate, 7-8 mm long, 3.5-4 mm wide, one side nearly straight, extending around the margin of the seed.

TWIGS: 2-5 mm diameter, greenish-brown with a rough and irregular surface; leaf scars rhomboidal on raised projec-tions; 1 bundle scar; pith brown, vacuolar, one-fifth of stem. Terminal buds conical, 5 mm long, 2.5 mm wide, scales brown, splotched with whitish resin spots, the lower scales mucronate and keeled; lateral buds smaller, scales thin, split at the upper end, the tips spreading.

TRUNK: Bark thin, furrowed, the ridges flat and somewhat plate-like or scaly; gray, tinged with brown; upper branches smooth, silvery-brown. Wood lightweight, weak, fine-grained, yellowish, with a narrow, light yellow sapwood.

HABITAT: Hillsides or valleys, often at the edge of boggy ground, stunted trees often in wet ground.

RANGE: Alaska to British Columbia, southeast to Northern Montana, northeastern Wyoming, Black Hills of South Dakota, east to Minnesota, Michigan, New York, and Maine, north to Labrador, and west to Yukon.

Picea glauca is the only native spruce in our area and is found only in the Black Hills. One vegetative specimen from a small and young tree in western North Dakota could not be accurately identified but is probably *P. glauca.*

This is an attractive tree and often covers a hillside in the Black Hills, the conical tops forming a geometric pattern. Typical of most spruces, the cones are mostly near the top. This, of course, does not prevent the squirrels from getting them for the seeds.

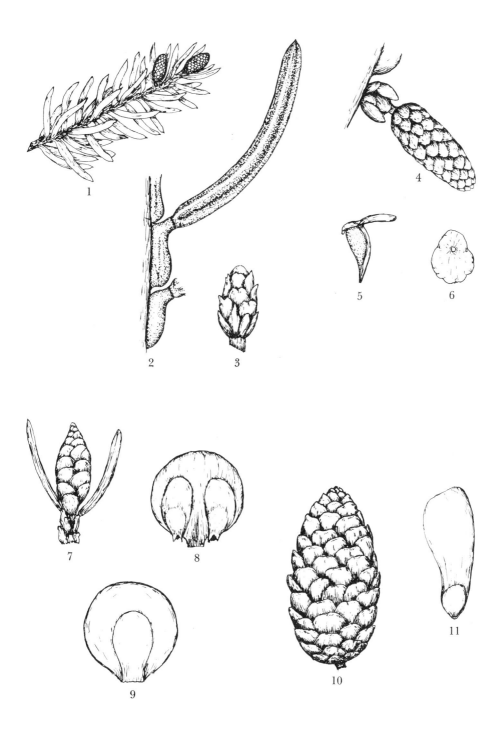

1. Branch with male strobile
2. Detail of leaf
3. Terminal leaf bud
4. Male strobile
5. Side view of peltate scale
6. Top view of peltate scale
7. Female strobile
8. Ventral view of scale of female strobile
9. Dorsal view of scale of female strobile
10. Mature cone
11. Seed

PINACEAE
Pinus contorta Dougl. ex Loud.
var. *latifolia* Engelm.

P. murrayana Balf. in A. Murr.; *P. contorta* var. *murrayana* Engelm.; *P. divaricata* (Ait.) Dumont var. *latifolia* (Engelm.) Boiv.; *P. tenuis* Lemmon

Lodgepole pine

A tree to 25 m high, with a central trunk and conical crown.

LEAVES: Spirally arranged, evergreen; needles in fascicles of 2, terminal on a dwarf spur branch and surrounded by the primary leaves, which appear as a brown, papery sheath around the base; needles 3-7 cm long, 1.5 mm wide, flattened on the inner surface and rounded on the outer, about 10 rows of minute white dots on the outer surface and fewer on the inner; margin minutely serrulate; the scale leaves (primary) ovate to lance-ovate, 4-6 mm long, brown, and thin. Needles remain on the tree for several years, falling only when the dwarf branch breaks and falls.

STROBILES: Late June; monoecious. Male strobiles in clusters of 25-50 at the base of new growth; the strobiles cylindric, 8-15 mm long, 5-6 mm diameter with several ovate, acuminate scale leaves at the base; face of strobile scales oval, pale brown, the margins erose and scarious; the 2 pollen sacs 1 mm long and attached along the abaxial side of the stalk. Ovulate strobiles near the tip of new growth, cylindric, 9-12 mm long, 8-9 mm diameter, the scales dark purple with brownish tips and a thin, rounded bract adherent to the abaxial side; scales ovate with long, often recurved, acuminate tip, each scale 4 mm long, 1 mm wide with 2 ovules embedded on the adaxial side, the pore located at the end of a small projection extending toward the cone axis.

CONES: Cones opening at the end of the second year or remaining closed for several years; cones asymmetrical, 3.5-5 cm long, 3-3.5 cm wide when open, usually slightly curved, the exposed portion of the scales on the more rounded side large and often nipple-like, those of the other side small and flattened, the umbo ending in a hard prickle curved toward the cone apex. Two seeds per scale, total seed length 12-14 mm, the wing 3.5-4 mm wide, the seed proper 4-4.5 mm long, 2-2.5 mm wide, and 1.4 mm thick, dark brown; the straight side of the wing rather thick, the rounded side thin and papery.

TWIGS: 4-5 mm diameter, red-brown the first year and gray the second year; leaves usually distributed along the branch, but often clustered at the end of each annual growth; older twigs rough with the old dwarf branch scars; pith greenish, irregular in section, continuous, two-thirds of the stem. Embryonic stem at the tip of a branch (naked bud) 10-16 mm long, 5-7 mm diameter, reddish-brown, and consists of thin, lance-ovate scale leaves and embryonic needles.

TRUNK: Bark gray or somewhat orange-brown, thin with shallow fissures or with small flakes which are often loose. Wood lightweight, hard, strong, coarse-grained, pale red-brown, with a wide, white sapwood.

HABITAT: Mountain sides and valleys, either moist or dry. In our area found scattered among ponderosa pine in one small area of the Black Hills.

RANGE: Alaska to southwestern Saskatchewan, south to Colorado, west to California, and north in the mountains to British Columbia.

Lodgepole pine and ponderosa pine both have the needles in fascicles of 2, but lodgepole pine needles are only 3-7 cm long, while those of ponderosa pine are 10-15 cm long. The cone of lodgepole pine is always asymmetrical and small. The trees, at a distance, appear yellow-green and ponderosa pine trees have a blue-green color.

Our plants are var. *latifolia,* the needles being much wider and the trees taller than var. *contorta* of the west coast.

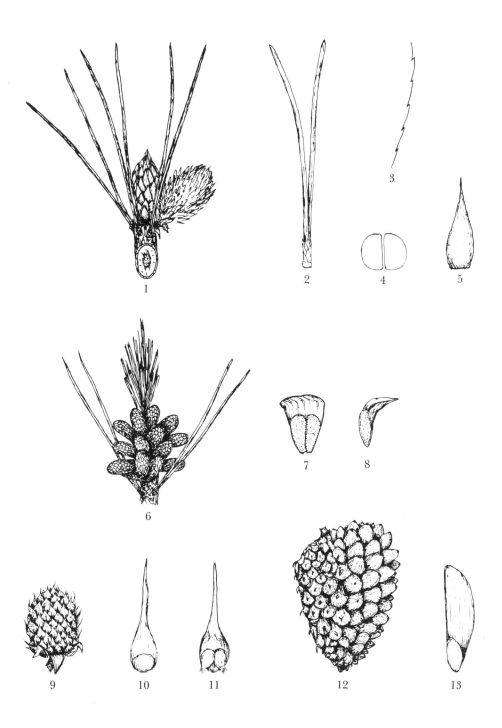

1. Winter twig with bud and first-year cone
2. Needle leaf
3. Margin of needle leaf
4. Transverse section of needles
5. Primary scale leaf
6. Cluster of male strobile
7. Abaxial view of male scale
8. Side view of male scale
9. Female strobile
10. Abaxial view of female scale
11. Adaxial view of female scale
12. Mature cone
13. Seed

PINACEAE
Pinus flexilis James

Apinus flexilis (James) Rydb.

Limber pine

A tree to 12 m high, either with a central trunk or with several low branches close to the ground.

LEAVES: Spirally arranged, evergreen, somewhat flexible; needles in clusters of 5 at the end of a dwarf branch and surrounded at the base by a sheath of primary leaves; primary leaves lanceolate, 8-15 mm long, light brown, thin, scale-like with scarious or fimbriate margins; needles somewhat triangular in section, 4-7 cm long, 0.75-1.5 mm thick, with several rows of minute white dots on each surface; margin entire; deciduous after several years but not until the dwarf branch dies and falls from the tree.

STROBILES: June; monoecious. The scales at the base of the male strobile broadly ovate, brown, thin with a scarious margin; male strobiles cylindric, 7-10 mm long, 4-5 mm wide, the top of the peltate scales yellow-brown, ovate with a pointed, toothed tip directed toward the cone apex, the face of the scale 1.5-2 mm long, 1 mm wide; the two pollen locules on the abaxial side of the stalk. Ovulate strobiles cylindric, 1-1.5 cm long, 6-8 mm wide, red-purple, the scales thick and resinous, 3 mm long, 2.5-2.75 mm wide, abruptly acuminate and reflexed; a reniform, reddish bract 1-1.5 mm long and 3 mm wide on the abaxial side; 2 ovaries on the adaxial side near the cone axis; ovules obovate, reddish, 1.5 mm long, the pore on the end of a tube extending toward the cone axis.

CONES: Cones often still green in color in June of the second year, but mature by the end of the season; mature cones 9-11 cm long, 5-6 cm wide, yellow-brown, resinous, the end of the scales thin, rounded and without a spine, the basal scales often recurved; 2 seeds per scale. Seeds irregularly ovoid, 10-11 mm long, 6.5-7 mm wide, 4.5-5 mm thick, wingless, light brown, often mottled with darker brown.

TWIGS: 7-8 mm diameter, young branches puberulent, green or brown the first year and light gray by the second year, older branches fairly smooth and light gray. Embryonic stem tip ovoid, pointed, pale red-brown, resinous, 6-10 mm long, 4-6 mm diameter.

TRUNK: Bark of young trees smooth, silvery gray, that of old trees gray-brown to nearly black, thick with shallow furrows and broad, nearly flat, scaly ridges. Wood hard, strong, fine-grained, red-brown, with a wide, white sapwood.

HABITAT: Mountainous areas, usually under dry conditions. In our area mainly on dry, eroded hills.

RANGE: Southern British Columbia and Alberta, south to northern New Mexico, west to California and north to Oregon and Idaho. Occurs in a large stand in Slope County, North Dakota, and in Kimball County, Nebraska, and a few trees in the Cathedral Spires area of Harney Peak, Custer County, near the Pennington County line, in South Dakota.

Pinus flexilis in our area grows in open stands and the largest trees about 8 meters high. In the North Dakota population the trees are short and rather dense, and even the small trees no more than 2 meters high may produce a full crop of cones.

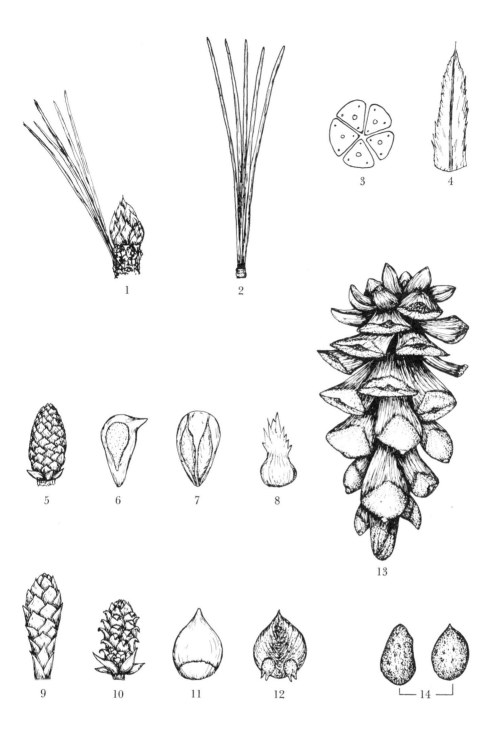

1. Winter terminal bud
2. Needle leaves
3. Transverse section of needles
4. Primary scale leaf
5. Male strobile
6. Side view of male scale
7. Abaxial view of male scale
8. Face (end) view of male scale
9. Immature female strobile
10. Female strobile
11. Abaxial view of female scale
12. Adaxial view of female scale
13. Mature cone
14. Seeds

PINACEAE
Pinus ponderosa Dougl. ex Loud.

P. ponderosa var. *scopulorum* Engelm.;
P. scopulorum Lemmon

Ponderosa pine, western yellow pine

A tree to 35 m high, with a straight trunk and high, open crown.

LEAVES: Spirally arranged, evergreen; needles in fascicles of 2-3, often more or less whorled at the tip of a branch. Needles 10-15 cm long, 1-1.5 mm wide, half-round in section and minutely toothed; terminal on a dwarf branch, the base enclosed in a sheath 1-2.5 cm long composed of several primary leaves; these primary leaves vary from 7 mm long, dark brown, rigid, ovate with an abruptly acuminate tip to scales 2.5 cm long, thin, pale brown with scarious or fimbriate margins and united around the needles; the sheath contracts, becoming gray and wrinkled and is deciduous by early winter.

STROBILES: May-June; monoecious. Male strobiles subtended by several brown, papery scales with hyaline margins; 10-20 strobiles clustered at the base of new growth, the new needles 1-2 cm long at the time of anthesis. Strobiles cylindric, 2-3 cm long, 6-8 mm wide, the top of the peltate scales circular with an irregular margin, brown, 1.5 mm across; the pollen locules 2.5 mm long, parallel and attached to the scale stalk. Ovulate strobiles near the tip of new growth, obovoid, 10-12 mm long, 8-10 mm thick, terminal on a thick stalk 5-7 mm long; scales ovate with a long point, thick, pink or purple, soft, resinous, ridged on the ventral (adaxial) side, the dorsal (abaxial) side with a thin, broadly ovate bract; 2 ovules on the ventral side with the pore toward the strobile axis.

CONES: Maturing in the summer of the second season; open cone 6-9 cm long, 6-7 cm wide, tip of scales rounded and often with a short spine, not especially resinous, brown; 2 seeds per scale. Seeds including the wing, 18-23 mm long, 6-8 mm wide; the seed itself 6-7 mm long, 5-5.5 mm wide and 3.5-4 mm thick, ovoid, brown, often mottled; wing papery, the straight edge heavy and the rounded edge thin, extending down along the seed.

TWIGS: 5-7 mm diameter, brown, covered with needles and occasionally the old, lanceolate leaf scales; dwarf branch scars oval; pith brown, hard, continuous, one-half of stem. Terminal bud naked, conical, 12-15 mm long, 5-7 mm wide, contains the brown primary leaves with the embryonic needles in their axils.

TRUNK: Bark of young trees dark gray-brown with deep furrows, the ridges long, wide and flat; bark of old trees scaly, gray-brown mixed with cinnamon brown. Wood hard, strong, fine-grained, red-brown, with a wide, white sapwood.

HABITAT: Rocky hillsides and at lower altitudes in the mountains, in dry or moist soil. It is the most common pine in the Black Hills and is the pine of the "pine ridge" region of northern Nebraska and southern South Dakota.

RANGE: Southern British Columbia, south to Baja California, east in northern Mexico, north through western Texas, western Nebraska, and South Dakota, and northwest into Montana.

Pinus ponderosa and its varieties is one of the principal lumber trees of western United States. It grows in pure stands in many places, but the stands are open. This should let light in for an abundance of understory plants, but these are seldom found in a pure stand of the pine.

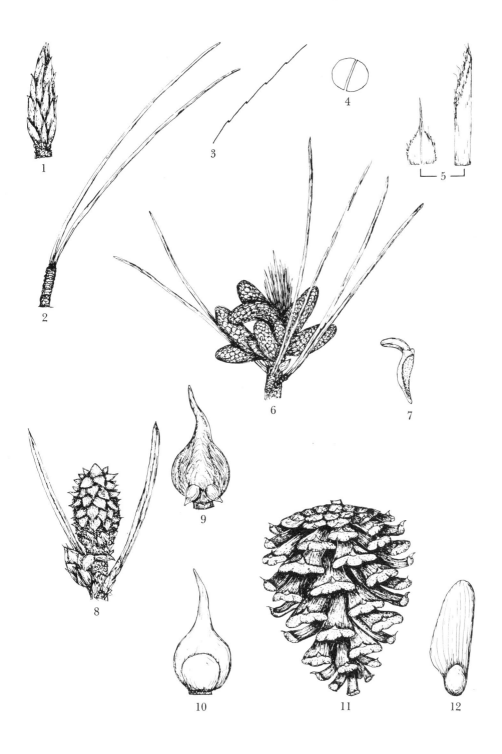

1. Winter terminal bud
2. Needle leaves
3. Margin of needle
4. Transverse section of needles
5. Primary scale leaves
6. Cluster of male strobile

7. Scale of male strobile
8. Female strobile
9. Adaxial side of male scale
10. Abaxial side of male scale
11. Mature cone
12. Seed

CUPRESSACEAE
Juniperus communis L.

Dwarf juniper

A decumbent shrub to 1.5 m high and 2-4 m across, usually dense.

LEAVES: Evergreen, simple, in whorls of 3; linear, 10-15 mm long, about 1 mm wide, curved and ascending, trough-shaped on the ventral side, a broad white stripe in the trough, the tip sharp pointed and the base jointed; entire, glabrous, somewhat gibbous on the ventral side at the base; no stipules.

STROBILES: June-July; dioecious. Axillary on growth of the season. Male strobile ovoid, 3-4 mm long, sessile or stalked, the stalk up to 0.5 mm long and with 3-5 acuminate bracts at the base; ament scales brown, stalked, broadly ovate with an acuminate point, the 4-6 sessile stamens attached to the lower edge of the basal side. Female strobiles 2-3 mm long, with the appearance of a minute cone with green, ovate, acuminate scales diminishing in size toward the tip, all of which are sterile; at the tip are 3 urn-shaped structures with an opening at the apex and the ovule at the bottom. These structures are entirely distinct at the time of pollination but fuse to form a berry-like cone which matures in the second year.

CONES: Late summer. Cones ovoid to ellipsoid, 8-10 mm long, 6.5-8.5 mm diameter, dark blue with a glaucous bloom, the cone in 3 fused sections each with 1 seed. Seeds broadly ovoid, light brown, about 4.5 mm long and 3 mm thick, somewhat ridged on two sides, the surface rough and with 3-9 resin glands, the seed apex usually apiculate.

TWIGS: 1.2-1.4 mm diameter, green, soft in the first season, turning brown and hardening in the winter or in the second season; a ridge-like resin gland extends from the leaf base to the node below; pith white, continuous, about one-fifth of stem. Buds naked.

TRUNK: Bark gray-brown exfoliating into short, thin strips, the inner bark red-brown. Wood fine-grained, durable, reddish, with a wide, white sapwood; the heartwood often reddish, only in concentric bands, that is, with alternating rings of red and white wood.

HABITAT: Wooded hillsides or rocky, exposed slopes and ledges; moist or dry soil.

RANGE: British Columbia, east to Newfoundland and Nova Scotia, south through New England to the mountains of North Carolina, west and north across Illinois to western South Dakota and Nebraska, south to New Mexico, west to California, and north to Idaho.

The varieties of *Juniperus communis* are apparently based on the habit of growth which in itself is a variable thing. If, as most manuals indicate, var. *communis* is a plant up to 3 m high, having a central trunk, and var. *montana* (*saxatalis*) is a trailing, mat-forming plant, we have neither of them in our area. Our plants have decumbent trunks and range from 50 to 100 cm high, rarely to 150 cm and those are usually spindly shade plants. This puts them in var. *depressa* Pursh, even though the needles are definitely curved at the base, which, according to some manuals, should not be.

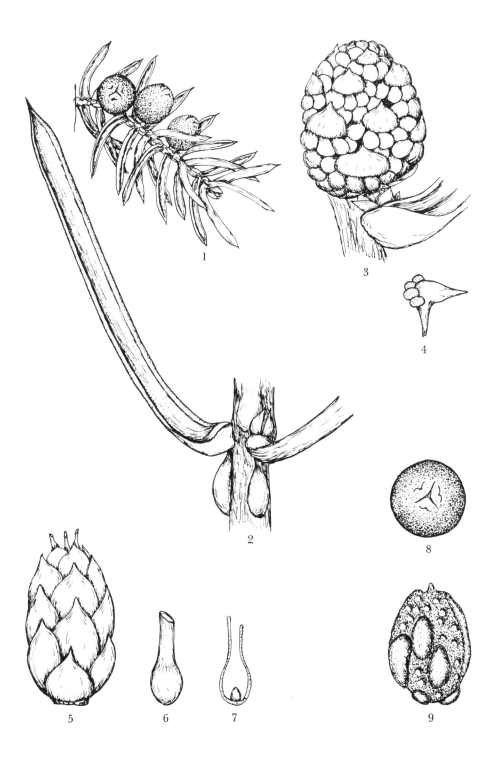

1. Twig with fruits
2. Detail of leaf
3. Male strobile
4. Scale of male strobile
5. Female strobile
6. Modified scale containing ovule
7. Sectional view of scale showing the ovule
8. End view of fruit
9. Seed

CUPRESSACEAE
Juniperus horizontalis Moench

J. prostrata Pers.; *J. sabina* L. var.
prostrata Loud.; *J. virginiana* L. var.
prostrata Torr.; *Sabina prostrata*
Antoine; *Sabina horizontalis* Rydb.

Creeping juniper

A creeping shrub, rooting along the branches and forming mats up to 7 m across and usually no more than 20 cm high.

LEAVES: Evergreen, opposite, scale-like or somewhat oblong, 1-2 mm long and 0.5 mm wide, longer on rapidly growing shoots; tip sharply pointed, base sessile and somewhat clasping; margin entire; the gland on the dorsal side shorter than the distance from the gland to the leaf tip.

STROBILES: May; dioecious. Male strobile cylindric, pale brown, 4-5 mm long, 2 mm thick, terminal on short branches; the peltate scales in 4 rows of 4 scales each; top of scale 1.5 mm across, nearly circular with a point on the apical side; the 4 pollen sacs attached to the stalk of the scale directly below the basal margin, sessile, pale yellow. Female strobile terminal on short, usually reflexed branches, the strobile consists of 2-6 ovoid ovaries about 0.25 mm long; these are soft, succulent modified scales, distinct to the base but fuse later, the pore being toward the outside and near the top.

CONES: August-September. Fertilized in the spring of one season and reach maturity in the fall of the next season. Peduncles usually curved, leafy, 3-5 mm long; cone globose, 8-10 mm diameter, scales fused, dark purple with a heavy bloom, 2-3 seeds per fruit. Seeds irregularly ovoid, 4-5 mm long, 2.5-3 mm wide, ridged on 2 sides, red-brown with a large light-colored basal scar, surface muricate.

TWIGS: About 1 mm diameter, usually covered with leaves, the fast growing stems up to 2 mm diameter with leaves farther apart and definitely decurrent on the stem; twigs green toward the end and brown farther back; pith whitish, continuous, one-fifth of stem. Buds naked.

TRUNK: Bark red-brown, thin, scaly, and exfoliating into flakes or flat strips. Wood soft, open-grained, red-brown, with a wide, white sapwood.

HABITAT: Open, prairie hillsides, eroded areas, open woods, or along rock ledges; usually in dry habitats.

RANGE: Alaska and northern Mackenzie, east to Newfoundland, south to New York, west to South Dakota, western Nebraska, Colorado, Wyoming, and northwest to British Columbia.

The trailing evergreen stems rooting at the nodes should be sufficient for the identification of this juniper; no other species is similar. Occasionally a plant appears that has the trunk erect for a few centimeters. If the plant is an herbarium specimen with fruits, the recurved leafy pedicel is usually a good character.

J. horizontalis is one of the finest plants for the prevention of soil erosion since it forms a solid mat covering a large area. It is seldom planted for that purpose because the land cannot be used for anything else; even grasses are eventually killed. But, fortunately enough, in nature the plant grows in places where the vegetation is already thin and the soil is in danger of eroding.

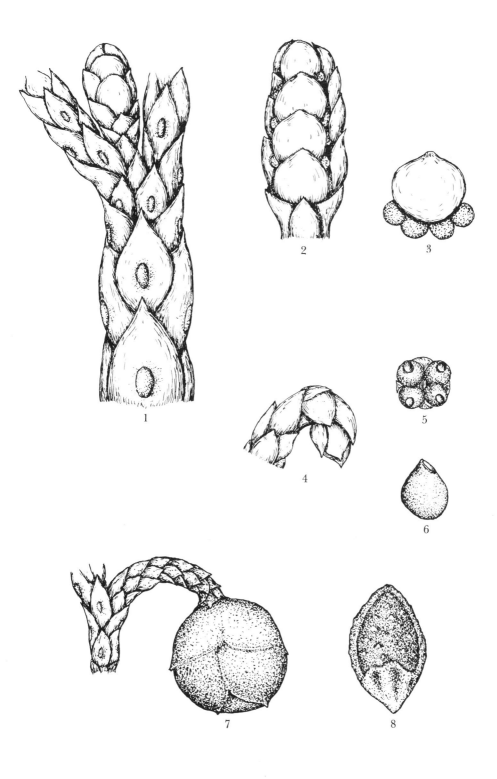

1. Twig with male strobile
2. Male strobile
3. Scale of male strobile with pollen sacs
4. Female strobile
5. The four ovaries at strobile end
6. One ovary
7. Mature cone
8. Seed

CUPRESSACEAE
Juniperus virginiana L.

Red cedar

Tree to 20 m high, conical with central trunk, or rounded and branched low, branches spreading, the small branches often drooping.

LEAVES: Evergreen, simple, opposite. Leaves on mature twigs closely appressed, imbricated, 4-ranked, broadly deltoid, 1.5 mm long, acute tip, broad base, entire; thick, rigid, resinous, glabrous, sessile. Leaves of fast growing twigs subulate, 6-12 mm long, sharp pointed, spreading. The gland oval, shorter than the distance from the gland to the leaf tip.

STROBILES: April; dioecious. Male strobiles terminal, sessile, globose or cylindric, cone-like, 2.5-3 mm long, 4 vertical rows of 3 peltate scales each, the stalk 0.8-0.9 mm long; scales yellow-brown, 1.2-1.4 mm across, each bearing 4-6 yellow, globose anther locules along the lower margin. Female strobiles terminal, subglobose cone-like, 1.2 mm high, 1.8 mm thick; scales usually 4, green, spreading, thick, acute, sessile, fused at the base, thin tip; center of cone fleshy with 1-3 flattened, urn-shaped ovules, with the opening at the top.

CONES: September. Globose, 6-7 mm diameter, dark blue with a heavy bloom, resinous, nondehiscent, the scales fused; 1-3 seeds. Seeds ovate, 3-3.5 mm long, 2-2.8 mm thick, light brown, darker at the small end, ridged, somewhat pitted. Fruits maturing in the first year.

TWIGS: 1.2-1.3 mm diameter including the leaves, green when young, brown in the second year, flexible, glabrous; immature cones for the following year at the tip of branchlets. Pith pale brown, irregular, continuous, one-fifth of stem.

TRUNK: Bark thin, pale red-brown, shredding into long, thin, flat strips. Wood lightweight, brittle, durable, fragrant, red, with a white sapwood.

HABITAT: Dry hillsides or in semi-barren lands; occurs most often in limestone, gypsum, or clay soils; occasionally in river-bottom woodlands.

RANGE: Maine, west through southern Ontario, south to the eastern half of South Dakota with an extension into the southwestern corner of North Dakota, south in eastern Nebraska, most of Kansas and Oklahoma to western Texas, east inland through Alabama and Georgia, and northeast to Maine. Does not appear in the high mountains of New England or the southern Alleghenies.

J. virginiana and *J. scopulorum* Sarg. have been the subject of much study in the Central Plains states. The above range has usually been accepted for *J. virginiana*. That given for *J. scopulorum* would be from British Columbia and Alberta south through the Great Basin and the Rocky Mountains to Arizona and New Mexico and into western Texas, north across eastern Colorado, western Nebraska to western North Dakota, and west in Montana. Such a range would not overlap that of *J. virginiana* except in North Dakota.

J. scopulorum is differentiated by the blunt leaves, leaves not overlapping, the glands elliptic and longer than the distance from it to the leaf tip and the fruits maturing in the second season. *J. scopulorum* is usually a more rounded tree and *J. virginiana* is more conical. However, in southwestern North Dakota there is a columnar cedar which Fasset called *J. scopulorum* var. *columnaris*.

Haverbeke (U. of Nebraska studies No. 38, 1968) states that the whole population within the Missouri River basin is of hybrid origin with neither of the extreme parental types being found. He continues to say that the plants tend increasingly more toward *J. scopulorum* in a line from the southeast to the northwest.

● *Juniperus virginiana*
■ *Juniperus scopulorum*

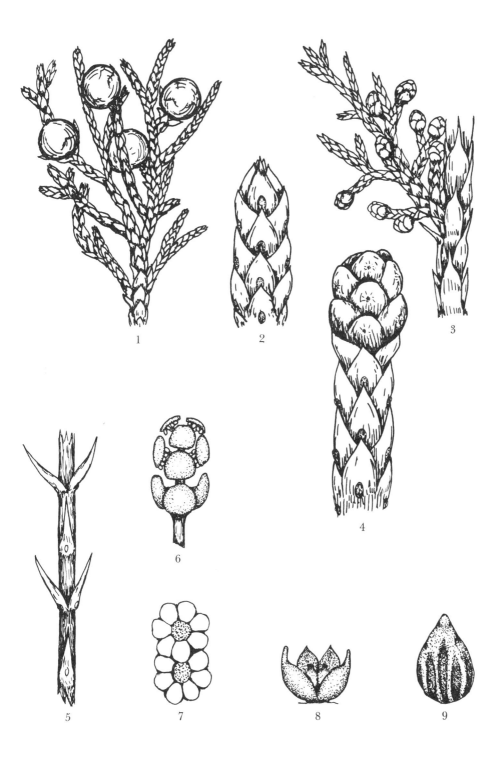

1. Fruiting branch
2. Leaves of female branch
3. Male branch in winter
4. Male winter cone and leaves
5. Juvenile leaves
6. Male cone
7. Stamen locules beneath peltate scale
8. Female strobile
9. Seed

LILIACEAE
Yucca glauca Nutt.

Yucca angustifolia Pursh

Soapweed, yucca, bear grass

A short-stemmed plant to 1 m high, the stem simple or branched at the base.

LEAVES: Simple, closely alternate, the bases completely covering the stem, evergreen and persisting for several years, the new leaves at the tip and the old dead leaves drooping at the base. Leaves linear with inconspicuous parallel veins, 4-6 dm long, 8-12 mm wide, stiff, radiating upward from the short caudex; glabrous, concave; gray-green with a narrow white margin from which white threads strip and curl; the margin rolled inward at the tip, forming a hard, sharp point; base white, broadened, and spoon-like.

FLOWERS: May-June; perfect. The inflorescence a raceme up to 1 m high, the peduncle about half of that length, occasionally the inflorescence is paniculate with a few short branches toward the base of the flowering portion; a few narrow, appressed leaf-like bracts 2-5 cm long widely spaced along the peduncle; raceme with 25-30 drooping flowers; pedicels 1.5-2 cm long, green, glabrous, recurved; sepals 3, greenish-white, 5-5.2 cm long, 2.6-2.8 cm wide, elliptical, tip acute; petals 3, white, 5.1 cm long, 3.5 cm wide, elliptical, the tip acute; stamens 5-6, filaments white, 1.7-1.9 cm long, slightly clavate and stocky, the upper end curved sharply outward, pubescent; anthers yellow, sagittate; ovary cylindric, 2.1-2.3 cm long, 1.7 cm wide, green, glabrous, 3-celled, each with 2 rows of ovules; style with 3 tumid divisions 1 cm long, green; stigmas capitate.

FRUIT: August-September. Capsules cylindric, erect, 5-8 cm long, 3-4.5 cm diameter, hard, brown, and wrinkled at maturity, the 3 carpels splitting to the base and each divided into 2 sections with an apiculate outcurved tip; many seeds tightly packed in 6 columns. Seeds roughly triangular, 9-11 mm long, 8-9 mm wide, 0.7-0.9 mm thick, flat, thin, black, semiglossy; a thin rim around the margin.

TRUNK: Short, stocky, up to 4 dm high, 8-10 cm thick, not including the dead leaves or their bases. Wood soft, fibrous, most of it decaying soon after the plant dies.

HABITAT: Dry, well-drained sandy or limestone soils in open areas; prairies, pastures, and waste ground.

RANGE: Montana, south through eastern Colorado to New Mexico, east to Texas, north through Kansas to Iowa and North Dakota.

The form with a paniculate inflorescence has been described as forma *guerneyi* McKelvey, but there is some question about its consistency or being different from the racemose plants. The two forms occur together in most of the western areas where either occurs.

A very similar species, *Y. arkansana* Trel. (*Y. glauca* var. *mollis* Engelm.) grows in southeastern Kansas. It has flatter, softer, and more flexible leaves, often bent and drooping. When the two species are grown from seed and side by side in a garden, *Y. arkansana* is a faster growing and more robust plant.

Yucca is usually termed a weed because it tends to take over a pasture during dry periods, but it is a definite aid in controlling wind erosion in sandy soils. Small mammals and birds, as well as reptiles, use the plant for shade and nesting sites, and the seeds are a staple food of the small mammals.

● *Yucca glauca*
■ *Yucca arkansana*

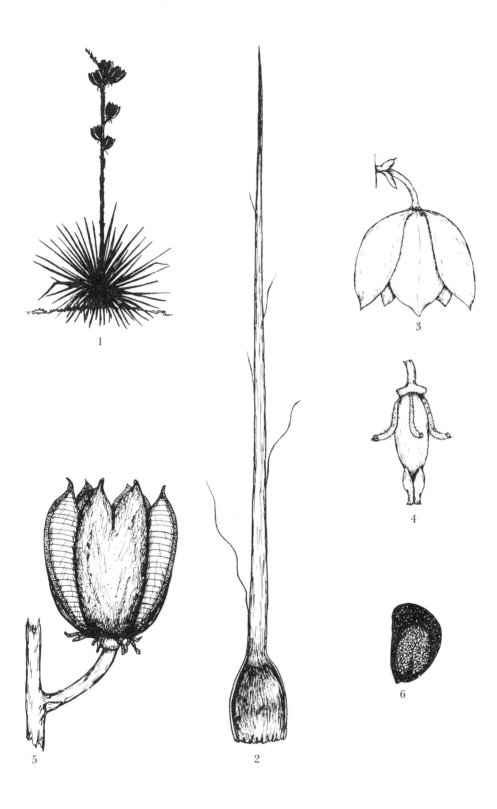

1. Plant in winter
2. Leaf
3. Flower
4. Stamens and pistil
5. Capsule
6. Seed

LILIACEAE
Yucca smalliana Fern.

Y. filamentosa of most authors, not L.

Yucca, Adam's needle and thread,
Spanish bayonet

A short-stemmed shrub to 2 m high, the leaves basal.

LEAVES: Simple, alternate, radiating upward from a short caudex, evergreen. Lance-linear, 4.5-6 dm long, 3-4 cm wide, parallel veins, stiff, gray-green; margin entire with long, white, usually curled threads peeling from it; the margins at the tip rolled in and united, forming a stiff spine-like point; sessile with a widened base, the base white and somewhat spoon-shaped; surface scabrous with minute projections.

FLOWERS: Early June; perfect. Panicle terminal on the caudex, 1-2 m high, 10-20 spreading branches with 4-8 flowers each; an acuminate bract at the base of each branch, 1-3 cm long; axis and branches finely puberulent, the main peduncle with many leaf-like bracts 4-8 cm long which tend to curve around the stalk; pedicels 1 cm long, green, puberulent, 2 acuminate bracts 5 mm long at the base of each. Flowers broadly campanulate; sepals 3, white, 3-4.5 cm long, 1.2-1.6 cm wide, elliptic, slightly keeled, thin, glabrous, cupped, distinct; petals 3, white, 3.5-5 cm long, 2-2.5 cm wide, ovate, thick at the base, glabrous, cupped, tip acute, distinct; stamens 6, filaments white, papillose, flattened, 2 mm wide, appressed to the ovary for 1.5 cm, then with a right angle bend outward for 5 mm; anthers yellow; ovary 3-celled, 14-17 mm long, 5-6 mm thick, pale green, glabrous; styles 3, fused, 6-8 mm long, white, glandular, tumid; stigma with 3 acute lobes.

FRUIT: July-August. Capsules cylindric, constricted at the middle and a low ridge along the back of each section, 4-6 cm long, 2-2.5 cm thick, hard, the 3 carpels splitting at the apex, each carpel with 2 sections, each with an outcurved tip; 6 columns of seeds, tightly packed. Seeds black, flat, roughly triangular, the longest dimension 7-10 mm, 5-5.6 mm wide, 0.3-0.5 mm thick; smooth, semi-glossy, a narrow rim around the margin.

TRUNK: Short, stout, concealed by the leaves or their old bases. Woody, but barely falls within the classification.

HABITAT: Escaped from cultivation usually along roadsides or around cemeteries; planted years ago as one of the reclamation plants on the strip-pit slag piles in southeastern Kansas and still persisting; usually grows in poor soil with plenty of light.

RANGE: Florida to Louisiana, north to Kansas (introduced), east to Tennessee and North Carolina.

Yucca smalliana is often planted for decoration around homes, parks, and cemeteries, and has escaped from there. It is not abundant in our area. The plant requires little attention and produces a long spike with a large, paniculate inflorescence of many white flowers. The flowers, capsules, and seeds are smaller and the leaves wider than the native *Yucca glauca*.

Although this plant has usually been called *Y. filamentosa* L., its characteristics as found in Kansas do not correspond with those given in the manuals for that species. It does, however, agree quite well with *Y. smalliana* as described by Fernald in the eighth edition of *Gray's Manual*.

18

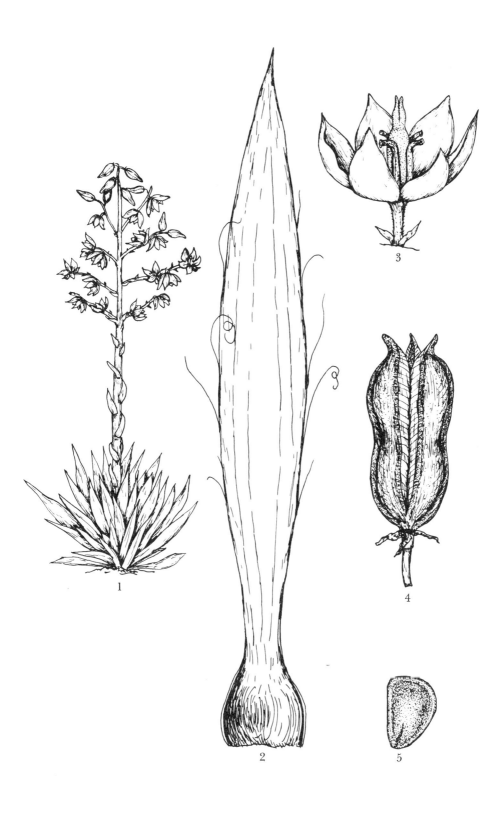

1. Flowering plant
2. Leaf
3. Flower
4. Capsule
5. Seed

LILIACEAE
Smilax bona-nox L.

S. *bona-nox* var. *hederaefolia* (Beyerich)
Fern.

Greenbrier, catbrier, Chinabrier,
bullbrier, smilax

Vine to 8 m long, supported by tendrils or sprawling over low bushes.

LEAVES: Alternate, simple, tardily deciduous. Shape variable, heart-shaped, broadly ovate or panduriform, 3-12 cm long, 3-10 cm wide; 3-5 veins from the base; margin smooth or bristly ciliate, slightly thickened or revolute; tip acute, cuspidate; base rounded or truncate, occasionally lobed; upper surface bright green and often with light blotches, glabrous; lower surface paler and usually with a few prickles on the midvein; petioles 1-2 cm long, often bent sharply and twisted, the blade somewhat decurrent; the tendrils are an extension of the stipules.

FLOWERS: Mid-May; dioecious. Staminate flowers in axillary umbels on new growth, 3-20 flowers; peduncle 1.7-2 cm long, green, glabrous; pedicels 6-7 mm long, green and glabrous. Tepals 6, oblong, obtuse, 5-6 mm long, 1.7-1.8 mm wide, yellow-green, strongly reflexed, a tuft of hair at the tip; stamens 5-6, filaments nearly white, 1.8-2 mm long; anthers 1.5-1.7 mm long. Pistillate flowers in axillary umbels on new growth, 8-20 flowers; peduncle 1.7-2 cm long, green, glabrous, flattened, and angular; pedicels 4-6 mm long, green, glabrous, a brown bract at the base of each; tepals 6, oblong, obtuse, yellow-green, 3-3.5 mm long, 0.7-1 mm wide, spreading, a small tuft of hair at the tip; ovary elliptic, 1.7-2 mm long, 1 mm thick, green, glabrous; stigma sessile, yellow-green, slightly lateral and curled above the ovary, 1.5 mm long, stout.

FRUIT: October. Globular clusters; peduncle flattened; 2-3 cm long, glabrous, minutely white-dotted; pedicels 4-6 mm long, radiating; berry globose, 6-7 mm diameter, black, smooth, semiglossy, usually 1-seeded. Seeds globose, 3.9-4.3 mm diameter, dark red, finely pitted, dull; a black scar at the attachment.

TWIGS: 1-1.5 mm diameter, green, minutely white-dotted, sharply angled, usually somewhat zigzag; prickles small, flat, broad-based, green or green with a dark tip. Buds conical, 2-3 mm long, pointed, green, concealed by the old leaf base.

TRUNK: Bark green, hard, often roughened with white scurf, the old leaf bases persistent; prickles stout, up to 8 mm long, flat, broad-based, green or green with a dark tip, spaced some distance apart along the stem and small twigs. Wood greenish-white, soft, pithy. The bark and wood are not truly such, since there is no cambium layer in the plant.

HABITAT: Woods and thickets, stream banks, wooded pastures, fence rows; in rich loam or rocky soil.

RANGE: Mexico and Texas, east to Florida, north to Maryland, west through Kentucky and southern Illinois to southeastern Kansas and Oklahoma.

Although var. *hederaefolia* is listed for our area, if it is present it intergrades with our variety *bona-nox*. No specimen examined could be separated into the eciliate, broadly cordate-ovate leaves without mottling group which is designated as var. *hederaefolia*.

This species is often confused with S. *hispida*, along side of which it is often found. S. *bona-nox* has flat, green prickles; many fine, intermeshed branchlets; and is commonly a tangled mass of fine branches and branchlets. None of these are characteristics of S. *hispida*.

This impenetrable mass of branches furnishes good cover for small mammals and birds for the larger animals cannot penetrate it. Mammals also use the fruit for food.

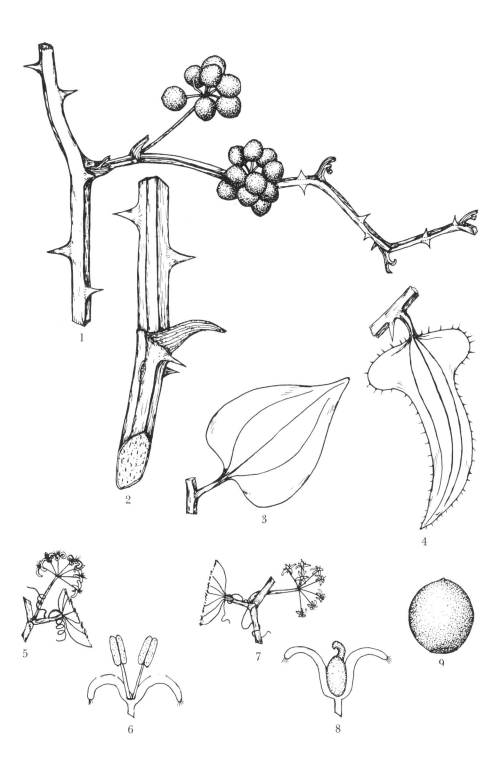

1. Winter twig with fruit
2. Detail of twig
3. Leaf
4. Leaf variation
5. Staminate flower cluster
6. Diagrammatic staminate flower
7. Pistillate flower cluster
8. Diagrammatic pistillate flower
9. Seed

LILIACEAE
Smilax hispida Muhl.

S. tamnoides L. var. *hispida* (Muhl.)
Fern.

Bristly greenbrier, smilax, Chinaroot,
catbrier, hellfetter

A vine to 13 m long, climbing high
by the use of tendrils.

LEAVES: Alternate, simple, tardily
deciduous. Ovate, 6-15 cm long, 4-13 cm
wide; 5 main veins from the base; margin
minutely serrulate; tip acute to acuminate, often cuspidate; base cordate or
broadly cuneate; upper surface dark
green, glabrous, and lustrous; lower surface pale and glabrous; petiole 1-2.5 cm
long, glabrous, grooved above, the blade
decurrent; petiole often twisted and bent
sharply at the summit; stipules adnate
for a short distance, then may be prolonged into a tendril.

FLOWERS: Mid-May; dioecious.
Staminate flowers in axillary umbels of
10-15 flowers on new growth; peduncle
2.5-3.5 cm long, green, glabrous; pedicels
0.7-1 cm long, glabrous. Tepals 6, linear,
6-8 mm long, 1.6-1.8 mm wide, greenish
to bronze, recurved, finely pubescent at
the tip; stamens 6, filaments 2.8-3.5 mm
long, flattened, green; anthers 1.8 mm
long; vestigial pistil in the center. Pistillate flowers in axillary umbels of 10-20
flowers on new growth; peduncle 3-3.5 cm
long, green, glabrous, flattened; pedicels
4.5-5.3 mm long, green, glabrous. Tepals
6, narrowly ovate, 4 mm long, 1.4-1.6 mm
wide, the inner three slightly narrower,
yellow-green to bronze, spreading but not
sharply reflexed, finely pubescent at the
tip; ovary green, 1.7-1.9 mm long, 1.5
mm wide, ovoid, slightly flattened on one
side, glabrous; style minute; stigmas 2-5,
stout, 1.3 mm long, yellow-brown, recurved.

FRUIT: October. Peduncles flattened, especially on one side, 4.5-5 cm
long, green; pedicels radiating from the
summit of the peduncle, 8-10 mm long,
enlarged at the base; the length varies
enough so the fruit cluster may be close
and compact or open and loose; pedicels
often brown scurfy. Fruits globose, 7-9
mm diameter, blue-black or with a greenish cast, smooth, often slightly indented
at the apex; 1-2 seeds. Seeds globose or
flattened on one side, 4.6-5 mm diameter, dark red, smooth, glossy, a black
scar at the point of attachment.

TWIGS: 1-1.5 mm diameter, green
with minute white dots, terete or nearly
so, flexible, few or no prickles on the
outer branches; young prickles yellowish,
older ones black. Buds green, conical,
pointed, one scale, hidden by the old
leaf base.

TRUNK: Neither true bark nor
wood present since there is no cambium
in the stem. The outer "bark" green,
smooth, hard; densely set with prickles
toward the base of the plant and gradually diminishing in number toward the
stem tips; prickles nearly black, 2-10 mm
long, straight, slender, terete or flattened,
the base hardly enlarged. "Wood" greenish-white, soft, pithy.

HABITAT: Rich woods, along
stream banks, in thickets, wooded pastures or upland in open woods; in rich
or rocky soil.

RANGE: Connecticut through
southern Ontario, southward across Minnesota to South Dakota, south to eastern
Kansas and Texas, east to Georgia, and
north to New York.

S. hispida is sometimes misidentified
as *S. rotundifolia,* a southeastern species.
This is possibly due to the statement
appearing in some manuals that the
prickles of *S. hispida* are terete and those
of *S. rotundifolia* are flattened. This may
be true of the very short prickles, but
occasionally *S. hispida* has prickles up to
2.8 mm wide and 1 mm thick. Also,
young prickles or those on fast growing
sprouts of *S. hispida* are light colored—a
characteristic of *S. rotundifolia* mature
prickles.

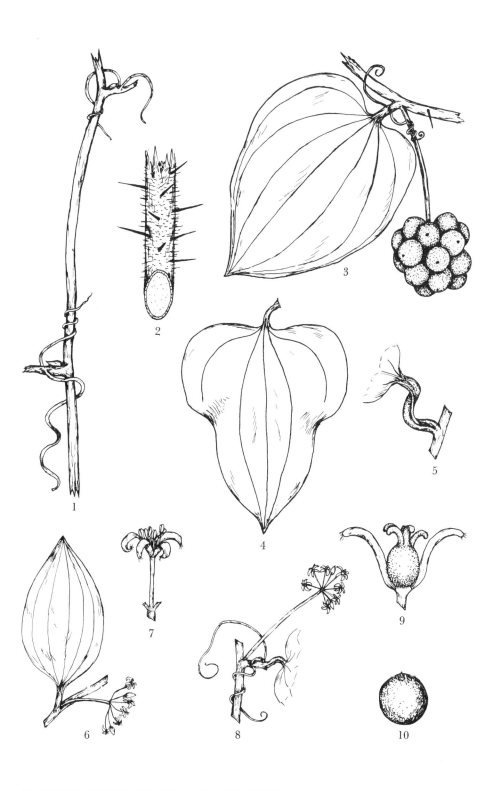

1. Winter twig
2. Detail of twig
3. Leaf and fruit
4. Leaf variation
5. Detail of petiole
6. Staminate flower cluster
7. Staminate flower
8. Pistillate flower cluster
9. Cutaway of pistillate flower
10. Seed

SALICACEAE
Salix alba L. var. *vitellina* (L.) Stokes

White willow

A tree to 25 m high, diffusely branched, the finer branches often drooping.

LEAVES: Alternate, simple, firm. Lanceolate, widest about the middle, 9-15 cm long, 12-20 mm wide; tip long-tapered, often falcate, base cuneate; margin serrate, 3-5 teeth per cm, the gland tip pointed forward; young leaves appressed pubescent, later glabrous on both surfaces and glaucous below; petiole 10-12 mm long, usually glandular at the summit, grooved above, pubescence heaviest above and at the broadened base; stipules narrowly lanceolate, 3 mm long, pubescent and usually with a tuft of long hairs at the base; early deciduous.

FLOWERS: May, with the leaves; dioecious. Staminate catkins on short peduncles about 1 cm long and with 2-4 leaves; catkin 4-6 cm long, 1 cm diameter, densely flowered, axis pubescent; scales narrowly ovate, 1.5 mm long, 0.5-0.75 mm wide, yellow, with long straight hairs on the outside; stamens 2, filaments slender, 4-4.5 mm long, white, pubescent at the base; anthers yellow; 2 glands, yellow, the front gland broadly triangular, back one narrow. Pistillate catkins 4-5 cm long, 4-5 mm wide, terminating a leafy shoot of the season, flowers crowded; scales lanceolate, 3-3.5 mm long, yellow-green, pubescent especially at the base and long ciliate at the tip; gland broad, shorter than the pedicel, green with a yellowish tip; pedicel 1 mm long; ovary lanceolate, 2.5-3 mm long, somewhat flattened, green, glabrous; style 0.75 mm long, the stigma with 2 lobes, greenish-brown.

FRUIT: Early June. Catkins 5-8 cm long, scales deciduous; capsule narrowly conic, 4-6 mm long, 1.5-1.7 mm diameter, glabrous, yellowish or light brown, the 2 valves splitting to the base but often curling outward unevenly. Seeds greenish brown, 1.4-1.7 mm long, 0.5-0.7 mm thick, cylindric, smooth, a slight mucro at the tip and a dense ring of long and short white hairs at the base.

TWIGS: 1.4-1.9 mm diameter, somewhat flattened, flexible, light golden yellow, puberulent only near the nodes, the lenticels elliptic, vertical, yellow or light orange with a dark marginal line; leaf scars crescent-shaped, enlarged at the middle and the rounded ends; 3 bundle scars; stipule scars small, oval; pith white, continuous, one-half of stem. Buds golden yellow, sparsely pubescent, tips usually turned toward the stem: floral buds ovoid, 8-10 mm long, 3-3.7 mm wide, flattened on the inner surface, tip rounded, sides and tip with a flattened rim; leaf buds similar but smaller, 3-6 mm long, 1.6-2.5 mm wide.

TRUNK: Bark light brown, with shallow furrows and broad, flat, tight ridges. Wood lightweight, soft, brownish, with a wide, white sapwood.

HABITAT: Escaping in many places, usually found in moist soils along streams, but may be in sandy areas around old farmsteads.

RANGE: Introduced from Europe and planted extensively in the United States. In our area it is especially common in North Dakota.

Most of our plants are var. *vitellina*, but occasionally var. *alba* is planted and escapes. It has pubescent branchlets and the young bark is brown or greenish brown.

During the winter, the brilliant yellow branches stand out sharply in contrast with the brown of other trees. This cannot be a positive identification character since several other species of cultivated willows also have bright yellow branches.

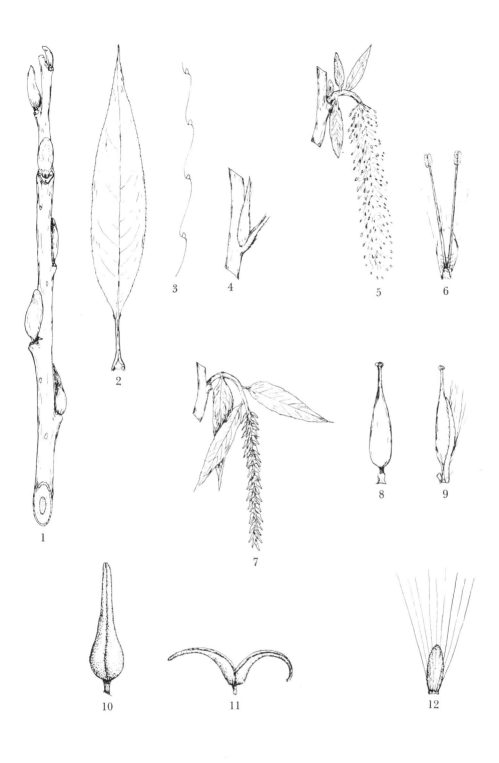

1. Winter twig
2. Leaf
3. Leaf margin
4. Stipule
5. Staminate catkin
6. Staminate flower
7. Pistillate catkin
8. Ventral view of pistil with gland
9. Lateral view of flower
10. Fruiting capsule
11. Open capsule
12. Seed

SALICACEAE
Salix amygdaloides Anderss.

S. wrightii Anderss.; *S. amygdaloides* var.
wrightii (Anderss.) Schneid.; *S. nigra*
Marsh. var. *amygdaloides* Anderss. in DC.

Peach-leaved willow, almond willow,
Wright willow

A tree to 20 m high, the branches fairly low, often somewhat drooping.

LEAVES: Alternate, simple. Lanceolate to ovate-lanceolate, 5-13 cm long, 18-30 mm wide; finely serrate, 6-7 teeth per cm with incurved callous tips; tip acuminate, base acute, often uneven; upper surface light yellow-green, lustrous, glabrous; lower surface pale and glaucous, glabrous; petiole slender, 12-18 mm long, glabrous, often twisted; stipules absent or reniform, 5-7 mm long, 8-12 mm wide, serrate, usually caducous.

FLOWERS: Early May, with the leaves; dioecious. Staminate catkins on short leafy branches 3-4 cm long; catkins 3-5 cm long, densely flowered, the axis pubescent; flowers sessile; scale ovate, 1.8-2.1 mm long, brownish, villous at the base and inside, deciduous; stamens 4-6, filaments white, 1-2 mm long, pubescent on the lower half; anthers yellow; 2 glands at the base of the scale. Pistillate flowers in catkins 5-8 cm long on a leafy branch 2-4 cm long; catkin erect or drooping, 60-80 flowers, the axis pubescent; scales oblong-ovate, 1.8-2 mm long, brownish, villous at the base and inside; pedicels 1-1.3 mm long, green, glabrous; ovary lanceolate, 2-2.5 mm long, green, glabrous; style 0.5 mm long; stigma 2-lobed, each divided, yellow-brown.

FRUIT: Late May. Ovoid with a tapering tip, 5.5-7 mm long, yellow at maturity, the two valves spreading, glabrous; pedicels 2 mm long. Seeds 1 mm long, 0.5 mm thick, cylindric, slightly constricted at the base, a minute groove on one side, greenish; a silky tuft of hairs 5-7 mm long at the base, breaks off easily as a ring.

TWIGS: 1.5 mm diameter, orange to red-brown, yellow-brown in age, shiny, glabrous, flexible; leaf scars narrow, nearly straight; 3 bundle scars; stipule scars small; pith white, continuous, one-fifth of stem. Buds broadly ovoid, 2-4 mm long, gibbous on the dorsal side, brown, lustrous.

TRUNK: Large, straight, and often leaning. Bark thick, brown, irregularly fissured, and with broad, flat ridges, often shaggy. Wood soft, lightweight, weak, not durable, light brown, with a wide, white sapwood.

HABITAT: Banks of streams and ponds, low woods, roadside gullies, and prairie sloughs.

RANGE: Quebec, west across southern Canada to British Columbia, south to Oregon, Utah, and Arizona, east to Texas, and northeast to Kentucky and Vermont.

S. amygdaloides and *S. nigra* are the only large willows native to the Central Plains States and should not be confused with the introduced *S. fragilis*, *S. babylonica*, or *S. alba*. These latter trees have bright golden-yellow branches. *S. amygdaloides* is the most common large willow of the western part of the Central Plains and is not as common in the eastern section. It is partially replaced in the southeastern section by *S. nigra* which does not extend far into the northwest of our area.

Salix X *glatfelteri* Schneid. is a hybrid between this and *S. nigra*.

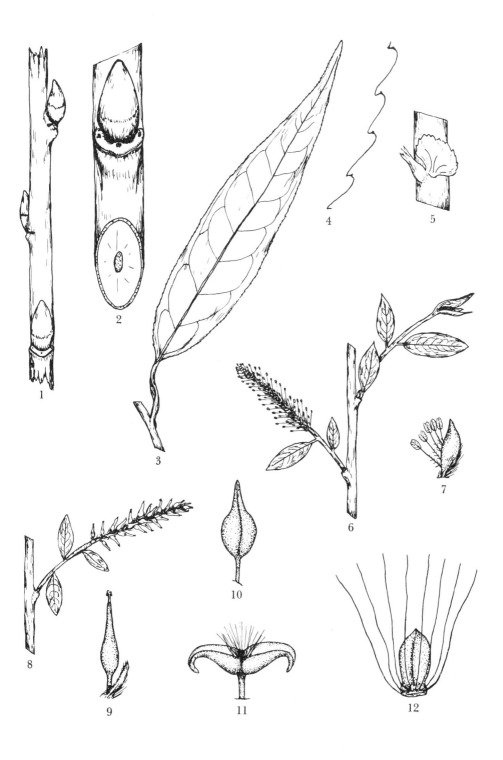

1. Winter twig
2. Detail of twig
3. Leaf
4. Leaf margin
5. Stipule
6. Staminate catkin
7. Staminate flower
8. Pistillate catkin
9. Pistillate flower
10. Mature capsule
11. Open capsule
12. Seed

SALICACEAE
Salix bebbiana Sarg.

var. *bebbiana* = *S. rostrata* Richards.;
S. cinerascens (var.) *occidentalis* Anderss.;
S. livida (var.) *rostrata* Dippel; *S. livida*
var. *occidentalis* Gray; *S. depressa* L. ssp.
rostrata Hiitonen. var. *perrostrata* (Rydb.)
Schneid. = *S. perrostrata* Rydb.; *S.
rostrata* Richards. var. *perrostrata* Fern.

Long-beaked willow

A shrub to 4 m high, with clustered trunks.

LEAVES: Alternate, simple, somewhat coriaceous. Elliptic, narrowly ovate or narrowly obovate, 4-6 cm long, 15-25 mm wide, tip short acuminate to acute, base cuneate; margin entire to shallowly toothed, 2-4 teeth per cm with a small dark gland at the tip; dark green and more or less villous above, often giving the effect of being gray-rugose; lower surface much lighter and somewhat glaucous, lightly or densely villous, the hairs more spreading than above; veins prominent beneath; petiole 5-8 mm long, pubescent, slightly grooved or terete, no glands at the summit; stipules foliaceous on sprouts, smaller on regular twigs, ovate to reniform, 2-6 mm long, 1-3 mm wide, early deciduous.

FLOWERS: Late May; dioecious. Staminate catkins 1.5-2.5 cm long on a short, leafy stem of the season, usually only two small leaves; catkin densely flowered, the axis pubescent; scales ovate to obovate, blunt, pubescent, greenish to pale brown, 1.5-2.5 mm long; stamens 2 with separate filaments about 5 mm long, glabrous; anthers yellow. Pistillate catkins 1.5-4 cm long on a shoot of the season with 2 leaves, densely flowered, the axis pubescent; scales ovate to oblong, greenish, pubescent; pedicel about 2 mm long, pubescent; ovary green, rather abruptly narrowed to a long neck, total length about 2.3 mm and about 0.5 mm wide, silky pubescent; style about 0.2 mm long; stigmas 2, spreading, 2-lobed.

FRUIT: June. Catkins ascending, 3-5 cm long, 1.5 cm wide; capsule conic-ovoid, 6-8 mm long, pedicels about 3 mm long, 2-3 times as long as the persistent scale; capsule puberulent or silky pubescent, becoming yellow-brown at maturity and splitting into 2 valves. Seeds obovoid, dark purple or green, smooth, 1 mm long, a ring of white, silky hairs 5-8 mm long around the base; apex rounded and often mucronate.

TWIGS: 1-1.2 mm diameter, flexible, red-brown to gray-brown, tomentulose, the lenticels vertical, broadly elliptic, light-colored but not obvious; leaf scars raised, crescent-shaped, with a narrow, yellowish rim on the lower side; pith white, continuous, about one-half of stem. Buds flattened with a flange on the sides and end, broadly ovate, light orange-brown, much lighter than the twig, glabrate, the scale thick; flower buds 3-4.5 mm long, 1.7-2.4 mm wide, glabrate or puberulent; leaf buds smaller and more rounded.

TRUNK: The shrub often with a gnarled appearance; bark gray, with shallow furrows and flat ridges. Wood medium hard, fine-grained, light brown, with a narrow, white sapwood.

HABITAT: Valleys and stream banks or on low hillsides, moist or dry.

RANGE: Alaska, east across Canada to Newfoundland and Labrador, south to Maryland, west to Iowa and northwestern Nebraska, south through Colorado to New Mexico and Arizona, north to eastern Oregon, Washington, and British Columbia. Apparently not in California.

Var. *perrostrata* is a more glabrous plant with thinner, entire leaves, the trunks in small clusters or single and grows in drier places.

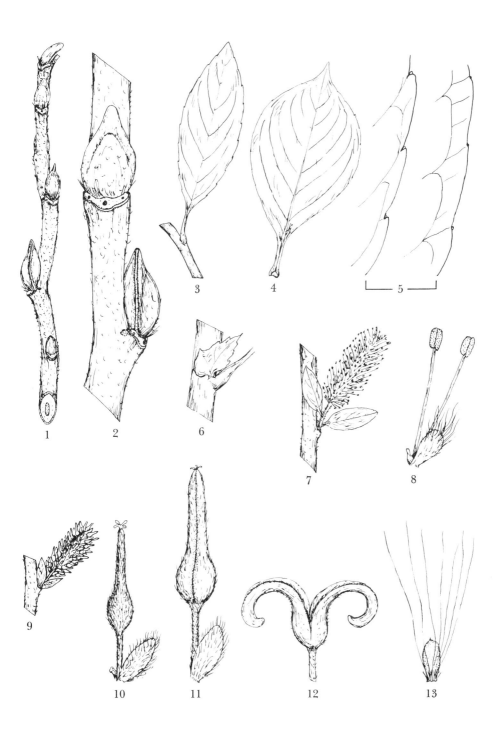

1. Winter twig
2. Detail of twig
3. Leaf
4. Leaf variation
5. Leaf margins
6. Stipule
7. Staminate catkin

8. Staminate flower
9. Pistillate catkin
10. Pistillate flower
11. Mature capsule
12. Open capsule
13. Seed

SALICACEAE
Salix candida Fluegge ex Willd.

S. candidula Nieuwl.

Hoary willow

A shrub to 1.5 m high, with many trunks from one base.

LEAVES: Alternate, simple. Oblanceolate or oblong, 7-12 cm long, 8-14 mm wide, acute, base cuneate and often uneven; dark green, rugose with the veins impressed, glabrate or permanently lightly tomentose above; densely tomentose below; margin revolute; young leaves with long, silky hairs in addition to the tomentum and with minute, wide-spaced, glandular teeth usually not seen in mature leaves; petiole 5-7 mm long, broadly grooved above, wide base, tomentose; stipules ovate, 2-3 mm long, entire or glandular serrulate, tomentose.

FLOWERS: May, with or before the leaves; dioecious. Catkins on short, bracteate peduncles or sessile, from buds of the previous season. Staminate catkins 1.5-2 cm long, 1-1.5 cm wide, densely flowered; one gland, stout, greenish; stamens 2, the filaments often united at the base, 3-3.5 mm long, white, glabrous; anthers red at first, yellow at anthesis; scale dark red or brownish, darker at the tip, oblanceolate, rounded, long silky pubescent on both sides. Pistillate catkins 2-2.5 cm long, 1 cm wide, densely flowered; scale obovate, 1-1.5 mm long, brown to nearly black, rounded, long silky pubescent; 1 gland, red-brown, shorter than the pedicel; pedicel 0.5-1 mm long; ovary narrowly ovoid, 2-2.5 mm long, densely woolly; style 1-1.25 mm long, red-brown, divided near the summit and with 2-4 linear, red stigma lobes.

FRUIT: Mid-June. Catkins 3-4 cm long, 1.5 cm wide; capsule narrowly ovoid, 6-9 mm long, blunt tip, densely tomentose, stramineous beneath the tomentum; pedicel 1 mm long; scales persistent. Seeds obovate and apiculate, 1.2 mm long, 0.5 mm thick, green, a single ring of long, fine silky hairs at the base.

TWIGS: 1-2 mm diameter, red-brown or green-brown, partly or densely tomentose, none of the hairs short or porrect, lenticels small, elliptic; leaf scars crescent-shaped, narrow; stipule scars broadly oval; 3 bundle scars; pith pale tan or white, continuous, one-third of stem. Buds ovoid, 2-5 mm long, 1-2 mm wide, blunt, slightly flattened, red-brown, glabrate or densely tomentose.

TRUNK: Bark gray or gray-brown with a greenish tinge, tight, roughened by old leaf scars and tan, oval lenticels, occasionally minutely fissured. Wood medium hard, white, with no noticeable line of sapwood.

HABITAT: Swamps and bogs, usually in open areas.

RANGE: Alaska, south to Idaho, southeast to Colorado, north to North Dakota, southeast and east across Indiana to Pennsylvania, north to Labrador and west to British Columbia.

S. candida and *S. humilis* are often confused, but *S. candida* grows right in the wet soil of a bog and *S. humilis* grows on drier land. Herbarium specimens may be difficult as some of the differences are quite subtle. The leaves of *S. candida* are dark green and usually rugose above with the veins indented, while those of *S. humilis* are yellow-green and usually smooth. The pubescence is quite different, that of the lower surface dense, cobwebby, and the individual hairs cannot be identified on *S. candida;* open, not cobwebby, and the individual hairs can be traced to their base on *S. humilis.* The upper surface of *S. candida* leaves is loosely covered with curly, cobwebby hairs matted and lying close to the surface, while those of *S. humilis* are straight, separate, and somewhat porrect.

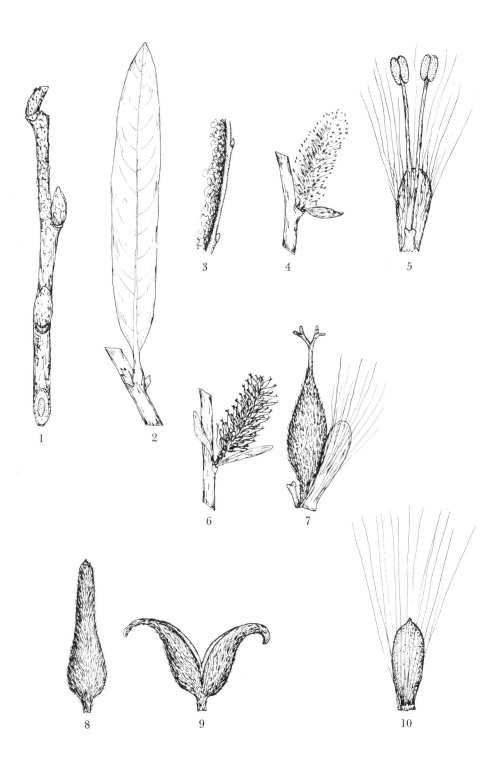

1. Winter twig
2. Leaf
3. Leaf margin, underside
4. Staminate catkin
5. Staminate flower
6. Pistillate catkin
7. Pistillate flower
8. Mature capsule
9. Open capsule
10. Seed

SALICACEAE
Salix caroliniana Michx.

S. amphibia Small; *S. occidentalis* Bosc
in Koch; *S. longipes* Shuttlew. ex
Anderss.; *S. wardii* Bebb; *S. longipes* var.
wardii (Bebb) Schneider

Carolina willow, Ward's willow, coastal-
plain willow

A tree to 10 m high, with one main trunk, occasionally clustered.

LEAVES: Alternate, simple. Lanceolate, 7-11 cm long, 1.5-2 cm wide; finely serrate, 10-12 teeth per cm with a light-colored callous tip; leaf tip acuminate, base rounded or cuneate; upper surface dark green, glossy, glabrous; lower surface silvery-white, glaucous, glabrate; young leaves tomentose; petiole 5-10 mm long, tomentose; stipules ovate to reniform, 5-15 mm long, 2-6 mm wide, acute, serrate, tomentose to glabrate, occasionally foliaceous and persistent.

FLOWERS: April, with the leaves; dioecious. Staminate catkins terminal on short, new leafy branches; catkins 4-5 cm long, axis villous, flowers sessile, somewhat in whorls; scales ovate, acute, green-yellow, villous inside and a few hairs outside; 2 yellow glands, one usually larger than the other; stamens 4-7, filaments 2 mm long, green, glabrous, the anthers yellow. Pistillate flowers in terminal catkins on short, new leafy branches; catkins 4.5-5.5 cm long, axis pubescent; pedicel 1.5-1.8 mm long, green, glabrous; scales greenish or light brown, glabrous outside, villous inside and on the margin; ovary green, ovoid, 2-3 mm long, tapered, often slightly enlarged at the tip, glabrous; style 0.2 mm long or absent; stigma 2-lobed, each lobe divided, yellowish. Often a small shoot is produced in the axil of the upper leaf of the branch on which the catkins are borne, apparently making the catkin lateral.

FRUIT: Mid-June. Pedicels 1.5-2 mm long, glabrous; capsule ovoid, 6-7 mm long, 2.5 mm thick, abruptly narrowed above the middle, glabrous, brown at maturity; scales persistent. Seeds cylindric, 1 mm long, 0.5 mm thick, green, constricted near the base, slightly grooved on one side, apex mucronate; long silky hairs at the base, which are 2-3 times as long as the seed, break off easily either as a ring or singly.

TWIGS: 2 mm diameter, red-brown to gray-brown, pubescent or glabrous; leaf scars narrowly crescent-shaped, enlarged at the ends and center; 3 bundle scars; stipule scars small; pith pale brown, continuous, one-fifth of stem. Buds 2-3 mm long, pubescent or glabrate, acute; 1 scale, dark brown with a lighter base.

TRUNK: Bark dark brownish-gray, shallowly furrowed, the ridges thin, brown, and firm. Wood soft, not durable, nearly white, with a narrow, white sapwood.

HABITAT: Gravel bars, sand bars, rocky stream beds and banks close to water; rocky prairie sloughs; usually in the bed of the stream.

RANGE: Pennsylvania, west to Missouri and eastern Kansas, south to Texas, east to Florida, and north to Maryland.

S. caroliniana may often be confused with *S. eriocephala,* especially along the Missouri River area in Kansas and southern Nebraska where the two ranges overlap. *S. caroliniana* has glabrous or pubescent twigs, definitely glaucous under leaf surface, deciduous yellow catkin scales, 3 or more stamens and usually no style. *S. eriocephala* has densely pubescent twigs, slightly glaucous under leaf surface, persistent dark brown scales, only 2 stamens and a style of about 0.5 mm.

S. caroliniana is not common in our area but is scattered with a few plants or a small population in one place, then it may be several miles before the next tree appears.

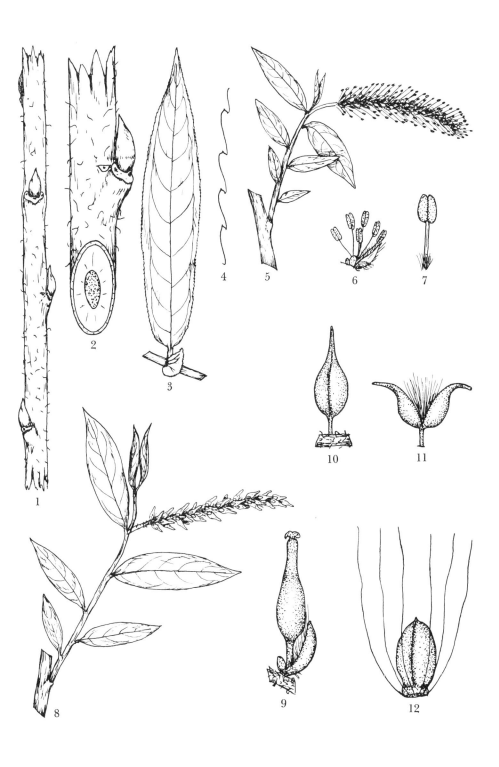

1. Winter twig
2. Detail of twig
3. Leaf
4. Leaf margin
5. Staminate catkin
6. Staminate flower

7. Stamen
8. Pistillate catkin
9. Pistillate flower
10. Mature capsule
11. Open capsule
12. Seed

SALICACEAE
Salix discolor Muhl.

Incl. var. *latifolia* Anderss. and var.
overi Ball

Pussy willow

A small tree to 5 m high, the trunks single or clustered, the branches high.

LEAVES: Alternate, simple. Elliptic, oblong or narrowly obovate, 4-8 cm long, 1.5-3 cm wide, acute to short acuminate, base cuneate to narrowly rounded; margin irregularly toothed with 2-3 per cm or merely undulate; teeth if present on young leaves, rounded and with an inturned gland; upper surface dark green, semilustrous, glabrous; lower surface lightly glaucous, glabrate; young leaves pubescent and occasionally with a few fulvous hairs; petiole 6-10 mm long, circular in section, puberulent, the blade often slightly decurrent; stipules lanceolate, 0.5-1 mm long, promptly deciduous.

FLOWERS: May, before the leaves; dioecious. Catkins sessile from winter buds. Staminate catkins 2-4 cm long, 1-1.5 cm wide, densely flowered; scales ovate, dark brown and often with a greenish base, long pubescent, 1.5-2 mm long; 1 gland, stout, blunt, greenish; stamens 2, distinct, filaments white, 3-4 mm long, glabrous; anthers yellow. Pistillate catkins 1.5-3 cm long, 1 cm wide; scales ovate, dark brown, densely pubescent, 1.5-2 mm long; pedicel 0.5 mm long, green with short appressed pubescence; gland yellow-green, stout, about as long as the pedicel; ovary narrowly conical with a long tapered tip, 2-2.5 mm long, green, appressed pubescent; style 1-1.5 mm long, green, glabrous or sparsely pubescent; stigma knob-like or with 2-4 lobes, greenish.

FRUIT: June. Catkins drooping, 6-8 cm long, 2.5-3 cm wide, capsules wide spreading; pedicels 2-3 mm long, pubescent; capsules ovoid with a long neck, 10-13 mm long, the neck being 8-9 mm long, finely pubescent; scales persistent. Seeds dark green, cylindric, 1.25-1.6 mm long, 0.5 mm diameter, blunt tip, a ring of short, stiff hairs and an outer ring of long, silky hairs around the base.

TWIGS: 1.5 mm diameter, flexible, red-purple to red-brown, dull, glabrous or glabrate, the lenticels oval, small, yellow; leaf scars crescent-shaped with rounded ends, pubescent on the upper edge; 3 bundle scars; stipule scars small and circular; pith white to brownish, continuous, one-half of stem. Buds ovoid, 6-9 mm long, 3-3.5 mm wide, gibbous, acute, often twisted to the side and somewhat appressed to the stem, sparingly appressed puberulent.

TRUNK: Bark gray-brown, tight with longitudinal, fine ridges, slightly fissured and ridged on old trunks. Wood soft, brownish, with a wide, white sapwood.

HABITAT: Along creeks and margins of bogs and swamps, either in the open or in wooded areas.

RANGE: Alberta, southeast across Montana and South Dakota to northeastern Missouri, east to Kentucky, north to Labrador, and west across southern Canada.

S. discolor is often confused with *S. bebbiana*, a species with smaller, dull leaves usually pubescent above and the flower scales are greenish or pale brown, also the leaves are more often obovate. Occasionally it is confused with *S. scouleriana*, but that plant is rather open, appears as a single trunk and somewhat scraggly, growing in higher and drier places, often toward the tips of the hills in the Black Hills.

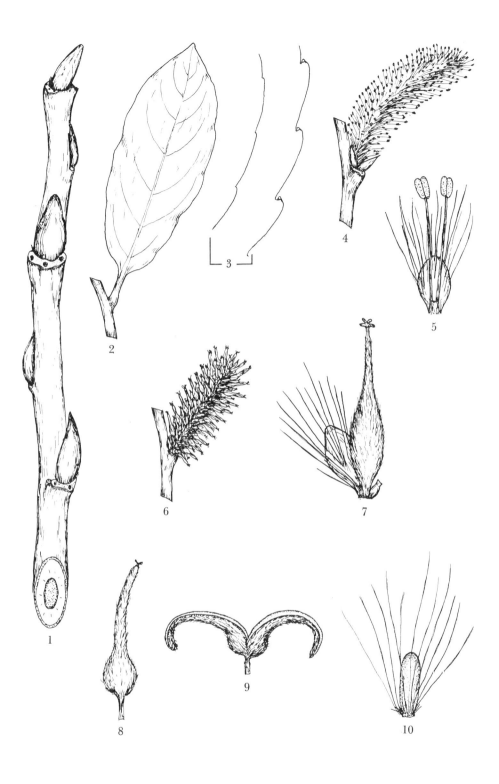

1. Winter twig
2. Leaf
3. Leaf margin variation
4. Staminate catkin
5. Staminate flower

6. Pistillate catkin
7. Pistillate flower
8. Mature capsule
9. Open capsule
10. Seed

SALICACEAE
Salix eriocephala Michx.

S. *missouriensis* Bebb; S. *rigida* Muhl.
var. *vestita* Anderss.

Missouri willow, heartleaf willow

A tree to 7 m high, with one main trunk, rarely clustered.

LEAVES: Alternate, simple. Lanceolate to elliptic, 4-9 cm long, 1-2 cm wide, finely glandular serrate, 6-11 teeth per cm; tip acuminate, base obtuse or rounded; upper surface dark green, lustrous, pubescent, becoming glabrate, the midrib pubescent on both surfaces; lower surface pale and lightly glaucous, glabrous or with a few hairs, the veins slender, often united near the margin, usually yellowish and obvious on both sides; petiole stout, 1-1.2 cm long, pubescent, yellowish; stipules foliaceous, semicordate, ovate or reniform, serrate with incurved teeth, dark green, glabrous above, lightly glaucous beneath, deciduous or persistent.

FLOWERS: Early April, before the leaves; dioecious. Staminate catkins from buds of the previous year, often showing the white hairs in early winter; catkins 4-7 cm long including the short peduncle, flowers crowded, axis woolly; 2 leafy bracts on the peduncle; pedicels 0.5-1 mm long; scale dark purple, or greenish with a dark tip, ovate, 1.4-1.5 mm long, pubescent, persistent; 1 greenish gland 1 mm long, narrow or wide; stamens 2, filaments white, glabrous, 3.5-4.8 mm long, often united near the base; anthers yellow. Pistillate catkins 3.5-7 cm long including the peduncle, axis woolly; pedicels 1.8-2.3 mm long, green, glabrous; scale dark purple-brown, oblong, 1.4-1.5 mm long, long pubescent; 1 gland, greenish-yellow, 1 mm long, narrow or wide; ovary lance-ovate, 2.8-3.3 mm long, green, glabrous; style 0.3-0.5 mm long, green; stigma 2-lobed, yellowish.

FRUIT: May. Catkins 5-7 cm long; pedicels 2-2.4 mm long; capsules 4-5 mm long, ovate with a long neck, glabrous, light brown, the 2 valves recurling sharply at maturity; scales persistent. Seeds cylindric, 1-1.3 mm long, 0.3-0.4 mm thick, constricted slightly at the base, a minute ridge or a groove on the sides; basal ring of short, appressed hairs and long hairs 2-3 times as long as the seed.

TWIGS: 1 mm diameter, light green or brown, covered with thick white or gray pubescence in the first year; red-brown, glabrous or pubescent in the second year; leaf scars narrowly crescent-shaped; 3 bundle scars; large stipule scars; pith white to cream, continuous, one-fourth of stem. Buds flattened, 4-7 mm long, acute, red-brown, hoary tomentose.

TRUNK: Bark thin, smooth, light gray with a slight tinge of red, the lenticels large, horizontal, elliptic or diamond-shaped; older bark shallowly fissured, the ridges wide, flat, and tightly appressed. Wood durable, dark brown, with a narrow, pale sapwood.

HABITAT: Sandy or rocky wet areas, along streams or in wet meadows and pastures.

RANGE: Kentucky, northwest to Minnesota and North Dakota, south to northeastern Kansas, and east to Missouri.

The range of S. *eriocephala* is rather small compared to most of the other willows, and in general, is only a wide band along the Mississippi and the Missouri rivers, starting in southern Missouri and extending into Minnesota and North Dakota. It is not abundant, but may be found occasionally along the smaller tributaries of these large rivers. It may form a thicket-like cluster in the silt at the upper end of a small lake where a creek enters the lake.

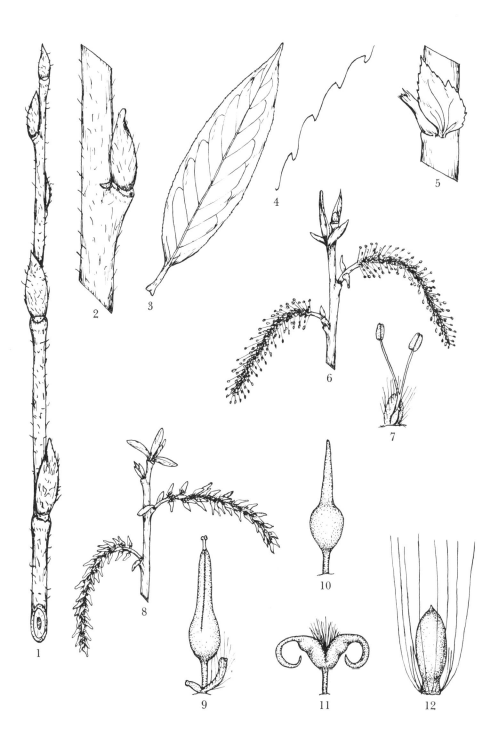

1. Winter twig
2. Detail of twig
3. Leaf
4. Leaf margin
5. Stipules
6. Staminate catkin
7. Staminate flower
8. Pistillate catkin
9. Pistillate flower
10. Mature capsule
11. Open capsule
12. Seed

SALICACEAE
Salix exigua Nutt. ssp. *exigua*

S. longifolia Muhl. var. *exigua* Bebb;
S. fluviatilis Sarg. var. *exigua* Sarg.;
S. argophylla Nutt.; *S. longifolia* var.
argophylla Anderss.

Sandbar willow, coyote willow

A colonial shrub to 4 m high, spreading from underground suckers.

LEAVES: Alternate, simple, the midrib usually raised on the upper side, the lateral veins inconspicuous. Lanceolate, 5-12 cm long, 5-10 mm wide; tip and base long tapered; margin with low distant teeth, 1-4 per cm; both sides permanently silvery, appressed pubescent; sessile or with a petiole about 1 mm long and winged to the base; stipules lanceolate, often lobed on one side at the base, pubescent, early deciduous.

FLOWERS: Late May-June; dioecious. Staminate catkins on new leafy branches 1-3 cm long from lateral buds of the previous year; catkins 1.5-2.5 cm long, 5-6 mm wide, erect or ascending, axis densely villous; scale pale greenish-yellow, ovate to obovate, densely pubescent on both sides; 1 linear yellow gland, 0.75 mm long, between the stamens and the scale, and another gland between the stamens and the catkin axis; stamens 2, filaments distinct, white, pubescent below the middle, 2.5-3 mm long; anthers light yellow. Pistillate catkins on leafy branches of the season 2.5-3 cm long; catkins 3-4 cm long, 4-5 mm wide, ascending or slightly drooping; capsules nearly parallel to the axis, sessile or nearly so; scales ovate, 2.5-3 mm long, pale greenish-yellow, rounded, often wrinkled, pubescent on both sides, densely crinkly ciliate; 1 gland between the ovary and the axis, broadly triangular; ovary green, narrowly pear-shaped, 4.5-5 mm long, 1 mm wide, the basal portion about 2 mm long, sparsely pubescent; style obsolete; stigmas brownish with 4 short, linear segments.

FRUIT: July. Catkins 3-5 cm long; capsules narrowly ovoid with the neck more than half the length, 6-7 mm long, 1.5 mm wide, blunt, glabrous or slightly pubescent, pedicel 1 mm long or less; mature capsule yellow-brown, the 2 valves widely spread and recurved.

Seeds dark green, cylindric, apiculate, constricted near the base, 0.75-1 mm long, a ring of long, fine, white hairs around the base, easily broken off.

TWIGS: 1 mm diameter, flexible, red-brown, densely puberulent, tips often die back from one to many nodes; lenticels scattered, same color as the twig, elliptic, slightly raised; leaf scars broadly V-shaped or crescent-shaped, narrow; 3 bundle scars; pith pale brown, continuous, one-third to one-half of stem. Buds ovoid with sides nearly parallel, flattened, blunt, puberulent, red-brown, 3-4 mm long, 1-1.75 mm wide.

TRUNK: Bark light gray-green to brown, smooth but with raised lenticels; occasionally a larger tree exists with gray-brown bark and shallow furrows. Wood soft, fine-grained, the growth rings obvious, no definite line of heartwood and sapwood.

HABITAT: Stream banks, valleys, marshy areas, ditches, and occasionally on drier ground.

RANGE: Southern British Columbia to Montana, south through the Black Hills and western Nebraska to Texas, west to Arizona, and north through the Intermountain Region.

This subspecies intergrades freely with ssp. *interior* and it is often difficult to determine the subspecies accurately.

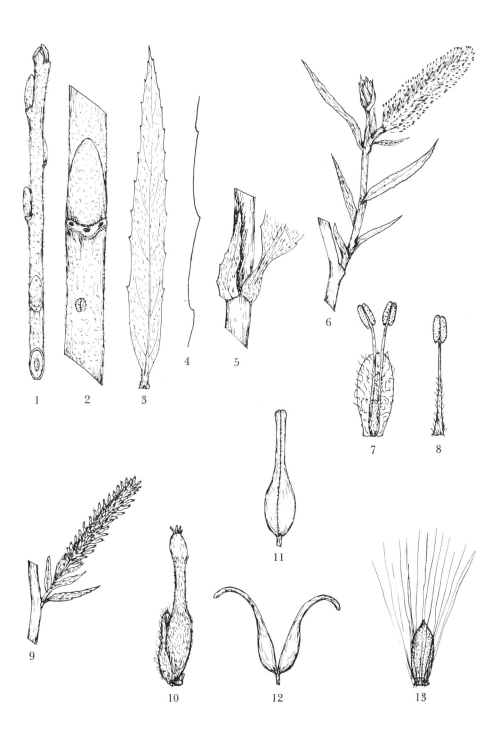

1. Winter twig
2. Detail of twig
3. Leaf
4. Leaf margin
5. Stipule and leaf base
6. Staminate catkin
7. Staminate flower

8. Stamen
9. Pistillate catkin
10. Pistillate flower
11. Mature capsule
12. Open capsule
13. Seed

SALICACEAE
Salix exigua Nutt. ssp. *interior* (Rowlee) Cronq.

S. interior Rowlee; *S. interior* var.
wheeleri Rowlee; *S. exigua* var.
luteosericea (Rydb.) Schneid.; *S.
luteosericea* Rydb.; *S. longifolia* Muhl.
var. *interior* M. E. Jones

Sandbar willow

A colonial shrub or small tree to 5 m high, usually with central trunk.

LEAVES: Alternate, simple. Linear-lanceolate, often falcate, 6.5-12 cm long, 5-8 mm wide; remotely dentate with shallow, glandular teeth, 1-4 per cm, often prolonged on sprouts; tip long acuminate, base acuminate; upper surface light yellow-green, slightly lustrous; lower surface paler, glabrous or glabrate; the lateral veins branch from the midrib at an acute angle but curve and run nearly parallel to the leaf margin before ending in a tooth; young leaves soft and silky; petiole 3-6 mm long, grooved, glabrous; stipules linear, 1.5 mm long, pubescent, caducous.

FLOWERS: Early May; dioecious. Staminate catkins 3-4 cm long on a short, leafy branch; axis silky; scales yellowish, 2.3-2.5 mm long, entire or erose, glabrous outside, pubescent inside and on the margin; stamens 2, filaments 4 mm long, pubescent toward the base; anthers yellow; 2-3 yellow glands at the base of the scale. Pistillate catkins 2.5-3 cm long, axis pubescent; pedicels 0.5-0.7 mm long, green, glabrous; scales yellowish, 2.5-3 mm long, glabrous outside and pubescent inside; ovary 3 mm long, lanceolate, sparingly pubescent, green; stigma sessile, 4-lobed, yellow-brown.

FRUIT: June. Catkins 3.5-5 cm long; pedicels 0.6-0.7 mm long; capsules 6-7.3 mm long, narrowly ovoid, crowded on the axis, pale brown, glabrous; the 2 valves spreading at maturity. Seeds cylindric, 0.9-1 mm long, 0.4 mm thick, dark green, a slight ridge on one side, constricted near the base, a ring of silky hairs 2-4 times the length of the seed around the base.

TWIGS: 0.8 mm diameter, light yellow to orange, smooth, flexible; leaf scars narrow, nearly straight; 3 bundle scars; stipule scars small; pith white, continuous, one-third to one-half of stem.

Buds 2-3 mm long, ovoid, narrow, chestnut brown, tip acute.

TRUNK: Slender, straight, with small branches. Bark thin, smooth, yellow-brown on small trees; dark brown, often red-tinged, and with shallow fissures on large trees. Wood soft, lightweight, close-grained, weak, brittle, light brown, with a pale brown sapwood.

HABITAT: Stream banks and lakeshores, roadside ditches, and prairie sloughs; in rocky or sandy soil; often covering a sandbar in a river.

RANGE: New Brunswick to Alaska, southeast to Wyoming, and south to New Mexico, east to Louisiana, north to Arkansas, east to Kentucky and Virginia, and north to Maine.

This is a common willow throughout our range and might be found in any locality. It forms thickets which become quite dense, and as the trees grow larger the thickets may be hard to penetrate. It is one of the first plants of a woody nature to inhabit a newly made sandbar in a river. Occasionally a plant is found which has more pubescence on the twigs and leaves; these have been called var. *wheeleri* Rowlee, but can hardly be separated. The narrow-leaved form, var. *pedicellata* (Anderss.) Ball, is commonly found, but just where the line is between the width of the leaves is questionable, although 5 mm wide is usually given as the widest for var. *pedicellata*. Gall-infected plants usually remain permanently silvery pubescent and have been identified as ssp. *exigua*.

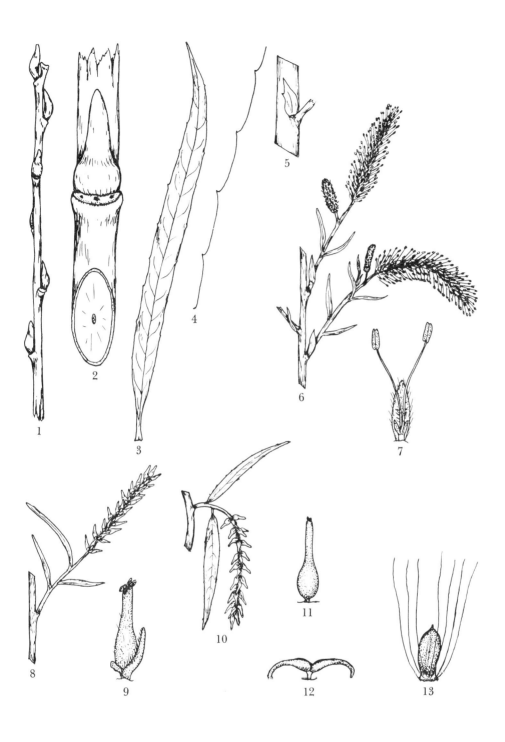

1. Winter twig
2. Detail of twig
3. Leaf
4. Leaf margin
5. Stipule
6. Staminate catkin
7. Staminate flower
8. Pistillate catkin
9. Pistillate flower
10. Fruiting catkin
11. Mature capsule
12. Open capsule
13. Seed

SALICACEAE
Salix humilis Marsh.

var. *microphylla* (Anderss.) Fern. = *S. tristis* Ait. and *S. humilis* var. *tristis* (Ait.) Griggs
var. *hyporhysa* Fern. = *S. humilis* var. *rigidiuscula* (Anderss.) Robins. & Fern.

Dwarf gray willow, sage willow, upland willow

A colonial shrub to 1.5 m high, trunks single or clustered.

LEAVES: Alternate, simple. Oblanceolate, 6-9 cm long, 7-20 mm wide; margin entire or undulate-crenate, revolute; tip rounded or acute, mucronate, base tapered; upper surface light yellow-green, glossy or dull, slightly pubescent with straight hairs, occasionally slightly rugose; lower surface silvery white pubescent or glabrate; leaves usually crowded near the summit of the branches; petiole 5-10 mm long, pubescent; stipules half-round to crescent-shaped, 1.2 mm long, pubescent; or linear and leaf-like, 4-7 mm long.

FLOWERS: Early April, before the leaves; dioecious. Staminate catkins axillary from buds of the previous year, sessile, 10-15 mm long, 8-15 mm wide, including the spreading stamens; scales obovate, 1.7 mm long, green at the base with a red or nearly black tip, long-hairy; 1 gland, 0.9 mm long, green to red, cylindric, enlarged toward the base; stamens 2, filaments white, 5-7 mm long, glabrous; anthers red-orange. Pistillate catkins axillary on growth of previous year, catkins sessile, 1.4-2.4 cm long, 6-9 mm wide; scales obovate, 1.5 mm long, green at the base, reddish or nearly black at the tip, long-hairy; 1 gland, 0.6 mm long, green or reddish; stalk of ovary 1.6 mm long, pubescent; ovary green, long conical, 2.5 mm long, 0.7 mm wide, pubescent; stigma nearly sessile, 2-lobed, reddish. Pistillate scales and glands slightly smaller than those of the staminate flowers.

FRUIT: Early May. Catkins 2-3 cm long; pedicels 1.5-2 mm long; capsule ovoid, 5-6 mm long, 1.8-2 mm wide at the base, yellow-brown, pubescent, a tuft of long, silky hairs at the base inside; the 2 valves strongly recurled at maturity. Seeds cylindric, 1.2-1.5 mm long, 0.4-0.5 mm wide, constricted at the base, a slight ridge on one side, dark greenish black, surrounded by a basal ring of long, white, silky hairs.

TWIGS: 1 mm diameter, yellow beneath the gray pubescence, flexible; usually crowded toward the end of the trunk; leaf scars narrowly crescent-shaped, enlarged at the ends and center; 3 bundle scars; stipule scars small; pith white becoming brown, continuous, one-fifth of stem. Buds ovoid, 4-6 mm long, flattened, reddish with white pubescence.

TRUNK: Bark gray-brown, rough, with crooked lines of cracked bark. Wood white, soft.

HABITAT: Dry, upland prairies, roadsides, fence rows; rocky or sandy soils; often abundant in sandy prairies; browsed heavily in pastures.

RANGE: Maine to North Dakota, south to eastern Oklahoma and Louisiana, east to Florida, and north to New England.

The varieties are not always completely separable but the extremes appear quite different.

var. *microphylla*: shrub to 1 m high, leaves 1.5-4 cm long, catkins 6-9 mm long, fruiting catkins 1-1.5 cm long.

var. *humilis*: shrub 1-1.5 m high, leaves 5-9 cm long, catkins 1-3 cm long, fruiting catkins 2-5 cm long, lower leaf surface densely hairy.

var. *hyprohysa*: same as var. *humilis* but the lower leaf surface becoming glabrate or glabrous.

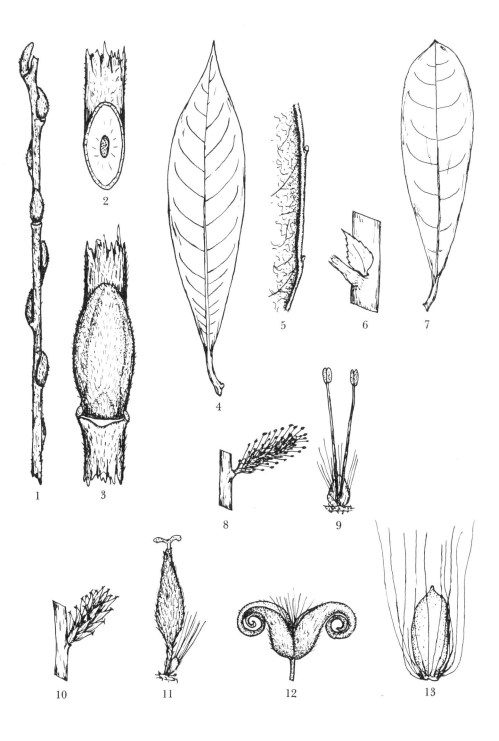

1. Winter twig
2. Twig section
3. Detail of twig
4. Leaf
5. Leaf margin, under side
6. Stipule
7. Leaf variation
8. Staminate catkin
9. Staminate flower
10. Pistillate catkin
11. Pistillate flower
12. Mature capsule
13. Seed

SALICACEAE
Salix lucida Muhl.

Shining willow

A tree to 4 m high, the trunks single or clustered.

LEAVES: Alternate, simple. Lance-ovate, usually broadest just below the middle, 5-8 cm long, 12-20 mm wide; tip acuminate to long acuminate and falcate, base broadly cuneate to somewhat rounded; margin finely crenate-serrate, 8-10 teeth per cm, a rounded gland on the infacing side; upper surface yellow-green, semiglossy; lower surface slightly paler, not whitened; both surfaces glabrous; petiole stout, 6-8 mm long, flattened or shallowly grooved above, thinly puberulent above, glabrous below, base broad, 2 to several glands at the summit, some raised on stalks; stipules flabellate, 2-3 mm long, 3-4 mm wide with coarse glandular teeth, the glands rounded. Young leaves pubescent with rufous hairs and golden, globular glands on the teeth.

FLOWERS: Early June; dioecious. Staminate catkins 2-3 cm long, 1 cm wide, densely flowered, on a puberulent branch 2-3 cm long, with 3-5 leaves; scales yellow, 3-4 mm long, oblong, often lobed on the end and curved around the stamens, lightly pubescent; 2 small, pyramidal, orange glands; stamens 3-5, filaments 4-5 mm long, pubescent at the base; anthers yellow. Pistillate catkins on short, leafy branches; catkin 2-2.5 cm long, 5 mm diameter, compact, capsules ascending and appressed to the axis, the axis densely pubescent; scales yellow-green, oblong, 2-2.5 mm long, often clasping the base of the ovary, pubescent on both sides; gland short, broad, yellow-green, shorter than the pedicel; pedicels 0.5 mm long, glabrous; ovary narrowly ovoid, 3-3.5 mm long, 0.8 mm wide, green, glabrous; style 1 mm long, yellowish, divided at the summit into two 2-lobed, red-brown stigmas.

FRUIT: Early June. Catkins 2.5-3 cm long, crowded; capsules ovoid, with a long neck, 7-8 mm long, 1.5 mm wide, light brown, glabrous; pedicels 1 mm long; scales deciduous. Seeds brownish, cylindric, 1.4-1.6 mm long, 0.4-0.5 mm wide, smooth, a ring of short and long hairs at the base.

TWIGS: 1.2-1.4 mm diameter, flexible, glabrous, glossy, yellow-brown, the lenticels small, inconspicuous, slightly lighter color; leaf scars crescent-shaped with rounded ends; 3 bundle scars; stipule scars oval; pith greenish, continuous, over one-half of stem. Buds ovoid, 4-6 mm long, 2-2.3 mm wide, rounded at the tip, flattened on the margins and tip, red-brown and darker than the twig, glabrous, short-stalked; end bud sharply curved.

TRUNK: Bark gray-brown, smooth, tight, lenticels broadly oval, horizontal, and orange-brown. Wood soft, close-grained, white, with no line of heartwood and sapwood.

HABITAT: Stream banks, moist hillsides, and open meadows along streams and on flood plains.

RANGE: Northern Manitoba to southern Labrador and Newfoundland, south to Delaware, west across Ohio and Iowa to the Black Hills and North Dakota; also in Colorado.

Salix lucida is rare in our area and apparently not many specimens have been collected. The leaves have glands at the summit of the petiole, the blade long acuminate and often falcate, definitely green on the underside. The first two of these characters should differentiate it from *S. rigida*, var. *watsonii*.

1. Winter twig
2. Leaf
3. Leaf margin
4. Summit of petiole
5. Stipule
6. Staminate catkin
7. Staminate flower

8. Pistillate catkin
9. Pistillate flower, side view
10. Pistillate flower, ventral view
11. Mature capsule
12. Open capsule
13. Seed

SALICACEAE
Salix monticola Bebb ex Coult.

S. padophylla Rydb.; *S. cordata* Muhl.
var. *monticola* L. Kelso; *S. padifolia*
Rydb., not Anderss.; *S. barclayi* Anderss.
var. *padophylla* L. Kelso; *S. barclayi* var.
pseudomonticola L. Kelso; *S. pseudomonticola* Ball; *S. pseudomonticola* var.
padophylla (Rydb.) Ball

Serviceberry willow

A shrubby tree to 3 m high, the trunks single or clustered.

LEAVES: Alternate, simple, soft, coriaceous, young leaves nearly glabrous but often more pubescent above than below, tinged with red on both the blade and the petiole. Broadly oblong or somewhat ovate, 4-6.5 cm long, 2-3 cm wide, tip acute or short acuminate, base rounded or subcordate; margin finely crenate-serrulate with a rounded, usually yellow-green gland on the tip, 7-9 teeth per cm; dark green above, glaucous below, both sides glabrous; petiole 8-13 mm long, nearly terete, puberulent above; stipules broadly ovate, up to 12 mm long and as wide, glabrous, glaucous below, tardily deciduous.

FLOWERS: May, before the leaves; dioecious. Staminate catkins sessile, 3-3.5 cm long, 1.5 cm diameter, densely flowered, axis woolly; scales obovate, 1.5 mm long, 0.75 mm wide, reddish at the base and brown at the tip, densely long pubescent on both sides, the hairs twice the length of the scale; 1 gland, greenish yellow, 0.75 mm long; stamens 2, filaments distinct, slender, 4-5 mm long, white, glabrous; anthers yellow or red, often drying purple. Pistillate catkins sessile or on a short peduncle; catkins 2.25 cm long, 1 cm wide, soon becoming longer; scales obovate, 1.5 mm long, 0.75 mm wide, dark brown, densely long pubescent on both sides; pedicels 1 mm long, glabrous; ovary narrowly conical, 2.5-3 mm long, 0.75 mm wide, green, glabrous; style 1 mm long, greenish-yellow, glabrous; stigma 4-lobed, yellowish.

FRUIT: Mid-June. Catkins erect or drooping, 6-7 cm long, 1.5 cm wide; scales persistent; capsules ovoid, abruptly narrowed to a neck, total length 8-9 mm, width 2.8-3 mm, slightly flattened, glabrous, brown. Seeds dark green, cylindric, 0.8-1 mm long, 0.5-0.6 mm wide, constricted, and with a ring of long and short hairs at the base.

TWIGS: 1.2-1.5 mm diameter, brownish-red to slightly purple, glabrous except for a few hairs at the nodes, slightly glossy, flexible; lenticels small, elliptic; leaf scars crescent-shaped, enlarged at the center and ends; 3 bundle scars; stipule scars large, circular; pith greenish, continuous, about one-half of stem. Flower buds usually toward the tip of the branch, ovoid, 8-10 mm long, 2.3-3 mm wide, yellowish to red-brown with a tan tip, gibbous at the base and flattened toward the tip, glabrous, the tip blunt and often spreading outward; leaf buds similar, smaller.

TRUNK: Gray-green or gray-brown, smooth or slightly cracked, branches often pruinose. Wood fine-grained, soft, no sapwood-heartwood line.

HABITAT: Moist meadows and stream banks, usually in the open and at higher altitudes in the mountains and the Black Hills.

RANGE: Central Alaska, across northern Manitoba to northwestern Quebec; in the Rocky Mountains from Idaho and Montana south to New Mexico; and in the Black Hills.

This is the only native willow of our area consistently with short, broad, cordate or subcordate leaves. It is not common but is scattered along streams in the Black Hills, always in quite moist situations. The trees are small and obviously heavily browsed by deer during the winter months.

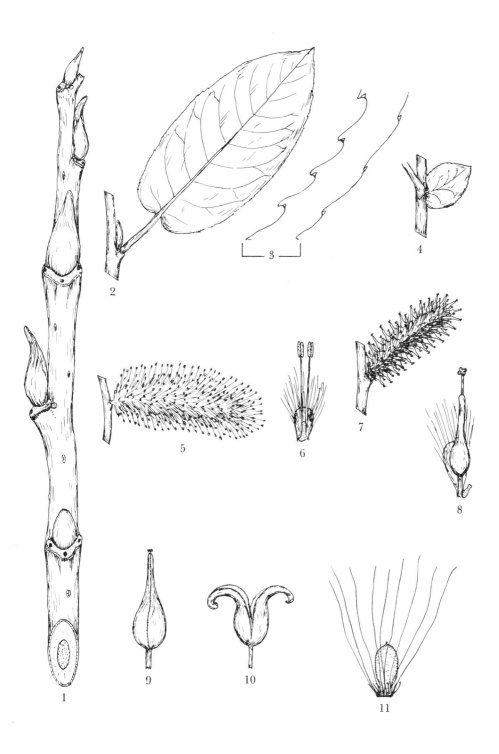

1. Winter twig
2. Leaf
3. Leaf margins
4. Stipule
5. Staminate catkin
6. Staminate flower

7. Pistillate catkin
8. Pistillate flower
9. Mature capsule
10. Open capsule
11. Seed

SALICACEAE
Salix nigra Marsh.

S. falcata Pursh; Incl. var. *altissima* Sarg.
and var. *lindheimeri* Schneid.

Black willow

A tree to 20 m high, the trunk often leaning.

LEAVES: Alternate, simple. Lanceolate, 5-13 cm long, 10-15 mm wide; margin finely toothed with incurved, callous-tipped teeth, 10-12 per cm; leaf tip long acuminate, often falcate, base rounded to cuneate; upper surface bright yellow-green, semiglossy, glabrous; lower surface paler green, pubescent when young, becoming glabrous; petiole 6-10 mm long, glabrate or glabrous; stipules foliaceous, 4-4.3 mm long, 2.3-2.5 mm wide, glandular serrate and persistent, or small, rounded, and deciduous.

FLOWERS: April, with the leaves; dioecious. Staminate catkins terminal on short, new leafy shoots; catkins 4-5 cm long, axis pubescent, flowers sessile, somewhat in whorls; scales yellow, 1-1.5 mm long, pubescent; 2 glands at the base of the scale; stamens 4-6, filaments 2-2.2 mm long, stout, greenish, pubescent at the base; anthers yellow. Pistillate catkins terminal on short, new, leafy shoots; catkins 4-5 cm long, axis pubescent; pedicels 1.5 mm long, green, glabrous; scale 1-1.5 mm long, yellow-green, pubescent; 1 large or 2 small glands; ovary green, 2.5 mm long, glabrous, ovoid with a conical, tapering neck; style obsolete or about 0.2 mm long; stigma 4-lobed. Occasionally a small shoot is produced from the axil of the uppermost leaf on the floral branch, apparently making the catkin lateral. The shoot is grown after the catkin is fully formed.

FRUIT: June. Pedicels 1.5 mm long, glabrous; capsule ovoid-conical, 4-5 mm long, glabrous, light brown, the 2 valves spreading widely at maturity. Seeds cylindric, 1-1.4 mm long, 0.4-0.7 mm thick, green, a slight ridge on one side; a ring of long silky hairs at the base with a few short, stiff hairs appressed against the seed.

TWIGS: 1 mm diameter, light red-brown, becoming darker and more gray with age, glabrous or pubescent, flexible, brittle; leaf scars narrowly crescent-shaped; 3 bundle scars; stipule scars small; pith pale brown, continuous, one-fifth of stem. Buds narrowly conical, 3 mm long, 1 scale, shiny red-brown, usually with a yellowish base.

TRUNK: Bark thick, gray-brown, fissured, the ridges broad, flat, and often loose and shaggy. Wood lightweight, soft, weak, not durable, pale brown, with a wide, white sapwood.

HABITAT: Any moist area, stream banks, lakeshores, pasture sloughs, and roadside ditches; rich loam, rocky or sandy soil.

RANGE: New Brunswick, west to Minnesota, southwest to Texas, east to Florida, and northeast to New England.

Salix nigra varies, as do all willows, and has been divided into several varieties. The characteristics of two of these are a matter of degree. These two varieties do not separate out geographically and are here placed in synonymy. They are: var. *altissima* Sarg. with more pubescent young leaves and var. *lindheimeri* Schneid. with narrower leaves. Var. *venulosa* (Anderss.) Bebb has bright yellow twigs which is not a characteristic of var. *nigra*. This variety is southwestern and enters our area sparingly in southwestern Kansas. It is synonymous with *S. goodingii* Ball and *S. nigra* var. *vallicola* Dudley.

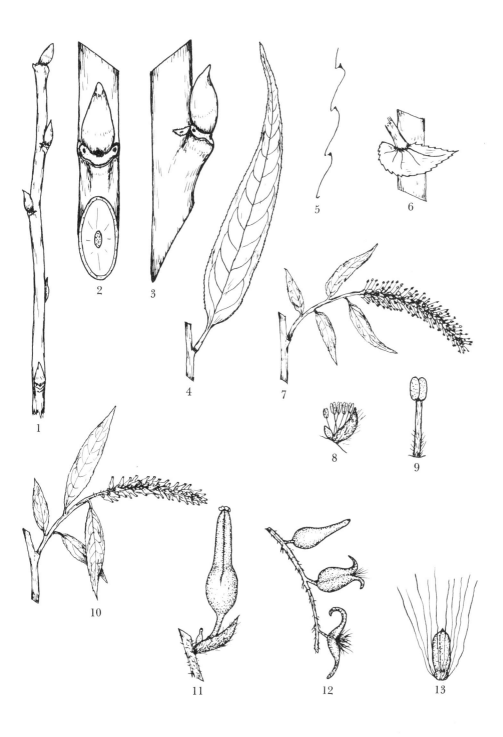

1. Winter twig
2. Detail of twig
3. Detail of twig
4. Leaf
5. Leaf margin
6. Stipule
7. Staminate catkin
8. Staminate flower
9. Stamen
10. Pistillate catkin
11. Pistillate flower
12. Mature capsules
13. Seed

SALICACEAE
Salix petiolaris J. E. Smith

Meadow willow

A shrub-like tree to 2 m high, often branched low.

LEAVES: Alternate, simple, lateral veins usually not obvious below. Lanceolate, 3-7 cm long, 10-17 mm wide; tip acuminate, base narrowly cuneate or slightly rounded; margin finely serrate, 3-5 teeth per cm, the teeth sharp and pointing forward; upper surface dark green, semiglossy, glabrous; lower surface glaucous, glabrous or sparingly pubescent, often some of the hairs fulvous; young leaves sericeous; petiole 5-10 mm long, glabrous or puberulent, broadly grooved above, the blade often decurrent to give a semiwinged margin; stipules narrowly linear, often with a few minute teeth, pubescent, caducous.

FLOWERS: May, the leaves one-fourth grown; dioecious. Catkins subsessile or on a short stalk with 1-3 leaves. Staminate catkins 2-2.5 cm long, 10-13 mm wide; scales vary from yellow with a brown tip to red-brown with a dark brown tip, obovate, 1.5 mm long, 1 mm wide, pilose; stamens 2, distinct, filaments slender, 5-6 mm long, white, pubescent at the base; anthers yellow; the gland broad, green with a yellow tip. Pistillate catkins 10-15 mm long, 6-7 mm wide; pedicels 0.5-2 mm long, pubescent; scales brown, ovate, 1-1.5 mm long, obtuse, long pilose; ovary ovoid, 1.5-2 mm long, short, silky pubescent; style short or absent; stigmas 2, each 2-lobed, red-brown; gland about 0.4 mm long and nearly as wide.

FRUIT: June. Catkins erect, 15-18 cm long, 10-14 mm wide; pedicels 1-2 mm long, pubescent; capsules conical, 6-7 mm long, 2 mm wide, short, silky, appressed pubescent; scales persistent. Seeds cylindric, 1.5-1.6 mm long, 0.5-0.6 mm wide, greenish, a ring of short, stiff hairs and one of long, silky hairs at the base.

TWIGS: 1.3-1.5 mm diameter, flexible, red-brown or yellow-brown, nearly glabrous or pubescent with both white and fulvous hairs, semiglossy; lenticels small, not noticeable, yellow-orange; leaf scars crescent-shaped with round ends; 3 bundle scars; pith greenish, continuous, about one-third of stem. Buds red-brown, 2-3.5 mm long, 1.8 mm wide, glabrous, semiglossy, microscopically striate, often with a light-colored base, a yellow band, and a dark red-brown outer half, the lower half of the scale lined with a green, scale-like membrane.

TRUNK: Bark of young branches gray-green or red-brown, often drying to a purple color; bark of old trunks brown, either smooth or flaky, branches often somewhat pruinose. Wood fine-grained, soft, white, with no distinction of heartwood and sapwood.

HABITAT: Stream banks and wet meadows, often associated closely with other species of willow.

RANGE: New Brunswick, across southern Canada to Saskatchewan, south to eastern Montana, and south to Colorado and from western Nebraska, east to New Jersey, and north to Quebec.

This willow is browsed heavily by deer and the trees are often dwarfed to dense, stubby shrubs only 50-75 cm high. It is quite similar to S. phylicifolia ssp. planifolia, but the leaves of S. petiolaris are sharply toothed, lanceolate, and narrow with an acuminate tip, while those of S. phylicifolia ssp. planifolia are usually entire, wider, and the tip acute or short acuminate.

Salix X subsericea (Anderss.) Schneid., a hybrid of S. gracilis var. textoris and S. sericea grows in one location east of Custer in the Black Hills. S. gracilis var. textoris is apparently a synonym of S. petiolaris.

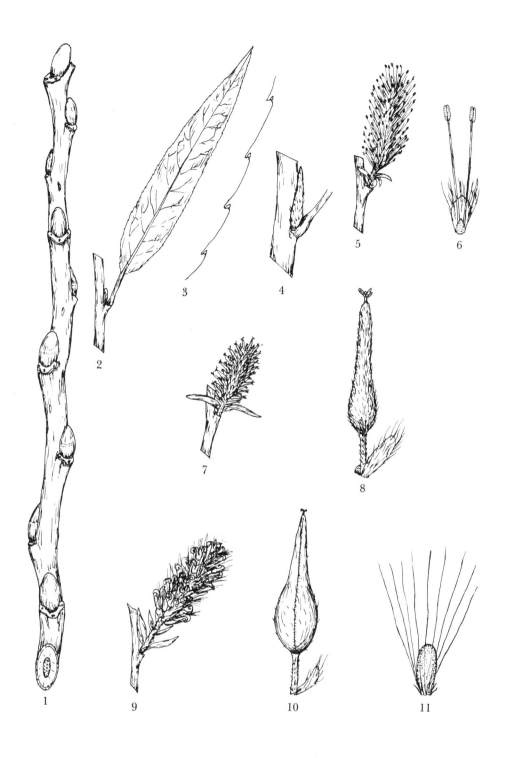

1. Winter twig
2. Leaf
3. Leaf margin
4. Stipule
5. Staminate catkin
6. Staminate flower
7. Pistillate catkin
8. Pistillate flower
9. Fruiting catkin
10. Mature capsule
11. Seed

SALICACEAE
Salix phylicifolia L. ssp. planifolia (Pursh) Hiitonen

S. planifolia Pursh; S. nelsoni Ball; S. planifolia var. nelsonii Ball ex E. C. Smith; S. chlorophylla Anderss. var. nelsonii Floderus

Planeleaf willow

A shrubby tree to 3 m high, the trunks clustered.

LEAVES: Alternate, simple, lateral veins not obvious beneath. Elliptic to oblong or slightly larger toward the outer end; 3.5-6 cm long, 12-25 mm wide; tip acute to short acuminate, base cuneate to slightly rounded; margin entire or occasionally with a few low teeth, 3-5 per cm; upper surface dark green, glabrous, semi-glossy, lower surface glaucous, pubescent when young; petiole 5-8 mm long, broad-based, slightly flattened above, glabrous or finely puberulent; stipules narrowly ovate, 2-3 mm long, toothed, early deciduous.

FLOWERS: May, with or before the leaves; dioecious. Staminate catkins sessile, 2-2.5 cm long, 1 cm thick, erect; pedicels 0.5 mm long, pubescent; scales obovate, rounded, 1.5 mm long, 0.75 mm wide, black, densely pilose, the hairs twice the length of the scale; 1 gland, broad, greenish-yellow, 0.75 mm long; stamens 2, distinct, filaments slender, 3-4 mm long, white, glabrous; anthers yellow or gray-brown. Pistillate catkins sessile, erect, 1.5-3 cm long, 7-8 mm wide; scales black, ovate, 1.5-1.75 mm long, 0.75 mm wide, densely pilose, rounded; ovary narrowly ovoid, 2.5-3 mm long, 1 mm wide, silky pubescent, green; style slender, 1 mm long, glabrous; stigma yellow-brown, 2 linear lobes.

FRUIT: June. Catkins erect, 4-6 cm long, 1.5-2 cm wide, dense; pedicels 0.5 mm long; capsules ovoid with a long neck, total length 7-8 mm, width 2.3 mm, finely pubescent, ascending; scales persistent. Seeds dark green, cylindric, 1.5 mm long, 0.4 mm diameter, somewhat angular, apiculate, a short, stiff, inner row and an outer, long, silky row of hairs at the base.

TWIGS: 1.3-1.5 mm diameter, flexible, red-brown to orange-brown, semi-glossy, glabrous or slightly pubescent at the nodes; lenticels elliptic, small, yellow-orange; leaf scars crescent-shaped with round ends; stipule scars large, oval; pith pale tan, continuous, about one-half of stem. End bud sharply curved; lateral buds ovoid, red-brown, puberulent with the more dense pubescence at the base, flattened against the twig, rounded on the outer side, 5-7 mm long, 1.6-2.3 mm wide, scale lined with a thin, greenish, scale-like membrane; leaf buds similar, 2-3 mm long, 1-1.4 mm wide.

TRUNK: Bark light gray, tight but slightly cracked toward the base, smooth except for the many raised, horizontal, orange-colored lenticels. Wood soft, fine-grained, white, with no distinction of heartwood and sapwood.

HABITAT: Stream banks, open meadows, or along low hillsides in moist soil.

RANGE: Across Canada from the south coast of Mackenzie to Baffin Island, Labrador, and Newfoundland, south to New England, and west across northern United States to North Dakota, south to the Black Hills and Colorado, and northwest to Idaho and Alberta.

Salix phylicifolia ssp. planifolia is a northern tree or shrub, the range extending into northern Canada. It is also found in the higher mountain elevations in the United States. The species varies greatly and has been divided into many different subgroups but ssp. planifolia is the only one which enters our area.

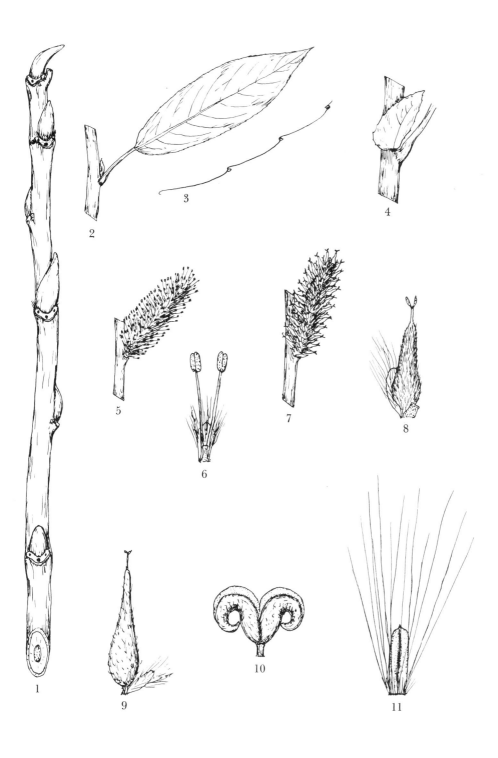

1. Winter twig
2. Leaf
3. Leaf margin
4. Stipule
5. Staminate catkin
6. Staminate flower
7. Pistillate catkin
8. Pistillate flower
9. Mature capsule
10. Open capsule
11. Seed

SALICACEAE
Salix rigida Muhl. var. *watsonii*
(Bebb) Cronq.

S. lutea Nutt.; *S. lutea* var. *platyphylla* Ball; *S. lutea* var. *watsonii* Jeps.; *S. cordata* Muhl., not Michx.; *S. cordata* var. *lutea* (Nutt.) Bebb; *S. cordata* var. *watsonii* Bebb; *S. watsonii* (Bebb) Rydb.; *S. cordata* var. *platyphylla* (Ball) L. Kelso; *S. flava* Rydb.; *S. ligulifolia* (Ball) Ball

Yellow willow

A tree or large shrub to 5 m high, the trunks clustered.

LEAVES: Alternate, simple, vigorous young leaves usually reddish. Elliptic or lanceolate, 4-6 cm long, 1-1.5 cm wide; tip acuminate or acute, base rounded to broadly cuneate; margin finely crenate-serrate with the gland on the apical side, 8-10 teeth per cm; young leaves pubescent, especially above; upper surface dark yellow-green, glabrous or somewhat puberulent along the midrib toward the base, lower surface glabrous; petiole 6-8 mm long, usually red, flattened and puberulent above, rounded and glabrous below; stipules about 3 mm long, reniform with one end acute, glandular serrate.

FLOWERS: Late May; dioecious. Staminate catkins 2.5-3 cm long, 5-7 mm wide, ascending or drooping, on a very short branch from lateral buds of previous year, usually with 2-3 leaf-like bracts which are often deciduous after anthesis; axis tomentose; scales ovate, dark brown, 1 mm long, pubescent with long curly hairs; stamens 2, filaments 4 mm long, distinct or united about one-third of the length, greenish-white, glabrous; anthers yellow; gland 0.75 mm long, broad base and knob-like apex, greenish-yellow. Pistillate catkins ascending, on a short stem with 2-3 leafy bracts; scales persistent or deciduous by the time the capsule is mature; catkin 2-3 cm long, about 1 cm wide; capsule narrowly lanceolate, 2 mm long, glabrous, stalk 0.5 mm long; style about 0.3 mm long; stigma divided into 2-4 red-brown lobes; gland and scales as in the staminate flower.

FRUIT: June. Capsule ovoid, abruptly narrowed to a long beak, stramineous, glabrous, 5 mm long, 1.5 mm wide, the 2 valves splitting to the base and curling; pedicel 1 mm long; about 10 seeds per capsule. Seeds cylindric, 1-1.2 mm long, greenish-brown, rounded apex, often mucronate, narrowed toward the base with an easily detached ring of long silky hairs at the base.

TWIGS: 1.3-1.8 mm diameter, flexible, red-brown to greenish-red, glabrous, somewhat pruinose; lenticels elliptic, vertical, same color as the twig; leaf scars crescent-shaped with enlarged ends; 3 bundle scars; stipule scars large, oval; pith greenish or pale tan, continuous, one-third to one-half of stem. Flower buds ovoid, flattened on the twig side, 8-11 mm long, 2.4-2.8 mm wide, appressed to the stem, often with the tip turned outward, greenish-red or purple-red, the greenish buds usually puberulent and the purplish buds glabrous; leaf buds similar but smaller, 4-5 mm long, 1.9-2.1 mm wide. "Pine cone" galls often present.

TRUNK: Bark gray-brown, tight, finely fissured, exposing the pale brown inner bark, lenticels horizontal, orange-colored, prominent, and causing a roughness to the bark of young stems. Wood soft, fine-grained, no distinction between heartwood and sapwood.

HABITAT: Stream banks and meadow flood plains, usually in moist soil.

RANGE: Alberta to Manitoba, south to western Kansas and New Mexico, west to Arizona and California, and north to eastern Washington.

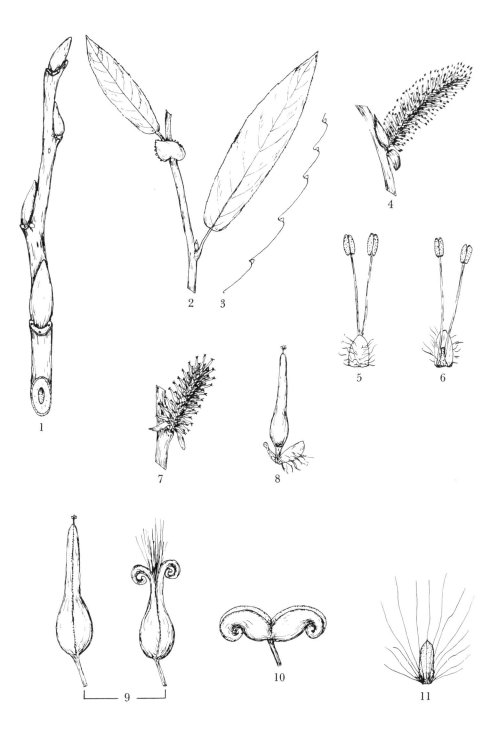

1. Winter twig
2. Leaves
3. Leaf margin
4. Staminate catkin
5. Dorsal side of staminate flower
6. Ventral side of staminate flower
7. Pistillate catkin
8. Pistillate flower
9. Mature capsules
10. Open capsule
11. Seed

SALICACEAE
Salix scouleriana Barratt

S. flavescens Nutt.; *S. nuttallii* Sarg.; *S. stagnalis* Nutt.; *S. capreoides* Anderss.; *S. brachystachys* Benth.

Scouler willow

A solitary, shrubby tree to 3 m high, open and somewhat scraggly.

LEAVES: Alternate, simple, firm, coriaceous; young leaves densely pubescent, hairs often fulvous. Elliptic to obovate, 3-7 cm long, 12-30 mm wide; tip acute, obtuse, or rounded, occasionally short acuminate; base cuneate; margin entire or minutely toothed; upper surface dark green, lower surface glaucous; the pubescence varies, some specimens glabrous above and pubescent below, other specimens are the opposite, always some of the hairs fulvous; petiole 5-8 mm long, blade somewhat decurrent, densely puberulent above with a few hairs below, flattened above; stipules ovate-reniform, up to 8 mm long, toothed, glabrate, early deciduous.

FLOWERS: May, before the leaves; dioecious. Catkins sessile or on a peduncle to 1 cm and with a few bract-like leaves; staminate catkins 2-4 cm long, 1-1.3 cm wide; scales dark brown, usually with a lighter base, rounded, 2-2.5 mm long, 1 mm wide, long pubescent; gland broad, stocky, yellowish; stamens 2, filaments slender, 6-7 mm long, glabrous or a few hairs at the base; anthers yellow. Pistillate catkins 2-3 cm long, 1.2-1.4 cm wide, the axis pubescent; scales dark brown, rounded, 2-2.5 mm long, long hairy; gland short, broad, yellowish; pedicel 1-1.2 mm long, pubescent; capsule slender with a long neck, 4-4.5 mm long, 1 mm wide, short pubescent; style 0.5 mm long; stigma usually divided into 2-4 red segments.

FRUIT: Late June. Catkins 3.5-4.5 cm long, 1-1.5 cm wide. Capsules ovoid with a long neck, 6-7 mm long, 2-2.5 mm wide, stramineous, densely pubescent. Seeds greenish, cylindric, 0.8-1 mm long, 0.4-0.5 mm wide, apiculate tip and truncate base, a ring of short and long silky hairs at the base.

TWIGS: 1.3-1.6 mm diameter, red to red-brown or greenish, first year twigs puberulent, glabrous by the second year; flexible, the lenticels small, elliptic, whitish margin; leaf scars narrowly crescent-shaped with rounded ends; 3 bundle scars; stipule scars small; pith greenish, continuous, one-third of stem, pith on older twigs light tan. Buds ovoid, blunt, flattened, semikeeled on the sides, 2-3 mm long, 1-1.5 mm wide, red-brown or yellowish, scurfy pubescent, often densely pubescent at the base, nearly the same color as the twig.

TRUNK: Bark gray-green, smooth with few lenticels; inner bark reddish. Wood soft, fine-grained, brown heartwood, and a wide, white sapwood.

HABITAT: In our area usually on hillsides in dry situations, occasionally along a stream bank or meadow; often near the top of the peaks in the Black Hills.

RANGE: Alaska, east to northern Manitoba, south to Montana, the Black Hills, Colorado, and New Mexico, west to California, and north to British Columbia.

The dry, hillside habitat and the single, rather open, shrub-like tree should be good field characteristics. In the herbarium, the obovate leaves (often only toward the stem tips), the entire leaves with some fulvous hairs, and the sessile catkins make a good combination for identification.

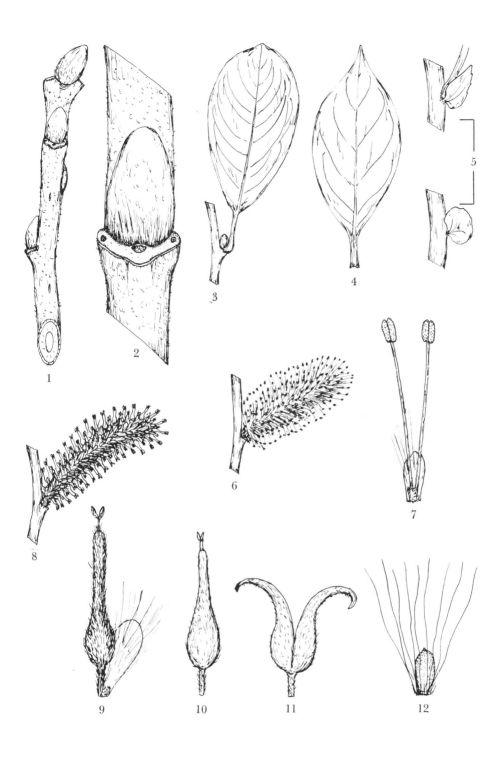

1. Winter twig
2. Detail of twig
3. Leaf
4. Leaf variation
5. Stipules
6. Staminate catkin
7. Staminate flower
8. Pistillate catkin
9. Pistillate flower
10. Mature capsule
11. Open capsule
12. Seed

SALICACEAE
Salix serissima (Bailey) Fern.

Autumn willow

A shrub to 3 m high, trunks either single or clustered.

LEAVES: Alternate, simple. Elliptic, oblong, or narrowly obovate, 6-8 cm long, 1-2 cm wide, short acuminate, base cuneate or narrowly rounded; margin finely serrate, 14-16 teeth per cm with a yellowish gland at the tip; thick, firm, leathery; upper surface dark yellow-green, lustrous, glabrous; lower surface light green and glabrous, the veinlets conspicuous and usually dark-colored; petiole 5-8 mm long, broadly grooved above, an occasional gland on the margin and with or without glands at the summit; stipules rarely present, flabellate, 1.5 mm long, 2 mm wide, glandular toothed. Young leaves often red.

FLOWERS: May-June; dioecious. Catkins on short, leafy shoots 1.5-2.5 cm long; staminate catkins 1.5-3 cm long, densely flowered, axis tomentose; scales obovate, 2-2.5 mm long, light yellow-brown, pubescent and long ciliate; 2 glands often united, conical, truncate, yellow; stamens 5, filaments slender, 4-6 mm long, pubescent on the lower fourth; anthers rich yellow. Pistillate catkins 2-2.5 cm long, 8 mm wide, axis tomentose, densely flowered; scales light yellow, pubescent, 2 mm long, obovate; glands small, yellow-orange; pedicels 0.75 mm long; ovary green, ovoid with a long neck, 4 mm long, glabrous; style 0.5 mm long; stigma with 2 lobes.

FRUIT: Late June to July. Catkins 4-5 cm long, 2-2.2 cm wide, axis tomentose, the capsules crowded; pedicels 1-2 mm long, glabrous; capsules ovoid with a long neck, 7-10 mm long, 2-2.5 mm wide, glabrous, pale brown and somewhat lustrous; scales persistent. Seeds cylindric, apiculate, 2 mm long, 0.3 mm wide, pale brown, smooth, a ring of both long and short silky hairs around the base.

TWIGS: 1-1.5 mm diameter, yellow-brown, glabrous, flexible, lustrous, lenticels small; leaf scars raised, U-shaped, enlarged at the ends and middle; 3 bundle scars; pith white, continuous, one-third of stem. Buds ovoid and slightly flattened, 5-7 mm long, yellow-brown, glabrous.

TRUNK: Bark yellow-brown, not fissured but with large lenticels causing a roughness, otherwise smooth. Wood soft, white, with no definite sapwood line.

HABITAT: Open woods along streams, roadside ditches; usually in moist, sandy, or rocky soil.

RANGE: Alberta to Newfoundland, south through New England to Pennsylvania, west through Indiana to Minnesota and the Dakotas, and south into Colorado. Not common in our area.

Salix is a complicated group; and even though a great deal of work has been done on it, more needs to be done. As in most plants, the specimens from any one local region are not sufficient to straighten out the whole genus, or even any one species. These species vary greatly within themselves, and identification must be made by using a group of characters rather than by just one outstanding characteristic. Hybrids are fairly common and will cause confusion because they are quite variable and in most cases have not been described.

In a number of species, as in *S. exigua,* our area lies between or in the overlapping ranges of two subspecies, and the plants may have characteristics of both. Too, many of the herbarium specimens are purely vegetative and this is not always sufficient for identification.

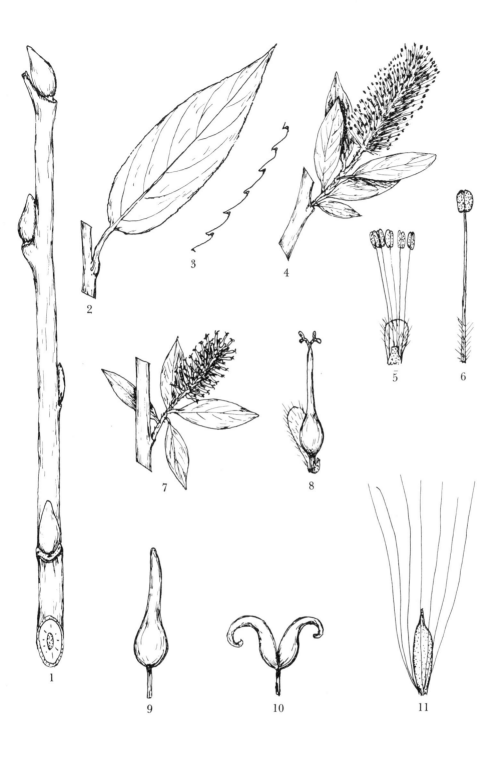

1. Winter twig
2. Leaf
3. Leaf margin
4. Staminate catkin
5. Staminate flower
6. Stamen
7. Pistillate catkin
8. Pistillate flower
9. Mature capsule
10. Open capsule
11. Seed

SALICACEAE
Populus acuminata Rydb.

Smooth-barked cottonwood

A tree to 20 m high, with a high, rounded crown.

LEAVES: Alternate, simple. Ovate, 8-12 cm long, 4.5-6 cm wide, tip acuminate, occasionally long acuminate, base broadly cuneate to somewhat rounded; margin coarsely toothed, 2-4 teeth per cm, teeth rounded and incurved with a small gland on the incurved tip, sparingly ciliate, the margin thickened and somewhat translucent; both sides green and glabrous, the lower side only slightly paler; thick and leathery; petiole 3-6 cm long, glabrous, often flattened laterally, and usually with 2 small glands at the summit, upper surface narrowly grooved; stipules ovate, long acuminate, 3-3.5 mm long, 1-1.5 mm wide, promptly turn brown and fall; young leaves heavily resinous.

FLOWERS: Early May, before the leaves; dioecious. Staminate catkins develop from lateral buds; catkins drooping, 6-10 cm long, 1 cm wide, flowers crowded, the axis green and glabrous; pedicels 2.5-3 mm long, flared to a broad funnel-shaped disk, 2.5 mm wide, beneath which are the 10-15 stamens; filaments 1-2 mm long, slender, white; anthers 1-1.5 mm long, yellow or reddish; the bract at the base of the pedicel green, flabellate, 6-7 mm long, including the brown fimbriate margin, often covering the flower. Pistillate catkins 8-12 cm long, pedicels 1.5-2 mm long, glabrous, slender; bract flabellate with a fimbriate margin, early deciduous; disk flared, glabrous, loose around the ovary, and with a thin, undulate margin; ovary globose, 2-2.5 mm diameter, glabrous, muricate, the two stigmas V-shaped, irregular, 1.5 mm long, yellowish.

FRUIT: June-July. Catkins 10-14 cm long, not crowded; pedicels slender, 2-3 mm long; capsules ovoid, glabrous, muricate, 5-7 mm long, 2-valved; disk saucer-shaped, thin, flaring with an irregular margin. No mature seeds were found during this study.

TWIGS: 2-2.5 mm diameter, red-brown to yellow-brown, glabrate, lenticels elliptic, yellowish; leaf scars bean-shaped, light brown, with 3 bundle scars; pith greenish, 5-angled, two-thirds of stem; stipule scars crescent-shaped. Buds 10-15 mm long, 3.5-4 mm diameter, greenish-brown, scales tight, resinous, the surface with a few short hairs and some scurf; lateral buds smaller.

TRUNK: Bark gray, thick, with shallow furrows and long, flat-topped ridges; branches somewhat smooth, gray-green. Wood lightweight, soft, light brown, with a wide, white sapwood.

HABITAT: Either moist or dry ground, along streams or on hillsides, usually in poor soil.

RANGE: Alberta, south to Arizona and southwestern New Mexico, north to Colorado, western Nebraska, South Dakota, North Dakota, and west to Montana.

Populus acuminata is usually considered a hybrid between *P. angustifolia* and either *P. deltoides* Marsh. var. *occidentalis* Rydb. or *P. fremontii* Wats. Hitchcock, et al. discuss this possibility but make no definite decision. Certainly the tree is distinctive, but that would not preclude a hybrid origin. The number of trees located during this study was small, and the author was unable to find mature fruiting capsules or seeds.

In our area it occurs sporadically with only a few trees in each location. Some of these trees were large and were estimated to be 50-75 years old, others were quite small.

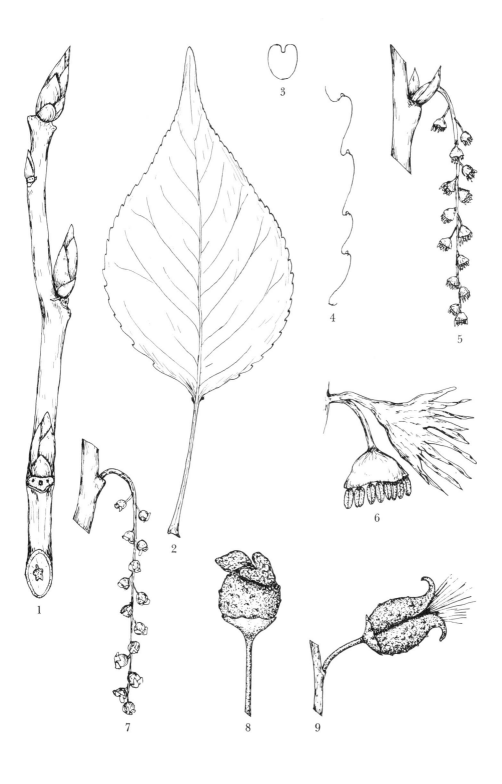

1. Winter twig
2. Leaf
3. Transverse section of petiole
4. Leaf margin
5. Staminate catkin
6. Staminate flower and bract
7. Pistillate catkin
8. Pistillate flower
9. Mature capsule

SALICACEAE
Populus alba L.

Silver poplar

A tree to 18 m high, widely branched.

LEAVES: Alternate, simple. Ovate, 4.5-8.4 cm long, 4-8 cm wide; the margin with 3-5 small, blunt lobes, irregularly dentate or sinuate; tip obtuse or rounded, base truncate or rounded; upper surface dark blue-green, lustrous, some pubescence on the veins; lower surface densely white tomentose; petiole 2-4 cm long, flattened, white tomentose; stipules narrowly deltoid, 2-2.5 cm long, white tomentose.

FLOWERS: Early April, before the leaves; dioecious. Staminate catkins from winter buds, pendulous, 5-10 cm long; bracts obovate, 3.5 mm long, dentate, long ciliate; stamens 6-10, filaments 1 mm long, anthers purplish; pistillate catkins from buds of the previous year, pendulous, 4-7 cm long, the axis tomentose; scales brown, 2-3 mm long, toothed, long ciliate, caducous; pedicels 0.5 mm long, glabrous, surrounded by a thin, loose sheath which broadens to a cup-shaped disk around the base of the ovary; ovary glabrous, 2-3 mm long, 1-1.5 mm wide, ovate, green, a slight crease on 2 sides; stigmas 2, sessile, each with 2 linear, terete lobes 1 mm long.

FRUIT: May-June. Catkins 5-10 cm long, drooping; capsules ovoid, 5-7 mm long, glabrous, granular, 2-valved, yellow-brown. Seeds light brown, 1-1.5 mm long with a dense ring of long, silky, white hairs at the base.

TWIGS: 1.5-2 mm diameter, rigid, flexible, white tomentose, greenish or brown beneath the tomentum; leaf scars wide, crescent-shaped or 3-lobed; 3-5 bundle scars; pith white, continuous, faintly 5-angled, one-third of stem. Buds ovoid, 6-7 mm long, acute, not viscid or resinous, the scales brown with a thin, white tomentum.

TRUNK: Bark on young limbs greenish-white, the lenticels horizontal, becoming enlarged and corky; bark of old trunks dark gray-brown, with shallow fissures and long, flat ridges. Wood light, soft, yellowish, with a wide, light sapwood.

HABITAT: Planted mainly around old farmsteads or on city lots, in any type of soil, the shoots often persisting for years, long after the original tree is gone.

RANGE: Naturalized from Europe. Escaping or persisting in parts of the United States.

Populus alba has a wide, rounded crown and furnishes good shade around a yard, but it may be objectionable because of its habit of sending up shoots from the roots which at times may cover an area several meters across. Cutting the tree does not necessarily destroy it since this may cause the sprouts to grow more abundantly. It grows rapidly under good conditions, but is not ordinarily long-lived. The largest tree known in our area has a trunk 11 feet and 4 inches in circumference at 54 inches above ground.

Although trees were examined throughout our area no staminate trees were located. The drawings were made from an old herbarium specimen and the description adapted from other authors.

This tree is often called the silver maple, but it is not a maple and the common name should be dropped in favor of silver poplar. The common name silver maple should be applied to *Acer saccharinum* L. if it is used at all.

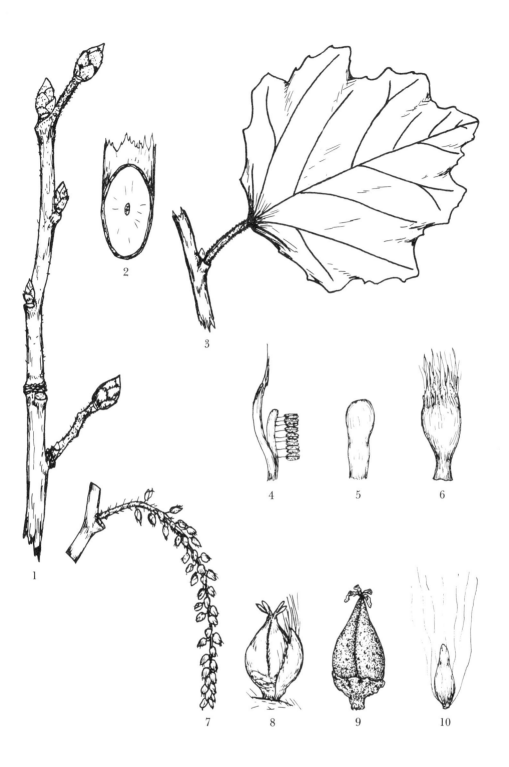

1. Winter twig
2. Section of stem
3. Leaf
4. Staminate flower with bract
5. Dorsal view of staminate flower
6. Dorsal view of bract
7. Pistillate catkin
8. Pistillate flower with bract
9. Mature capsule
10. Seed

SALICACEAE
Populus angustifolia James

P. canadensis Moench var. *angustifolia*
Wesmael in DC.; *P. balsamifera* L. var.
angustifolia S. Wats.; *P. fortissima*
Nels. & Macbr.

Black cottonwood, narrow-leaved
cottonwood

A tree to 16 m high, with long trunk and high cylindric or conical crown.

LEAVES: Alternate, simple, thick, coriaceous, the lateral veins occasionally not prominent. Lanceolate to narrowly ovate, 5-10 cm long, 13-25 mm wide; gradually tapered to the tip, cuneate or narrowly cuneate at the base; margin finely crenulate-serrulate, the tips inturned and with a yellowish callous on the inturned face, 7-9 teeth per cm; yellow-green, slightly lighter below, glabrous on both sides; petiole 10-15 mm long, flattened on the sides, channeled above, rounded below, puberulent when young; stipules lanceolate, 4-6 mm long, 1-1.25 mm wide, deciduous.

FLOWERS: May; dioecious. Staminate catkins 4-6 cm long from lateral buds on wood of the previous year; the axis glabrous; the bract subtending the flower flabellate, about 2 mm wide with a fimbriate margin, the base narrowed abruptly to a claw about 1.25 mm long; flowers sessile, somewhat turbinate and terminating in a flat disk, 12-16 stamens suspended under the disk; filaments filiform, 1-1.5 mm long, the anthers yellow and about the same length as the filaments. Pistillate catkins 6-10 cm long, 7-8 mm diameter, on a short peduncle often with 1-2 leaves; pedicels 2-2.5 mm long, green, glabrous, expanded into a disk 2-2.5 mm across and 0.5-0.75 mm deep, enclosing the basal one-fourth of the ovary; ovary ovoid, green, muricate, 2-2.2 mm long, 2 mm wide; style less than 1 mm long; stigmas 2, Y-shaped, about 2.5 mm long, yellow-green.

FRUIT: Late June. Catkins pendent, 9-12 cm long, not crowded; capsules ovoid, 7-7.5 mm long, 4 mm across, lightly muricate, yellowish; pedicels 2-3 mm long; scales early deciduous, the disk persistent. Seeds cylindric or slightly larger at the outer end, greenish, puberulent, 1.8-2 mm long, 0.8-0.9 mm across, a dense ring of curly hairs and long, silky hairs at the base.

TWIGS: 1.5-2 mm diameter, twigs of the season red-brown, later grayish yellow-brown, glabrous; the lenticels narrow, vertical, and inconspicuous; leaf scars reniform to elliptical, smooth, large, pale yellow-brown; 3 bundle scars; pith greenish becoming pale brown in one-year-old twigs, angular, continuous, one-fourth of stem. End bud narrowly ovoid, 9-11 mm long, 3-3.5 mm wide, red-brown, resinous, the basal scales broad and somewhat keeled; lateral buds 7-10 mm long, 2 mm wide, slightly compressed, often curved outward.

TRUNK: Bark thin, olive-green, smooth with dark, horizontal breaks; on older trees, light gray, fissured and with long, flat, irregular ridges. Wood soft, open-grained, light brown, with a wide, white sapwood; warps badly when cut into lumber.

HABITAT: Stream banks and margins of marshy areas, usually in rocky soil; occasionally in roadside ditches.

RANGE: Black Hills and western Nebraska, south to New Mexico and Mexico, west to eastern California, north to southeastern Oregon, central Idaho, Montana, and Alberta.

Populus angustifolia is abundant in our area only in local situations. It is quite common in Spearfish Canyon and again in the deeper valleys around Deadwood in the Black Hills. The locations in Nebraska are along the streams in the pine covered hills of northern Sioux and Dawes counties, but the stands are scattered.

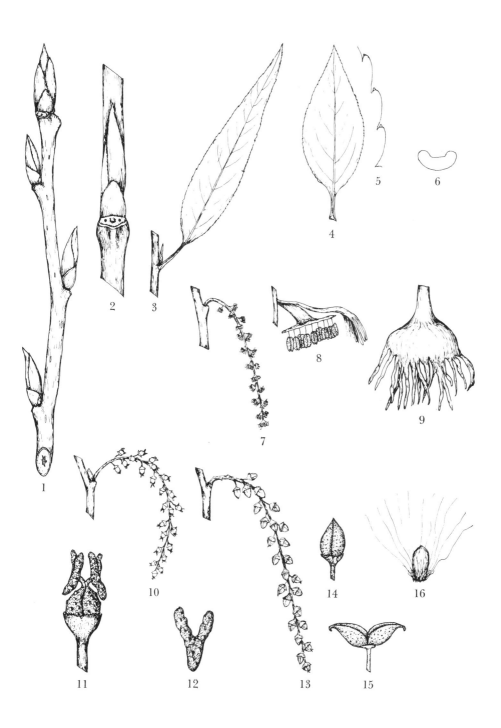

1. Winter twig	9. Bract
2. Detail of twig	10. Pistillate catkin
3. Typical leaf	11. Pistil
4. Leaf variation	12. Face view of one stigma
5. Leaf margin	13. Fruiting catkin
6. Section of petiole	14. Capsule
7. Staminate catkin	15. Open capsule
8. One flower with bract	16. Seed

SALICACEAE
Populus balsamifera L.

P. tacamahacca Mill.

Balsam poplar

Tree to 20 m high, usually with central trunk and cylindric crown.

LEAVES: Alternate, simple, thick, coriaceous, with a rusty appearance, the veins hardly raised beneath. Broadly ovate, 6-9 cm long, 5-6 cm wide; tip acuminate, base truncate, rounded or semicordate; margin finely low toothed, 3-5 teeth per cm, abruptly rounded and often with a callous tip on the incurved side, ciliate; upper surface dark yellow-green, glabrous, lower surface glaucous but rusty appearing, glabrous or with a few hairs on the midvein; petiole 2-4 cm long, shorter than the blade, flattened above, rounded below, finely puberulent, often twisted, the base abruptly broadened; stipules linear or lanceolate, 4-6 mm long, 1 mm wide, ciliate, promptly deciduous.

FLOWERS: May; dioecious, from lateral winter buds. Staminate catkin 5-7 cm long, 1 cm wide, flowers crowded, axis puberulent; bract flabellate, 3.5-4 mm long, 4-4.5 mm wide, greenish-yellow with long, brown fimbriate margin; pedicels 1.5 mm long, expanding to a circular disk, 2.5-3 mm across, flattened on the underside and with 25-30 stamens, the filaments white, slender, 1.3 mm long; anthers yellow, 0.75 mm long. Pistillate catkins 6-15 cm long, densely flowered, axis green and pubescent; bracts brown, flabellate, 4-4.5 mm long, 4.5-5 mm wide, early deciduous, the margin dark, fimbriate, ciliate; pedicels green, glabrous, 1.5-2 mm long, disk cup-shaped, 2.5-3 mm across and 1.5 mm deep and covers the lower part of the ovary; ovary 2.5-3 mm long, 2-3 mm wide, muricate, green, glabrous; style short, the 2 stigmas Y-shaped, 2.5-3 mm long, yellowish.

FRUIT: June. Catkins 9-12 cm long, not crowded; capsules ovoid, 8-10 mm long, 5-6 mm thick, occasionally with 3 valves instead of the usual 2, green becoming brown, glabrous, irregularly muricate, disk persistent, 18-25 seeds per capsule. Seeds 2.5 mm long, 1 mm wide, roughly ellipsoid, compressed, yellow-brown, apiculate, slightly constricted at the base just above the ring of variable length, silky hairs.

TWIGS: 2.3-2.6 mm diameter, brown to gray-brown, often with a waxy coating which peels off in large flakes, glabrous; the lenticels inconspicuous, light-colored and narrowly elliptic; leaf scars reniform, red-brown, smooth; 3 bundle scars; pith brown, slightly 5-angled, continuous, one-third of stem. Buds narrowly ovoid, sharply acute, 15-18 mm long, 4-5 mm diameter, slightly flattened, the scales greenish red-brown, resinous, glabrous or finely puberulent, ciliate, the hairs often inconspicuous because of the resin. Most of the buds are at the end of spur branches, with few lateral buds.

TRUNK: Bark of main branches olive, bark of trunk gray-brown, with shallow grooves and firm, flat ridges. Wood soft, nearly white, with little differentiation between heartwood and sapwood.

HABITAT: Creek banks, moist hillsides, sandhill potholes, and knolls.

RANGE: Alaska, eastward across southern Keewatin to Labrador and Newfoundland, south to New York, west to Minnesota, North Dakota, and the Black Hills, south across western Nebraska to Colorado, and northwesterly to Montana and Mackenzie. This apparently is the eastern and northern form of *P. trichocarpa,* and the exact western range is rather indefinite due to the introgression. A great many of the North Dakota plants have the 3 valves of *P. trichocarpa,* but the capsules are ovoid and smooth instead of globular and pubescent as in *P. trichocarpa.*

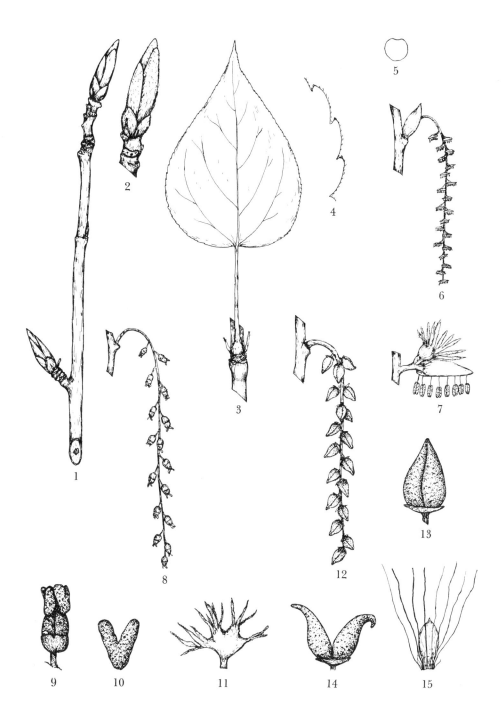

1. Winter twig
2. Detail of end bud
3. Leaf
4. Leaf margin
5. Section of petiole
6. Staminate catkin
7. Staminate flower with bract
8. Pistillate catkin

9. Pistillate flower
10. Face view of stigma
11. Pistillate flower bract
12. Fruiting catkin
13. Capsule
14. Mature capsule
15. Seed

SALICACEAE
Populus deltoides Marsh.

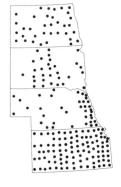

P. *balsamifera* L. var. *virginiana* Sarg.;
P. *virginiana* Foug.; P. *monilifera* Ait.;
P. *canadensis* Moench var. *virginiana*
(Foug.) Fiori

Cottonwood, eastern cottonwood

A tree to 30 m high, wide and spreading in open areas, but with a tall, central trunk when crowded.

LEAVES: Alternate, simple, thick, coriaceous. Broadly ovate to deltoid, 6-12 cm long and equally as broad; margin serrate except across the base and the very tip, the teeth coarse and incurved, 2-4 teeth per cm; tip short acuminate, base truncate or cordate; upper surface dark green, lustrous, glabrous; lower surface paler, glabrous, and somewhat lustrous; becoming brilliant yellow in autumn; petioles 5-7 cm long, flattened from the sides, glabrous; stipules narrowly ovate, 3-4 mm long, promptly deciduous.

FLOWERS: April, before the leaves; dioecious. Staminate catkins 8-12 cm long, from lateral buds of the previous year, pendulous, densely flowered with 50-70 reddish flowers; bracts 1.5-2 mm long, flabellate, green with a brown fimbriate margin, early deciduous, bract attached either on the pedicel or at its base; disk 8-9 mm across, the many stamens clustered beneath it; filaments white, 1-1.5 mm long, glabrous; anthers red. Pistillate catkins lateral, pendulous, 3-6 cm long, 25-50 green flowers; bracts 1.5-2 mm long, green with brown fimbriate margin, caducous; pedicels 2-4 mm long; ovary green, ovoid, 2.5-3 mm long, 2 mm wide, glabrous, subtended by a cup-shaped disk which surrounds the pedicel and the lower portion of the ovary; stigmas 3-4, sessile, roughly heart-shaped with divergent lobes and erose margins.

FRUIT: May. Catkin 12-20 cm long, 2-2.5 cm wide, 30-50 capsules; pedicels 6-8 mm long; capsules ovoid 8-12 mm long, 5-7 mm thick, glabrous, green with small white dots, drying to straw-color. Seeds oblong, 4-4.2 mm long, 1-1.2 mm wide, flattened, brown, puberulent with minute hooked hairs; apex pointed, base with a dense tuft of long, silky hairs.

TWIGS: 2.5-4 mm diameter, green-brown to yellow-brown, becoming grayish; glabrous, brittle, often ridged, the lenticels prominent; leaf scars broad, shield-shaped; 3 bundle scars; fruit scars oval; pith white, 5-angled, continuous, one-third of stem. Buds 1.5-2 cm long, conical, pointed, glossy, glabrous, resinous.

TRUNK: Bark on young trees smooth, yellow-green; on older trees ashygray to dark gray, thick, deeply furrowed, the ridges wide and flat. Wood lightweight, soft, weak, light brown, with a wide, white sapwood. Warps badly when cut into lumber.

HABITAT: Grows well in almost any moist situation; common along streams, ditches, and rich bottom lands; one of the first trees to become established on new ground along streams.

RANGE: Maine, across southern Canada to North Dakota, south to Texas, east to Florida, and north to New England.

The plants of the eastern part of the Central Plains are mostly var. *deltoides*. Westward it passes freely into var. *occidentalis* Rydb. (P. *sargentii* Dode) so that many of the cottonwoods in the central portion of our range are difficult to place as to variety. Var. *occidentalis* has slightly pubescent buds, coarser teeth on the leaf (1-2 per cm), and the twigs and branches have a yellowish cast.

It is a fast-growing tree and is used in windbreaks but is objectionable as a street tree because of the weak branches, the cottony seeds, the large red staminate catkins, and the quantity of leaves which do not decompose readily.

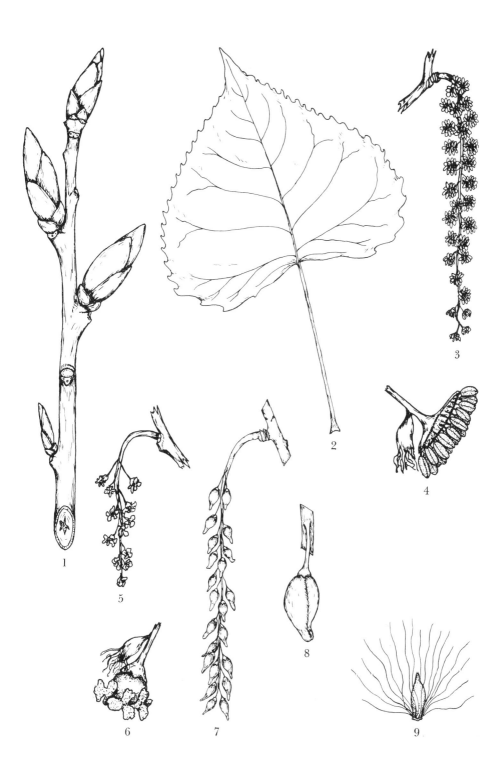

1. Winter twig
2. Leaf
3. Staminate catkin
4. Staminate flower with bract
5. Pistillate catkin
6. Pistillate flower with bract
7. Fruiting catkin
8. Capsule
9. Seed

SALICACEAE
Populus tremuloides Michx.

P. *aurea* Tidestrom; *P. tremuloides* (var.)
aurea Daniels; *P. cercidiphylla* Britt.;
P. tremuloides var. *cercidiphylla* Sudw.

Quaking aspen, trembling aspen

A tree to 15 m high, usually in small stands, but often covering a large area.

LEAVES: Alternate, simple, somewhat coriaceous, veins not prominent below. Circular to broadly ovate, often broader than long, 3-6 cm long and as wide, abruptly short acuminate, base subtruncate to rounded; crenate-serrate, 4-5 low teeth per cm, the marginal vein often thickened, the margin long, white, ciliate until midsummer; upper surface dark green, lower surface much paler and yellow-green, both surfaces glabrous; petiole 2.5-4 cm long, flattened from the sides, glabrous, often with glands at the summit; stipules narrowly lanceolate, 5-7 mm long, ciliate with long hairs, immediately deciduous.

FLOWERS: May-June; dioecious. Staminate catkins pendent, 2-3 cm long, 20-30 flowered, axis pale yellow-green, pubescent; bracts obovate, brown, deeply cut into 3-8 narrow segments, long ciliate; pedicels 1 mm long, pubescent, expanded to a cup-shaped, glabrous disk with entire or minutely toothed margin, oblique, 1.5-2 mm across; stamens 6-10 attached near the base of the disk, filaments white, 0.5 mm long, anthers dark purple, 0.5 mm long. Pistillate catkin erect or pendent, 2-3 cm long, many flowered, axis pubescent, green; bracts green-brown at the base, obovate, deeply cut into 3-6 narrow, long, ciliate segments; pedicels 0.5 mm long, glabrous or pubescent, disk cup-shaped, 1.5-2 mm across, undulate at the top, green, glabrous, deciduous or persistent; ovary 1 mm long, ovoid, green; style short, stigmas 2, each with a bulbous base and 2 ascending lobes, red-purple, 1-1.2 mm long.

FRUIT: June. Catkins 4-6 cm long, fruits not crowded, drop from the tree about as soon as the capsules start to open. Capsules green, muricate, conical, 5-7 mm long, 2-2.3 mm wide, grooved on two sides, glabrous; pedicels 2-2.5 mm long, disk usually persistent. Seeds yellow-green, narrowly obovoid, 1 mm long, 0.5 mm wide; a ring of short, straight hairs and a ring of long, silky hairs at the base.

TWIGS: 1.5-2 mm diameter, yellow-brown to reddish, glabrous, rigid, semi-glossy, the lenticels small and elliptic; leaf scars large, usually longer than the stem diameter, shallow U-shaped to half-round, light-colored; 3 bundle scars; pith greenish, continuous, slightly 5-angled, one-third of stem. Older twigs roughened with many leaf and fruit scars. End bud ovoid, acutely pointed, 5-7 mm long, 2.3-2.5 mm diameter, somewhat 3-angled, red-brown, glossy, glabrous, lightly resinous, the scales ciliate or with a narrow, scarious, erose margin.

TRUNK: Color of bark varies from olive-brown to nearly white, the Black Hills trees often white enough to be confused from a distance with *Betula papyrifera;* marked with rough, dark, horizontal scars; bark of old trees light gray, with shallow furrows and flat-topped ridges. Wood soft, lightweight, open-grained, light brown, with a wide, white sapwood.

HABITAT: Dry hillsides or moist valleys, usually in rocky or sandy soil, common in recently burned areas.

RANGE: Alaska, across northern Manitoba to Labrador and Newfoundland, south to Virginia, west to northeastern Missouri, northern Nebraska, and Wyoming, south to New Mexico and Mexico, west to California, and north to British Columbia.

1. Winter twig
2. Leaf
3. Leaf margin
4. Section of petiole
5. Stipule
6. Staminate catkin
7. Staminate flower with bract
8. Pistillate catkin
9. Pistillate flower with bract
10. Pistil
11. Fruiting catkin
12. Capsule
13. Mature capsule
14. Seed

SALICACEAE

Populus trichocarpa T. & G.
ex Hook.

P. hastata Dode; *P. trichocarpa* var.
hastata Henry in Elwes & Henry

Black cottonwood

A large tree to 30 m high, with high, spreading branches and often drooping branchlets.

LEAVES: Alternate, simple, veins not prominent below, thick, tough, coriaceous. Broadly ovate, 5-10 cm long and approximately as wide; abruptly acuminate and usually not toothed at the tip, base truncate to subcordate; margin coarsely or finely toothed, 2-4 teeth per cm, the marginal vein somewhat hyaline and lightly ciliate, basal side of each tooth long, abruptly rounded, with a gland on the forward facing side; upper surface yellow-green, lower side slightly paler, both surfaces glabrous except for some puberulence on the veins near the base; petiole 5-7 cm long, glabrous to finely puberulent, a narrow groove above, slightly flattened laterally, gradually widened to a broad base and often with two raised, brown glands at the summit; stipules lanceolate to oblong, 6-12 mm long, pubescent at the base and ciliate toward the tip.

FLOWERS: May; dioecious. Staminate catkins 2-3 cm long, pendent, bracts nearly circular, lacerate and ciliate, early deciduous; 30-50 stamens suspended on short filaments beneath the nearly circular disk. Pistillate catkins 5-12 cm long, axis densely pubescent, pedicels 0.5-1 mm long, pubescent; disk shallow, saucer-shaped, with irregular margin, pubescent; ovary globose, 3-4 mm diameter, usually 3-carpellate, pubescent, the 3 yellow-red stigmas V-shaped, divergent. (Fresh flower material not seen by the author.)

FRUIT: Mid-June. Catkins 8-15 cm long, not crowded; pedicels about 1 mm long; capsule broadly ovoid, 7-8 mm long, 5-6 mm wide, 3-carpellate, short pubescent or scurfy. Seeds pale brown, 2.5 mm long, 1 mm wide, glabrous, slightly wider at the apex and somewhat flattened, definitely apiculate, completely covered by the dense ring of hairs from the base, the hairs buff-colored, 7-9 mm long.

TWIGS: 5-7 mm diameter, red-brown with prominent, light-colored lenticels; nodes much enlarged and glabrous to short puberulent; leaf scars large, angular crescent-shaped; 3 bundle scars; pith 5-angled, continuous, greenish, one-third of stem; annual growth on twigs comparatively short. Buds ovoid, pointed, 1.5-2 cm long, 6-8 mm wide, scales short, puberulent, red-brown, the lower scales extending nearly around the bud, strongly keeled and often with 2 small lateral keels.

TRUNK: Bark of old trees light gray with shallow furrows and flat ridges, upper branches gray-green, definitely marked with old lenticels and branch scars, bark of young trees often streaked light and dark. Wood soft, open-grained, lightweight, brown, with a wide, white sapwood.

HABITAT: Usually in moist soil of valleys and around springy areas, but occasionally on drier ground.

RANGE: Alaska, south to Baja California; Utah, northeast to western North Dakota and Wyoming, and northwest to Alberta and British Columbia.

It is quite possible that our plants called *P. trichocarpa* are a hybrid form with *P. balsamifera*. The two ranges overlap in the area along the west edge of North Dakota and since the two species are closely related, hybridization would be quite possible. The trees grow in the same habitat as *P. deltoides* and have somewhat the same general appearance. These two also hybridize and it may be that our *P. trichocarpa* has been influenced in some manner by this.

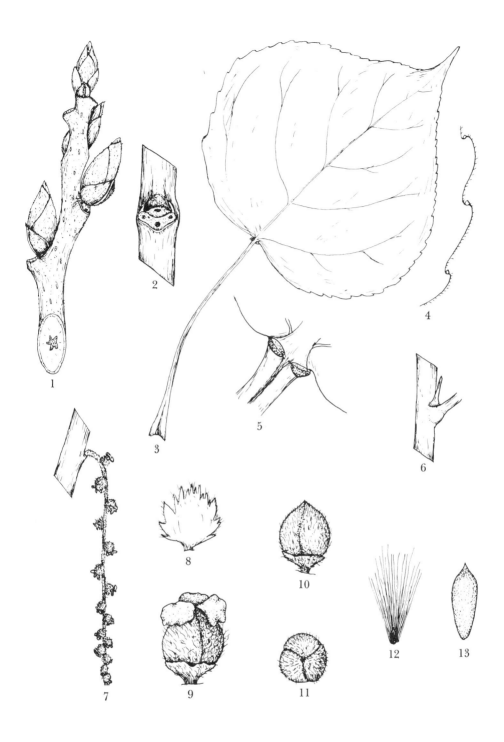

1. Winter twig
2. Detail of leaf scar and fruit scar
3. Leaf
4. Leaf margin
5. Summit of petiole
6. Stipule
7. Pistillate catkin
8. Flower bract
9. Pistillate flower
10. Mature capsule
11. End view of capsule
12. Seed
13. Seed with hairs removed

JUGLANDACEAE
Juglans microcarpa Berlandier

J. rupestris Engelm.

Texas walnut

A spreading tree to 15 m high, usually with large, low branches.

LEAVES: Alternate, pinnately compound, 9-23 cm long, 11-21 leaflets; thin, firm. Leaflets narrowly ovate to lanceolate, 3-9 cm long, 1-2.5 cm wide, the larger ones near the middle; serrate with low teeth, 3-5 per cm, the basal side long and nearly straight, the veinlets ending in the sinuses; tip acute to long attenuate and falcate; base unequally rounded to broadly acute; upper surface dark yellow-green, dull, glabrous except for a few hairs on the main veins; the lower surface slightly paler and sparingly pubescent; young leaves densely glandular below; petiole 2.5-4 cm long, with a shallow groove above and both glandular and stellate pubescence, the young petioles densely pubescent; terminal petiolule 0.5-1 cm long, the lateral leaflets sessile or with a petiolule to 3 mm; rachis with both sessile glands and glandular hairs; no stipules.

FLOWERS: May; monoecious. Staminate catkins from buds on wood of the previous year; pendent, 4-6 cm long, loosely flowered with 25-35 flowers, each about 3 mm across, yellow-green, the axis glandular and glandular pubescent, pedicels 3 mm long; bract below the flower ovate, 1 mm long, brown, tomentose outside and sparingly pubescent inside; calyx lobes 4-6, obovate to rotund, 1-1.5 mm long, green, glabrous; stamens 18-30, nearly sessile, yellow-green, glabrous. Pistillate flowers 1-4 in short spikes terminal on new growth; peduncle 8-10 mm long, glandular and stellate pubescent, the flowers sessile; involucre ellipsoid, 6-8 mm long, 3 mm diameter, green, glandular and stellate pubescent, the 2-3 lobes triangular; calyx lobes 3-4, lanceolate, 2-3 mm long, green, glabrous; style short; stigmas 2, linear, 7-9 mm long, papillate and coarsely rugose, greenish-brown, divergent.

FRUIT: October. Single or 2-3 in a cluster, globular, 2.8-3.2 cm diameter; husk dark brown, 5-7 mm thick, fibrous, nondehiscent, the surface dull and granular. Nut globose, 2-2.3 cm diameter, dark brown, hard, longitudinally and irregularly ridged with flat-topped ridges and shallow furrows; kernel sweet, edible.

TWIGS: 3.5-5 mm diameter, brown to gray-brown, with a vesture of stellate and long simple hairs and stipitate glands, some of the pubescence persistent into the third year; lenticels small, numerous, elliptic, light-colored; leaf scars large, 3-angled with rounded corners and concave sides, or heart-shaped, often longer than broad, the upper margin with a dense fringe of long, white hairs; 3 bundle scars; pith brown, partitioned, one-fourth to one-third of the stem. Buds gray-brown with stellate and simple pubescence; terminal bud ovoid, compressed, 7-11 mm long, 4-6 mm wide, obtuse, the outer scales keeled; lateral buds ovoid to globose, 3-6 mm long, 3-4 mm wide, densely pubescent; flower bud between the leaf bud and leaf scar, ovoid, 3 mm long, 2.5 mm wide, the scales small and numerous, pubescent.

TRUNK: Bark brown, deeply fissured, the ridges flat-topped and occasionally loose; bark of young trees and branches often silvery-gray.

HABITAT: Dry, rocky ravine banks and hillsides; flood plains in good soil.

RANGE: Southwestern Kansas, western Oklahoma, Texas, New Mexico, Arizona, and northern Mexico.

The plants of our area are probably intergraded with the more eastern *Juglans nigra*.

74

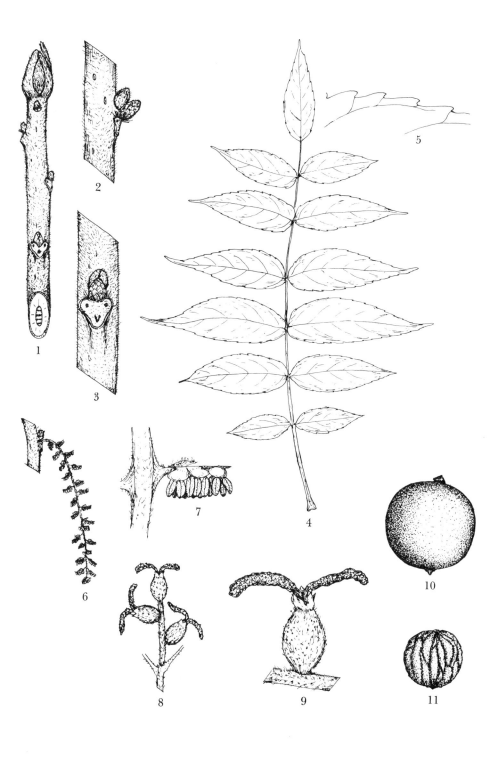

1. Winter twig
2. Detail of buds
3. Detail of buds
4. Leaf
5. Leaflet margin
6. Staminate catkin
7. Staminate flower
8. Pistillate cluster
9. Pistillate flower
10. Fruit
11. Nut

JUGLANDACEAE
Juglans nigra L.

Wallia nigra (L.) Alef.

Black walnut

A tree to 30 m high, with long, straight trunk when crowded.

LEAVES: Alternate, pinnately compound, 20-50 cm long, 9-21 leaflets. Leaflets ovate to ovate-lanceolate, 4-10 cm long, 2-5 cm wide, sharply serrate or biserrate, 3-7 teeth per cm; tip acuminate, base rounded, or unevenly oblique; upper surface dull yellow-green, glabrous, the lower surface paler and pubescent, especially on the veins; petiole 5-10 cm long, glandular and stellate pubescent; terminal petiolule 1-3 cm long, pubescent; lateral leaflets sessile or nearly so; rachis pubescent; terminal leaflet often poorly developed; no stipules.

FLOWERS: Early May, the leaves partly grown; monoecious. Staminate catkins single or in clusters on wood of the previous year, 6-9 cm long, densely flowered with 35-40 flowers; axis elongating after anthesis, glabrous or pubescent; pedicels 1-1.5 mm long with a bract at the upper end, the bract ovate to lanceolate, 1.4 mm long, brown, pubescent outside, dark brown and glossy inside; calyx lobes 4-6, obtuse, green, glabrous; anthers 30-40, sessile, 1.6-2 mm long, yellow-green, glabrous. Pistillate flowers terminal on new growth, in short spikes of 1-4 flowers, peduncle 1 cm long, pubescent; the 2 bracts at the base of the peduncle lanceolate, pubescent; flowers sessile; involucre ovoid, 6.5-7 mm long, 4.5-5.2 mm wide, stipitate glandular and stellate pubescent, 3-4 irregular lobes at the apex; calyx lobes 4, lanceolate, 2-3.5 mm long, pubescent, green; style green, 3.8 mm long, glabrous, surrounded by the calyx lobes; stigmas 2, divergent, yellow-green, 6-7 mm long, 2-3 mm thick, papillose.

FRUIT: October. Solitary or in clusters of 2-4; yellow-green, becoming dark brown at maturity, globose, 4-6 cm diameter, granular surface, somewhat puberulent; husk 5 mm thick, fibrous, indehiscent. Nuts black, globose, 3-4 cm diameter, slightly flattened laterally, irregularly fissured, the ridges sharp.

TWIGS: 4.5-5 mm diameter, rigid, stout, brown to gray-brown with tawny stellate or simple pubescence; leaf scars large, shield-shaped or 3-lobed; 3 groups of bundle scars; pith brown, angular, chambered with thin, close plates, one-fourth of stem. Terminal bud ovoid, 6-7 mm long, lateral buds 2-3 mm long, obtuse, slightly compressed, silky pubescent, the scales in pairs. Staminate catkin buds often below the regular leaf bud, rounded, compressed, 3.2-3.8 mm wide, 2.5-2.7 mm thick, the many small scales acute, silky pubescent.

TRUNK: Bark dark brown or blackish, deeply fissured, the ridges flat-topped, long or blocky. Wood heavy, strong, hard, durable, dark brown, with a yellowish sapwood.

HABITAT: Rich bottom lands, flood plains, stream banks; upland in good soil and often planted around farmsteads.

RANGE: Massachusetts, west to southeastern South Dakota, south to western Oklahoma and eastern Texas, east to Florida, and north to New York.

This is one of our finest lumber trees and most of those large enough for lumber have been cut. When crowded, it grows tall and straight, the bole often 15 m high without limbs of any size. The wood is of good quality and is used for furniture, panels, and gun stocks.

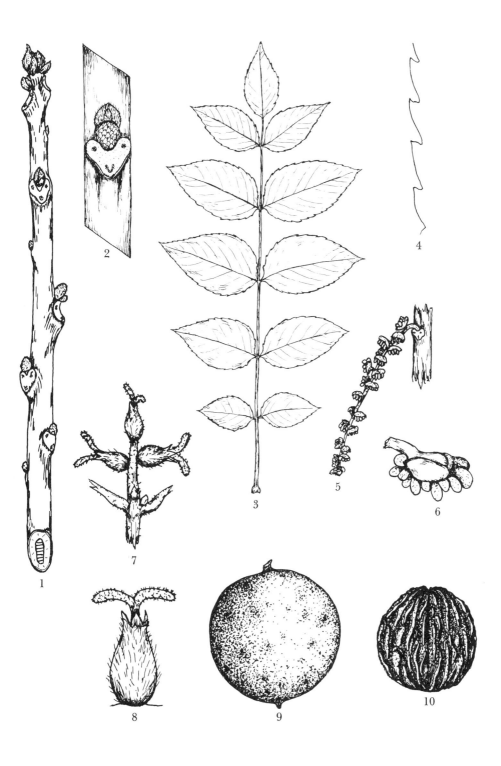

1. Winter twig
2. Detail of twig showing staminate flower bud below leaf bud
3. Leaf
4. Leaf margin
5. Staminate catkin
6. Staminate flower
7. Pistillate flower cluster
8. Pistillate flower
9. Fruit
10. Nut

JUGLANDACEAE
Carya cordiformis (Wang.) K. Koch

C. amara Nutt.; *Hicoria cordiformis* Britt.; *Hickoria minima* (Marsh.) Britt.; *Juglans cordiformis* Wang.

Bitternut, pignut, yellow-bud hickory

A tree to 25 m high, with central trunk, occasionally with large, low branches.

LEAVES: Alternate, pinnately compound, 15-30 cm long, 7-9 leaflets. Terminal leaflet 10-15 cm long, 5-6 cm wide, petiolule 3-8 mm long; lateral leaflets 4-10 cm long, 2-5 cm wide, sessile; margin serrate, 2-4 teeth per cm, sparingly ciliate; terminal leaflet obovate, laterals elliptic to lanceolate; tip acuminate, base of terminal leaflet attenuate and decurrent, laterals unequally rounded or cuneate; upper surface bright green, glabrous; lower surface paler, pubescent especially on the midrib, hairs fascicled or simple, often with a few scurfy scales; petiole 5-7 cm long, pubescent; stipules absent.

FLOWERS: Mid-May, after the leaves partly grown; monoecious. Staminate catkins in fascicles of 3 from lateral buds on wood of the previous year or at the base of new growth; peduncles 1-3 cm long with lanceolate, pubescent bracts 7-9 mm long at the apex; stalk of the catkin 5-15 mm long; catkin 8-12 cm long, pubescent, many-flowered, each flower with 3 pubescent bracts 1.2-1.5 mm long, the central one longest; stamens 4, filaments white, 0.3 mm long; anthers 1 mm long, yellow, pubescent at the tip. Pistillate flowers terminal on new growth, single or in a spike of 2-3; involucre 6-10 mm long, yellow, scurfy, ridged, the 4 acute lobes 1.7-2.1 mm long, green, and pubescent; ovary green, 2 mm long, barrel-shaped, ridged, pubescent at the base, surrounded by semitransparent scales; style absent; stigmas 2, stout, conical, rough with fleshy processes, green, occasionally broad and somewhat flabellate.

FRUIT: October. Obovoid to globose, slightly compressed, 2.6-3 cm long, 2-2.3 cm wide, 1.8-2 cm thick; husk 1-2 mm thick, yellow, scurfy, the 4 sections slightly winged and splitting halfway to the base at maturity. Nut smooth, pale brown, ovoid to obovoid, compressed, 1.8-2.6 cm long, 1.8-2.7 cm wide, 1.6-1.8 cm thick, a sharp tip at the apex, the surface with prominent vein lines; kernel bitter.

TWIGS: 2.5-4 mm diameter, rigid, greenish, becoming brown, glabrate, the lenticels prominent; leaf scars shield-shaped, large; bundle scars numerous, in 3 groups; pith brown, angled, continuous, one-fifth of stem. Terminal bud 8-15 mm long, compressed, oblique at the base, obtuse tip, scales valvate, yellow scurfy and keeled; lateral buds ovoid, 3-5 mm long, pubescent or scurfy, angled, often stalked; occasionally a tree is found with brown buds which are not scurfy, possibly a hybrid.

TRUNK: Bark gray, close, smooth, but in small plates on large trees. Wood heavy, strong, hard, elastic, durable, dark brown, with a wide, white sapwood.

HABITAT: Rich bottom lands, stream banks, flood plains, or on rocky hillsides.

RANGE: Quebec, west to Minnesota, south through eastern Nebraska to Texas, east to Florida, and north to New England.

C. cordiformis is one of the most common hickories in eastern Kansas, and its range extends north along the Missouri River into Nebraska. Forma *latifolia* (Sarg.) Steyerm. is described as having wider leaflets, but in our material both leaf shapes may be found on the same tree. It hybridizes with *C. illinoensis* (Wang.) K. Koch to produce *C.* X *brownii* Sarg., commonly called the "hickon," and with *C. ovata* to produce *C.* X *laneyi* Sarg.

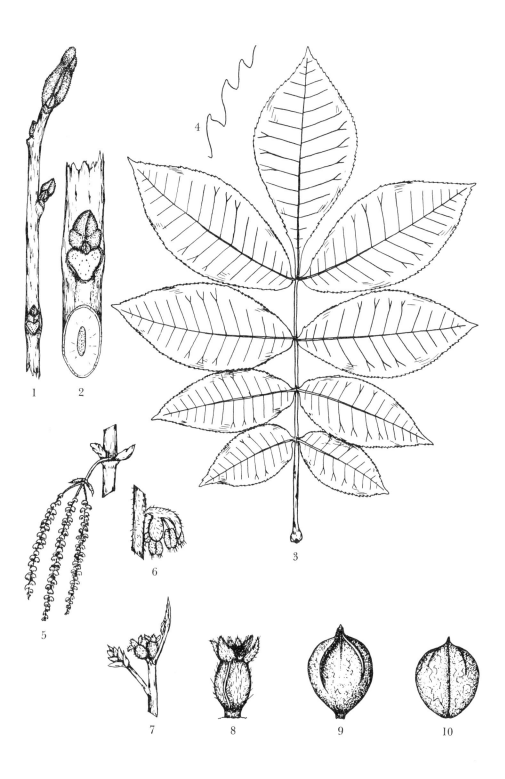

1. Winter twig
2. Detail of twig
3. Leaf
4. Leaf margin
5. Staminate catkins

6. Staminate flower
7. Pistillate cluster
8. Pistillate flower
9. Fruit
10. Nut

JUGLANDACEAE
Carya illinoensis (Wang.) K. Koch

C. pecan (Marsh.) Engelm. & Graebn.;
Hicoria pecan Britt.; *Juglans pecan*
Marsh.; *J. olivaeformis* Michx.

Pecan

A tree to 25 m high, with large, spreading branches and short trunk.

LEAVES: Alternate, pinnately compound, 30-50 cm long, 7-13 leaflets. Leaflets oblong-lanceolate to ovate-lanceolate, 6-10 cm long, 2.5-4 cm wide; serrate, 2-5 teeth per cm, often ciliate; tip long acuminate, often falcate, base unequal, the apical side more rounded and extended further along the petiolule; upper surface dark yellow-green, dull, glabrous, the lower surface paler, glabrous except for small tufts in the vein axils; petiole slender, 5-7 cm long, glabrous or pubescent; rachis glabrous; terminal petiolule 2-3 cm long, laterals 3-10 mm long, glabrous; stipules absent.

FLOWERS: April-May, with the leaves; monoecious. Staminate catkins slender, 8-12 cm long, fascicled in groups of 3 from buds of the previous year; flowers sessile or nearly so; calyx bract linear with 2 broad, lateral lobes, green and pubescent; corolla none; stamens 3-6, the anthers nearly sessile, yellow, pubescent at the tip. Pistillate flowers solitary or in short spikes of 2-4 flowers at the end of new growth; involucre oblong, 3 mm long, slightly 4-angled, narrowed at both ends, yellow-green with a scurfy pubescence, 4 lanceolate bracts 2-4 mm long at the upper end, one much longer than the other three; stigma 2-lobed from a globose base, yellow-green, the lobes with short processes.

FRUIT: October-November. Cylindric, 3-4.5 cm long, brown, 4-winged, the husk 2-2.5 mm thick, rough dotted, splitting to the base at maturity, often remaining on the tree after the nut falls. Nut oblong, 3-3.5 cm long, 1.4-1.7 cm thick, both ends pointed, light brown mottled with dark brown; shell thick or thin, smooth; the kernel sweet and edible.

TWIGS: 3-4 mm diameter, coarse, gray-brown with light lenticels, rigid, pubescent when young and becoming glabrous; leaf scars heart-shaped, large; 3-4 bundle scars; pith brownish, angled, continuous, one-third of stem. Buds ovoid, acute, compressed, scurfy, and pubescent with fascicled hairs, end buds 5-8 mm long, lateral buds 3-5 mm long.

TRUNK: Bark dark gray with shallow fissures and long, flat ridges which are often loose; branches gray and smooth. Wood hard, not strong, brittle, light brown, with a narrow, light sapwood.

HABITAT: Rich, moist bottom lands or uplands in good soil. Often planted around farm homes and in commercial groves.

RANGE: Ohio, west to Iowa and eastern Kansas, south to central Texas, east to Alabama, and north to Kentucky.

Carya illinoensis hybridizes readily with other hickories, producing a wide variety in the type of nuts. Some of them are sweet and edible, others bitter and not palatable; some with a hard, thick shell and others with a thin shell. The only one reported for Kansas is *Carya* X *brownii* Sarg., a hybrid between *C. illinoensis* and *C. cordiformis*, the nut of which is sweet and edible with a comparatively thin shell.

Recently several commercial groves of pecan have been planted in the southern part of our area. The trees grow rapidly, producing a good crop of nuts in about 12 years. The nuts have a high percentage of fat and are used extensively in candies and cookies. They are also a favorite food of the squirrel.

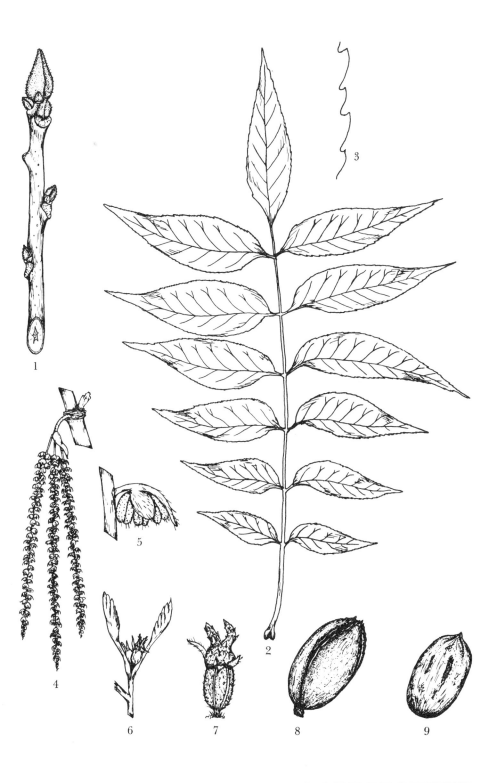

1. Winter twig
2. Leaf
3. Leaf margin
4. Staminate catkins
5. Staminate flower

6. Pistillate cluster
7. Pistillate flower
8. Fruit
9. Nut

JUGLANDACEAE
Carya laciniosa (Michx.) Loud.

Hicoria laciniosa (Michx. f.) Sarg.;
Juglans laciniosa Michx. f.; *Carya sulcata* Nutt.

Kingnut, big shellbark

A tree to 20 m high, with low, spreading branches.

LEAVES: Alternate, pinnately compound, 30-50 cm long, 7-9 leaflets. Leaflets obovate to oval, finely serrate, 3-5 teeth per cm, ciliate with simple or fascicled hairs; tip acuminate, base of the terminal leaflet attenuate, occasionally with 2 small, rounded lobes, the base of the laterals unequal, rounded on the basal side, obtuse to rounded on the apical side; upper surface dark yellow-green, glabrous; lower surface paler with soft, white or brown fascicled or simple hairs, often with stipitate glands on the midvein; petiole 5-7 cm long with a wide base, glabrate, often persistent on the tree after the leaflets have fallen; terminal petiolule 1-1.5 cm long, lateral leaflets nearly sessile, or on petiolules to 2 mm; stipules absent.

FLOWERS: Late April, after the leaves; monoecious. Staminate catkins ternate on old or new growth; peduncles 1.5-3 cm long with 2 subulate bracts at the apex, stalk of the catkins 5-8 mm long; catkin 10-18 cm long, slender, many flowered; axis with pale hairs, flowers nearly sessile, the central bract 3-4 mm long, linear, pubescent, occasionally with some reddish hairs, the lateral bracts ovate, pubescent especially at the tip; stamens 4, filaments short; anthers yellow-green, often red-tinged, pubescent at the tip. Pistillate flowers terminal on new growth, usually 2 with an undeveloped flower between, peduncles 2-3 mm long, with 2 broad-based lanceolate bracts; involucre 3.5-4 mm long, ridged, white pubescent, the 4 acuminate lobes green, pubescent, one usually longer than the others; the 2 stigmas sessile, yellow-green, conical, and spreading, the surface rough.

FRUIT: Late September. Solitary or in pairs, subglobose to obovoid, 5-6 cm long, 5-6 cm wide, 4-5 cm thick; 4-valved, indented at the sutures; surface glabrate, brown, with minute clear scales or orange spots; husk woody, 10-12 mm thick, splitting to the base. Nut ovoid, 4-ridged, 3.5-4 cm long, 3.3-3.7 cm wide, 2.5-2.6 cm thick; tip acute, point of attachment conical; pale brown at first, weathering to light red-brown; kernel sweet, edible.

TWIGS: 3-5 mm diameter, orange-brown, becoming dark gray-brown by the first winter; pubescent or glabrous, rigid; the lenticels narrow and elongated; leaf scars heart-shaped, large; 3-4 bundle scars; pith greenish, angled, continuous, one-third of stem. Terminal bud ovoid, 10-18 mm long; the outer scales keeled with a long pointed apex, tawny pubescent, slightly resinous; middle scales white pubescent, apiculate; inner scales silky, orange pubescent, accrescent in spring; lateral buds ovoid, 5-9 mm long.

TRUNK: Bark gray, thick, exfoliating into long, broad plates, curved outward at one or both ends. Wood heavy, hard, strong, elastic, dark brown, with a narrow, light sapwood.

HABITAT: Rich bottom lands, seldom in rocky areas unless on a stream bank.

RANGE: Southern Quebec, west to Ontario, south to Iowa, southeastern Nebraska, eastern Oklahoma, and northeastern Texas, east to Louisiana, Alabama, and Tennessee, and north to New York.

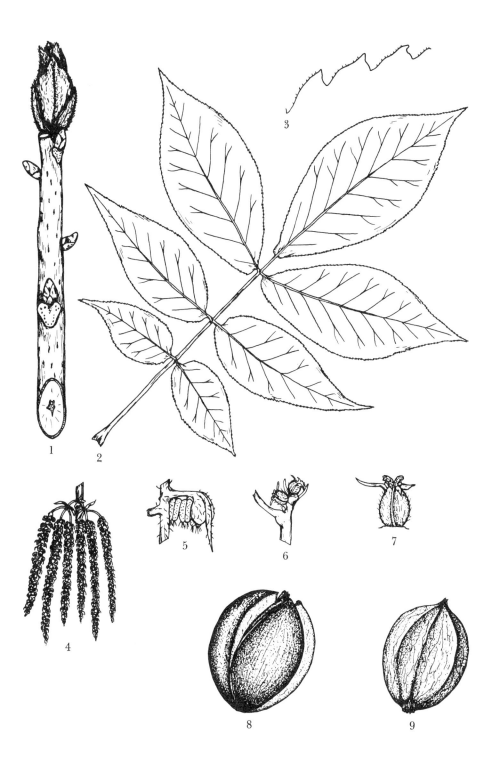

1. Winter twig
2. Leaf
3. Leaf margin
4. Staminate catkins
5. Staminate flower
6. Pistillate cluster
7. Pistillate flower
8. Fruit
9. Nut

JUGLANDACEAE
Carya ovata (Mill.) K. Koch

C. mexicana Engelm.; *Hicoria ovata*
Britt.; *Hicoria carolinae-septentrionalis*
Ashe; *Juglans ovata* Mill.

Shagbark hickory, shellbark hickory

A tree to 20 m high, usually with large, low branches.

LEAVES: Alternate, pinnately compound, 20-35 cm long, 5 leaflets. Leaflets 13-18 cm long, 5-7.5 cm wide, obovate to lanceolate, finely serrate, 3-4 teeth per cm; ciliate at least when young, a tuft of hairs remaining on each side of the tip of each tooth; tip of leaflets acuminate; base of the terminal leaflet acute and abruptly rounded to the petiolule, the base of the laterals unequal, the basal side more rounded; upper surface dark yellow-green, glabrous; lower surface paler, glabrous except on the veins and the tufts in vein axils, the hairs fascicled or simple; petiole stout, 8-12 cm long, broadly grooved above and enlarged at the base, scurfy glandular, glabrous or slightly hairy; terminal petiolule 5-15 mm long, the lateral leaflets sessile or nearly so; stipules absent.

FLOWERS: April-May, after leaves nearly grown; monoecious. Staminate catkins ternate, at the tip of old wood or new growth, peduncle 1-3 cm long, sparsely pubescent, green, 2 lanceolate bracts at the apex; catkins 6-12 cm long, many flowered, the flowers sessile or on pedicels 0.5 mm long; flower bract narrowly lanceolate to linear, 1.5-2.5 mm long, pubescent, especially at the tip, the 2 lateral bracts ovate to obovate, glabrate, green; stamens 4-5, filaments green, 0.5 mm long, the anthers 1.2-1.4 mm long, yellow and pubescent at the tip. Pistillate flowers terminal on new growth, solitary or 2-3; peduncles 1-2 cm long with 2 broad-based, lanceolate bracts; involucre ovoid, 8-9 mm long, ridged, green, and covered with transparent and reddish scales; the 4 lobes lanceolate with pubescent tips, one lobe much longer than the others; stigmas sessile, brown or green, irregularly cone-shaped, and erose.

FRUIT: September-October. Solitary or clusters of 2-3; globose, 3.3-3.5 cm long, 3.4-3.7 cm thick, indented at the sutures, the 4 valves splitting to the base at maturity; husk 9-12.4 mm thick, dry with a dark brown granular surface covered with yellowish scales. Nut 4-ridged, ovoid, slightly compressed, 2.4-2.6 cm long, 2.2-2.6 cm wide, 1.8-2 cm thick, light brown; kernel sweet and edible.

TWIGS: 3-6 mm diameter, brown and downy when young, becoming gray and glabrous; lenticels pale, elliptic; leaf scars large, heart-shaped; 3-4 bundle scars; pith greenish, angled, continuous, one-third of stem. Terminal bud 12-18 mm long, broadly ovoid, obtuse; the outer scales dark brown, tomentose or glabrous, keeled, the tip often with a long point; middle scales apiculate, white pubescent; inner scales orange, tomentose, persistent and accrescent when the bud opens, becoming 5-7 cm long, yellow-green tinged with red, and remaining attached until the leaves are half-grown. Lateral buds 6-9 mm long, ovoid, obtuse, the scales slightly keeled, pubescent or glabrous.

TRUNK: Bark thick, gray, separating into thick, long strips, free at one or both ends and curved outward. Wood heavy, hard, strong, tough, elastic, light brown, with a narrow, white sapwood.

HABITAT: Well-drained soils, loam or rocky hillsides, river banks or in hilly woods.

RANGE: Maine, across southern Canada to Minnesota, south through eastern Nebraska to Texas, east to Florida, and north to New England.

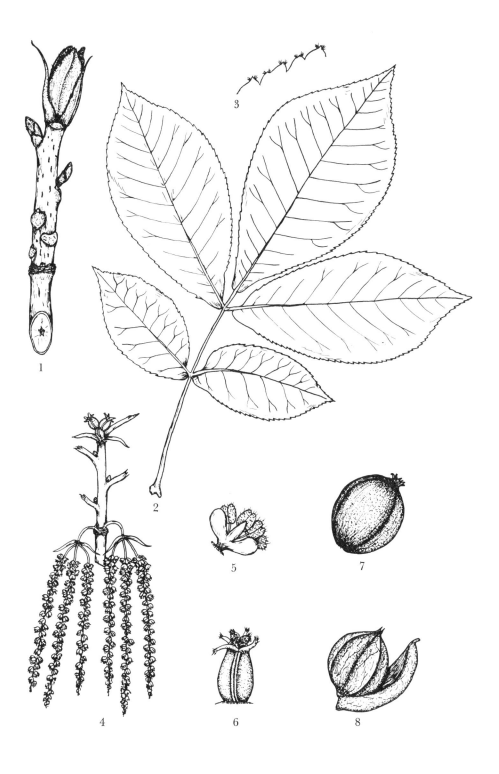

1. Winter twig
2. Leaf
3. Leaf margin
4. Staminate catkins with pistillate flowers at tip of stem
5. Staminate flower
6. Pistillate flower
7. Fruit
8. Nut

JUGLANDACEAE
Carya texana Buckl.

C. buckleyi Durand.; *C. villosa* Schneid.;
C. arkansana Sarg.; not *C. texana*
(Le Conte) C. DC.

Black hickory

A tree to 15 m high, with small, low branches.

LEAVES: Alternate, pinnately compound, 20-28 cm long, 5-7 leaflets. Leaflets oblanceolate to obovate, terminal leaflet on lower branches often broadly obovate; 6-14 cm long, 3-6 cm wide; serrate, 3-6 teeth per cm, ciliate with fascicled hairs; tip acuminate; base of the terminal leaflet attenuate, the laterals oblique, rounded on the basal side; upper surface dark green, glabrous; lower surface paler with rusty, fascicled hairs, and a few red or orange glands; midrib and rachis with fascicled hairs; young leaves with rusty scurf on both sides; petiole slender, 4-5 cm long, pubescent or glabrate; terminal petiolule 0.5-0.8 mm; lateral leaflets nearly sessile; stipules absent.

FLOWERS: April-May, leaves half-grown; monoecious. Staminate catkins ternate on the base of new growth or tip of old growth; peduncles 8-12 mm long with 2 lanceolate bracts at the summit; catkins 8-13 cm long with many sessile flowers; flower bract linear, 3-3.5 mm long, green, pubescent, glandular, the 2 lateral lobes ovate, 2.5 mm long, occasionally a small calyx lobe beneath the flower axis, lobes and bract with rusty scurf; stamens 4-6, filaments short, anthers 1 mm long, yellow, pubescent at the tip. Pistillate flowers terminal on new growth, usually 2 on a short peduncle; involucre ovoid, 2.5-3 mm long, ridged, covered with pale hairs, transparent scurfy scales and a few red, scurfy glands; the 4 lobes acute, 1-2 mm long, pubescent outside; the two large, erose, cone-shaped stigmatic points exserted, yellow at first, becoming brilliant red.

FRUIT: October. Sessile or with a 5 mm pubescent stalk; globose to obovoid, 3-3.6 cm long, 3-3.6 cm wide, 2.7-3 cm thick; husk 2-2.4 mm thick, 4-valved, the 2 lateral valves narrow, the surface with scattered rusty hairs and yellow or orange scales especially at the tip and base; slightly indented at the sutures and splitting to the base late in the season. Nut subglobose, 2.5-2.8 cm long, 2.7-2.9 cm wide, 2.2-2.3 cm thick; slightly angled or smooth, veiny, pale brown at first, weathering to red-brown, a small point at the apex and base; kernel sweet, edible.

TWIGS: 2.8-3.3 mm diameter, rigid, brown, rusty pubescent and scaly at first, but becoming glabrous; leaf scars large, triangular; 3 groups of bundle scars; pith brown, continuous, one-third of stem. Terminal bud 7.8-8 mm long, 3.9-4 mm wide, ovoid, obtuse to acute, covered with rusty pubescence and yellow-brown scurf, often a few white hairs at the apex; lateral buds 5 mm long, 2.5 mm thick.

TRUNK: Bark tight, thick, dark brown, shallow furrows on young trees; deeply furrowed and with blocky ridges on old trees. Wood hard, light red-brown, with a pale sapwood.

HABITAT: Dry uplands, upper rims of creek banks, rocky soils, and open woods pastures.

RANGE: Texas, Arkansas, Oklahoma, Missouri, southeastern Kansas, southern Illinois, and Indiana.

Carya texana trees of our area are scattered in wooded areas and no location was found where there were more than a few trees. The young trees are often mistaken for other hickories, especially *C. tomentosa*, but the blocky bark of the old trees is distinctive.

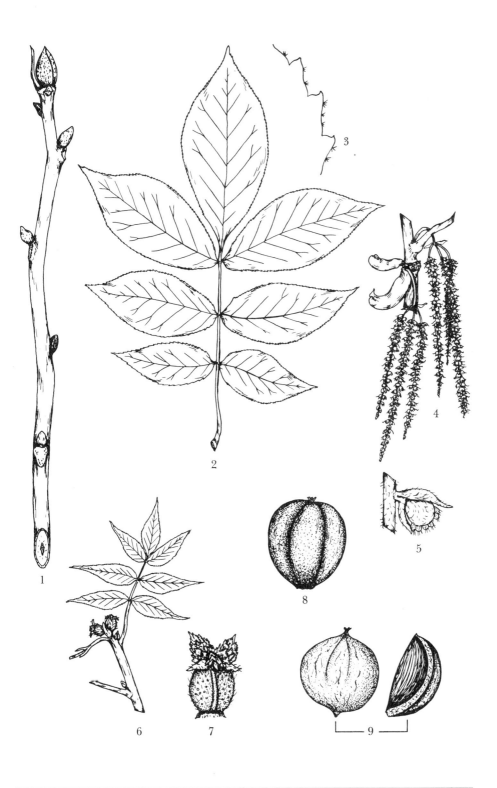

1. Winter twig
2. Leaf
3. Leaf margin
4. Staminate catkins
5. Staminate flower
6. Pistillate cluster
7. Pistillate flower
8. Fruit
9. Nut with portion of husk

JUGLANDACEAE
Carya tomentosa Nutt.

C. alba K. Koch, not Nutt.; *Hicoria alba*
(L.) Britt.; *Juglans alba* L.

Mockernut, white hickory, hognut

A tree to 20 m high, with large, low branches.

LEAVES: Alternate, pinnately compound, 20-35 cm long, 5-9 leaflets. Leaflets 13-20 cm long, 5-8 cm wide, obovate; serrate, 2-5 teeth per cm, ciliate with fascicled hairs; tip acuminate or short acuminate; base of terminal leaflet attenuate, that of the laterals nearly equilateral; upper surface dark yellow-green, glabrous except on the veins; lower surface paler, yellow-green, covered with fascicled hairs and red glands, the midvein with a few straight, loosely appressed hairs up to 2 mm long; rachis tomentose with rusty fascicled hairs and a few long, reddish, straight hairs; petiole 5-7 cm long, abruptly enlarged at the base, tomentose; terminal petiolule 5-10 mm long; the lateral leaflets nearly sessile; stipules absent.

FLOWERS: April-May, after the leaves are partly grown; monoecious. Staminate catkins ternate on the base of new growth or the tip of old wood; peduncle 1-1.5 cm long with 2 lanceolate bracts at the summit; catkin 8-13 cm long, many flowered, slender, the axis pubescent; flowers nearly sessile; the bracts green, pubescent with long gold-brown hairs and short white hairs; the central lobe lanceolate, 12-15 mm long, the 2 lateral lobes ovate, ciliate at the tip. Pistillate flowers terminal on new growth, usually 2 on a short, bracted peduncle; involucre 5-6 mm long, green, ridged, tomentose and scurfy with nearly transparent scales; the 4 lobes pubescent on both sides, one lobe much longer than the others; stigmas sessile, yellow-green to reddish, disk-like at the base with 2 spreading, erose lobes.

FRUIT: October. Compressed globose, 3.5-4.5 cm long, 3.7-4.1 cm wide, 3-3.5 cm thick, 4 valves, the laterals narrower than the others, winged toward the tip of the sutures and slightly indented toward the base; surface granular with minute clear glands; husk brown, 3-3.5 mm thick, woody, splitting to near the base. Nut 4-angled, ovoid, 3.3-3.8 cm long, 2.8-3.3 cm wide, 2.4-2.6 cm thick, pale tan at first, weathering to light brown; the shell thick and hard with the vein lines prominent; a sharp tip at the apex and a small, conical, basal area; kernel sweet, edible.

TWIGS: 4-6 mm diameter, brown at first, becoming gray, the twigs tomentose with brown, fascicled hairs; leaf scars heart-shaped, with the lateral margins concave, or broadly crescent-shaped; 3-4 bundle scars; pith brown, angled, continuous, one-third of stem. Terminal bud 12-15 mm long, 7-9 mm wide, ovoid, red-brown, pubescent; outer scales keeled, deciduous in autumn; inner scales silvery-brown tomentose, accrescent in spring; lateral buds 6-8 mm long, 3-4 mm wide, acute, brown, pubescent.

TRUNK: Bark thick, hard, gray-brown, furrowed, the ridges flat-topped and tight. Wood strong, heavy, hard, dark brown, with a wide, white sapwood.

HABITAT: Well-drained soils, bottom lands, or rocky hills and slopes.

RANGE: New Hampshire to southeastern Ontario, Michigan, Iowa, eastern Kansas, and eastern Texas, east to Florida, and north to Massachusetts.

In many herbarium specimens, *C. tomentosa* may be confused with an overly pubescent form of *C. laciniosa*, but neither the rachis nor the twigs of the latter are at all tomentose. In winter specimens, the buds of *C. tomentosa* are without the hard, brown outer scales, thus exposing the dense, red-brown pubescence of the inner scales. In the field, the tight, ridged bark of *C. tomentosa* cannot be confused with the loose, plate-like bark of *C. laciniosa*.

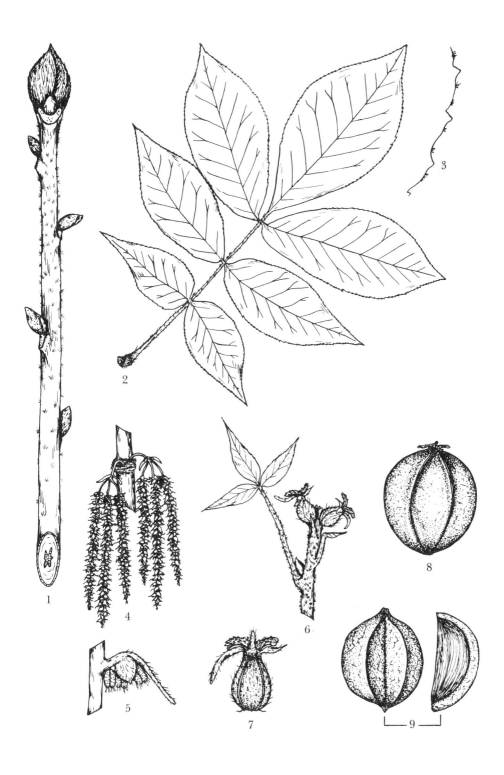

1. Winter twig
2. Leaf
3. Leaf margin
4. Staminate catkins
5. Staminate flower
6. Pistillate cluster
7. Pistillate flower
8. Fruit
9. Nut with portion of husk

BETULACEAE
Corylus americana Walt.

Hazelnut, filbert

A colony forming shrub to 2.5 m high, either open or densely branched.

LEAVES: Alternate, simple. Broadly ovate to oval; 8-12 cm long, 4.5-8 cm wide; biserrate with 2-3 of the larger teeth per cm, often with shallow lobes, ciliate; tip acute or acuminate, the base rounded to cordate; upper surface dark, dull green, thinly pubescent, older leaves often rugose; lower surface paler and more densely pubescent; petiole 5-12 mm long, pubescent and stipitate glandular; stipules 5-7 mm long, lanceolate, pubescent, often serrate, caducous.

FLOWERS: March, long before the leaves; monoecious. Staminate catkins, formed in the previous summer, winter over as a closed catkin 3.5-5 cm long, elongating in the spring to 5-8 cm long and 4 mm wide; peduncle 3-6 mm long; catkin scales brown, 1.3 mm across, apiculate, pubescent, ciliate, with 2 small ciliate bracts below the scale; stamens 4, each filament divided, bearing a half anther which is apically pilose. The pistillate flowers are from axillary buds formed in the previous summer, these buds are similar to the leaf buds, and, at the time of flowering, only the stigmas are exserted; 6-12 flowers per bud, the flowers in pairs subtended by a small bract and 2 minute, inner bracts, these accrescent in fruit; ovaries small, and one of the pair is sterile; style and stigma linear, 3-5 mm long, bright red, becoming darker and withered after exposure.

FRUIT: September. Clusters of 2-4; a short, broad, irregularly toothed bract, 2-2.5 cm wide, 1-1.5 cm long, at the base of each cluster; the 2 accrescent bracts adnate to the lower part of the nut, enclose the nut and extend beyond for 1-1.5 cm, the outer margin broad, sinuate, and coarsely toothed, the bracts downy pubescent and with a few stipitate glands. Nut subglobose, 1.5-1.7 cm diameter, brown with a velvety puberulence on the otherwise smooth surface; lower one-third of nut rough where it was adnate to the bract; apex broadly rounded with a minute tip; base subconical; shell hard, 1.5-2.3 mm thick, corky lined; kernel sweet, edible, white, with a thin, brown, membranaceous skin.

TWIGS: 1-1.2 mm diameter, tan to red-brown, pubescent and with some stipitate glands, lenticels obvious; leaf scars half-round; 3 bundle scars; pith green, becoming brown, continuous, one-third of stem. Buds ovoid, 2-3 mm long, scales red-brown, ovate, pubescent and ciliate.

TRUNK: Bark brown to gray-brown, fairly smooth, the outer, thin layer slightly fissured. Wood nearly white, medium hard, fine-grained, and with obvious rays.

HABITAT: Dry or moist areas, usually forming thickets along the edge of a woods, in fence rows, or on ravine banks; in low, rich soils, or on rocky hillsides.

RANGE: Newfoundland, west to Saskatchewan, southeast to eastern Oklahoma, east to Georgia, and north to New England.

Two forms have been described and may be found in our area. Forma *americana,* as described above, has gland-tipped hairs on the petiole, twigs and bracts; f. *missouriensis* (A. DC.) Fern. is without gland-tipped hairs. Since the amount of gland-tipped hairs varies greatly on forma *americana,* the distinction between the two forms is questionable.

The kernels are excellent and may be eaten raw or used in candy or cookies. Squirrels and other small mammals are also fond of them and will take them as soon as they loosen from the bracts.

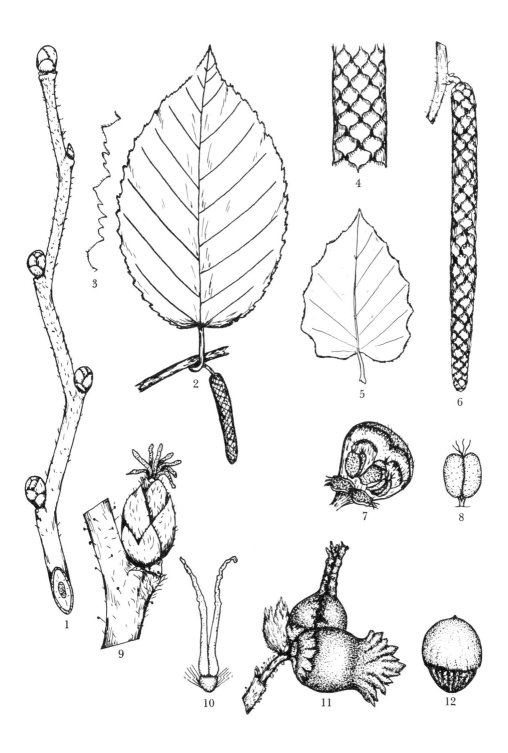

1. Winter twig
2. Leaf and staminate winter catkin
3. Leaf margin
4. Portion of winter catkin
5. Leaf variation
6. Staminate catkin
7. Bract with flower
8. Stamen
9. Pistillate flower cluster
10. Pistillate flower
11. Mature fruit with bracts
12. Nut

BETULACEAE
Corylus cornuta Marsh.

Corylus rostrata Ait.

Beaked hazelnut

An open shrub to 3 m high, usually in loose thickets.

LEAVES: Alternate, simple. Ovate to slightly obovate, 7-10 cm long, 5-7 cm wide, tip acute or short acuminate, the base rounded to cordate; sharply serrate or double serrate, 3-7 teeth per cm; often with shallow lobes 1-2 cm from tip to tip, and a main vein extending into the tip, the sinuses 2-4 mm deep; young leaves densely pubescent on both sides, the upper surface soon glabrate and dark green, the lower surface with a few hairs on the veins and with tufts in the vein axils; petioles 1-2 cm long, slightly grooved or flattened above, finely pubescent; stipules narrowly ovate to narrowly triangular, 7-8 mm long, 2.5-3 mm wide, pubescent, early deciduous.

FLOWERS: Late April; monoecious. Staminate catkins formed the previous summer and expanding at anthesis to 1.5-2.5 cm long and 4-5 mm wide, drooping, sessile or short-stalked; bracts triangular with a long point, purplish, pubescent, ciliate, and long ciliate at the tip; 2 bractlets, obovate, greenish, appressed to the apical side of the larger bract; 4 stamens, the filaments short and divided, each bearing one locule, the anthers yellow and bearded at the tip. Pistillate flowers from lateral buds on wood of the previous year, the buds resembling leaf buds but with the stigmas extruded; 3-4 flowers per bud, each subtended and surrounded by 2 bractlets; stigmas filiform, 4-4.5 mm long, red, extended long before the bud opens or the leaves emerge; ovary 0.5 mm long, obovate, green, pubescent at the base. As the bud opens, the axis elongates and the young fruits appear with a short pedicel.

FRUIT: August. Single or in clusters of 2-6, pedicels 5-10 mm long, reflexed, pubescent; body of the fruit ovoid, 1.5-2 cm long and as wide, the 2 bracts which enclosed the flower are accrescent and now enclose the nut and extend 3-4 cm beyond, becoming 5-8 mm wide toward the foliaceous tip; the bracts puberulent and bristly with long, stiff hairs, the beak straight or curved, greenish-brown, and somewhat ridged and tuberculed. Nut brown, ovoid, 12-15 mm long and as wide, finely puberulent and with longer hairs at the apex; basal scar 9-10 mm across and nearly flat.

TWIGS: 1.4-1.6 mm diameter, yellow-brown, flexible, dull, glabrous or with a few hairs at the nodes; bark of older twigs often flaky; leaf scars small, oval to nearly deltoid; 5-9 scattered bundle scars; stipule scars narrowly elliptic to narrowly deltoid; pith pale brown, firm, continuous, one-third of stem. Buds ovoid, acute, yellow-brown to red-brown, 3.5-5 mm long, 2-2.5 mm diameter, the scales pubescent and densely ciliate. Staminate catkins formed during late summer in a leaf axil; catkin oval, 7-8.5 mm long, 4 mm thick, erect on a pubescent stalk 2 mm long, the scales apiculate, densely ciliate, red-brown but gray with a dense pubescence.

TRUNK: Bark light gray-brown, irregularly fissured, scales tight. Wood hard, close-grained, rays obvious, heartwood pale tan, with a wide, nearly white sapwood.

HABITAT: Creek banks, rocky hillsides, borders of woods and brushy pastures, usually in slightly moist soil.

RANGE: Newfoundland to British Columbia, south into California (var. *california*); northern Idaho, but apparently not in Montana or the Intermountain Region; Colorado, northeastern Wyoming, North and South Dakota, east across northern United States to New England, and south into the mountains of Georgia. Listed in several manuals for Kansas, but no specimens were located.

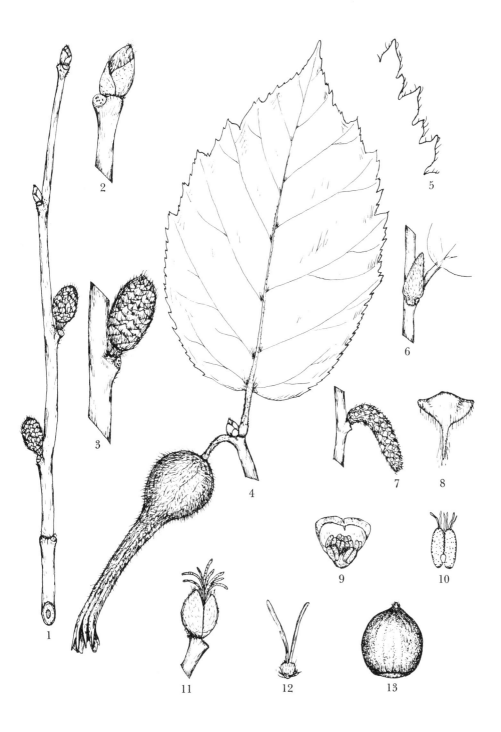

1. Winter twig
2. Detail of end of twig
3. Staminate winter catkin
4. Leaf and mature fruit with bracts
5. Leaf margin
6. Stipule
7. Staminate catkin

8. Staminate catkin scale
9. Stamens and bractlets
10. Stamen
11. Pistillate flower cluster
12. Pistillate flower
13. Nut

BETULACEAE
Ostrya virginiana (Mill.) K. Koch

Carpinus virginiana Mill.

Ironwood, hop hornbeam

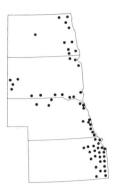

A tree to 8 m high, with wide, spreading branches.

LEAVES: Alternate, simple. Ovate to elliptic, 6-10 cm long, 3-5 cm wide; sharply biserrate, 3-5 teeth per cm, slender tip on each tooth; leaf tip acute to acuminate; base narrowly cordate to broadly cuneate; upper surface dull, dark green, glabrous or sparingly pubescent; lower surface paler and pubescent, the pubescence more dense on the veins and in vein axils; petiole slender, pubescent, 4-10 mm long; stipules lanceolate, 4-6 mm long, caducous.

FLOWERS: April-May with the leaves; monoecious. Staminate catkins formed in the previous season, 2-3 cm long, red-brown, usually in 3's; expand at anthesis to 6-7 cm long, drooping, the axis pubescent and flowers crowded; each scale of the catkin cupped, green with the outer margin brown, ciliate, apiculate, the body of scale glabrous, pubescent at the base inside; stamens 3-7, attached beneath the scale, filaments divided, each with a yellow half-anther 1 mm long, pilose at the tip. Pistillate catkins, solitary or in pairs on new growth, erect or nodding, opening shortly after the staminate flowers; peduncle 1.2-1.8 cm long, pubescent; catkin 2-3 cm long; outer bracts ovate, long acuminate, pubescent, deciduous; these bracts subtending 2 flowers, each flower enclosed in 2 white, ovate, woolly bractlets with the margins united; ovary 0.2 mm long, broadly ovoid, compressed, green, glabrous; style and stigma not noticeably different, filiform, 4-4.3 mm long, pinkish, finely pubescent, the 2 stigmatic ends long exserted.

FRUIT: June-July. A hop-like strobile consisting of the accrescent, inflated bracts which surrounded the ovary in flower; strobile 2.5-5 cm long, 2 cm thick, on a hairy, slender peduncle 2 cm long; bracts ovate, acuminate, pubescent, 1.5 cm long, 0.8 cm wide, stramineous; at maturity these bracts, with the margins fused and enclosing 1 seed, are deciduous from the strobile axis. Seeds ovate to elliptic, 7-8 mm long, 2.8-3.2 mm wide, 1.8-1.9 mm thick, olive-brown, slightly pointed, finely striate, lustrous.

TWIGS: 1-1.5 mm diameter, flexible, pubescent toward the tip, otherwise glabrate or glabrous, usually slightly zigzag, red-brown to dark brown; lenticels small, not obvious; leaf scars somewhat raised, half-round; 3 bundle scars; pith greenish or pale brown, continuous, one-fourth to one-third of stem. Buds ovoid, acute, 3-6 mm long, the scales red-brown, pubescent, often striate.

TRUNK: Bark thin, red-gray, fissured into narrow, plate-like, tight scales, or an occasional tree with loose scales. Wood heavy, strong, hard, red-brown, with a wide, white sapwood.

HABITAT: Woods on shaded, moist but well-drained slopes and stream banks, usually in rocky soil.

RANGE: Nova Scotia, west to Manitoba, south to the Black Hills, southeast to eastern Texas, east to Florida, and north to New England.

The species has been divided into var. *virginiana*, f. *virginiana* with glabrous or glabrate young branches; var. *virginiana*, f. *glandulosa* (Spach) Macbr. with stalked glands on the twigs; and var. *lasia* Fern. with the twigs permanently and heavily pubescent. All three variations may be in our area but they are not always clearly separable.

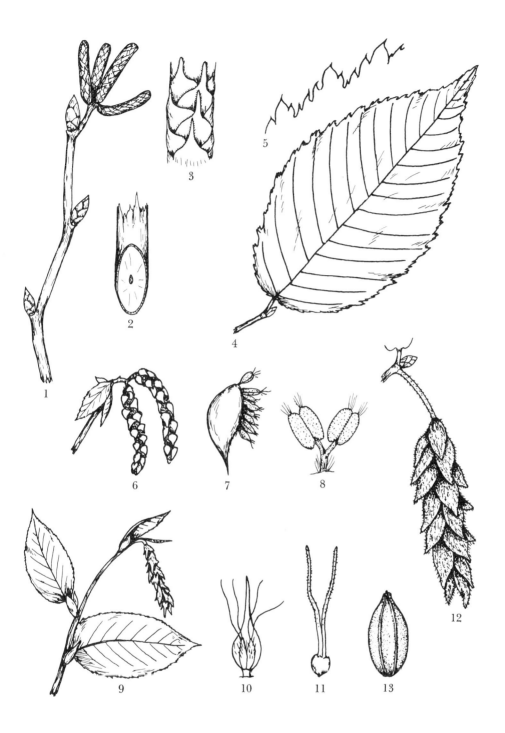

1. Winter twig with staminate winter catkins
2. Section of stem
3. Portion of staminate winter catkin
4. Leaf
5. Leaf margin
6. Staminate catkin
7. One staminate flower

8. Stamen
9. Pistillate catkin
10. Scale of pistillate catkin with two flowers enclosed in bracts
11. One pistillate flower
12. Fruiting catkin
13. Seed

BETULACEAE
Betula glandulosa Michx. var.
glandulifera (Regel) Gl.

Betula pumila L. var. *glandulifera* Regel

Dwarf birch, swamp birch, marsh birch

A freely branched shrub to 2 m high, usually in small colonies.

LEAVES: Alternate, simple, somewhat reticulate beneath. Obovate to nearly rotund, 15-20 mm long, 12-18 mm wide; tip rounded, base cuneate; margin crenate, teeth often gland-tipped, 4-6 teeth per cm, not toothed toward the base; glabrous on both surfaces, dark green above, pale green below, older leaves with dark glandular dots; firm, subcoriaceous; petiole 3-6 mm long, grooved above, finely puberulent, often rather stout; stipules ovate, 2-2.5 mm long, scarious, ciliate, promptly deciduous.

FLOWERS: Mid-June; monoecious, but some plants apparently dioecious. Staminate catkins sessile, cylindric, 15-18 mm long, scales peltate, 1.25 mm across, brown, ciliate; 2 brown bracts beneath the scale; 3 flowers beneath the peltate scale, each with a calyx of one scale-like lobe 0.75 mm across; stamens 2, filaments short, divided near the outer end, each division with 1 anther locule; anthers 1 mm long, yellow. Pistillate catkins on peduncles 5-10 mm long with a leaf-like, petioled, toothed bract 3-5 mm long near the base; catkin cylindric, 14-18 mm long, green; bracts 3-lobed, 2 mm wide, broader than long, center lobe the longest, each lateral lobe with 1 ovary beneath; ovary 1 mm across, flat, winged, brownish with green wing, nearly circular, 2 minute, pointed lobes at the outer end, each lobe with a red style and stigma 0.5 mm long.

FRUIT: August-September. Fruiting catkins erect, cylindric, 10-15 mm long, 5-6 mm wide, light brown, remain on the shrub until early winter then shatter, the center lobe of each bract reflexed or spreading. Fruit flat, irregularly circular to obovate, winged on two sides, 1.5 mm long, 1.2-2 mm wide, light brown; styles or their bases often remain attached.

TWIGS: 1-1.3 mm diameter, flexible, red-brown, pruinose, finely puberulent, densely covered with clear or brownish resin glands; leaf scars crescent-shaped with rounded ends; 3 bundle scars; stipule scars long and narrow; pith greenish, angular, continuous, one-sixth of stem. Winter staminate catkins 5-7 mm long, 2 mm diameter, scales red-brown with lighter margin, densely ciliate, resinous. Buds ovoid, acute, 3 mm long, 2 mm wide, red-brown, the outer scales glabrous or puberulent, ciliate, often with a gland-like structure near the tip, inner scales dark brown and glabrous.

TRUNK: Bark thin, fairly smooth, purplish-brown or gray-brown, with numerous, light, horizontal lenticels. Wood medium hard, fine-grained, light brown, with a wide, nearly white sapwood.

HABITAT: Marshes and wet areas along streams or in broad valleys, often growing with spruce and willow.

RANGE: New York, across southern Canada to British Columbia, south into northern Idaho and Montana, the Black Hills of South Dakota, north to North Dakota, east along the Great Lakes; overlaps with var. *glandulosa* in southern Canada.

In the winter condition, the twigs of *B. glandulosa* might easily be mistaken for those of *B. occidentalis* Hook., both having resinous glands. *B. occidentalis* is a taller shrub and the trunks are more liable to be clustered. The leaves are entirely different, those of *B. glandulosa* are small, under 2 cm long, and are obovate, those of *B. occidentalis* over 3 cm long and are ovate.

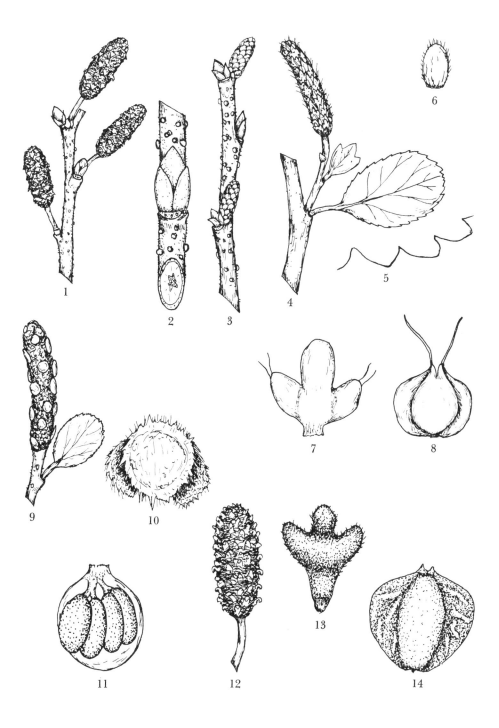

1. Winter twig with old fruiting cones
2. Detail of twig
3. Twig with winter staminate catkins
4. Leaf and pistillate catkin
5. Leaf margin
6. Stipule
7. Pistillate bract, dorsal view
8. Pistillate flower
9. Staminate catkin
10. Staminate bract, dorsal view
11. Staminate flower
12. Fruiting cone
13. Bract of cone
14. Seed

BETULACEAE
Betula nigra L.

River birch, red birch, black birch

A tree to 20 m high, with central trunk and drooping branchlets.

LEAVES: Alternate, simple. Ovate to narrowly deltoid, 4-9 cm long, 3.5-6.5 cm wide; coarsely serrate, 1 tooth per cm, with 3-5 small teeth on the long side of it; tip acute, the base broad, cuneate to subtruncate; upper surface dark green, lustrous, glabrous or sparsely pubescent, lower surface paler, pubescent, especially on the midrib; both surfaces pubescent on young leaves; petiole 8-14 mm long, tomentose; stipules ovate, acute, 3-4 mm long, pale green, white pubescent, caducous.

FLOWERS: Mid-April, with the leaves; monoecious. Staminate catkins formed the previous season and becoming 7-9 cm long at anthesis; the circular, peltate scales reddish-brown, ciliate, minutely apiculate; 3 flowers beneath each peltate scale; calyx with 2-4 lobes, ovate, 0.1 mm long; stamens 2, filaments 1 mm long, divided about the middle, each part with 1 anther locule, giving the impression of 4 stamens. Pistillate catkins from buds of the previous season, erect, 2-3 cm long, peduncle 1-4 cm long with 2 bracts near the middle; flowers in 3's, each with a 3-lobed, green, ovate, pubescent, ciliate bract; ovary broadly ovate, 0.5-0.6 mm across, flat, winged, pubescent; style short; stigmas 2, filiform, 2 mm long, divergent, green, with a red-brown tip.

FRUIT: Late May. Fruiting cone 2.4-3.5 cm long, 1 cm thick, erect or pendent; scales 7 mm long, the 3 lobes 3.5 mm long, densely pubescent; fruit broadly ovate, with wide, lateral wings, total width 5.2-5.4 mm, length 2.4-3.6 mm, wing width 2-2.3 mm, red-brown, pubescent around the nutlet and on the apical portion; wing thin, membranaceous.

TWIGS: 1-1.3 mm diameter, reddish-brown to gray-brown, finely pubescent, flexible, pendent, often zigzag; the older stems with many spur branches; lenticels oval, small, numerous, horizontal on older stems; leaf scars broadly triangular with a convex upper margin; 3 bundle scars, the center one U-shaped; pith green or brown, slightly angular or irregular, continuous, one-fourth of stem. Buds ovoid, acute, 5-6 mm long, 2 mm thick, the 2-4 outer scales red-gray and pubescent. Winter staminate catkins 2-2.5 cm long, 3 mm wide.

TRUNK: Bark pale salmon color, peeling into thin, papery layers; bark of old trunks soft, dark yellow-brown, shallowly furrowed with flat-topped ridges. Wood strong, lightweight, pale brown, with a wide, white sapwood.

HABITAT: Along streams or in low, wet areas.

RANGE: New Hampshire, west to southeastern Minnesota, south through eastern Kansas to eastern Texas, east to Florida, north to Massachusetts. Does not occur in the eastern mountains.

Betula nigra is the only native birch in the southern part of our area and is not easily mistaken for any other tree. The salmon-colored bark should distinguish it from the cultivated birches, most of which have a white bark. It is commonly one of the first trees to establish itself around the pools resulting from mining or removal of sand and gravel. In our area it is restricted to the southeastern corner of Kansas where, locally, it becomes quite common.

The wood is used for pulp in papermaking, furniture, wooden trinkets and bowls, and to some extent for panels. However, our trees are not sufficiently numerous to be of commercial value.

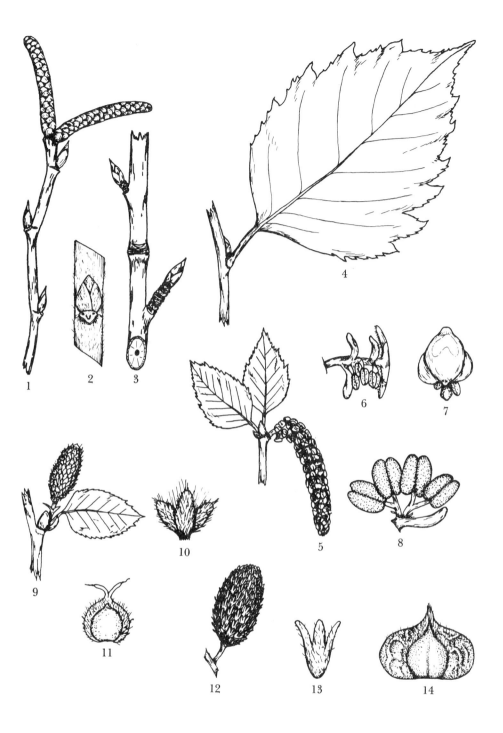

1. Winter twig with staminate winter catkins
2. Detail of twig
3. Twig with spur branches
4. Leaf
5. Staminate catkin
6. Peltate scale with flowers, side view
7. Peltate scale, dorsal view
8. Staminate flower
9. Pistillate catkins
10. Bract of pistillate catkin
11. Pistillate flower
12. Fruiting cone
13. Bract of cone
14. Seed

BETULACEAE
Betula occidentalis Hook.

B. *alba* L. ssp. *occidentalis* Regel; *B. microphylla* var. *occidentalis* M. E. Jones; *B. papyrifera* Marsh. var. *occidentalis* Sarg.; *B. fontinalis* Sarg.; *B. rhombifolia* Nutt.

Western birch

A shrub to 5 m high, the trunks usually clustered and with low branches.

LEAVES: Alternate, simple, often clustered on the end of spur twigs. Ovate to broadly ovate, 3-4.5 cm long, 2-3 cm wide, larger on sprouts; tip acute, base cuneate to rounded; serrate except at the base, 3-5 teeth per cm, often irregular in size; upper surface dark green, lower surface pale yellow-green, both surfaces glabrate and without tufts in the vein axils; young leaves resinous, occasionally with definite resin dots; petiole 1-1.5 cm long, slender, glabrous or lightly pubescent, grooved above; stipules ovate-oblong, with a wide base, long ciliate, caducous.

FLOWERS: Early June; monoecious. Staminate catkins brown, pendent, 3-4 cm long, 5-6 mm wide; peltate scales 1.5-2 mm across, ovate with an acuminate tip, ciliate, stalk of scale 1.5 mm long; 2 smaller bracts on the underside of the scale; 3 flowers on the underside of each peltate scale, each flower with 1 calyx lobe and 2 stamens, the filaments, divided, each branch with 1 anther locule. The exact position of the floral parts is difficult to determine, and the appearance is more like 6 one-locular anthers on a short axis subtended by 3 ovate, ciliate, calyx lobes. Pistillate catkins cylindric, 12-15 mm long, 2-3 mm wide, erect, single, terminal on a very short branch of the season, peduncle 3-4 mm long, with a leaf-like, petioled bract and 2 sessile, lanceolate bracts (possibly a small leaf with 2 stipules); catkin bracts green, 3-lobed, the center lobe ovate, spreading, 2 mm long, the laterals triangular, ciliate on the lower margin; beneath the bracts are 1-3 ovaries, broadly ovate to circular, 0.3-0.7 mm across; styles 2, filiform, reddish, 1 mm long, with a few hairs at the base.

FRUIT: Late August. Fruiting cone cylindric, 2-2.5 cm long, 8-10 mm wide, peduncles 8-10 mm long; bracts brown, 5 mm long, 4 mm wide, ciliate, the center lobe lanceolate, spreading or recurved, the lateral lobes broad and angular. Fruits brown, 2 mm long, 3-3.5 mm wide, the seed portion 1-1.5 mm wide, wing thin and semitransparent.

TWIGS: 1-1.2 mm diameter, red-brown, pruinose, and with a few crystalline resin glands; lenticels elliptic, light-colored; leaf scars half-round, dark-colored; 3 bundle scars; pith greenish, angular, continuous, one-fourth of stem; spur branches short. Buds 3-4 mm long, 0.8-1.2 mm wide, ovate, acute, the scales red-brown, glabrous, often scurfy, ciliate with brownish hairs, the outer scales usually with an abrupt, spreading tip. Winter staminate catkins 1-1.5 cm long, 2.8-3 mm thick, the scales red-brown with a dark margin, long ciliate with pale brown hairs.

TRUNK: Bark dark gray-brown, tight, not peeling, dull, smooth except for the long, horizontal lenticels. Wood soft, pale brown, with a wide, white sapwood.

HABITAT: Stream banks and moist valleys, occasionally on hillsides.

RANGE: British Columbia to Saskatchewan, south to western Nebraska, Colorado and New Mexico, west to California, and north to Washington.

Usually this small tree has several trunks from one base, thus easily distinguished from other small species of birch. However, *B. papyrifera* often grows in clusters and the bark of the young trees is quite similar to that of *B. occidentalis*. In the winter condition, the two may be distinguished by the size of the winter catkins as given in the text.

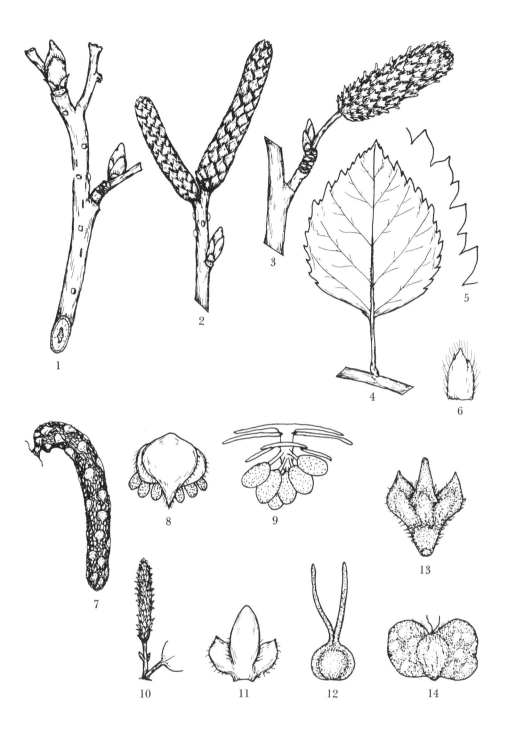

1. Winter twig
2. Twig with staminate winter catkin
3. Twig with fruiting cone
4. Leaf
5. Leaf margin
6. Stipule
7. Staminate catkin

8. Staminate bract, dorsal view
9. Staminate bract and flowers, side view
10. Pistillate catkin
11. Pistillate bract, dorsal view
12. Pistillate flower
13. Bract of fruiting cone
14. Seed

BETULACEAE
Betula papyrifera Marsh.

B. alba var. *papyrifera* Spach; *B. alba*,
ssp. *occidentalis* var. *commutata* Regel;
B. andrewsii A. Nels.; *B. subcordata*
Rydb.

White birch, paper birch, canoe birch

A tree to 15 m high, with central trunk
and small branches, the fine branches
drooping.

LEAVES: Alternate, simple. Ovate
to broadly elliptic, 5-8 cm long, 3-6 cm
wide; tip short acuminate, base cuneate
or somewhat truncate; sharply serrate,
double-serrate, or very shallowly lobed,
3-5 teeth per cm, lobes 1-1.5 cm from tip
to . tip, the sinuses 2-4 mm deep; lateral
veins nearly straight and extending directly to a lobe tip; upper surface dark
yellow-green, semiglossy, sparingly long
pubescent; lower surface much paler
green, glandular, sparingly pubescent,
and with tufts in the vein axils; petiole
1-2 cm long, flattened or slightly grooved
above, glandular, glabrous or puberulent,
and with a few long hairs; stipules narrowly ovate, ciliate, caducous.

FLOWERS: Early June; monoecious.
Staminate catkins sessile, pendent, terminal on twigs of the previous year, single
or clustered, 4-6 cm long, 5-8 mm thick;
the peltate scales ciliate, brown, 1.5-2 mm
across, on a stalk 2-2.5 mm long, the 2
smaller scales beneath it subtending the
3 flowers; each flower with one ovate
calyx lobe, 1-3 small bractlets and 2 stamens, the short filaments divided at the
base, each branch with 1 anther locule
1 mm long. Pistillate catkins single, erect,
terminal on a short branch of the season,
peduncles 5-10 mm long, green, glandular
and with 1-2 bracts; catkin cylindric 15-18
mm long, 2-3 mm diameter, the bracts
green, 3-lobed, the center lobe oblong,
2 mm long, rounded, usually reflexed, the
lateral lobes ovate, small, ciliate along
the basal edge; 2-3 ovaries beneath each
bract, ovary broadly ovate, about 0.4 mm
across and as long, flat with narrow,
lateral wings; styles 2, linear, 0.75 mm
long, red-brown, spreading, often a few
hairs at the base.

FRUIT: August-September, the
cones often remain into early winter.
Pedicels 1.5 cm long, fruiting catkins
cylindric, 2.5-3.5 cm long, 1 cm thick,
brown, drooping; the scales pale brown,
7 mm long, 6 mm wide, with 3 variable
lobes, the terminal one lance-shaped, the
laterals angular. Fruit pale brown with
a broad, rounded, undulate, membranaceous wing on each side, the whole fruit 6
mm wide, 4 mm long, the seed portion
elliptic, 3 mm long, 1.2 mm wide.

TWIGS: 1.5-1.7 mm diameter, red-
brown, flexible, usually pruinose, glabrous or finely puberulent with longer
pubescence around the buds and at the
nodes; lenticels light-colored, elliptic;
older branches with short spurs; leaf scars
kidney-shaped to half-round, dark-colored; 3 bundle scars; pith greenish, angular, continuous, one-fifth of stem. Buds
ovoid, acute, 4-7 mm long, 1.9-2.1 mm
wide, the scales glabrate, ciliate, red-
brown. Staminate catkins clustered at
the end of a twig, 2.5-4 cm long, 4-4.5
mm wide, greenish-brown, the scales
rounded, mucronate, the margin light
brown and ciliate.

TRUNK: Bark on young trees red-
brown, with horizontal lenticels; bark of
old trees white, peeling readily, the lenticels dark and · horizontal; inner bark
brown. Wood lightweight, hard, strong,
fine-grained, light brown or occasionally
pink tinged, and with a wide, white sapwood.

HABITAT: Moist hillside woods or
stream banks and meadows.

RANGE: Alaska to northeast Oregon, east to Colorado, Nebraska, Iowa,
West Virginia, and North Carolina, north
to Labrador, and west across southern
Canada.

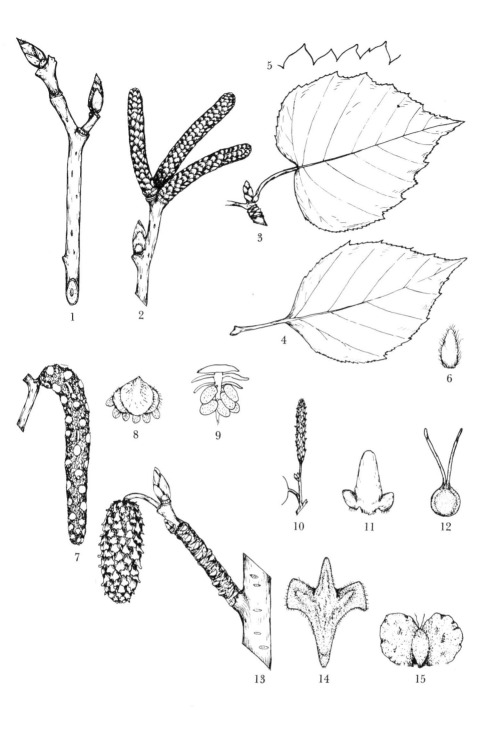

1. Winter twig
2. Twig with staminate winter catkins
3. Typical leaf
4. Leaf variation
5. Leaf margin
6. Stipule
7. Staminate catkin
8. Peltate scale, dorsal view
9. Peltate scale and flowers, side view
10. Pistillate catkin
11. Pistillate bract, dorsal view
12. Pistillate flower
13. Twig with fruiting cone
14. Fruiting bract
15. Seed

BETULACEAE
Alnus incana (L.) Moench ssp.
rugosa (DuRoi) R. T. Clausen

A. rugosa (DuRoi) Spreng.

Alder

A shrub with clustered trunks to 5 m high, the branches on the upper half.

LEAVES: Alternate, simple, with 10-12 nearly straight lateral veins on each side, ending in a large tooth (small lobe); leaf ovate to broadest at the middle, 5-10 cm long, 3-6 cm wide; tip acute, base broadly cuneate or rounded, the margin with small lobes 1-1.5 cm across, each with 2-5 small teeth; upper surface dark green and glabrate, lower surface light green, densely pubescent or nearly glabrous with pubescence on the veins; petiole 1-3 cm long, the narrow groove above becoming flattened toward the blade, glabrous or pubescent; stipules oblong, 6-10 mm long, 1.5-2 mm wide, sparsely or densely pubescent, caducous.

FLOWERS: Late April; monoecious. Inflorescence cymose, open. Staminate catkins 4-9 cm long, 7-9 mm diameter, with many globose flowers; the purple-red, peltate bracts on stalks about 2 mm long; beneath and adnate to this bract are 3-4 small, green, red-tipped, obovate bractlets; the 2-3 sessile flowers attached to the stalk of the peltate scale, each flower with 4 obovate perianth lobes 1 mm long, the lobes cupped, green with a reddish tip, and each bearing 1 stamen, the short filament adnate to the lobe, and the two anther locules separated but attached to the lobe for most of their length. Pistillate catkins sessile, cylindric, 5-6 mm long, 2-2.5 mm wide, the flowers crowded; each fleshy, broadly ovate bract 1 mm long and subtending 2 flowers, each with 2 small, green bracts; ovary green, obovate, flattened, about 0.25 mm long; the 2 red, linear styles 0.75 mm long, and extending beyond the fleshy bract.

FRUIT: September. Cone-like, 14-18 mm long, 9-13 mm thick, ovoid, brown when mature and with spreading scales; scales fan-shaped, flat, 4-ribbed, obscurely 4-lobed with a fifth lobe above, 4.5 mm long, and 4 mm across; 2 seeds on each scale. Nutlets obovate, 3 mm long, and the same width across the top, dark brown, flat, thin, slightly winged on the sides and apex, the style persistent.

TWIGS: 1.8-2.3 mm diameter, red-brown, somewhat scurfy, glabrous except at the nodes, the lenticels light-colored, prominent, oval; leaf scars small, triangular, red-brown; 5 bundle scars; pith angular, continuous, greenish-brown, about one-third of stem. Buds 10-12 mm long, 3-3.3 mm diameter, slightly angular, gray-purple, sparsely pubescent and scurfy, glutinous inside; scales valvate, ciliate; bud stalk 3-4 mm long. Winter staminate catkins 2-2.5 cm long, 5-7 mm diameter, usually in 3's, dark purple-gray; the pistillate catkins 4-5 mm long, 2 mm diameter, purple-gray, attached at the base of the inflorescence.

TRUNK: Bark gray-brown, smooth with small, slightly raised, pale, oval lenticels. Wood soft, pale brown, with a narrow, yellow-brown sapwood.

HABITAT: Swampy ground, usually sandy soil, either along streams or in boggy areas. In the mountains it lines the small streams, growing in rocky soil.

RANGE: Saskatchewan, east to southern Labrador, south to West Virginia, and northwesterly to North Dakota.

Our plants are var. *rugosa* as described above. Further west and in the Rocky Mountains is var. *occidentalis* (Dippel) C. L. Hitchc. (*A. tenuifolia* Nutt.), with slightly narrower leaves and the twigs and leaves more pubescent. *A. incana* var. *rugosa* is the only alder of our northern area.

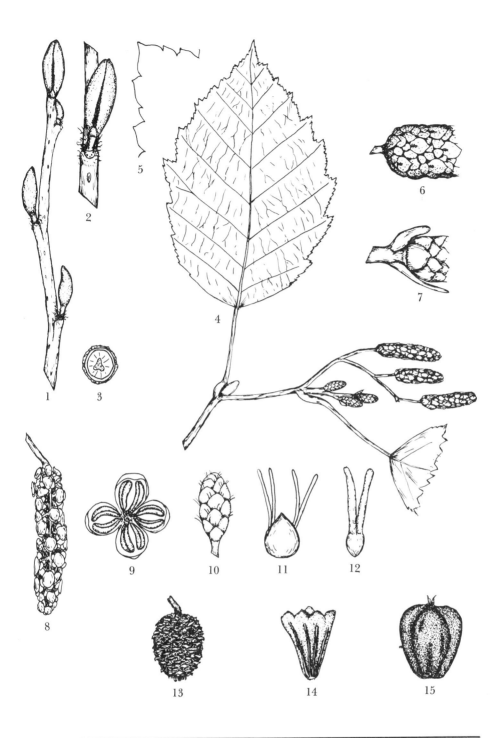

1. Winter twig
2. Detail of bud
3. Transverse section of twig
4. Leaf with winter catkins; staminate (large), pistillate (small)
5. Leaf margin
6. Portion of winter staminate catkin
7. Portion of winter pistillate catkin
8. Staminate catkin
9. Staminate flower with perianth lobes
10. Pistillate catkin
11. One bract, dorsal view
12. Pistillate flower
13. Cone-like fruit
14. One scale of fruit
15. Seed

BETULACEAE
Alnus serrulata (Ait.) Willd.

A. noveboracensis Britt.; *A. rubra*
(Marsh.) Tuckerman; *Betula serrulata*
Ait.

Alder, black alder, smooth alder

A shrub to 4 m high, trunks clustered, with most of the branches above the middle.

LEAVES: Alternate, simple. Elliptic to obovate, 5-9 cm long, 3-5 cm wide; tip obtuse to acute; base cuneate; finely serrate, 3-6 teeth per cm; upper surface dark green, essentially glabrous but pubescent on the midrib; lower surface paler, sparingly pubescent, and with axillary tufts; petiole 8-16 mm long, pubescent; stipules elliptic, 7-8 mm long, 3-4 mm wide, pubescent, caducous.

FLOWERS: March, long before the leaves; monoecious. Staminate catkins formed in the previous season, terminal, clustered, expanding at anthesis to 3-5 cm long, 5-8 mm wide, becoming pendent; catkin composed of many peltate, nearly circular bracts, 1.4 mm wide, with 4 bractlets beneath; each peltate bract with 3 sessile flowers; calyx lobes 4, ovate, distinct, 1-1.2 mm long, 0.9 mm wide, minutely toothed; stamens 4, attached to the calyx lobes, filaments short, not divided; anthers 1 mm long, yellow. Pistillate catkins clustered, formed in the previous season just below the staminate clusters; because of the drooping nature of the staminate cluster and the erect position of the pistillate, it may appear as if the pistillate flowers are terminal; catkin 4-6 mm long, 2-2.5 mm thick; bracts fleshy, 2 flowers per bract; no calyx; pistil 0.6-1 mm long, ovary 0.3 mm long, flattened, ovate, green, glabrous; style short, red; stigmas 2, exserted, filiform, red.

FRUIT: September. Woody, cone-like structure, 15 mm long, 8 mm wide, cylindric or ovoid, scales brown, widely separating; peduncle 4.5-5 mm long, scurfy with small scales. Nutlet irregularly circular, ovate or quadrate, 2-2.3 mm long, 2-2.2 mm wide, brown, variously ridged, one surface nearly flat, the other convex, marginal wing narrow, style persistent.

TWIGS: 1.4 mm diameter, flexible, brown with fine white pubescence, older twigs gray-brown and with small, light-colored lenticels; leaf scars half-round; 3 bundle scars; pith green, continuous, slightly angled, one-third of stem. Buds 2.5-5 mm long, ovoid, obtuse, 2 main outer scales, pubescent; bud stalk tomentose, 1-1.5 mm long. Staminate catkins 1.5-2 cm long, pistillate catkins 4-5 mm long.

TRUNK: Bark red-brown, with horizontal lenticels; bark with narrow fissures starting as a vertical row of dots and eventually opening to a narrow crack. Wood lightweight, not strong, light brown.

HABITAT: Low, wet areas, stream banks, lakeshores.

RANGE: Nova Scotia, southwest to New York, Ohio, Indiana, Missouri, southeast Kansas, Oklahoma, and to eastern Texas, east to Florida, and north to New England.

Alnus serrulata is the only alder in the southern part of our range; and there, in only one location in southeastern Kansas. The only plant with which it might be confused would be a small *Betula nigra*. In the winter, *Alnus* can be distinguished by the clustered, purple-brown catkins, staminate and pistillate together, and the old fruiting cones which often remain until the next summer. In the summer, the clustered trunks and the obovate leaves of alder should distinguish it from the single trunk and triangular leaves of the black birch.

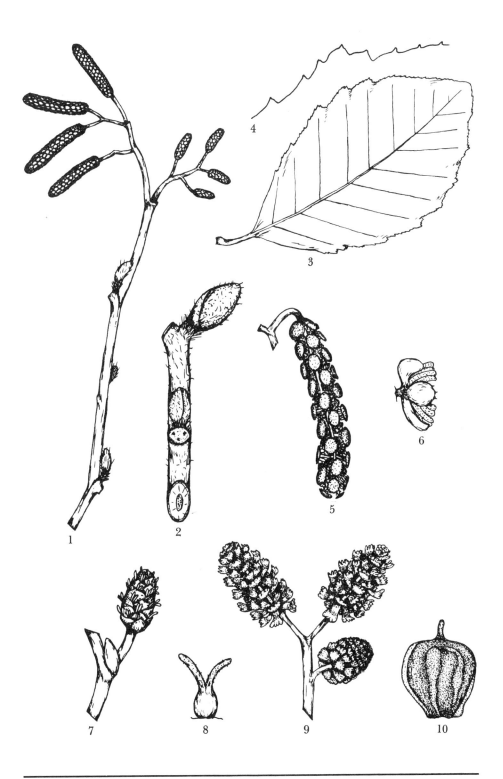

1. Winter twig with staminate catkins (large) and pistillate catkins (small)
2. Detail of nonflowering twig
3. Leaf
4. Leaf margin
5. Staminate catkin
6. Staminate flower
7. Pistillate catkin
8. Pistillate flower
9. Cone-like fruit
10. Seed

FAGACEAE
Quercus alba L.

White oak

A spreading tree to 25 m high, usually with a short, thick bole and large, low branches.

LEAVES: Alternate, simple, subcoriaceous. Obovate, 13-23 cm long, 8-13 cm wide; 5-9 lobes with rounded or emarginate ends, often divided to near the midrib; lobes pointing outward and toward the apex; margin entire with a marginal vein beneath; base of leaf acuminate to acute; upper surface bright green, glabrous, semilustrous; lower surface pale, glaucous, glabrous but often pubescent along the midrib, the hairs stellate; petiole stout, 1-1.5 cm long, glabrous or pubescent; stipules linear, 6-10 mm long, pubescent, caducous; young leaves densely stellate pubescent on both sides.

FLOWERS: May, with the leaves; monoecious. Staminate catkins at the base of new growth or from lateral buds near the tip of old growth, catkins 6-10 cm long, loosely flowered, the axis sparingly pubescent with fascicled hairs; flowers sessile; calyx lobes 3-6, brown, 1.2-1.5 mm long, rounded or acute tip, pubescent and long ciliate; stamens 5-7, filaments white, 0.8 mm long; anthers yellow. Pistillate flowers axillary on new growth, usually in pairs; sessile or with pubescent peduncles 2-7 mm long; ovary globose, 2.5-3 mm diameter, scales green with a red tinge, pubescent; styles 3, short, stout, pubescent, rusty-red; stigmas short, stout, red-brown.

FRUIT: October, of the first year. Acorn sessile or with a 3-5 mm stalk; cup gray, the base rounded, 8-12 mm high, 18-22 mm wide, encloses one-third to one-half of the nut; scales gibbous at the base with thin, acute tips, giving a warty appearance to the cup; exposed portion of the scale 2.7-2.9 mm wide, 3-3.5 mm long, ciliate or erose margin, upper scales small and appressed to the nut; inside of cup densely brown tomentose. Nut ellipsoid to ovoid, 21-30 mm long, 15-19 mm wide, glabrous, light brown; base broad, the scar 8-12 mm across, nearly flat; apex rounded; often indented around the small, blunt tip; kernel sweet, edible.

TWIGS: 2.5-3 mm diameter, rigid and somewhat brittle, bright green, tomentose at first, becoming reddish and glabrate in the first winter, finally ashy-gray to brown; prominent light lenticels; leaf scars half-round to crescent-shaped; 10 or more bundle scars; pith white to brown, angled, continuous, one-third of stem. Buds clustered at the tip, broadly ovoid, 3-6 mm long, the scales numerous, obtuse, dark red-brown, glabrous or pubescent, often ciliate; lateral buds smaller.

TRUNK: Bark thick, light gray, shallow fissures, the ridges short, flat, and plate-like. Wood heavy, strong, hard, durable, light brown, with a narrow, light brown sapwood.

HABITAT: In open areas on well-drained soil, but also in lowlands along streams; commonly in rocky soil.

RANGE: Maine, across southern Canada to Minnesota, south to Texas, east to Florida, and north to New England.

Q. alba when grown in the open has a short, thick trunk, stout horizontal branches often gnarled with age, and a broad, open crown. In a wooded area the trunk is fairly long and straight without branches and is considered to be a good lumber tree.

The acorns vary in shape and those examined from the northern part of Kansas were, in general, short and broadly ovoid; those of the southern part were long and elliptic.

Specimens with leaf sinuses less than halfway to midrib have been described as *Q. alba* f. *latiloba* (Sarg.) Pal. & Steyerm.

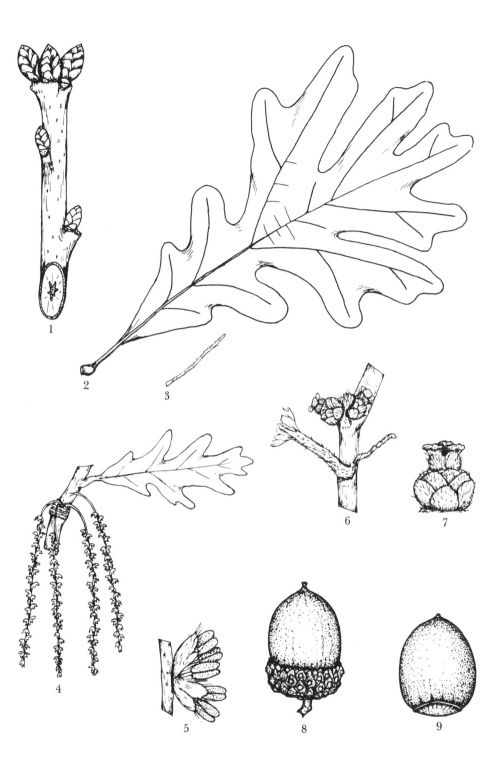

1. Winter twig
2. Leaf
3. Stipule
4. Staminate catkins
5. Staminate flower

6. Pistillate spike
7. Pistillate flower
8. Acorn
9. Nut

FAGACEAE
Quercus borealis Michx. f. var. maxima (Marsh.) Ashe

Q. *rubra* L.; Q. *maxima* (Marsh.) Ashe

Red oak

A tree to 25 m high, the bole long and the crown high and rounded.

LEAVES: Alternate, simple, coriaceous. Oval to obovate, 13-23 cm long, 10-15 cm wide; 5-11 lobes, the sinuses wide and rounded, lobes tapered to the outer end which is bristle-tipped, usually with a few coarse teeth near the lobe tip; base cuneate to truncate; upper surface dark green, lustrous, glabrous; lower surface paler, glabrous and with small axillary tufts of tawny hairs; young leaves densely hairy below with white, fascicled hairs, and slightly pubescent above, with a few red, fascicled hairs scattered among the white; petiole stout, 2.5-4 cm long, glabrous; stipules 7-10 mm long, linear with a broadened end, long pubescent, caducous.

FLOWERS: Mid-April, the leaves half-grown; monoecious. Staminate catkins at the base of new growth or from lateral buds on wood of the previous season; catkins 7-12 cm long, loosely flowered with 30-35 flowers, the axis white tomentose; calyx with 2-5 lobes, divided to near the base, green with a thin, brown edge, glabrous except for a tuft of hairs at the apex; tip often erose and rounded; stamens 4, filaments white, 2 mm long; anthers yellow, 2.2 mm long. Pistillate flowers axillary near the tip of new growth; the short stalk densely tomentose with fascicled hairs; scales broadly ovate, green at the base, red in the middle and with a brown, scarious tip, glabrous and often fringed; ovary depressed globose, 0.7 mm diameter, green, glabrous; styles 3, greenish, 2 mm long; stigmas linear, red to yellow-brown.

FRUIT: October of the second year. Acorn sessile or on a short stalk; cup saucer-shaped, the base nearly flat; 7-9 mm high, 2.5-3 cm wide, upper margin incurved; encloses one-fourth to one-third of the nut; scales tight, light red-brown with darker margin, appressed puberulent; largest scales 5 mm long, 3.5 mm wide, nearly flat, occasionally gibbous, upper scales sparingly ciliate; inside of cup nearly glabrous but with a few hairs at the bottom and around the scar. Nut broadly ovoid, 2-2.5 cm long, red-brown, finely puberulent with stellate and straight, appressed hairs; the base broad with a flat scar 10-12 mm across; apex rounded, not depressed, and with a conical tip; kernel white to pinkish, bitter.

TWIGS: 2.5-2.6 mm diameter, rigid, outer end somewhat flattened or ridged, lustrous green at first, becoming reddish-brown in the first year and dark brown to ashy in the second year, glabrous; new twigs at flowering time densely pubescent; leaf scars half-round; 10 or more bundle scars; pith cream-colored, 5-angled, continuous, one-third of stem. Buds clustered at the tip, 3.6 mm long, ovoid, acute, the scales red-brown, lustrous, glabrate and ciliate; lateral buds smaller.

TRUNK: Bark dark brown to gray, with shallow or deep fissures and firm, broad ridges; inner bark bright red, rich in tannic acid. Wood heavy, hard, strong, light red-brown, with a narrow, light sapwood.

HABITAT: Rich, moist loams, stream banks, and flood plains, or on rocky hillsides.

RANGE: Nova Scotia, west to Ontario, south to Minnesota, Nebraska, Oklahoma, and Louisiana, east to Georgia, northeast to Virginia and New England.

Q. *borealis*, var. *maxima* is easily confused with Q. *shumardii*. Q. *borealis* has the lobes of the leaf tapered to a narrow end, a deeply fissured bark, and a shallow acorn cup; whereas, Q. *shumardii* has enlarged ends on the lobes, smoother bark, and deeper acorn cup. The wood is used in furniture and interior finish, but has a tendency to check.

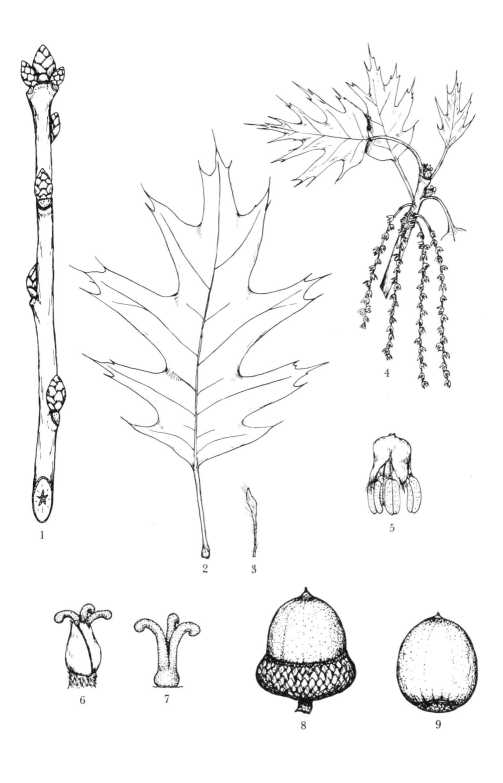

1. Winter twig
2. Leaf
3. Stipule
4. Staminate and pistillate flowers
5. Staminate flower
6. Pistillate flower
7. Pistil
8. Acorn
9. Nut

FAGACEAE
Quercus imbricaria Michx.

Shingle oak, laurel oak

A tree to 20 m high, with central trunk and comparatively small branches.

LEAVES: Alternate, simple, coriaceous. Lanceolate to narrowly obovate, 10-15 cm long, 2.5-5 cm wide; margin entire with a marginal vein beneath, often undulate; tip acute, mucronate, the base cuneate; upper surface dark green, lustrous, glabrous; lower surface paler and brownish, pubescent with both red and white stellate hairs; petiole stout, 6-12 mm long, pubescent or glabrous; stipules linear or with a broad outer end, 7-10 mm long, pubescent, promptly deciduous. Young leaves often with a caudate tip.

FLOWERS: Early May, with the leaves; monoecious. Staminate catkins from the base of new growth or from lateral buds on old growth; catkin 5-7.5 cm long, loosely flowered with 30-35 sessile flowers, axis tomentose; calyx with 3-5 distinct lobes, 1-1.2 mm long, brownish, glabrous with a tuft of hairs at the tip; stamens 2-4, filaments white, 1 mm long; anthers yellow, 1.2 mm long; receptacle hairy. Pistillate flowers axillary on new growth; peduncles tomentose, 4-7 mm long, 1-3 flowered; 1 linear, pubescent calyx lobe; lower scales of involucre obtuse, reddish, long ciliate; upper scales green, acuminate; ovary green, depressed globose, 1.2 mm diameter, 0.8 mm long, glabrous; styles 3, linear, 2 mm long, recurved, reddish-green; stigma red.

FRUIT: September of the second year. Acorns solitary or in pairs on stout peduncles 8 mm long; cup slightly turbinate, cup-shaped, 9-11 mm high, 13-16 mm wide, red-brown; encloses one-third to one-half of the nut; scales ovate or deltoid with rounded or truncate tip, appressed pubescent, the edges scarious and erose; exposed portion of scales 3.2 mm long, 3 mm wide, the upper scales smaller; upper row of scales with pubescent margins appressed to the nut; inside of cup glabrous, red-brown, lustrous.

Nut broadly ovoid to subglobose, 12-14 mm long, 12-14 mm diameter, puberulent, dark brown, often with light-colored stripes; base broad, scar 7-8.5 mm across, rounded; apex dome-shaped, the tip small and conical; kernel bitter.

TWIGS: 2-2.5 mm diameter, rigid, glabrous, dark green to light brown, lustrous; leaf scars small, half-round; 10 or more bundle scars; pith brown, 5-angled, continuous, one-fourth of stem. Buds clustered at the tip of the twig, 2.9-3.2 mm long, ovoid, acute, lustrous, obscurely angled; scales tight, imbricated, light brown, often ciliate.

TRUNK: Bark thick, light brownish-gray, with shallow fissures and long, flat ridges. Wood heavy, hard, light red-brown, with a narrow, light sapwood.

HABITAT: Rocky upland soil or rich river flood plains and stream banks.

RANGE: Pennsylvania, west to Wisconsin, southwest to Iowa and eastern Kansas, southeast to Arkansas, northern Georgia, and north in the Appalachians to New Jersey.

Q. imbricaria is the only oak of our area with entire leaves. It is a good shade tree for street planting, but is subject to fungus diseases as it becomes older. The leaves remain green longer than most oaks, finally turning orange to rusty brown and remaining on the tree in early winter.

The wood is inferior to that of other oaks, but is used to some extent in construction, furniture, and shingles. It grows to a height of 20 meters with an open, rounded crown. In open areas, it branches low with nearly horizontal lower branches.

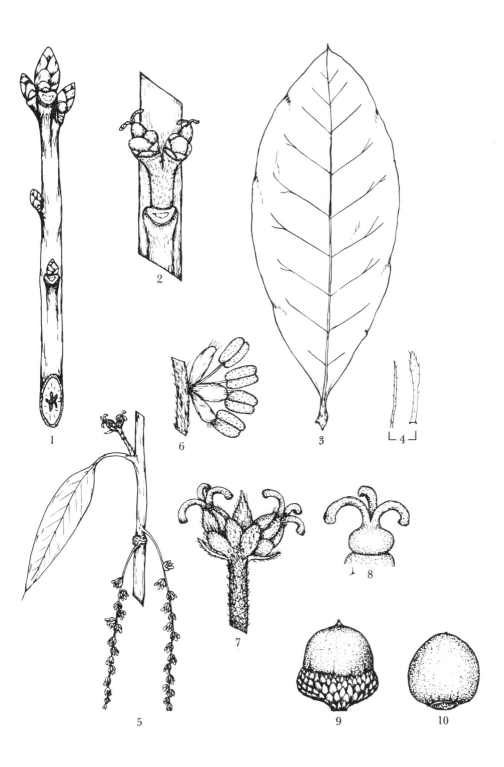

1. Winter twig
2. Twig with first-year acorns
3. Leaf
4. Stipules
5. Staminate and pistillate flowers
6. Staminate flower
7. Pistillate flower
8. Pistil
9. Acorn
10. Nut

FAGACEAE
Quercus macrocarpa Michx.

Bur oak, mossy-cup oak

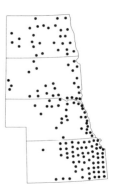

A spreading tree to 28 m high, with massive bole and low, large branches.

LEAVES: Alternate, simple, coriaceous. Obovate, 15-25 cm long, 6-15 cm wide; 5-7 lobes, the sinuses often extending to near the midrib, end of main lobes wider than the base and with a marginal vein; apex rounded, the base cuneate or rounded; young leaves densely pubescent; upper surface of mature leaves dark green, lustrous, with sparse pubescence; lower surface whitened and heavily pubescent; petiole 1.3 cm long, stout, the base enlarged; stipules linear, 6-8 mm long, pubescent, caducous.

FLOWERS: Late April, with the leaves; monoecious. Staminate catkins from the base of new growth or from buds near the tip of old growth, 7-10 cm long, yellow-green, the axis pubescent; flowers sessile; calyx lobes 3-5, distinct, 2 mm long, lance-ovate, green with a brown tip, pubescent; stamens 6-10, filaments white, 1 mm long; anthers yellow, 1 mm long. Pistillate flowers in clusters of 1-4, axillary on new growth; sessile or with a stalk to 3 mm; scales of involucre pubescent, green with some red; ovary globose, 2-2.5 mm long; style short or absent; stigmas 3, linear, yellow-green, often united to a disk-like structure at the top of the ovary.

FRUIT: October of the first year. Acorns sessile or with a stalk of 3-4 mm; cup gray, 2.5-3 cm high, 3.5-4.5 cm wide, cup-shaped, enclosing one-half to two-thirds of the nut; in the southern part of the range the fringe often covers the nut; main scales gibbous, slightly keeled with an acute tip, or occasionally long acuminate; appressed pubescent; exposed portion of scale 5-8 mm long, 4-5 mm wide, smaller toward the rim; scales near the upper edge of the cup small with elongated tip to 1 cm and the scales on the rim reduced to a coarse filament 5-14 mm long, forming a thick, curled fringe around the nut; cup silky pubescent in-side. Nut broadly ovoid, 2.5-3.5 cm long, 2.5-3 cm diameter, brownish, puberulent; base broad, scar 1.8-2.2 cm wide, nearly flat; apex rounded, often depressed, the tip small and conical; kernel white, with a brown, membranaceous coating; sweet, edible.

TWIGS: 3-5 mm diameter, rigid, yellow-brown, becoming ashy-gray or brown; tomentose when young, pubescent in the first winter; small branches often with corky ridges; leaf scars large, shield-shaped; 10 or more bundle scars; pith pale brown, 5-angled, continuous, one-third of stem. Buds clustered at the tip of the twig, 3-5 mm long, broadly ovoid, red-brown, with many pubescent scales.

TRUNK: Bark thick, gray-brown, deeply furrowed, the ridges long and flat-topped. Wood hard, heavy, strong, durable, pale brown, with a narrow, pale sapwood.

HABITAT: Common on rich bottom land soils but also on rocky hillsides in open areas and poor soil.

RANGE: New Brunswick, west to Manitoba, south through Kansas to Texas, east to Alabama, northeast to Virginia and New England.

The identification characters for *Q. macrocarpa* are distinct enough so that in any specific area the tree is not confused with other species. However, the trees in eastern Kansas are quite different from those of the Black Hills and of the northwestern part of North Dakota. In general, the trees of the northern states have smaller leaves and fruits, the leaves being less deeply lobed and the acorn cup less fringed and covering a smaller portion of the nut. Maze (Brittonia 20: 321-333) suggests that the northern plants are a result of past hybridization with *Q. gambellii* Nutt. Rydberg has described this northern taxon as *Q. mandanensis*.

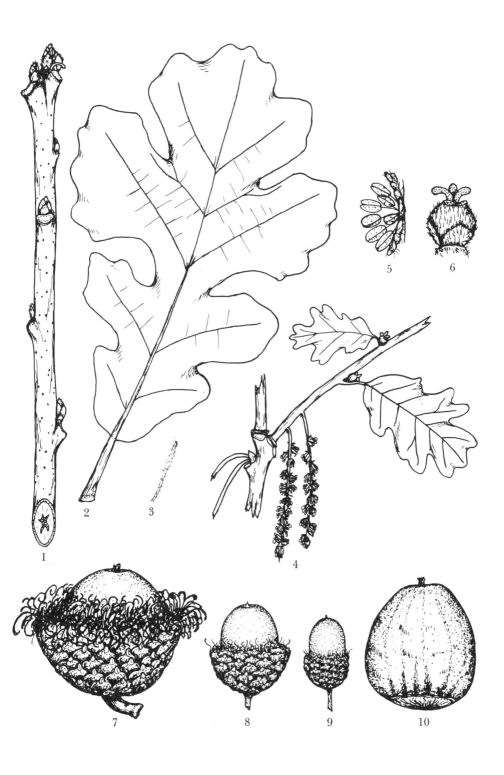

1. Winter twig
2. Leaf
3. Stipule
4. Staminate and pistillate flowers
5. Staminate flower
6. Pistillate flower
7. Acorn, Kansas
8. Acorn, eastern Nebraska
9. Acorn, western South Dakota
10. Nut

FAGACEAE
Quercus marilandica Muenchh.

Black Jack oak, scrub oak

A tree to 8 m high, usually with crooked trunk and contorted branches, the crown low and irregularly rounded.

LEAVES: Alternate, simple, coriaceous. Broadly obovate, abruptly expanded to a broad, 3-lobed, somewhat truncate end, 10-15 cm long, 7-12 cm wide, the lobes and the few teeth with bristle tips; base narrow, rounded; upper surface dark yellow-green, lustrous, sparsely and minutely pubescent with fascicled hairs; lower surface brownish with rusty, fascicled hairs and axillary tufts, the midrib often densely pubescent; petiole stout, 7-12 mm long, rusty pubescent; stipules oblanceolate, 6-8 mm long, pubescent.

FLOWERS: Mid-April, with the leaves; monoecious. Staminate catkins single or in clusters of 2-4 in the leaf axils at the base of new growth; catkins 5-12 cm long, axis pubescent, 30-50 sessile flowers; calyx lobes 3-6, distinct, 2 mm long, green with a brown end, the outside glabrate with a tuft of hairs at the broadly acute tip, pubescent on the inside; stamens 4-6, filament 0.5-1 mm long; anthers yellow, 1-1.5 mm long. Pistillate flowers axillary near the end of new growth, clustered on a short, tomentose peduncle; flowers broadly ovoid, 1-1.5 mm long, the involucre scales acute, rusty tomentose with a few bright red hairs near the tip; ovary green, glabrous; the 3 styles 1.5 mm long, green at the base and red at the outer end, recurved; stigma subcapitate, bright red.

FRUIT: October of the second year. Acorn sessile or with a stalk to 2 mm; cup turbinate, 9-11 mm high, 14-18 mm broad, red-brown, enclosing one-third to one-half of the nut; exposed part of the largest scales 5 mm long, 3 mm wide, appressed pubescent; scale tips truncate, those near the rim with transverse wrinkles; inside of the cup densely pubescent with fulvous hairs. Nut ovoid to subglobose, 13-17 mm long, 9-13 mm

diameter, yellow-brown, puberulent; base broad, scar 5-8 mm across, slightly rounded; apex rounded, the tip conical; kernel yellow, not palatable.

TWIGS: 2.8-3 mm diameter, rigid; young twigs light red-brown, scurfy, becoming glabrous and red-brown to ashy-brown; leaf scars small, half-round; 10 or more bundle scars; pith brown, 5-angled, continuous, one-fourth of stem. Buds clustered at the tip, 5-6 mm long, ovoid, acute, slightly angled, the scales light red-brown with a few rusty hairs; lateral buds smaller. Overwintering, fertilized pistillate flowers in leaf axils near the end of the twig.

TRUNK: Bark thick, black, furrowed, the ridges squarish, plate-like and flat-topped. Wood hard, heavy, strong, dark brown, with a wide, light sapwood.

HABITAT: Dry sandy or clay soil, more common in the open or along the border of a woods in rocky sandstone areas, but often in pure, dense stands.

RANGE: New York, west to Wisconsin, south to Kansas and Texas, east along the gulf to Florida, and north to New Jersey.

Black Jack is a shrubby tree with contorted branches and an irregularly rounded crown, seldom over 8 meters high. It hybridizes freely with several other species of oak and, although the author found no record of it hybridizing with *Q. stellata*, some trees certainly have characteristics of both species. It often grows in mixed stands with that species.

The leaves are tenacious, turning brown and clinging to the tree throughout the winter. It affords good cover for squirrels, and the heavy crops of small acorns provide food for them and other small mammals and many birds. The wood is used mainly for fuel, fence posts, and railroad ties.

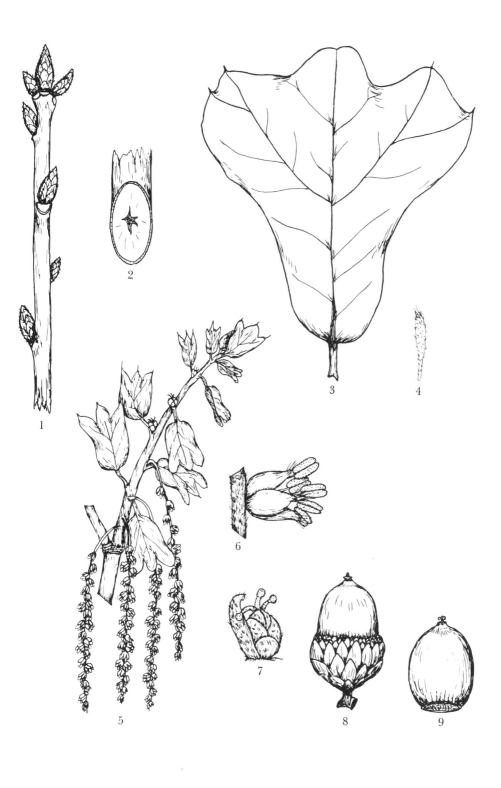

1. Winter twig
2. Section of twig
3. Leaf
4. Stipule
5. Staminate and pistillate flowers
6. Staminate flower
7. Pistillate flower
8. Acorn
9. Nut

FAGACEAE
Quercus muehlenbergii Engelm.

Q. prinoides Willd. var. *acuminata*
(Michx.) Gl.

Chestnut oak, chinquapin oak,
yellow oak

A spreading tree to 18 m high, often with large, low branches and rounded crown.

LEAVES: Alternate, simple, coriaceous. Oblong, lanceolate or obovate, 10-18 cm long, 2.5-10 cm wide; 8-14 teeth on each side, the teeth variable, rounded, sharp, and spreading or incurved, each tooth mucronate; a marginal vein beneath; lateral veins nearly parallel, each directly into a tooth; tip acute to acuminate; base acute to rounded; upper surface dark yellow-green, glossy, glabrous; lower surface whitened and with fascicled hairs; petiole 1.5-3 cm long, pubescent; stipules narrowly oblanceolate, 5-7 mm long, pubescent, caducous.

FLOWERS: Early May, the leaves partly grown; monoecious. Staminate catkins from the base of new growth or from lateral buds on growth of the last season; catkin 7-10 cm long, pubescent axis, loosely flowered, the flowers sessile; calyx lobes 4-6, united toward the base, 1 mm long, green with brown ends, glabrate, but with long crinkly cilia; stamens 4-8, filaments white, 1 mm long; anthers yellow, 1 mm long; receptacle hairy. Pistillate flowers in short spikes, axillary on new growth, sessile or with short, tomentose stalks; ovary globose, 1.4-1.6 mm long, involucral scales green with red tips, pubescent; stigmas 3, discoid, sessile, yellow-green.

FRUIT: September of the first year. Acorn sessile or on a stalk to 5 mm; cup brownish-gray, 6-10 mm high, 8-17 mm wide, enclosing one-fourth to one-half of the nut; scales small, exposed portion 4 mm long, 2 mm wide, gibbous about the middle, tip truncate to acute, tomentose, the margin scarious and ciliate; cup densely pubescent inside or with long hairs at the base and glabrous near the rim. Nut ovoid, 16-22 mm long, 11-17 mm diameter, light brown, puberulent near the apex; scar 6-8 mm across, rounded or flat; apex narrowed and

rounded, tip cylindric or conical; nut often shows longitudinal stripes of yellow and brown, especially during maturation; kernel sweet, edible.

TWIGS: 2-3.5 mm diameter, rigid, glabrous, greenish-brown at first, then orange-brown, finally gray-brown; often slightly flattened; leaf scars half-round; 10 or more bundle scars; pith white, 5-angled, continuous, one-fifth of stem. Buds clustered at the tip of the twig, 3-4.5 mm long, ovoid, acute; scales chestnut-brown with a light, scarious, hairy or fringed margin; lateral buds smaller.

TRUNK: Bark thin, silvery-gray or ash-color, with shallow fissures, the ridges short and flaky. Wood hard, heavy, strong, durable, dark brown, with a narrow, pale sapwood.

HABITAT: Rocky limestone hillsides and rocky stream banks, bottom lands and along the border of a woods.

RANGE: Vermont, west to Ontario, south to Minnesota, eastern Nebraska, and Texas, east to Alabama, and northeast to Virginia and New England.

The shape and size of the leaves of *Q. muehlenbergii* are so variable that they have led to many misidentifications. The leaves may be narrowly oblong, or broadly obovate, with the teeth from rounded to sharply pointed. It has often been identified as *Q. michauxii* Nutt., which has mostly simple hairs, sufficiently dense to give a velvety feeling, on the leaf underside.

It is not uncommon to see *Q. prinoides*, var. *prinoides* growing on a pasture hilltop as a shrub 1 meter high, becoming taller as it extends downward on the hillside and grading into *Q. muehlenbergii* on the steep slope with no apparent difference excepting the size. The number of lateral veins is not always a good differentiating character, and a vegetative herbarium specimen is often difficult to identify.

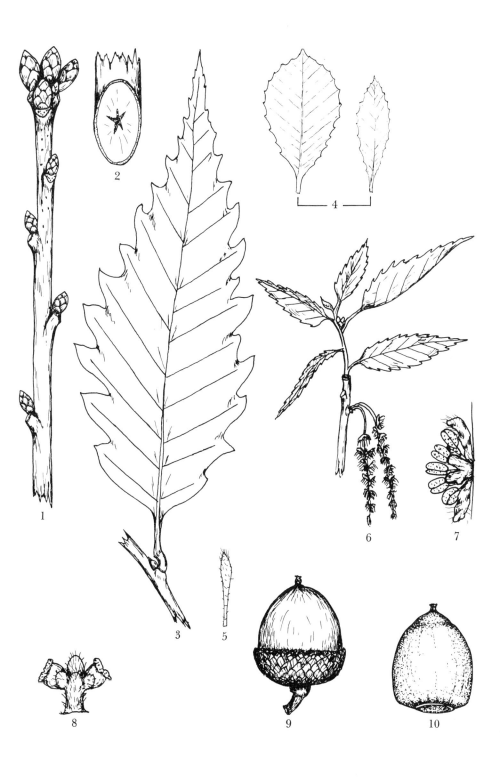

1. Winter twig
2. Section of twig
3. Leaf
4. Leaf variation
5. Stipule
6. Staminate catkins
7. Staminate flower
8. Pistillate spike
9. Acorn
10. Nut

FAGACEAE
Quercus palustris Muenchh.

Pin oak, swamp oak, Spanish oak,
water oak

A tree to 20 m high, with central trunk and conical crown.

LEAVES: Alternate, simple, coriaceous. Obovate to ovate, 10-15 cm long, 5-10 cm wide; 5-7 lobes, the sinuses deep and rounded; lobes generally tapering and at right angles to the midrib; upper surface dark green, lustrous, glabrous; lower surface paler and with pale to rusty hairs in the vein axils; leaves scarlet in autumn, often turning brown and remaining on the tree during the winter; unfolding leaves densely covered with fascicled hairs on both sides; petiole 2-5 cm long, slender, glabrous; stipules linear or oblanceolate, 6-7 mm long, pubescent, long hairs at the tip.

FLOWERS: Mid-April, with the leaves; monoecious. Staminate catkins mainly from buds at the tip of old growth, but some on the base of new growth; catkins 5-7 cm long, axis pubescent, loosely flowered, flowers sessile or with a short pubescent stalk; calyx broadly and shallowly campanulate with 3-5 thin, brownish lobes ciliate on the irregularly acute tip; stamens 5-6, filaments 0.3 mm long, greenish; anthers yellow, 1.6 mm long; receptacle hairy. Pistillate flowers axillary on new growth; peduncles short, tomentose; scales of involucre tomentose, the inner scales thin and appressed to the styles; ovary depressed globose, 0.3-0.5 mm diameter; styles 3, flattened linear, curved outward, green but pinkish toward the tip, persistent over winter; stigmas bright red, recurved, flattened.

FRUIT: October of the second year. Acorn sessile or on a 3-5 mm stalk; cup flat, saucer-shaped, 6-8 mm high, including the turbinate base, 14-17 mm wide, enclosing only the base of the nut; cup scales red-brown, appressed pubescent, narrow, deltoid, flat, with a truncate, acute, or rounded tip and thin, dark brown, ciliate margin; exposed portion of larger scales 3.5 mm long, 3 mm wide; cup pubescent inside. Nut broadly ovoid to dome-shaped, 12-14 mm long, 13.5-15.5 mm diameter; light brown, puberulent; base broad, scar 9-11 mm across, rounded; apex broadly rounded, with a conical tip; kernel bitter.

TWIGS: 2.3-2.5 mm diameter, rigid, dark red and tomentose at first, becoming red-brown in the first winter and gray-brown in the second winter; leaf scars half-round; 10 or more bundle scars; pith pale brown, 5-angled, continuous, one-fifth of stem. Buds clustered at the tip of the twig, 3-4 mm long, ovoid, red-brown, the scales glabrous or with hairs at the tip.

TRUNK: Bark thick, gray-brown or black, smooth except for a few shallow fissures; bark on very old trunks often with shallow fissures and broad, flat-topped ridges. Wood heavy, hard, strong, light brown, with a narrow, pale sapwood; often with many small knots.

HABITAT: Moist rich soil, river bottoms, wet lands. Does well on uplands and is used extensively in street planting.

RANGE: Massachusetts, west through southern Ontario to Wisconsin, southwest to Iowa, Kansas and Oklahoma, south to Louisiana, northeast to Kentucky and North Carolina, and north to Rhode Island.

A general characteristic of *Q. palustris* is the central trunk with small branches, the lower ones drooping, the central ones nearly horizontal, and the upper branches ascending. Its symmetry and graceful branches, as well as the red autumn foliage, make it a desirable tree for street and yard planting. It is not an important lumber tree in our area, but is used to some extent for interior finish, railroad ties, and clapboards.

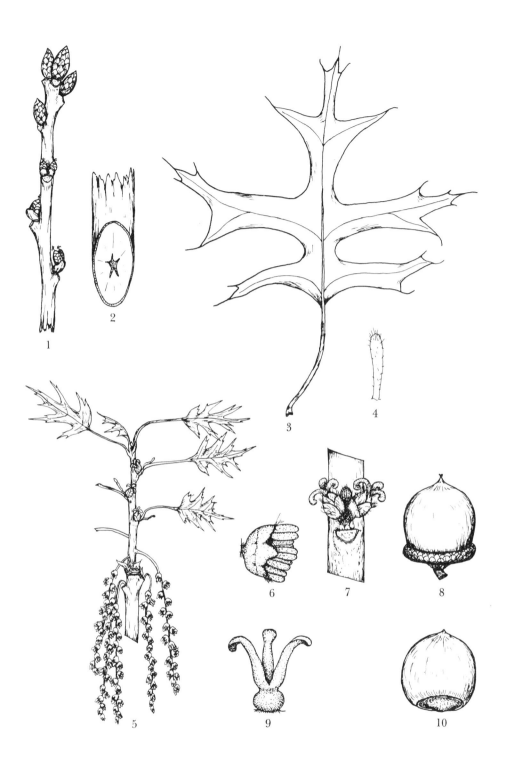

1. Winter twig
2. Section of twig
3. Leaf
4. Stipule
5. Staminate and pistillate flowers
6. Staminate flower
7. Pistillate flower
8. Acorn
9. Pistil
10. Nut

FAGACEAE
Quercus prinoides Willd. var.
prinoides

Dwarf chestnut oak, dwarf chinquapin
oak, scrub oak

A shrub or bushy tree 1-2 m high, with several trunks from one base.

LEAVES: Alternate, simple, coriaceous. Elliptic to obovate, 4-10 cm long, 2-4 cm wide; 4-8 mucronate teeth on each side or the margin sinuate; a lateral vein extending to each tooth; teeth short, with broad, rounded sinuses; leaf tip acute, base cuneate, uneven; upper surface dark yellow-green, semilustrous and with scattered fascicled hairs; lower surface paler, tawny, velvety with fascicled hairs; petiole 7-12 mm long, pubescent; stipules 3-5 mm long, linear or with broadened tip, sparingly pubescent, caducous.

FLOWERS: May; monoecious. Staminate catkins at the base of new growth, from lateral buds or axillary to the bud scales; catkin 3-5 cm long, 25-30 flowers, not crowded, axis pubescent, flowers nearly sessile; calyx cup-shaped, lobes 4-6, distinct or united near the base, 1 mm long, rounded or acute, variable in size and shape, green with brownish tip, pubescent; receptacle hairy; stamens 4-6, filaments white, 0.5 mm long; anthers yellow, 1 mm long, occasionally with a few lateral hairs. Pistillate flowers axillary, near the end of new growth, 2-4 flowers on a short, stout, pubescent stalk; involucre 0.9-1 mm long, 1-1.1 mm wide, green, brownish toward the base, the scales rounded, 0.5-0.6 mm long, densely pubescent; stigmas 3, disklike, sessile, greenish or brown, extending laterally beyond the edge of the involucre.

FRUIT: September of the first year. Acorn sessile or nearly so; cup gray-brown, 8-11 mm high, 15-16 mm wide, enclosing one-third of the nut; scales small, the exposed portion 2-2.2 mm long, 1.8-2 mm wide, gibbous at the center, with an acute, dark tip, densely puberulent; cup heavily pubescent inside. Nut ovoid, 1.5-2 cm long, 1.3-1.5 cm thick, dark red-brown, often purplish; the base broad, apex rounded and slightly de-

pressed, tip cylindric. Nut puberulent toward the apex, the basal scar 6-7 mm across, slightly rounded; shell thin; kernel sweet, edible.

TWIGS: 2-2.4 mm diameter, rigid, red-brown, pubescent when young; leaf scars small, half-round; 5 bundle scars; pith tan, 5-angled, continuous, one-fourth of stem. Buds globose or ovoid, 1.5-3.5 mm long, 1.4-2.5 mm wide, obtuse; clusters of 2-4 at the end of a twig; scales many, imbricated, pubescent, red-brown to grayish, the tips rounded and scarious; the largest bud at the tip faintly 5-angled.

TRUNK: Bark on young trunks gray, smooth except for noticeable, horizontal lenticels; on old trunks gray with shallow fissures and broad, thin plates. Wood hard, yellowish.

HABITAT: Dry, rocky hills and pastures, usually on poor soil; roadside banks, borders of woods, and in open areas.

RANGE: Maine to Minnesota, south through Nebraska to Texas, east to Alabama, and north to Tennessee and Virginia.

Q. prinoides var. *prinoides* in its extreme form is a clump-forming bush about 1 meter high. Its leaves are small, with only 4-8 teeth on each side. From this it grades into *Q. muehlenbergii*, which is a tree 25 meters high, with large leaves and 8-14 teeth on each side. The question always arises as to exactly where to draw the line between them. Var. *prinoides* often forms thickets several meters across and is excellent for soil conservation and wildlife management.

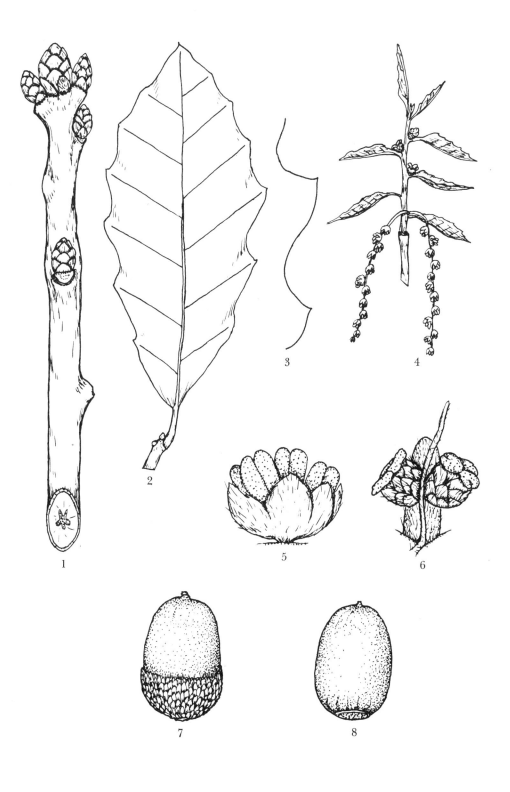

1. Winter twig
2. Leaf
3. Leaf margin
4. Staminate and pistillate flowers

5. Staminate flower
6. Pistillate flowers and stipule
7. Acorn
8. Nut

FAGACEAE
Quercus shumardii Buckl.

Incl. var. *schneckii* (Britt.) Sarg.

Shumard's oak

A tree to 25 m high, with high, spreading crown.

LEAVES: Alternate, simple, coriaceous. Obovate, 15-20 cm long, 10-13 cm wide; 5-7 lobes, each with small lobes or teeth, bristly-tipped; lobes wider at their apex than at the base, sinuses rounded, narrow; base acute to truncate, often uneven; upper surface dark green, lustrous, glabrous; lower surface paler green with tufts of tawny tomentum in the vein axils; petiole 3-4 cm long, glabrous; stipules narrowly linear, 3-7 mm long, pubescent, promptly deciduous.

FLOWERS: Mid-April, with the leaves; monoecious. Staminate catkins on the base of new growth or from the ring of bud scale scars at the end of old growth; catkins 9-15 cm long, 30-40 flowered, not crowded, axis pubescent, flowers sessile; calyx broadly campanulate, 3-5 lobes divided halfway to the base, green with brownish, ciliate tip; receptacle hairy; stamens 4-6, filaments white, 1.4 mm long; anthers yellow, 1.6 mm long. Pistillate flowers axillary on new growth, 1-1.7 mm long, pedicel 1-4 mm long; involucral scales green tinged with brown, pubescent at the tips; styles 3, linear, 1.5 mm long, green at the base and red toward the tip; stigmas linear, bright red, recurved.

FRUIT: October of the second year. Acorns with a short stalk up to 5 mm; cup shallow, the base flat or slightly turbinate, 7-10 mm high, 1.9-2.5 cm wide, enclosing one-fifth to one-fourth of the nut, the rim curved in against the nut; scales thin with wide base and abruptly narrowed to a rounded or truncate, erose tip, light red-brown with darker margin, densely appressed pubescent; exposed portion of the scales 4.4 mm long, 3 mm wide; inside of cup thinly pubescent with a dense ring of hairs around the scar. Nut red-brown, puberulent, ovoid, 1.8-2.4 cm long, 1.8-2.1 cm diameter, the base broad, the scar 9-11 mm across, nearly flat; apex rounded, the tip conical; shell thick; kernel edible.

TWIGS: 2.5-3 mm diameter, red-brown, somewhat flattened, rigid, glabrous; leaf scars half-round to triangular; many bundle scars; pith white, 5-angled, continuous, one-fourth of stem. Buds clustered at the stem tip, ovoid, acute, 3.5-7 mm long; scales closely imbricated, gray-brown, glabrous with pubescent tip; lateral buds smaller. Dormant, immature acorns may be present in leaf axils near the end of the twig.

TRUNK: Bark of old trees with shallow fissures and broad, long, flat ridges. Inner bark reddish to gray. Wood heavy, hard, light red-brown, with a narrow, light sapwood.

HABITAT: Rich moist soils, stream banks, in heavy woods or open areas; upland on rocky hillsides.

RANGE: Texas, east to Florida, north to Pennsylvania, west to Illinois and Kansas, and south to Oklahoma.

The differences between this and *Q. borealis* have been discussed under *Q. borealis*, the two often confused. *Q. shumardii*, var. *schneckii* (Britt.) Sarg. has been listed as a variation in Kansas, but our material cannot be separated into two varieties. Var. *schneckii* is described as having deeper, narrower acorn cups, the trees growing in upland, drier soil, and the leaves being more uniform on the upper and lower branches.

The wood is of good quality and is used for furniture, interior finish, construction, and clapboards. It is usually sold as red oak.

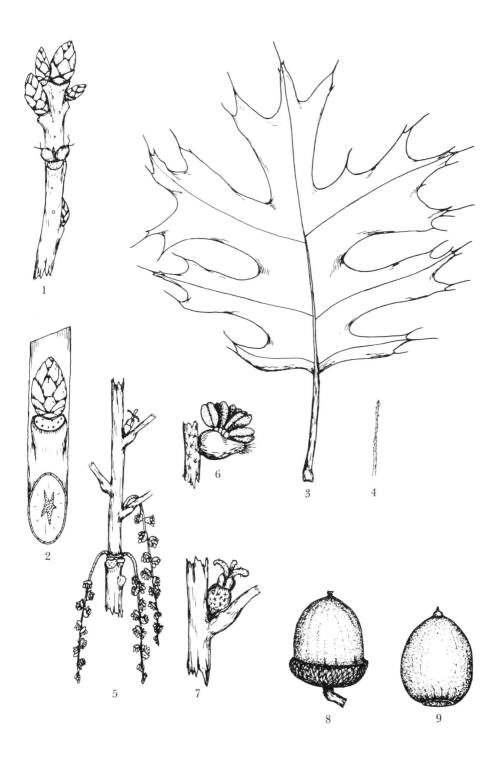

1. Winter twig with first-year fruit
2. Detail of bud
3. Leaf
4. Stipule
5. Staminate and pistillate flowers
6. Staminate flower
7. Pistillate flower
8. Acorn
9. Nut

FAGACEAE
Quercus stellata Wang.

Post oak

A tree to 18 m high, with a broad, cylindric crown.

LEAVES: Alternate, simple, coriaceous. Obovate, 12-20 cm long, 8-15 cm wide; 3-7 lobes, the terminal lobe rounded, the upper laterals broad with a truncate or retuse end; the lower lobes tapered toward a rounded tip; base of the leaf rounded; upper surface dark green, lustrous, slightly granular, scabrous, and with fascicled rusty or white hairs; lower surface pale, covered with fascicled rusty or white hairs; petiole stout, 5-10 mm long, pubescent; stipules linear, 6-8 mm long, pubescent, caducous.

FLOWERS: April, with the leaves; monoecious. Staminate catkins from lateral buds of the previous season or from the cluster of buds at the end of the twig; catkins 8-10 cm long, axis pubescent, flowers somewhat crowded and nearly sessile; calyx with 4-6 lobes, distinct or united at the base, 1.2-1.4 mm long, pubescent, obtuse, yellow with brown tips; stamens 5-8, filaments short; anthers yellow and pubescent. Pistillate flowers clustered in the leaf axils on new growth, 2 to several in each cluster, short-stalked; ovary globose, 1.8-1.9 mm long, surrounded by the pubescent, green involucral scales; style short or absent; stigma lobes 3, broad, yellow-green, clavate.

FRUIT: October of the first year. Acorns sessile or on a short stalk; cup gray-brown, 6-8 mm high, 11-14 mm broad, base rounded, enclosing one-third to one-half of the nut; scales thin, ovate, the exposed portion 1.7-1.8 mm wide, 1-2.6 mm long, red-brown, pubescent, slightly gibbous, tip rounded and scarious; cup pubescent inside. Nut ovoid, 11-13 mm long, 10-13 mm diameter, finely puberulent, base broad, scar 4-5 mm wide, flat, sometimes slightly raised; apex rounded, often with a concave ring near the end; tip small, conical, blunt; shell thin; kernel bitter.

TWIGS: 3-4 mm diameter, rigid, pubescent, orange to reddish at first, becoming gray or dark brown; leaf scars moderate size, broadly crescent-shaped; 10 bundle scars; pith brownish, 5-angled, continuous, one-fifth of stem. Buds clustered at the tip, 3-6 mm long, broadly ovoid with obtuse tip, chestnut brown, scales pubescent and at least some with cilia; lateral buds 3.5-4.5 mm long with an acute tip, pubescent.

TRUNK: Bark dark brown to black, deeply fissured, the ridges flat-topped, broad, and blocky. Wood heavy, hard, durable, brown, with a wide, light sapwood.

HABITAT: Dry rocky or sandy soils; hillsides and open pasture woods.

RANGE: Massachusetts, south and west to Ohio, Illinois, and Kansas, south to Texas, east to Florida, and north to New York.

Var. *stellata* as described above is the common form found in Kansas. An occasional specimen has the lobes narrow and rounded, none of them broadened at the outer end, tending toward var. *margaretta* (Ashe) Sarg.

The post oak is a common tree in parts of eastern Kansas, especially in the sandstone areas, but apparently does not extend further north in our area. It is not an important lumber tree, but may be used for railroad ties, fence posts, ornamental rail fences, and fuel. The leaves turn dark red in autumn, then brown, and may remain on the tree well into the winter.

The oaks are a valuable food plant for wild game, the nuts being eaten by wood ducks, prairie chickens, quail, wild turkeys, blue jays, woodpeckers, raccoons, squirrels, chipmunks, gophers, ground squirrels, and wood rats. Deer may also eat the nuts, but browse mostly on the twigs and leaves of the young trees.

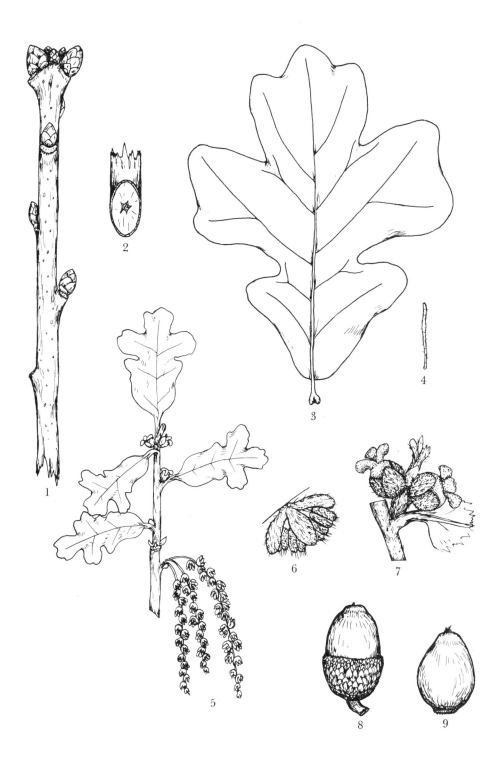

1. Winter twig
2. Section of twig
3. Leaf
4. Stipule
5. Staminate and pistillate flowers
6. Staminate flower
7. Pistillate flowers
8. Acorn
9. Nut

FAGACEAE
Quercus velutina Lam.

Q. leiodermis Ashe; forma *velutina*=
f. *macrophylla* (Dippel) Trel., f. *dilaniata*
Trel. and f. *pagodaeformis* Trel.

Black oak

A tree to 18 m high, usually with the branches on the upper two-thirds.

LEAVES: Alternate, simple, coriaceous. Obovate, 13-20 cm long, 8-15 cm wide, with 5-7 shallow or deep lobes and a marginal vein, the sinuses rounded; upper lobes commonly wider at the outer end, lower lobes tapering, the coarse teeth and lobes with bristle-tips; base rounded or cuneate; upper surface dark green, lustrous, often with a granular appearance and a few rusty hairs on the midrib; lower surface pale and tawny with scattered rusty, fascicled hairs and axillary tufts; petiole stout, 3-6 cm long, densely rusty pubescent, or glabrate; stipules narrowly oblanceolate, 7-11 mm long, tawny pubescent, with a dense tuft at the tip, caducous.

FLOWERS: Late April, with the leaves, monoecious. Staminate catkins mainly from buds on old wood, some on the new growth; catkins 7-10 cm long, axis pubescent, flowers sessile; calyx 4-lobed, united at the base, green to red-brown, acute, glabrate except for a heavy tuft at the apex; stamens 4-8, filaments white, 0.5 mm long; anthers yellow, 1 mm long; receptacle hairy. Pistillate flowers axillary on new growth; peduncles short, tomentose; involucral scales green, tinged with red, pubescent; ovary 2 mm long, ovoid; styles 3, short, green, pubescent at the base; stigmas red, divergent and recurved.

FRUIT: October of the second year. Acorn sessile or on a stalk 3-4 mm long; cup turbinate, red-brown, 10-17 mm high, 16-26 mm wide, enclosing one-half to three-fourths of the nut; scales thin, light red-brown with a darker margin, velvety puberulent, tip acute to truncate; basal scales 4.8-5 mm long, 3.8-4.5 mm wide; scales at the bulge 6-6.2 mm long, 2.7-2.8 mm wide; scales on the upper half of cup with free, spreading tips, giving a coarsely fringed appearance; cup pubescent inside. Nut ovoid, red-brown, 15-19 mm long, 13-19 mm diameter, puberulent; base broad, scar 7-12 mm wide, slightly convex; apex broadly or narrowly rounded, tip small, conical; kernel bitter.

TWIGS: 3.4-3.6 mm diameter, rigid, often slightly ridged; young twigs fulvous, scurfy pubescent, becoming glabrate and red-brown by the first winter; young branches with orange inner bark; leaf scars medium-sized, half-round; 10 or more bundle scars; pith cream-colored to pale brown, 5-angled, continuous, one-fourth of stem. Buds clustered at the twig tip, 9-12 mm long, ovoid, obtuse, strongly 5-angled, hoary pubescent; lateral buds 3-5 mm long.

TRUNK: Bark nearly black and deeply furrowed, the ridges rounded, short, and blocky; inner bark thick, yellow, bitter, with an abundance of tannic acid. Wood hard, heavy, strong, red-brown, with a narrow, pale sapwood.

HABITAT: Dry uplands in poor soil, rocky hilltops, and hillsides.

RANGE: Maine, west to Minnesota, south through eastern Nebraska to Texas, east along the Gulf Coast to Florida, and north to New England.

Several forms have been described on the basis of leaf lobe variations and pubescence. Our material shows a complete intergradation of all these forms, and they are here considered as not separable from the species.

Q. velutina is the only oak of our area with densely pubescent, strongly angled buds and reflexed tips of the acorn scales. Tannic acid may be extracted from the inner bark and a yellow dye may be made from the inner bark. The wood is used mostly for short lumber such as flooring, railroad ties, and barrels.

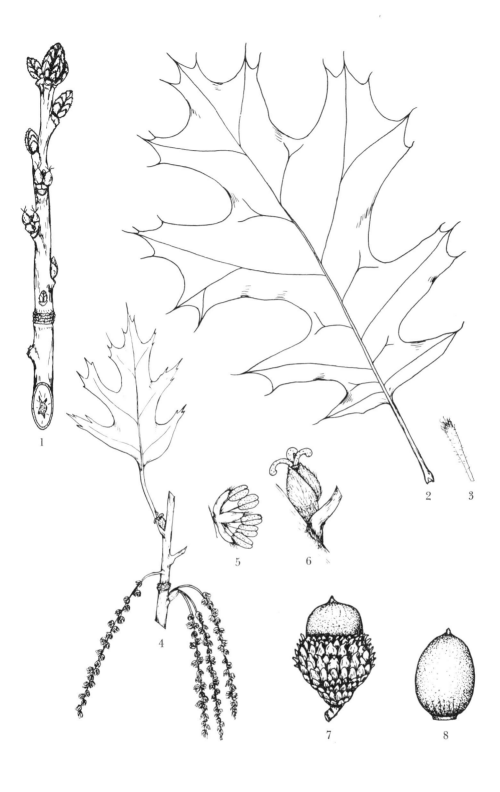

1. Winter twig with first-year acorns
2. Leaf
3. Stipule
4. Staminate and pistillate flowers
5. Staminate flower
6. Pistillate flower
7. Acorn
8. Nut

ULMACEAE
Ulmus alata Michx.

Winged elm, wahoo

A tree to 20 m high, with spreading crown and many slender branches.

LEAVES: Alternate, simple, the lateral veins parallel and terminating in a tooth. Narrowly ovate to elliptic, 5-7 cm long, 2-3 cm wide, tip acute to acuminate, the base cuneate, narrowly rounded or subcordate; margin biserrate or triserrate, 2-3 major teeth per cm, tip of teeth blunt; upper surface dark green, semiglossy, glabrous except for a few short hairs on the midrib; lower surface slightly paler, downy pubescent, the veins more heavily pubescent; petiole stout, 2-4 mm long, terete, puberulent; stipules lanceolate to oblanceolate, 6-8 mm long, broad-based, thin, papery, brownish, early deciduous.

FLOWERS: March; perfect. Racemes from as long as the bud scales to 15 mm, with 3-5 flowers; pedicels 2-3 mm long, glabrous, a small, brown, linear or spatulate, ciliate bract at the base; calyx broadly and symmetrically turbinate, 1.5 mm broad, 2 mm long including the stipe-like base, the 5-6 obovate lobes about 1.5 mm long and 1 mm wide, brown, ciliate, especially at the base and tip; no corolla; stamens 5, inserted near the base of the calyx, filaments 1.5-3 mm long, white; anthers about 1 mm long, red; ovary 1 mm long, ovate, greenish, flattened, appressed pubescent; styles 2, divergent, 0.5 mm long, stout, flat, stigmatic on the puberulent inner surface.

FRUIT: Early May. Racemes 1-1.5 cm long with 3-4 fruits; fruits ovate, 7-8 mm long, 3-3.5 mm wide, flat with a narrow wing, brown, pubescent, densely tawny ciliate, the hairs up to 1.5 mm long, the base of the fruit abruptly narrowed to a stipe, the apex with 2 tips 1.5 mm long, curved inward, touching or overlapping; fruit 1-seeded. Seeds attached apically, ovate, 3.2-3.6 mm long, 2.3-2.6 mm wide, 1.3-1.4 mm thick, brown, the apex acute, base truncate and slightly concave, the surface wrinkled.

TWIGS: 0.8-1 mm diameter, red-brown, pubescent, especially in a line upwards from the buds, the hairs straight or curved, the lenticels oval, orange, often raised; corkiness starts on the first-year wood or not until the twig is 3-4 years old; leaf scars half-round to nearly circular, pale red-brown; 3 bundle scars; pith white or pale green, continuous, one-fourth of stem. Leaf buds narrowly ovoid, 3.5-4.5 mm long, 1.7 mm wide, 1.4 mm thick, flattened laterally, acute, the scales red-brown, scurfy puberulent, the margin dark and often with a few cilia; the terminal bud is the largest; flower buds obovoid, rounded, red-brown, 4 mm long, 2 mm wide.

TRUNK: Bark gray-brown, irregular, with shallow furrows and flat-topped ridges, similar to the bark of *U. americana*. Wood hard, heavy, not strong, fine-grained, hard to split, light brown, with a wide, white sapwood.

HABITAT: Moist lowlands or along rock ledges or on hillsides in dry soil.

RANGE: Virginia, south to Florida, west to eastern Texas, north to southeastern Kansas, east across Missouri and southern Indiana to Kentucky.

Only one herbarium specimen and no living plants could be located in our area and for this reason the descriptions and drawings were made from fresh specimens in Oklahoma, just south of the Kansas line, where it is a common tree.

This is one of the two species of elm in our area which has a densely pubescent and long ciliate fruit. The other is the northern *U. thomasi*, the fruits of which are 14-18 mm long as compared to 8-11 mm for *U. alata*.

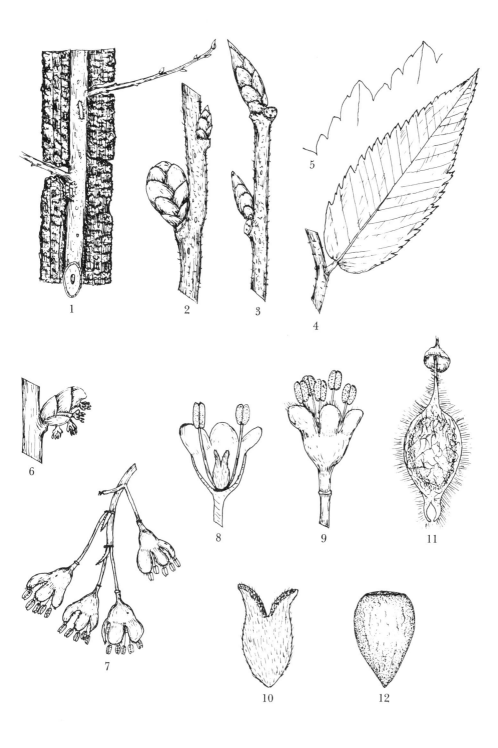

1. Winter twig
2. Twig with flower and leaf buds
3. Twig with leaf buds
4. Leaf
5. Leaf margin
6. Flower bud opening
7. Flowering raceme
8. Flower cutaway
9. Flower
10. Pistil
11. Fruit
12. Seed

ULMACEAE
Ulmus americana L.

American elm, white elm

A spreading tree to 25 m high, attaining a massive bole.

LEAVES: Alternate, simple. Ovate to elliptic, 4-12 cm long, 2-6 cm wide; coarsely biserrate or triserrate, teeth incurved, average 1.5 teeth per cm, a lateral vein to the tip of each tooth; leaf tip acuminate or acute, the base obliquely cuneate or rounded; upper surface dark green, glabrous or scabrous, dull or lustrous; lower surface paler, pubescent or glabrate; petiole 2-6 mm long, glabrous or pubescent; stipules lance-linear, 5-7 mm long, pubescent, caducous.

FLOWERS: March, before the leaves; perfect. Clustered from buds of the previous year; pedicels 1-1.5 cm long, glabrous; calyx tube obliquely campanulate, 1-2 mm long, 2.5-3 mm wide, greenish or reddish, glabrous with a few hairs at the tip of the lobes; lobes 5-6, brownish, 0.5 mm long, rounded; stamens 5-9, filaments white, 3-4 mm long; anthers 1 mm long, red; ovary flattened ovate, 1-1.5 mm long, glabrous with a hairy margin, green; styles 2, about 1 mm long, divergent, green, pubescent, the inner surface stigmatic and with papillose hairs.

FRUIT: Early May. Clusters of 8-15; pedicels 15-18 mm long; fruit ovate, 11-14 mm long, 9-12 mm wide, flattened; a 1-seeded samara with the wing all around, a deep notch at the apex with incurved, sharp points, the sinus between usually open and the points not touching but occasionally overlapping; conspicuously reticulate, glabrous with a ciliate margin.

TWIGS: 1.5-2.5 mm diameter, red-brown to gray-brown, glabrous or pubescent, flexible; leaf scars half-round; 3 bundle scars; pith white, continuous, one-sixth of stem. Leaf buds narrow, acute, 2-3 mm long; flower buds broadly ovoid, acute, flattened, glabrous or slightly pubescent, 3-4 mm long, the scales brown, 2-ranked.

TRUNK: Bark thick, dark ashy-gray, furrowed, the ridges short, flat-topped, or somewhat angular, anastomosed. Wood heavy, hard, strong, hard to split, light brown, with a wide, white sapwood.

HABITAT: Adaptable to nearly any type of soil except wet; appears most often in rich bottom lands and stream banks but is also found on rocky hillsides in the open. Commonly planted as a shade tree.

RANGE: Newfoundland, west to Saskatchewan, south to Kansas and Texas, east to Florida, and north to New England.

Ulmus americana is one of our finest shade trees; but in the last several years this species has been attacked by the Dutch Elm disease, and many of the trees in cities and along the streams have died. In wooded areas it may have an unbranched bole up to 8 meters, but in open areas it may branch within 2 meters of the ground, with large, nearly horizontal branches. It produces an abundance of seed and the ground may soon be covered with elm seedlings. This is objectionable when the trees grow next to a house, since the fruits fill the eve troughs and even germinate there.

The wood is used extensively for agricultural implements, furniture, flooring, wall panels, chopping bowls, barrels, boxes, and fiber for roofing felt.

Four forms have been separated, but the distinguishing characteristic is the degree of pubescence on the leaves and twigs. The forms are: f. *pendula* (Ait.) Fern., glabrous leaves and pubescent twigs; f. *laevior* Fern., glabrous leaves and twigs; f. *alba* (Ait.) Fern., scabrous leaves and pubescent twigs (often mistaken for *U. rubra*); f. *intercedens* Fern., scabrous leaves and glabrous twigs.

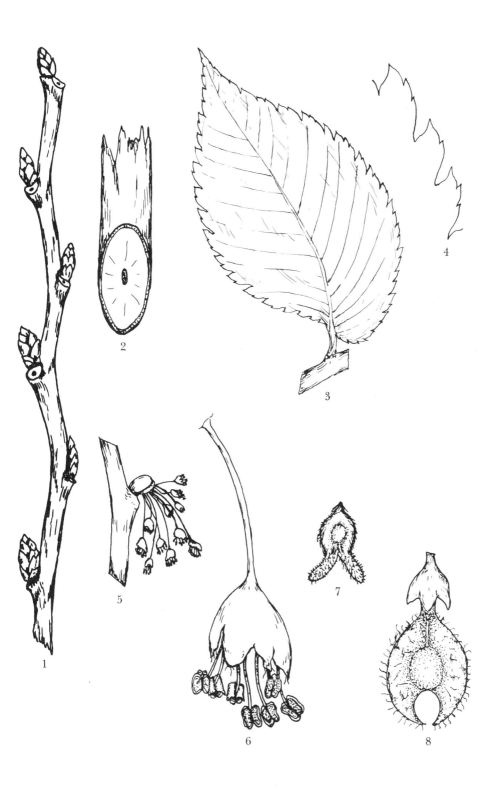

1. Winter twig
2. Section of twig
3. Leaf
4. Leaf margin
5. Flower cluster
6. Flower
7. Pistil
8. Fruit

ULMACEAE
Ulmus pumila L.

Siberian elm

Tree to 15 m high, with a broad crown and many fine branchlets.

LEAVES: Alternate, simple. Ovate to elliptic, 4-7 cm long, 2-3 cm wide; serrate or biserrate, 3-4 teeth per cm; leaf tip acute or acuminate; base oblique or evenly rounded; upper surface dark green, glabrous, semilustrous; lower surface paler, glabrous with some pubescence in the vein axils; petiole 5-8 mm long, pubescent; stipules broad-based, lanceolate, 2-4 mm long, caducous.

FLOWERS: March, before the leaves; perfect. Clustered from the buds of the previous year; pedicels 0.5 mm long, calyx tube campanulate, 1.4-1.6 mm long, green, pubescent; calyx lobes 4-5, reddish-brown, 0.8-0.9 mm long, tip rounded, pubescent; corolla none; stamens 4-5, filaments white, 2-3 mm long; anthers 1 mm long, red-brown; ovary 0.9-1 mm long, green, oval, glabrous, flattened, winged, the wing extending upward along the outer margin of the styles; styles 2, divergent, linear, 0.5-0.6 mm long, the inner surface stigmatic, with white, papillose hairs.

FRUIT: Early May. Clustered, 2-8 fruits, pedicels 2-4 mm long; fruit a flat samara, circular, 11-14 mm diameter, glabrous, light brown, the wing all around, notched at the apex with rounded, overlapping tips. Seed area 4-5 mm across.

TWIGS: 1.5 mm diameter, flexible, slightly pubescent, gray-brown, drooping; leaf scars half-round to irregularly elliptical; 3 bundle scars; pith white, continuous, one-fourth of stem. Leaf buds ovoid, 2-3 mm long, obtuse, the scales dark brown, ciliate; flower buds globose, 3.5-4 mm diameter, dark red-brown, lustrous, the scales long-ciliate.

TRUNK: Bark gray, with shallow furrows and long, flat ridges. Wood hard, difficult to split, red-brown, with a narrow, light brown sapwood.

HABITAT: Planted in many places, yards, streets, shelterbelts; escapes readily and is established in many places.

RANGE: Introduced from Asia. Escaped throughout central United States.

The Siberian elm is usually called Chinese elm, a name which probably belongs to *U. parvifolia* Jacq. Siberian elm is commonly planted around homes and along streets. It is a rapid-growing tree and reaches a height of 15 meters, the crown wide and spreading. The pendent branches give it a graceful appearance, especially in the summer when the long, drooping branchlets sway in the wind. However, these branches are not strong and may break in a heavy wind. It is more resistant to disease than the white elm. Because of this and the rapid growth it is often used in shelterbelts where it is protected from the wind by other trees.

During the winter, *U. pumila* can often be identified from a distance by the gray, misty appearance caused from the abundance of fine, light gray branchlets. At this season it stands out sharply in contrast to other elms with which it may be growing.

Three other elms are often planted in yards and parks in our area and might be confused with *U. pumila*. All have small leaves with single- or double-toothed margins. *U. parvifolia* Jacq. is a fall-flowering tree, with the flowers in short racemes and the axis being less than 1 cm long. *U. serotina* Sarg., another tree with fall flowers, has long racemes with the axis from 2-3 cm long. *U. procera* Salisb. flowers in the spring and the flowers are clustered from a bud. It often has corky-winged branches and the fruits are longer than broad, neither of which is characteristic of *U. pumila*.

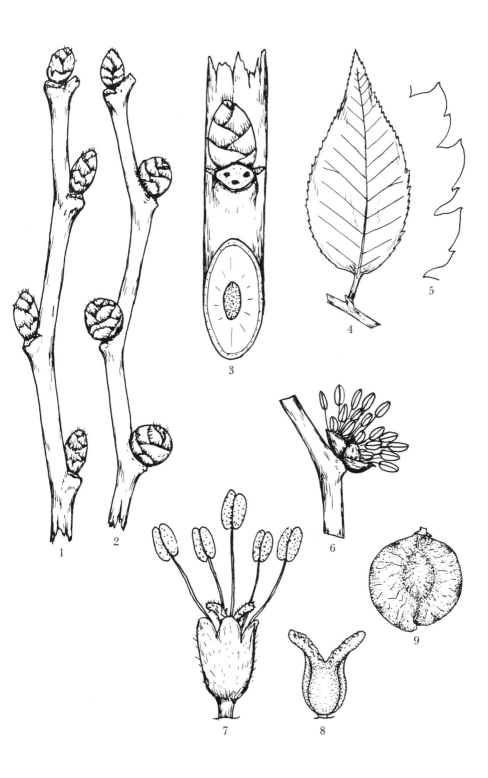

1. Winter twig, leaf buds
2. Twig, flower buds
3. Detail of twig
4. Leaf
5. Leaf margin

6. Flower cluster
7. Flower
8. Pistil
9. Fruit

ULMACEAE
Ulmus rubra Muhl.

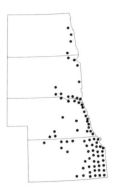

U. fulva Michx.

Red elm, slippery elm

A tree to 20 m high, with a high, rounded crown.

LEAVES: Alternate, simple. Ovate to oval, 10-13 cm long, 5-7 cm wide; coarsely double serrate, the teeth incurved, the larger ones up to 1 cm across; leaf tip acute, the base oblique, rounded on one side; upper surface dark green, often rugose, scabrous, the hairs with a definite pustulate base; lower surface paler and pubescent and with small axillary tufts; petiole stout, 3-5 mm long, pubescent; stipules ovate-oblong, 10-12 mm long, 5-7 mm wide, caducous, pubescent, usually a tuft of rufous hairs at the tip.

FLOWERS: March, before the leaves; perfect. Flowers in clusters of 8-12 from buds of the previous year; pedicels 1.7-1.9 mm long, green, pubescent; calyx tube campanulate, 2.6-4 mm long, pubescent with pale hairs; calyx lobes 5-9, green, 0.5-0.7 mm long, rounded, fringed with red-brown hairs; corolla none; stamens 5-8 inserted on the base of the calyx, filaments 4-5 mm long, white, glabrous; anthers 1.3-1.5 mm long, dark red; ovary ovate, flattened, 1.2-1.4 mm long, 1 mm wide, pubescent in the center, slightly winged; styles 2, pubescent, 1.8-2 mm long, divergent, the stigmatic surface on the inner side, reddish papillose.

FRUIT: Late April or early May, the leaves about half-grown. Dense clusters of 3-10 fruits; pedicels 2-4.5 mm long; fruits circular to obovate, 1.8-2 cm long, 1.3-1.5 cm wide; a 1-seeded samara with the wing all around, a notch at the apex with the tips rounded and overlapping; wing glabrous, obscurely reticulate, 5-8 mm wide; seed area 5-6 mm diameter, pubescent. Seed chestnut-brown, flattened ovoid.

TWIGS: 1.8-2 mm diameter, grayish or brown, pubescent, flexible; leaf scars half-round; 3 bundle scar groups; pith white, continuous, one-fifth of stem. Leaf buds narrowly ovoid, obtuse; flower buds broadly ovoid to obovoid, 4-5 mm long, dark brown, rusty tomentose, the scales 2-ranked.

TRUNK: Bark thick, gray-brown, with shallow fissures and long, flat, often loose, plates; inner bark mucilaginous. Wood hard, heavy, strong, easy to split, dark red-brown, with a narrow, light sapwood.

HABITAT: Stream banks and bottom lands, usually in rich, moist soil, but occasionally on rocky hillsides in poor soil.

RANGE: Maine, west across southern Canada to Minnesota and North Dakota, south to Texas, east to Florida, and north to New England.

U. rubra is more resistant to disease than *U. americana* but has not been used as much for yard planting. At a distance, the two species can often be distinguished if growing side by side; *U. rubra* is more yellow-green and has a dull, harsh appearance to the leaves. Also, the ridges of bark are longer and less anastomosed, which gives the appearance of vertical lines on the trunk. This does not always hold true, as some red elm bark is quite blocky. The leaf buds on winter twigs of *U. rubra* and *U. americana* are often quite similar; both may have red-brown hairs. The flower buds are quite different; those of *U. rubra* are definitely rounded and densely covered with red-brown hairs.

The wood is used for fence posts, agricultural implements, window sills, railroad ties, and wagon hubs. It is easily split, but is strong and durable. It was formerly used for rail fences.

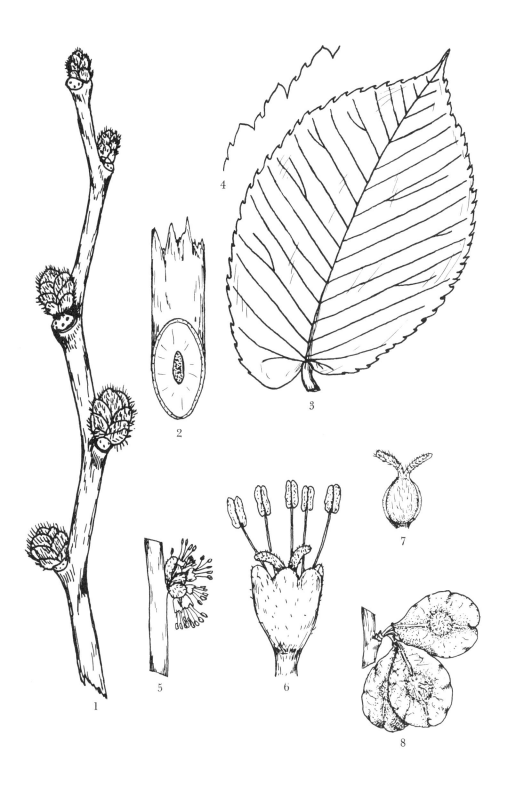

1. Winter twig
2. Section of twig
3. Leaf
4. Leaf margin
5. Flower cluster
6. Flower
7. Pistil
8. Fruit

ULMACEAE
Ulmus thomasi Sarg.

U. racemosa Thomas, not Borkh.

Rock elm, cork elm

A tree to 20 m high, with a spreading crown.

LEAVES: Alternate, simple, the midrib impressed above, the lateral veins prominent beneath, straight and nearly parallel and terminating in a tooth. Broadly elliptic or oval, widest about the middle, 8-10 cm long, 4-6 cm wide; tip acute or abruptly acuminate, the base uneven, definitely rounded or cordate on the apical side and cuneate on the basal side; margin biserrate or triserrate, the teeth averaging 1.5 per cm; upper surface dark green, glabrous or with a few hairs on the midrib, lower surface pale green, pubescent, rather densely so on the veins; petiole 4-6 mm long, terete, glabrous or pubescent; stipules lanceolate, 8-10 mm long, the base broad, 3 mm across, and abruptly narrowed to the lanceolate end, almost immediately deciduous.

FLOWERS: Early April, before the leaves; perfect. Racemes puberulent, 1-3 cm long, 3-9 flowers; pedicels 2-4 mm long, with a brown, ciliate, obovate bract at the base; calyx tube obconic, equilateral or oblique, 2.5-3 mm long, 2 mm wide, with 7-9 brown, ovate or obovate lobes about 1.5 mm long; no corolla; stamens 7-9 inserted near the base of the calyx tube, filaments 4-5 mm long; anthers red-purple, 1 mm long; ovary flattened oval, 1 mm long, 0.5 mm wide, pubescent, brownish with a narrow wing; styles 2, stout, 0.75 mm long, green, divergent, the stigmatic surface on the inner side, white and hairy.

FRUIT: May. Racemes pendent, 2-5 cm long. Fruit obliquely ovate, broadly winged, 14-18 mm long, 11-13 mm wide, seed portion 9-12 mm long, 6-8 mm wide, brown, flat, densely pubescent over all parts, densely ciliate with hairs about 0.5 mm long, 2 points at the apex about 1 mm long and usually overlapping; the base abruptly narrowed and stipe-like; 1 seed in each fruit. Seeds easily separated from the wing, ovate, flat, yellowish, 6-8

mm long, 4-5 mm wide, 2 mm thick, with a flat rim on one margin, apex acute, the base truncate and concave.

TWIGS: 1.5-1.8 mm diameter, red-brown or yellow-brown, velvety pubescent on young twigs, often nearly glabrous by spring; lenticels light-colored; larger twigs and branches usually with corky wings or protuberances; leaf scars shield-shaped to half-round or nearly triangular, brown with 3-5 bundle scars; stipule scars narrowly triangular; pith greenish-white, continuous, one-half of stem. Buds narrowly ovoid, 5-7 mm long, 2.3-2.8 mm diameter, slightly flattened, acute, red-brown, the pubescence somewhat appressed, outer scales without cilia, the inner scales ciliate, thin, translucent, greenish at the base; the embryonic leaves conduplicate.

TRUNK: Bark dark gray, with shallow furrows and flat-topped ridges with vertical sides, similar to that of *U. americana*; bark of branches rough and warty, not symmetrically winged on two sides. Wood hard, fine-grained, of good quality, pale brown, with a wide, white sapwood.

HABITAT: Wooded hillsides and creek banks, common in the loess soil of northeastern Nebraska; rich flood plains of streams.

RANGE: Southern Quebec, west to Minnesota and southeastern South Dakota, south to northeastern Kansas, east across Missouri to Tennessee, and north through West Virginia and western New England.

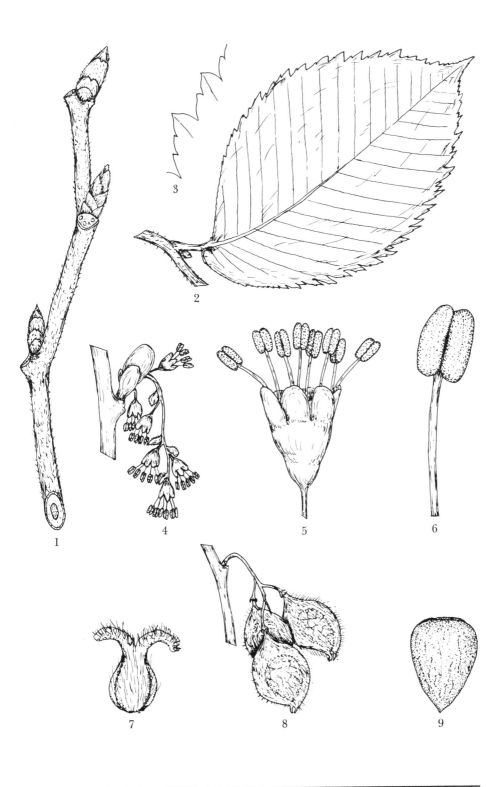

1. Winter twig
2. Leaf
3. Leaf margin
4. Flower cluster
5. Flower
6. Stamen
7. Pistil
8. Fruit
9. Seed

ULMACEAE
Celtis laevigata Willd.

C. mississippiensis Bosc; *C. smallii* Beadle

Sugarberry, southern hackberry

A tree to 20 m high, with high, rounded crown.

LEAVES: Alternate, simple, 3 principal veins. Ovate to ovate-lanceolate, 4-9 cm long, 1.2-3.5 cm wide, more than twice as long as wide; margin entire or with a few teeth; tip long acuminate, often falcate; base unequal, obtuse or semicordate; upper surface light yellow-green, glabrous; lower surface pale green, with a few hairs in the vein axils; thin or subcoriaceous; petiole slender, 12-14 mm long, glabrous; stipules lanceolate, 5-6 mm long, membranaceous, densely curly ciliate, puberulent on the dorsal side, promptly deciduous.

FLOWERS: Mid-May, with the leaves; monoecious. Staminate flowers axillary at the base of a short new branch, solitary or in clusters of 2-3; pedicels 4 mm long, glabrous; calyx lobes 5-6, distinct, 3 mm long, narrowly ovate, green with a brownish tip; no petals; stamens 5-6, filaments white, 2.5 mm long, hairy at the base; anthers yellow-green; the center of the flower pubescent. Pistillate flowers axillary, toward the apex of the same new branch, solitary or in 2's; pedicels 4-6 mm long, glabrous; calyx lobes 4-6, ovate, 2.4-2.5 mm long, green, united at the base, scarious tip, the margin ciliate with a tuft of hairs at the tip; vestigial stamens often present; ovary green, ovoid, 2.1-2.4 mm long, glabrous; style short, thick; stigmas 2, white, papillose on the inner margin, 2.2-2.5 mm long.

FRUIT: September-October. Pedicels glabrous, 8-15 mm long, shorter or longer than the subtending petiole; fruit globose, 6-8 mm diameter, orange-red to brownish-red, smooth. Seeds globose, 5.1-6 mm long, 4.4-4.9 mm wide, cream-colored, reticulate, usually 4 main ridges from the acute base, the apex rounded with a small tip.

TWIGS: 1-1.6 mm diameter, light red-brown, lustrous, glabrous, flexible; leaf scars half-round; 3 bundle scars; pith white, chambered or continuous, one-third of stem. Buds ovoid, 1-2.5 mm long, acute, appressed to the twig, glabrous, chestnut-brown.

TRUNK: Bark light gray with broad, flat furrows and short, narrow, warty ridges. Wood lightweight, soft, weak, light yellow, with a wide, white sapwood.

HABITAT: Rich bottom lands, stream banks, flood plains, and rocky hillsides near streams.

RANGE: Virginia, west to Illinois and Kansas, southwest to New Mexico and Mexico, east to Florida, and north to Tennessee.

Three varieties of *C. laevigata* are found within our area, but they are often not distinguishable: Var. *laevigata*, with glabrous petioles, the blade smooth, membranaceous, and without teeth; var. *smallii* (Beadle) Sarg., with glabrous petioles, the blade membranaceous, smooth, and with many teeth; and var. *texana* Sarg., with pubescent petiole, the blade coriaceous, scabrous, and without teeth. The leaves of most of our specimens are smooth on the surface, with a few, or no, teeth.

This species is resistant to the witches'-broom disease and is considered a good shade tree in the southern part of our range. It is not hardy farther north.

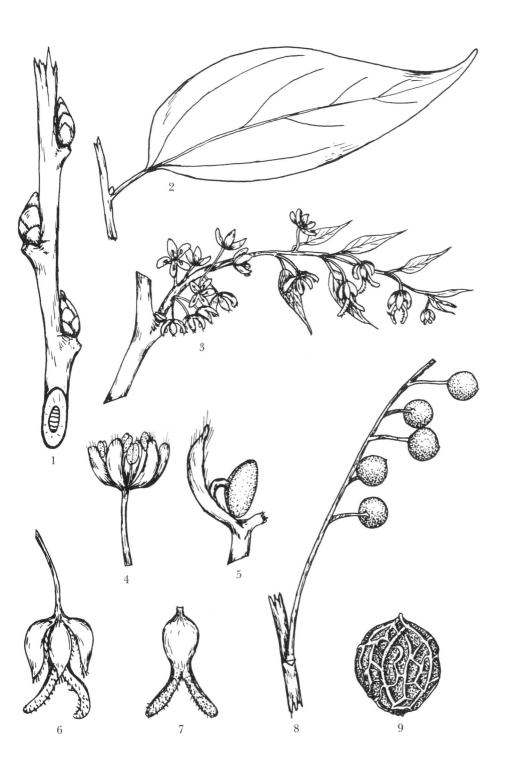

1. Winter twig
2. Leaf
3. Floral branch
4. Staminate flower
5. Calyx lobe and stamen
6. Pistillate flower
7. Pistil
8. Fruiting branch
9. Seed

ULMACEAE
Celtis occidentalis L.

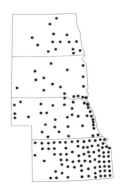

Including var. *crassifolia* (Lam.) Gray;
var. *pumila* (Pursh) Gray;
var. *canina* (Raf.) Sarg.

Hackberry

A tree to 25 m high, the larger branches often 8-10 m above ground.

LEAVES: Alternate, simple, with 3 principal veins. Ovate to ovate-lanceolate, 5-12 cm long, 3-6 cm wide; coarsely serrate, 2-4 teeth per cm; tip acuminate, base oblique or obliquely subcordate; upper surface glabrate or scabrous, the lower surface paler and pubescent; young leaves densely tomentose; petiole slender, 10-12 mm long, glabrous or pubescent; stipules narrowly ovate, 5-6 mm long, erose ciliate, membranaceous, glabrous or pubescent; promptly deciduous.

FLOWERS: April-May, with the leaves; monoecious. Staminate flowers single or in clusters of 2-3 at the base of a short, new branch; pedicels 5-7 mm long, green, glabrous; calyx lobes 5-6, deeply divided, ovate, 2-2.5 mm long, yellow-green with brownish tips, glabrous except for a fringe at the tip, often toothed and usually cupped or folded around a filament; no petals; stamens 5, filaments 2 mm long, flattened, and with a tuft of hairs at the base; anthers yellow. Pistillate flowers single or in pairs, axillary, apical to the staminate flowers on the same branch; pedicels green, 6-7 mm long, glabrous; calyx lobes 5-6, deeply divided, ovate, 2-2.3 mm long, green with brownish tip, glabrous except for ciliate margin, often toothed; a tuft of hairs at the base of the vestigial stamens; ovary 1.9-2.1 mm long, ovate, green, glabrous; style 0.8-1 mm long; the 2 stigmas linear, 1-1.8 mm long, rough, yellow, spreading or recurved.

FRUIT: September-October, often remaining on the tree until midwinter. Pedicels glabrous, slender, 14-16 mm long, longer than the subtending petiole; globular drupe, 8-10 mm diameter, brownish to dark purple, the thin pulp edible and sweet. Seeds cream-colored, reticulate, globose, 7.3-7.8 mm long, 6.1-6.8 mm wide, 4 main ridges, the apex rounded with a small tip.

TWIGS: 1-2 mm diameter, gray-brown, glabrate or pubescent, flexible; leaf scars small, crescent-shaped; 3 bundle scars; pith white, closely chambered, one-third of stem. Buds ovoid, 2-3 mm long, light brown, acute, flattened, the scales 2-ranked.

TRUNK: Bark thick, light gray-brown, deeply furrowed, the ridges short, narrow, and warty, or the bark with shallow furrows and plate-like ridges. Wood nearly white, soft, weak, with a narrow, white sapwood.

HABITAT: Commonly in rich, moist soil along stream banks or on flood plains; grows well on rocky hillsides or in open hilly pastures.

RANGE: Quebec to Manitoba and Idaho, southeast to Utah, Kansas, and Oklahoma, east to Alabama and Georgia, and north to New England.

The genus *Celtis* is a complicated group and is in need of further research. The trees vary to such an extent that specimens taken from different parts of the same tree might well be identified as two different varieties. The three commonly recognized varieties are: var. *occidentalis* with scabrous leaves and dark brown fruits; var. *pumila* (Pursh) Gray with thin, smooth leaves, the base uneven, and brown to purplish fruits; var. *canina* (Raf.) Sarg., with thin, smooth, narrow leaves with symmetrical base, and brown to purplish fruits. Definitely these varieties can be found, but there is still the multitude of intermediate forms with which one has to contend.

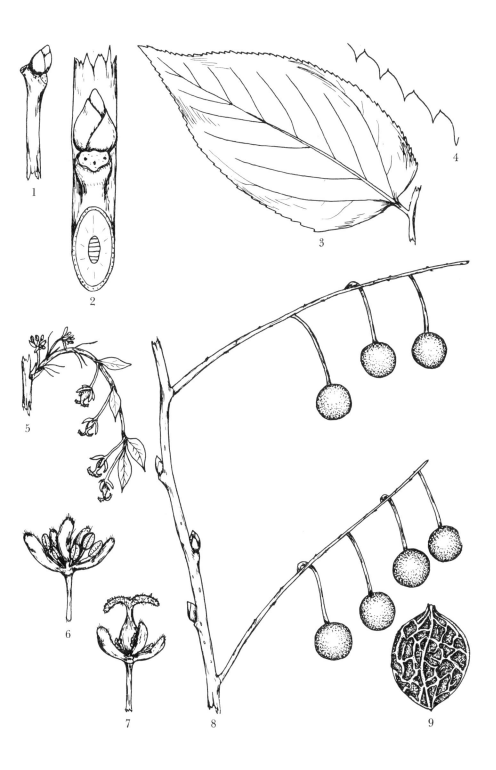

1. Winter twig
2. Detail of twig
3. Leaf
4. Leaf margin
5. Floral branch
6. Staminate flower
7. Pistillate flower
8. Fruiting branch
9. Seed

ULMACEAE
Celtis reticulata Torr.

C. laevigata Willd. var. *reticulata* (Torr.) Benson; *C. rugosa* Rydb.; *C. rugulosa* Rydb.; *C. reticulata* (var.) *vestita* Sarg.

Netleaf hackberry

A scraggly tree to 7 m high, the limbs usually crooked.

LEAVES: Alternate, simple, usually with 3 veins from the base, strongly reticulate below. Ovate 4-7 cm long, 2.5-4 cm wide; tip acute, obtuse, or short acuminate; base cordate or occasionally oblique; margin entire or with a few teeth, 1-4 per cm, ciliate; gray-green above and yellow-green below; young leaves densely pubescent on both sides, becoming scabrous above and pubescent below; thick and rigid; petiole stout, 3-8 mm long, grooved above, pubescent; stipules lanceolate, 4-5 mm long, 1-1.2 mm wide, tomentose, promptly deciduous.

FLOWERS: Late April when the leaves are just starting; monoecious. Axillary on a short branch of the season; pedicels slender, glabrous, green, 5-8 mm long, erect or drooping. Staminate flowers usually toward the base of the branch, the 5 calyx lobes distinct, ovate, 3 mm long, 1.5 mm wide, cupped, green, thin, spreading and glabrous or pubescent, the margin fimbriate and ciliate; no petals; 5 stamens, anthers sessile or nearly so, 1.5 mm long, yellow-green; center of flower pubescent. Pistillate flowers toward the tip of the same branch, calyx lobes as in the staminate; ovary ovoid, 1.5 mm long, 1 mm wide, green, glabrous; style short, the 2 stout stigmas 1.5-2 mm long, curved, divergent, white, and finely papillate; rudimentary stamens often present.

FRUIT: August-September. Pedicels 10-14 mm long, longer than the subtending petiole, fruit globose, 8-10 mm diameter, red-brown, smooth, the thin pulp sweet and edible. Seeds globose, 5.5-5.8 mm long, 4.8-5.3 mm wide, reticulate, cream-colored, hard.

TWIGS: 0.75-1 mm diameter, pubescent, red-brown; lenticels elliptic, light-colored, and abundant; leaf scars half-round, raised; 3 bundle scars; pith white, chambered, two-thirds of stem.

Buds ovoid, 3-3.5 mm long, 1.5-2 mm wide, pubescent.

TRUNK: Bark gray, warty, the younger branches light gray, with spots of warty bark. Wood soft, coarse-grained, pale yellow, with a wide, white sapwood.

HABITAT: Dry, rocky hillsides and ravine banks, occasionally in sandy soil.

RANGE: Southern Nebraska, south through Kansas and Colorado to Texas, west to Arizona and northern Mexico.

This species is as confusing and apparently as misunderstood as any of the genus. Kearney and Peebles give *C. douglasii* Planch. and *C. laevigata* Willd. var. *reticulata* (Torr.) Benson as synonyms and give the range from Oklahoma and Colorado southwestward. Benson gives the same synonyms and extends the range to include Washington, Idaho, and the Great Basin. This is the range given by Hitchcock et al. for *C. douglasii*, but they mention no synonyms, which, if there are any, is unusual. Correll and Johnston give the same range as Benson and do not mention *C. douglasii*, but they divide the species into two unnamed forms—large leaf and small leaf. None of these manuals mention the species entering our area. If the western Kansas *Celtis* is *reticulata*, it is almost identical to the thick-leaved *C. tenuifolia* Nutt. var. *georgiana* (Small) Fern. & Schub., but the range of that taxon is much farther east.

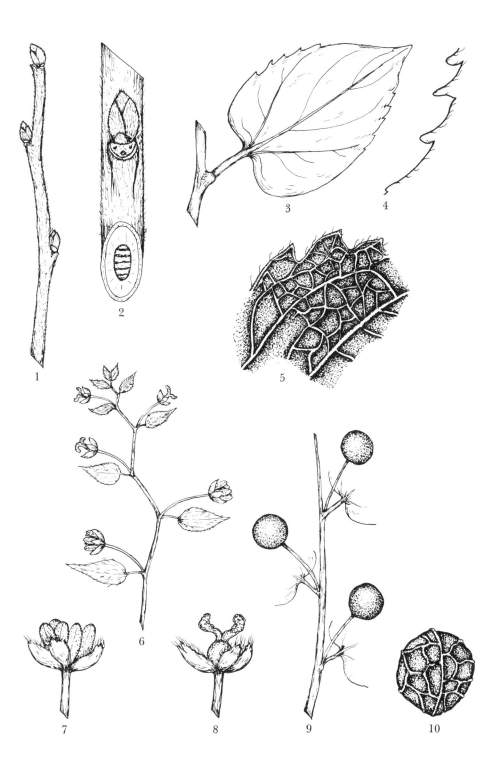

1. Winter twig
2. Detail of twig
3. Leaf
4. Leaf margin
5. Under surface of leaf
6. Floral branch
7. Staminate flower
8. Pistillate flower
9. Fruiting branch
10. Seed

ULMACEAE
Celtis tenuifolia Nutt.

C. pumila of auth., not Pursh

Dwarf hackberry

A small tree to 8 m high, often somewhat scraggly.

LEAVES: Alternate, simple, 3 main veins, the leaves one-half to three-fourths as broad as long. Ovate, 5-6 cm long, 3-4 cm wide; margin entire or serrate, often toothed near the apex, 2-4 teeth per cm; tip acuminate, base unequal, at least one side rounded; upper surface dark green, scabrous, with short, stiff hairs bent toward the apex; lower surface pubescent with bent hairs, or glabrate; thin to coriaceous; petioles 6-10 mm long, pubescent; stipules lanceolate to oblanceolate, 5-6 mm long, membranaceous, pubescent, promptly deciduous.

FLOWERS: April-May, with the leaves; monoecious. Staminate flowers axillary, near the base of a short new branch; pedicels 6-7 mm long, glabrous; calyx lobes 5, ovate, 3 mm long, divided to near the base, acute or erose tip, crinkly ciliate, often toothed, pubescent on the dorsal side, glabrous inside; stamens 5, filaments white, 3-4 mm long; anthers yellow. Pistillate flowers axillary toward the tip of the same new shoot, solitary; pedicels 9-10 mm long, glabrous; calyx lobes 5, ovate, 3 mm long, divided to near the base, green, entire or toothed, crinkly ciliate, pubescent on the dorsal side, glabrous inside; vestigial stamens 3-5; ovary dark green, 1.5 mm long, glabrous, glossy; style 0.5 mm long; stigmas 2, divergent, 3.2 mm long, 1 mm thick, the inner surface white hairy.

FRUIT: September-October. Pedicels 5-10 mm long, pubescent; fruit globose, 7.5-8 mm long, 6.5-7 mm thick, orange, brown, or deep red, smooth. Seeds cream-colored, globose, 4.9-5.1 mm long and about as thick, reticulate, 4 main ridges from the base; base acute, apex rounded with a small tip.

TWIGS: 1-1.5 mm diameter, pubescent when young, red-brown, gray by the second year, somewhat rigid, the lenticels small and light; leaf scars half-round with 3 indistinct bundle scars; pith white, finely chambered, one-third of stem. Buds ovoid, 2-3 mm long, grayish pubescent.

TRUNK: Bark light gray, furrowed, the ridges short, warty, and with vertical sides. Wood light, soft, nearly white, with a wide, white sapwood.

HABITAT: Rocky, open woods and bluffs, usually exposed areas. Occasionally on stream banks.

RANGE: Pennsylvania to Indiana and Kansas, south to eastern Oklahoma and Louisiana, east to Florida, and north to New Jersey.

Celtis tenuifolia is quite variable, and two varieties of it have been described: var. *tenuifolia*, leaves thin, upper surface smooth, lower surface glabrous or glabrate, petioles, pedicels, and twigs glabrous or with a few hairs; var. *georgiana* (Small) Fern. & Schub., leaves thick and coriaceous, upper surface scabrous, lower surface pubescent, petioles, pedicels, and twigs pubescent.

The trees usually grow in rocky, limestone soil, often in the shallow soil over bedrock or along a rock ledge in a ravine. In such locations it appears as a somewhat stunted and gnarled shrub-like tree. At the base of the same bluff or bottom of the ravine, it may grow to a height of 8-10 meters. The fruits vary also, from pale brown or orange to dark red. The flesh is thin and sweet, and raccoons eat large quantities of them.

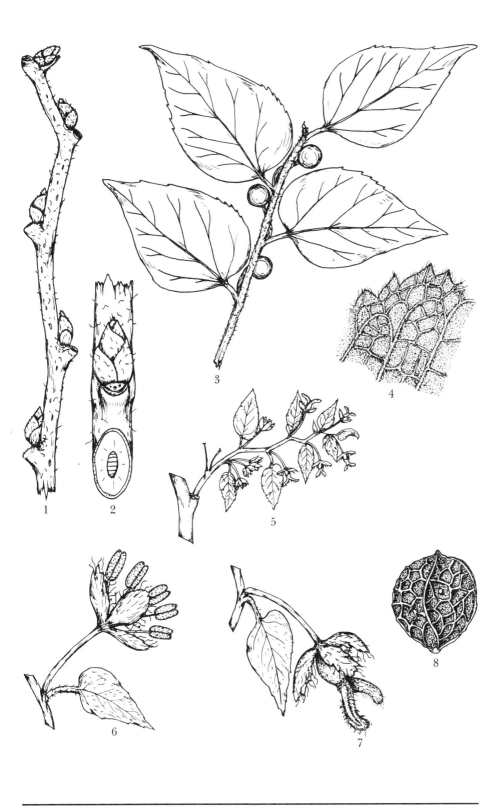

1. Winter twig
2. Detail of twig
3. Fruiting branch with leaves
4. Under surface of leaf
5. Floral branch
6. Staminate flower
7. Pistillate flower
8. Seed

MORACEAE
Morus alba L.

White mulberry, black mulberry

A spreading tree to 8 m high, with low, rounded crown.

LEAVES: Alternate, simple. Ovate, 6-10 cm long, 3-6 cm wide; margin variable, coarsely serrate to deeply lobed and serrate, 2-3 teeth per cm; tip acute to short acuminate, base cuneate, truncate, or cordate; upper surface dark blue-green, glabrous, often lustrous; lower surface sparingly pubescent, the veins with porrect pubescence and with tufts in the axils; petiole 2.5-5 cm long, a narrow groove above, exudes a milky juice when broken; stipules ovate, acute, thickened at the base, pubescent, with long, white hairs, caducous.

FLOWERS: Early May, with the leaves; monoecious or dioecious. Staminate flowers axillary in catkins on new growth; catkins 2.5-4 cm long, densely flowered, the axis and pedicels pubescent; calyx lobes 4-5, distinct, ovate, 1.5 mm long, thin, green with red tip, pubescent; stamens 4, filaments tapered 2.7 mm long, white; anthers yellow; filaments inflexed in bud and enfolded by a calyx lobe; as the flower opens, the filament snaps straight, the pollen sacs split, and the pollen is expelled. Pistillate catkins 5-8 mm long, axillary on new growth, densely flowered, axis and pedicels pubescent; ovary glabrous, ovoid, 2 mm long, green, compressed, enclosed by 2-4 green, ciliate, calyx lobes; style linear, 0.5-1 mm long; stigmas 2, linear, red-brown, divergent, papillose on the inner surface.

FRUIT: May-June. A cylindric multiple fruit 1.5-2.5 cm long, 0.8-1 cm diameter, consisting of obconic drupes 3.8-4 mm long, 1.4-1.5 mm wide, the calyx lobes becoming fleshy and thick; bright red when immature, becoming black, purple, or nearly white, sweet, juicy, and edible. Nutlet ovoid, 2.4 mm long, 1.8 mm wide, 1.4 mm thick, light brown, with a thin, soft shell.

TWIGS: 1.8-2.5 mm diameter, glabrous, flexible, orange-brown or dark green with a reddish cast; the lenticels prominent; leaf scars half-round; many bundle scars; pith white, continuous, one-third of twig. Buds ovoid, 4-6 mm long, acute to rounded, the scales 2-ranked, yellow-brown with a dark margin, glabrous or with a few marginal hairs.

TRUNK: Bark thin, dark brown tinged with red or yellow, shallow furrows, the ridges long and narrow. Wood soft, lightweight, durable, orange-brown, with a wide, light yellow sapwood.

HABITAT: Mainly in the open on prairie hills, in rocky ground; planted in windbreaks.

RANGE: Introduced from Asia, escaping over a large portion of eastern United States.

The name *Morus alba* is the cause of many misidentifications of this plant. Most people who might be trying to identify the tree know that there are white mulberries and purple-black mulberries and may assume that the white ones are *M. alba* and the blackish ones are *M. rubra*. Common names are also confusing; the names of white mulberry and black mulberry may both apply to *M. alba*. *M. rubra* fruits are black, but *M. alba* fruits may be black, lavender-purple, or white. The plant has also been misidentified as *M. nigra* L., which is introduced and not hardy in our area.

The leaves of *M. alba* are usually lobed and the veins beneath have some pubescence, and occasionally a few hairs on the surface. *M. rubra* leaves are seldom lobed excepting on sprouts, are larger, and the underside of the typical leaf is velvety hairy.

1. Winter twig
2. Detail of twig
3. Leaf
4. Leaf variation
5. Leaf margin
6. Staminate catkin
7. Staminate flower

8. Position of stamen in the bud
9. Pistillate catkin
10. Pistillate flower
11. Multiple fruit
12. Ventral view of seed
13. Dorsal view of seed

MORACEAE
Morus rubra L.

Red mulberry

A tree to 18 m high, usually with a long, straight trunk and high crown.

LEAVES: Alternate, simple; 3 main veins from the base, prominent below, the lateral veins curve and join the next vein above near the margin, and a small vein extends to each tooth. Broadly ovate, 10-18 cm long, 8-12 cm wide; tip abruptly acuminate, base rounded to sub-cordate; margin serrate, 2-3 teeth per cm; sides of teeth convex, a short mucro at the tip, the sinuses acute; occasionally with lobes; upper surface glabrous, often somewhat rugose; lower surface densely short pubescent on the surface as well as on the veins; petiole 2-2.5 cm long, terete, glabrous or short pubescent; stipules linear, 10-13 mm long, 1 mm wide, whitish, early deciduous.

FLOWERS: April-May; dioecious. Leaves of the staminate trees usually more developed at flowering time than are those of the pistillate trees. Staminate catkins 3-5 cm long, 10 mm wide, pendent, axillary on branches of the season, densely flowered; peduncle 1 cm long, pubescent; flowers sessile, calyx tube shallow, 1-1.5 mm deep, green, pubescent; lobes 4, ovate, 2-2.5 mm long, tinged with red, pubescent outside, ciliate toward the acute tip, each lobe enfolding a stamen in the bud; stamens 4, filaments white, 3-3.5 mm long with a broad base tapered to the small, light yellow anther; vestigial pistil in the center. Pistillate flowers axillary on branches of the season, catkins densely flowered, 8-12 mm long, 5-7 mm wide, erect or drooping, peduncle 3-5 mm long, pubescent; ovary green, enclosed in 2 pairs of green, apically ciliate scales, ovary broadly elliptic or obovate, 1.5-2 mm long, 1.25 mm wide, slightly compressed; style none or very short; stigmas 2, linear, whitish, 1.5 mm long, divergent.

FRUIT: June-July. Peduncle 9-12 mm long, green, pubescent; multiple fruit cylindric, 1.5-2.5 cm long, about 1 cm thick, black or deep purple, the calyx becoming enlarged, succulent, juicy and edible, the 4 lobes obvious, but fused. Seeds irregularly oval, flattened, smooth, 2-2.3 mm long, 1.4-1.6 mm wide, 1 mm thick, yellowish; 1 seed per individual fruit.

TWIGS: 2-3 mm diameter, red-brown to light greenish-brown, slightly zigzag, smooth, often a few hairs at the nodes; lenticels elliptic, narrow, light-colored, and prominent; leaf scars large, oval to irregularly circular; a circle of small bundle scars; pith white, continuous, one-half to one-third of stem. Buds ovoid, acute, 3-7 mm long, 2.4 mm diameter, slightly compressed, outer scales dark brown, often pubescent and minutely ciliate, inner scales greenish-brown with a dark margin.

TRUNK: Bark gray-brown with an orange tint, furrows shallow, the ridges flat, broad, and tight, or occasionally with loose scales. Wood medium hard, durable, red-brown, with a wide, white sapwood.

HABITAT: Rich, wooded flood plains and creek banks, usually in moist soil.

RANGE: Southern Ontario to Minnesota and South Dakota, south to Texas, east to Florida, and north to New England.

Morus rubra is our only native mulberry and grows along streams usually in rich, moist soil. Here it may reach 18 meters high and has a long central trunk. The amount of pubescence on the under leaf surface varies, but is usually rather dense, most of the hairs being porrect.

1. Winter twig
2. Leaf
3. Leaf margin
4. Staminate catkin
5. Staminate flower
6. Pistillate catkins
7. Pistillate flower
8. Multiple fruit
9. Face view of one simple fruit
10. Seed

MORACEAE
Maclura pomifera (Raf.) Schneid.

Toxylon pomifera Raf.

Osage orange, hedge, hedge apple,
bois d'arc

A tree to 12 m high, with crooked branches.

LEAVES: Simple, alternate, or clustered at the ends of short spurs; thick, subcoriaceous. Ovate to oblong-lanceolate, 4-12 cm long, 2-6 cm wide; margin entire; tip acuminate, base rounded; upper surface dark green, lustrous, glabrous with some pubescence on the midrib; lower surface paler, glabrate with midrib and veins pubescent; petiole 1-2.5 cm long, grooved above, pubescent, exuding a milky juice when broken; stipules minute, lanceolate, pubescent and long ciliate, mainly on sprouts.

FLOWERS: Mid-May, after the leaves; dioecious. Staminate flowers clustered at the end of a spur on 1- to 2-year-old wood; peduncle 1-1.5 cm long, green, pubescent; heads globose or cylindric, 1.3-2.3 cm diameter, many flowered; pedicels 0.2-1 cm long, green, glabrate; calyx lobes 4, distinct, acute, 1 mm long, yellow-green, pubescent on both sides; stamens 4, filaments 2-2.2 mm long, closely appressed to the calyx lobes, flattened, broad, nearly transparent; anthers yellow. Pistillate flowers axillary in dense, globose heads 1.5 cm diameter; peduncle stout, 2-2.5 mm long, green, glabrous or pubescent; flowers sessile on an obconic receptacle; calyx lobes 4, obovate, 3 mm long, thick, green, enclosing and closely appressed to the ovary, cucullate and ciliate near the tip; ovary circular, compressed, 1 mm across; style filiform, 2.8-3.2 mm long, green, glabrous; stigma filiform, yellowish, 4-6 mm long, papillose, style and stigma elongating and becoming 1.5-2 cm long after pollination.

FRUIT: September. Globose, 8-12 cm diameter, surface irregular, yellow-green, exuding a milky juice when broken; peduncle short, stout, glabrous or pubescent. Seeds completely covered by the accrescent, thickened calyx lobes and deeply embedded in the fruit. Seeds cream-colored, oval to oblong, 8-12 mm long, 4.5-6 mm wide, 2.7-3 mm thick, base truncate or rounded with 1-3 minute points, apex rounded and with a pointed tip; surface minutely striate or pitted, the margin with a narrow groove.

TWIGS: 1.8-2 mm diameter, greenish-yellow, becoming orange-brown, glabrous, rigid, milky; spines stout, straight, 1-1.5 cm long, usually lateral to a spur and above a leaf scar; spur branches often in pairs; leaf scars half-round; several bundle scars arranged in an oval; pith white, continuous, one-third of stem. Buds globose, often paired, larger one 1.7-1.9 mm diameter, red-brown, scales ciliate; smaller buds about half as large; buds on spurs inconspicuous, brownish.

TRUNK: Bark dark orange-brown, with shallow furrows, the flat ridges often peeling into long, thin strips. Wood hard, strong, heavy, flexible, durable, orange-brown, with a narrow, lemon-colored sapwood.

HABITAT: Rich woods along streams and flood plains; hillsides in open, rocky pastures. Planted as a windbreak along fields and roads.

RANGE: Native to Texas, Oklahoma, and Arkansas, probably introduced in Kansas; escaped and naturalized from New York to Nebraska and to some extent throughout the central states.

This tree was formerly planted in rows, especially near a farmstead or along a roadway, and formed a dense windbreak and shelter. Both song birds and game birds found shelter and food beneath the low branches. Squirrels tear the fruit apart and eat the seeds, often leaving a large pile of fruit pulp below the tree.

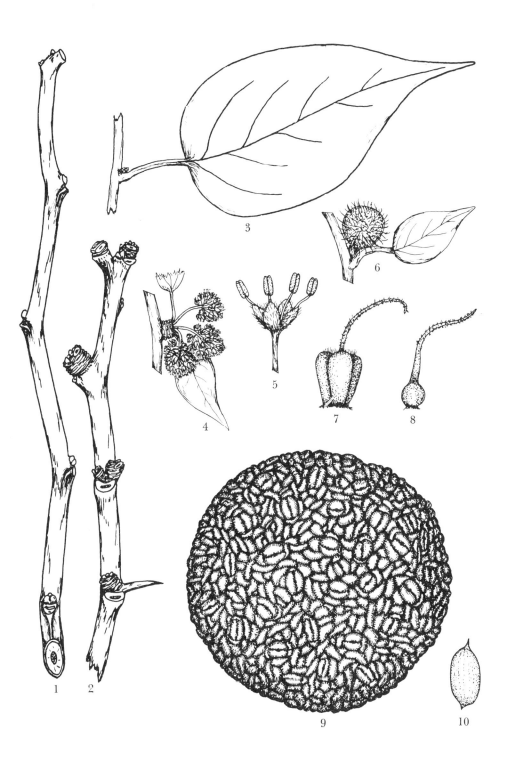

1. Winter twig
2. Winter twig with fruiting spurs
3. Leaf
4. Staminate flower cluster
5. Staminate flower

6. Pistillate flower cluster
7. Pistillate flower
8. Pistil with calyx lobes removed
9. Fruit
10. Seed

LORANTHACEAE
Phoradendron tomentosum (DC.)
Gray ssp. *tomentosum*

Phoradendron serotinum (Raf.) M. C.
Johnston; *Phoradendron flavescens*
(Pursh) Nutt.

Mistletoe

A bush-like plant, semiparasitic on trees, often forming clusters up to 60 cm in diameter.

LEAVES: Opposite, evergreen, simple. Elliptic to obovate, 3-4.5 cm long, 1.5-2.5 cm wide; entire, the tip rounded, base acute; both surfaces yellow-green, granular, pubescent with coarse, white, simple or fascicled hairs; thick, firm, fleshy or coriaceous; petioles stout, 5 mm long, thinly pubescent.

FLOWERS: March-July; dioecious. Staminate flowers in axillary, jointed spikes 2-3 cm long, each joint with 2 clusters of 7 flowers each, the clusters on opposite sides of the joint, the flowers of the adjacent joint in an alternate (decussate) position; calyx lobes 3, broadly deltoid, small, green, pubescent with coarse hairs; stamens 3, small, sessile on the calyx lobes. Pistillate flowers on jointed, axillary spikes 1.5-3 cm long; spikes green, pubescent, 3-6 flowers on each joint, each depressed into the thick axis of the spike; flowers short cylindric, green; calyx lobes 3, deltoid, 0.7-0.9 mm long, yellow-green, fleshy with a fringe of hairs at the base; ovary 1.5-1.7 mm long, 1.4-1.6 mm wide, globose, green, glabrous; stigma sessile, globose, hard.

FRUIT: November-December. Fruits on short spikes in the leaf axils, globose, 4-6 mm diameter, white translucent, stigma persistent as a black tip; sticky when crushed. Seed, with the covering, white, ovate, pointed; seed proper dark green, oval, compressed, 2.4-2.9 mm long, 1.8-2 mm wide, 0.9 mm thick, granular, a rounded knob at the apex.

TWIGS: 1.5-2.5 mm diameter, rigid, brittle, yellow-green, coarsely pubescent toward the tip with simple or fascicled hairs; leaf scars oval, small; 3 bundle scars; pith green, continuous, one-fourth of stem. Buds axillary near the stem tip, 1 mm long, green, pubescent, 2 scales.

TRUNK: Bark green with irregular, very shallow fissures, the inner bark brownish. Wood soft, white.

HABITAT: Semiparasitic on deciduous trees along streams, occasionally on upland trees and common on upland trees farther south. American elm is the most common host in Kansas.

RANGE: Mexico, north to Texas, Oklahoma, and southeastern Kansas, east to eastern Arkansas and Louisiana.

The time of anthesis and pollination appears to be quite variable. Wiens (in Correll and Johnston, *Manual of the Vascular Plants of Texas*) states in his key that the period is from November through March. This may well be true, but it is a strange situation to have anthesis at the same time the fruits are maturing. In this study, the author has examined Kansas material in the field during all seasons of the year and no developed stamens were found in the winter, November to February. The earliest found were in March, but fully developed stamens with apparently viable pollen were also found in July.

Although a few of our plants are nearly glabrous, they apparently are not *P. serotinum* var. *serotinum* as previously thought. That species is further east and south.

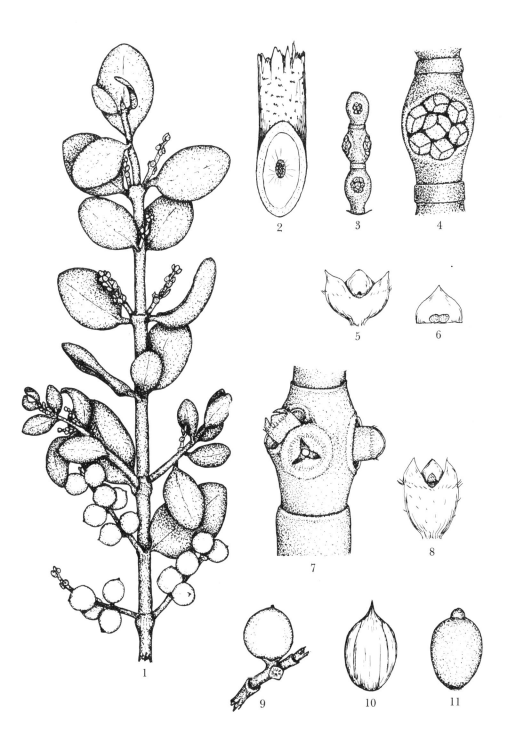

1. Winter branch with leaves and fruit
2. Section of twig
3. Staminate spike
4. Staminate flowers
5. One flower
6. Calyx lobe with stamen
7. Portion of pistillate spike with flowers
8. One flower
9. Fruit
10. Seed
11. Seed with papery coat removed

ARISTOLOCHIACEAE
Aristolochia tomentosa Sims

Isotrema tomentosa (Sims) H. Huber

Pipevine, dutchman's pipe

A twining vine to 20 m long, the trunk becoming 4 cm in diameter.

LEAVES: Alternate, simple. Heart-shaped, 6-15 cm long, 5-14 cm wide, 3 main palmate veins; margin entire; tip acute but not pointed, the base cordate; upper surface dark green, slightly pubescent; lower surface pale, tomentose; petiole 5-6 cm long, tomentose.

FLOWERS: Mid-May; perfect. Flowers single, opposite a leaf on new growth; pedicels green, soft hairy, curved, 3-3.5 cm long; ovary inferior, cylindric, green, 1.5-2 cm long, 3-4 mm wide, tomentose; calyx cylindric, abruptly enlarged at the upper end of the ovary, the enlargement 10-13 mm long, 9-11 mm thick, tomentose, pale green, slightly ridged, the inside glabrous and purple; the narrow part of the calyx 2.3-2.5 cm long, 5 mm thick, sharply curved upward, ridged, tomentose, pale green, the inside glabrous and spotted with purple; calyx lobes 3, broadly deltoid to ovate, 9-10 mm long, acute, yellow-green, the lower lobe appressed against the tube, tomentose outside, rough and wrinkled on the inner side; orifice 5-6 mm across, with an irregular, deep red-purple rim; corolla none; stamens 6, anthers sessile yellow, 3 mm long, entirely adnate to the stigma, 2 on each stigma lobe; style about 0.5 mm long; stigma 3-lobed, 3.5-4 mm long, stout, pale yellow.

FRUIT: September. Pedicels 3-5 cm long, tomentose; capsule gray-brown, cylindric, 4-7 cm long, 2.7-3.2 cm diameter, puberulent, 6-angled; the 6 valves with many seeds compressed in vertical columns. Seeds deltoid, 10.5-12.1 mm long, 10.9-13.2 mm wide, flat, thin, gray-brown, notched at the wide end; a narrow rim extending around the seed. Before maturity, the notch at the apex is filled with a white, spongy material which dries and cracks into a broken, scarious membrane; adnate to the apical surface of the immature seed is a thick,

white, spongy layer which dries at maturity and shrivels to a scarious membrane attached to the seed at its axial end only, resembling an undeveloped seed; apical surface of dry seed minutely granular, basal surface covered with a thin, scurfy, finely pitted film.

TWIGS: 2 mm diameter, flexible, tomentose, gray-green and often blotched, enlarged at the nodes; ends of the vine usually die back several internodes during winter; leaf scars narrowly U-shaped; 3 bundle scars; pith white, continuous, most of the stem. Buds barely visible in a dense, white tomentum and nearly surrounded by the U-shaped leaf scars.

TRUNK: Bark gray-brown, slightly fissured, often exfoliating into strips. Wood soft, somewhat pithy, pale brown, porous, with prominent rays.

HABITAT: Rich woodlands along streams, stream banks, and flood plains, occasionally on rocky hillsides or the base of a bluff.

RANGE: North Carolina, west to southern Indiana and eastern Kansas, south to eastern Texas, east to Florida, and north to South Carolina.

The flowers of *Aristolochia* are distinctive and are one of the principal characteristics for identification of the plant. During late summer and autumn, the sausage-like, pendent fruits are also a distinguishing characteristic, for no other plant of our area resembles *A. tomentosa* in these two respects. Only the larger vines are high climbing; most of them reach about 8 meters.

1. Winter twig
2. Cutaway of twig
3. Detail of twig
4. Leaves

5. Flower
6. Stamens and stigmas
7. Fruit
8. Seed

POLYGONACEAE
Eriogonum effusum Nutt.
var. *effusum*

Umbrella plant

A suffrutescent shrub to 40 cm high, the old inflorescence persistent over winter.

LEAVES: Alternate, simple. Elliptic, oblong, or oblanceolate; 2-3 cm long, 5-7 mm wide, tip rounded or obtuse with a minute mucro, base gradually narrowed; margin entire, either straight or undulate, often revolute, upper surface thinly tomentose but showing the green color, lower surface densely white tomentose and concealing the midrib; petiole 5-8 mm long with widened base, tomentose; no stipules.

FLOWERS: August-September. Inflorescence about half of the total plant height. Cymes, loose or compact, 12-20 cm long, branched and rebranched, tomentose, involucres with 5-12 flowers; 3 triangular bracts 1-1.5 mm long at the base of each branch or involucre, the bract tip brown and rigid; involucre 1.5 mm high, 1.2 mm wide, obconic, green, tomentose outside, glabrous within, the 5 triangular lobes 1 mm long, green with brown tip; pedicels slender, weak, 2 mm long, glabrous; calyx lobes 6, pink or white, the outer 3 obovate, 2 mm long, 1 mm wide, the inner three shorter and narrower; no petals; stamens 9, the filaments delicate, 1 mm long, flattened, ciliate; anthers pale yellow, the base of the ovary globose and smooth, the neck with 3 angles, the angles rough, almost barbed, the whole structure 1.5 mm long, 0.3 mm wide; styles 3, filiform, 0.75 mm long, white; stigmas capitate, yellowish.

FRUIT: Late September. Achene enclosed in a membranaceous, dry coating; achene body with 2 angles, gibbous on one or both sides, 1 mm long and the same width, red-brown, the neck 3-angled, 1 mm long, and usually darker brown than the body, the surface of the body granular striate, the base yellow-brown and somewhat hyaline; 1 achene per flower.

TWIGS: 1.5-2.5 mm diameter below the persistent inflorescence, red-brown, covered with a dense, white tomentum which often remains only in patches, stem glossy beneath the tomentum; leaf scars crescent-shaped, usually concealed by old petiole bases; 3 bundle scars; pith one-half to three-fourths of stem, pale brown, continuous. Buds ovate, compressed, 1-2 mm long, 0.75-1 mm wide; concealed beneath old leaf bases and embedded in a dense, white tomentum; scales red-brown, often with an acuminate tip, mainly glabrous with a long, dense, crinkly ciliate margin; the inner scales green.

TRUNK: Bark red-brown, exfoliated into long, papery shreds or broad, thin strips. Wood hard, fine-grained, white but often with a pinkish cast.

HABITAT: Dry, rocky or gravel soils in well-drained areas, either among prairie grasses or in partly denuded, eroded areas.

RANGE: Nebraska to Utah, south to Arizona and New Mexico, and north to eastern Colorado.

Var. *effusum* enters our area only in the westernmost counties of the Nebraska panhandle, but its range extends west and southwest from there, a far wider range than var. *rosmarinoides*, which is limited to western Kansas. In the southwest corner of the Nebraska panhandle var. *effusum* becomes quite abundant and varies in size from 15-40 centimeters high, including the inflorescence. In contrast with var. *rosmarinoides*, var. *effusum* is a much smaller plant with fewer branches and is more inclined to have several upright stems from ground level; the leaves are 5-7 mm wide, while those of var. *rosmarinoides* are only 1-2 millimeters. The range alone would be sufficient to separate the two varieties.

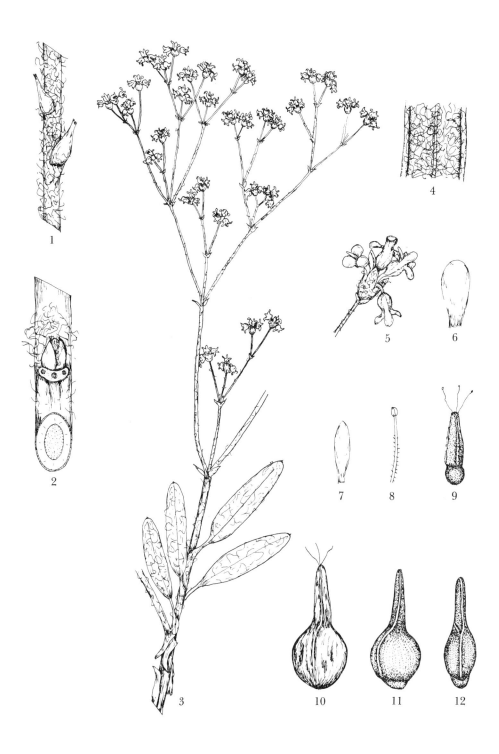

1. Winter twig
2. Detail of twig
3. Branch with leaves and inflorescence
4. Portion of underside of leaf
5. Involucre with flowers
6. Outer calyx lobe
7. Inner calyx lobe
8. Stamen
9. Pistil
10. Fruit
11. Achene, face view
12. Achene, edge view

POLYGONACEAE
Eriogonum effusum Nutt. var. *rosmarinoides* Benth. in DC.

E. microthecum Nutt. var. *helichrysoides* Gand.; *E. helichrysoides* (Gand.) Rydb.

Umbrella plant

A shrub to 60 cm high, with many fine branches, usually larger and more woody than var. *effusum*.

LEAVES: Alternate, simple. Linear, 3-4 cm long, 2 mm wide; entire and strongly revolute; tip acuminate and mucronate, base attenuate; upper surface gray-green with some cobwebby hairs, the midrib indented; lower surface densely white with cobwebby hairs, the lateral veins concealed; petiole 5 mm long, blade decurrent, forming a slight wing, densely pubescent, grooved above, the base abruptly flared; no stipules.

FLOWERS: July-August; perfect. Inflorescence cymose, 5-15 cm wide, terminal and with small bracts at the base of the branches. Involucre cylindric, 8-10 mm long, 4-5 mm wide, pubescent, green; lobes 5, rounded or acute, 2-3 mm long, pubescent; 3 of the lobes short with a red tip, 2 long and narrow with a yellow tip; 3-7 flowers in each involucre, each flower with a pedicel 6-10 mm long, green and glabrous; calyx campanulate, 3 mm long, the 6 lobes 2 mm long, white or pink with a green midvein, glabrous, the 3 outer lobes obovate to oval, 1.4 mm wide, the 3 inner lobes ovate, 1 mm wide; corolla none; stamens 9, inserted on the base of the calyx lobes; filaments white, pubescent with an orange gland at the base; anthers yellow; ovary 3-angled, narrowly ovate, 1-1.5 mm long; styles 3, filiform, 2.3-2.5 mm long, white, glabrous; stigma globular, greenish.

FRUIT: August-September. Fruit ovate, 3.2-3.4 mm long, 1.5 mm thick, 2-3 angled, abruptly narrowed to a long pointed tip, pale red-brown. Achenes ovate, slightly compressed, 2-ridged, abruptly narrowed to a tapering point, 2.6-2.9 mm long, 1.3-1.4 mm thick, red-brown, darker at the base and tip.

TWIGS: 3 mm diameter below the inflorescence, semirigid, red-brown, with gray, cobwebby pubescence on new growth; leaf scars indistinct, since the abscission layer is on the petiole; pith white to brown, continuous, one-half of stem. Buds 2.5 mm long, ovoid, brown, woolly, concealed by the persistent leaf base and the dense tomentum.

TRUNK: Bark red-brown to gray-brown, splitting into long, thin, loose shreds; bark of young branches smooth, glossy. Wood medium hard, brown, with a light sapwood.

HABITAT: Dry, limestone hillsides where the sod is broken, in chalk bluffs and open pastures.

RANGE: Kansas.

Hitchcock, et al. give var. *rosmarinoides* as synonymous with var. *laxiflorum* Benth., but our specimens from western Kansas will not fit into their key to varieties.

Var. *rosmarinoides* is a much larger plant with a heavier trunk than var. *effusum*, often 2-3 cm diameter at the level of the ground. The main branches are usually decumbent for a short distance, the flowering branches always erect. The plant clings tenaciously to the eroding slopes of the chalk bluffs, and erosion often exposes the root for 3-4 decimeters. This root then develops a shredded bark similar to that of the trunk.

The plant is very local in distribution but may be abundant in the few locations where it grows. It is most common in the chalk-bluff region of southern Logan County, Kansas, that county being in the west central part of the state. During winter months it is easily identified at some distance by the red-brown color and the dense mass of fine branches.

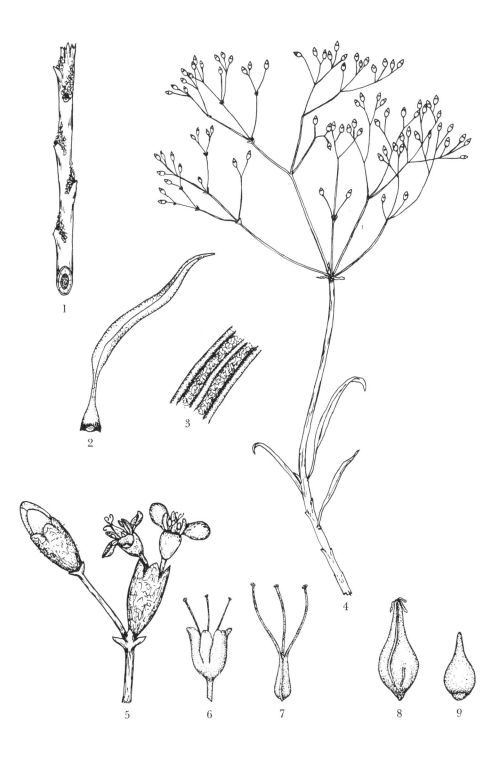

1. Winter twig
2. Leaf
3. Underside of leaf
4. Inflorescence
5. Involucres and flowers
6. Flower as it opens
7. Pistil
8. Achene
9. Seed

CHENOPODIACEAE
Atriplex canescens (Pursh) Nutt.

Calligonum canescens Pursh; *Obione canescens* Moq.; *Pterochiton canescens* Nutt.; *P. occidentale* Torr. & Frem.

Wingscale

A rigid, densely branched shrub to 1 m high.

LEAVES: Alternate, simple, usually in clusters. Oblong, obovate, or lanceolate, 2.5-4.5 cm long, 3-7 mm wide; margin entire, occasionally undulate; tip rounded, base acuminate; both sides gray-green, covered with silvery scales, the lateral veins hidden by the scales; petiole none or to 2 mm; no stipules.

FLOWERS: Late May; dioecious. Staminate flowers in globose clusters in terminal panicles on new growth, the lower panicles leafy; perianth with 5 lobes united at the base, incurved, 0.9 mm long, 0.8 mm wide, ovate, scaly, and coarsely pubescent outside, glabrous inside; stamens 5, filaments 1-1.3 mm long with yellow anthers; pistil, if present, vestigial. Pistillate panicles 5-10 cm long, terminal, leafy, on new growth; 2 fleshy bracts 1.6 mm long adnate to the ovary, green, pubescent, and scaly; ovary green, glabrous, ovoid, 0.6-0.8 mm long; style stout, 0.5 mm long, stigma of 1-2 linear, yellow lobes, 0.8-1.3 mm long.

FRUIT: July. Pedicels 3 mm long; the accrescent bracts of the ovary are fused, forming a 4-angled, fruit-like structure 0.8-1 cm long, the outer end laciniate or undulate, tan in color and the surface covered with scales. Seeds 2.3-2.9 mm long, 1.7-1.8 mm wide, 1.1-1.2 mm thick, ovoid with an extension on one side, light brown.

TWIGS: 1.5 mm diameter, gray to tan or yellow, pubescent when young, becoming glabrous, rigid; leaf scars small, crescent-shaped, with 2-3 bundle scars; pith greenish, continuous, one-fifth of stem. Buds ovoid, 1 mm long, yellow-brown to gray, pubescent or scaly.

TRUNK: Bark tight, slightly furrowed, gray, thin. Wood hard, brown, with a greenish sapwood.

HABITAT: Dry soil, usually in grassy lands where the sod is broken and on barren flats or slopes in chalk bluff areas.

RANGE: Alberta to Washington, south to Baja California, east to Texas, north across western Kansas to North Dakota and Montana.

Atriplex canescens forms a low bushy shrub to 1 meter high and about the same width, but most of them are smaller. It may be diffusely branched, or it may have one central trunk with the branches extending horizontally from it and with the tips turning more or less erect. The plant is found in association with *Eriogonum effusum* and *Eurotia lanata* in the drier, often eroded prairie hills, but may also be found in flat, sandy locations. It is hardy and makes a strong comeback after the cattle have grazed the branches down to the larger limbs. The stands in our area are not thick enough to be of much value in preventing soil erosion, but individual plants may break up the surface wind currents. It does, however, furnish good cover for small birds and mammals and it is quite common to find a cottontail or jack rabbit resting in the shade of a bush. The plants produce an abundance of seed, but apparently very few of these seeds find proper germinating conditions.

Atriplex nuttallii Coult. is quite similar in appearance and habitat, but is usually less than 30 cm high. It may be a herbaceous perennial or with a woody base, the winter buds up to 15 cm above ground. The fruiting bracts of the ovary do not enlarge but form only a small flange instead of the broad wing of *A. canescens*.

● *Atriplex canescens*
■ *Atriplex nuttallii*

1. Winter twig
2. Detail of twig
3. Leaf
4. Staminate panicle
5. Staminate flower
6. Pistillate panicle
7. Pistillate flower with bracts
8. Pistil
9. Fruiting panicle
10. Fruiting bracts
11. Seed

CHENOPODIACEAE
Atriplex confertifolia (Torr. & Frem.) Wats.

A. spinosa D. Dietr.; *A. subconferta*
Rydb.; *Obione confertifolia* Torr. &
Frem.; *O. spinosa* Moq. in DC.

Shadscale, spiny saltbush

A shrub 40-70 cm high, freely branched close to the ground, the branches rigid and spinescent.

LEAVES: Alternate, simple. Lateral veins obscured. Leaves thick, ovate, obovate, triangular or elliptic, 1-2 cm long, 8-12 mm wide, tip acute, obtuse or rounded, the base broadly to narrowly cuneate, decurrent on the petiole; margin entire; surface dull, scurfy, light gray-green; petiole 3-5 mm long, elliptic in section; no stipules.

FLOWERS: July; dioecious. Staminate flowers sessile in dense, axillary glomerules or on short, often leafy, terminal spikes. Perianth lobes 4-5, distinct nearly to the base, green, scurfy, oblong, 1.5 mm long, 1 mm wide, curved over the stamens in bud; stamens 4-5, filaments tapered, 1-1.5 mm long, white; anthers 0.5 mm long, yellow. Pistillate flowers sessile in axillary glomerules or in short terminal spikes; no perianth; 2 foliaceous bracts surround the ovary, the bracts ovate, 2-3 mm long, 1.5-2 mm wide, green, scurfy, thick, the margins united about one-third of their length and the faces in close contact above the ovary; pericarp thin, transparent; ovule broadly oval, flattened, 1 mm across, green; styles 2, filiform, green, 1.5 mm long, the stigmas extending from between the bracts.

FRUIT: August. Pistillate bracts accrescent, 8-12 mm long, 8-10 mm wide, gray-green, becoming yellow-brown, fused completely around the seed; seed brownish, circular, flat, the center slightly concave, 2 mm across, 0.6 mm thick, radicle end extending from one side, funiculus attached at the base of the radicle.

TWIGS: 2 mm diameter at 5 cm from the tapered, sharp-pointed tip; twigs rigid, yellow-brown, with many short, spinescent branches, young twigs densely scurfy; no obvious lenticels; leaf scars oval to nearly circular, raised; 3 bundle scars; pith greenish, firm, continuous, one-half of stem. Buds naked, consist of 4-8 embryonic, thick, scaly leaves 0.5-2 mm long.

TRUNK: Trunk irregular, appears to be composed of several fused branches. Bark yellow-brown, with shallow furrows and ridges, somewhat scaly. Wood hard, porous, light yellow-brown, with no sapwood-heartwood line.

HABITAT: Semiarid conditions, sandy, rocky or clay soil, often on steep, eroded hillsides in "badland" areas.

RANGE: Northern Oregon, east across central Montana to North Dakota, south to Wyoming, Colorado, and northern Mexico, and north to California.

Shadscale is a plant of the Great Basin Province and of the Sonoran Province. Since it appears only in North Dakota in our area it would be an eastward extension from the Great Basin. In areas where it is more abundant it is an important browse plant, and the condition of the plants observed in the field in North Dakota would substantiate that.

The branches are rigid, tough, and spinescent, and where it grows in pure stands it becomes a barrier to easy travel. After having brushed against it a few times, one wonders how cattle or sheep manage to get the leaves.

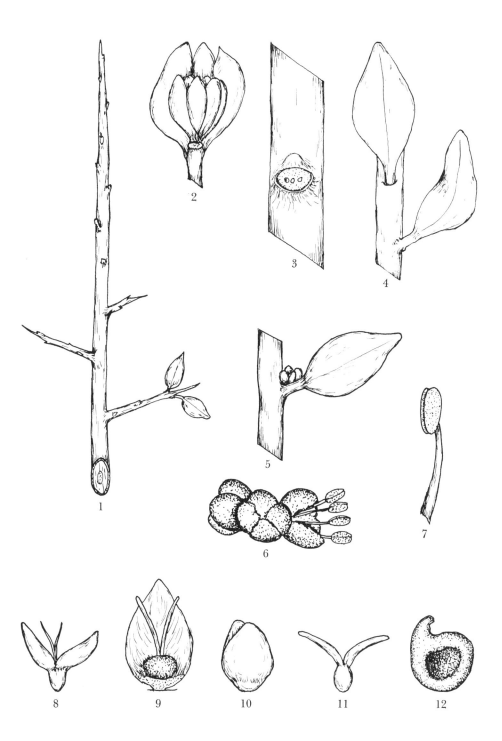

1. Winter twig
2. Naked terminal bud
3. Detail of twig
4. Leaves
5. Leaf with staminate flowers
6. Group of 3 staminate flowers

7. Stamen
8. Pistillate flower
9. Pistillate flower with one bract removed
10. Fruiting bract, face views
11. Fruiting bract, edge view
12. Seed

CHENOPODIACEAE
Eurotia lanata (Pursh) Moq.

E. *ceratodes* C. A. Mey. var. *lanata*
Kuntze; *Diotis lanata* Pursh

Winter fat, winter sage

A low shrub to 50 cm high, with several erect branches from near the ground.

LEAVES: Alternate, simple. Linear or narrowly lanceolate, 1.5-3 cm long, 2-2.5 mm wide; margin entire, strongly revolute; tip obtuse, base acuminate; both surfaces gray-green and densely pubescent with stellate and simple hairs, the midrib indented above, the lateral veins obscured; petiole 1-3 mm long, pubescent, the blade decurrent on the petiole; no stipules.

FLOWERS: Late May; monoecious or dioecious. The flowering portion is terminal on each of the many upright branches and may be 35 cm long. The staminate inflorescence consists of axillary clusters on the main branch or spikelike clusters with 6-8 flowers on short axillary branches; calyx lobes 4, distinct, obovate, 1.8 mm long, 1 mm wide, the tip nearly white with a longitudinal green stripe along the center, densely woolly outside, glabrous inside; corolla none; stamens 4, opposite the calyx lobes, filaments white, 2-2.3 mm long; anthers 1 mm long, orange-yellow. Pistillate flowers in axillary clusters of 1-4, flowers usually sessile; each flower surrounded by 2 ovate, green, pubescent bracts with only the stigmas exserted; perianth none, stamens none; ovary green, 0.5 mm long, obovoid, sparingly pubescent; style green, 0.3-0.7 mm long, glabrous; stigma lobes 2, linear, 1-1.4 mm long, greenish.

FRUIT: July-September. The two bracts surrounding the ovary in the flower accrescent, 5-6 mm long, white woolly, with fluffy, straight hairs. Fruit oval, flat, 2 mm long, with white pubescence; embryo annular, erect.

TWIGS: 1-1.2 mm diameter, gray-brown, densely pubescent when young; leaf scars indistinct, the abscission layer being on the petiole; pith green, continuous, one-sixth of stem. Buds naked, white woolly, 1 mm long and consist of a cluster of tiny leaves.

TRUNK: Bark gray-brown, exfoliating into short flakes. Wood hard, brownish-yellow, with a light yellow sapwood.

HABITAT: Dry clay or chalky soils, usually on broken ground but occasionally in short-grass prairies.

RANGE: Texas and New Mexico, north through the western part of our area and into Canada; also in Washington and California.

Eurotia lanata branches low to the ground and has many ascending branches, forming a rather dense, low shrub. The leaves become dry in the autumn but may remain on the plant most of the winter. The white woolliness gives a gray color to the plant and causes it to stand out among the other shrubs and grasses. In the autumn the long white hairs of the fruiting bracts fluff out and completely hide the fruit. At this season, a heavily productive plant will appear pure white and can be spotted from quite some distance, especially if the sun is shining.

Eurotia is common but not abundant in our area, but further west and southwest it occurs in nearly pure stands and covers large areas. It is highly nutritious and is a valuable forage plant for domestic animals and the large native mammals. In some sheep-grazing regions the plant has been so heavily grazed that it is in danger of becoming completely exterminated in those sections.

It is not used for decorative purposes but could well be, since the dense, white fruiting columns remain on the plant for several weeks. They are often cut and taken into the house for winter bouquets, eventually turning brownish, as they often do on herbarium specimens.

166

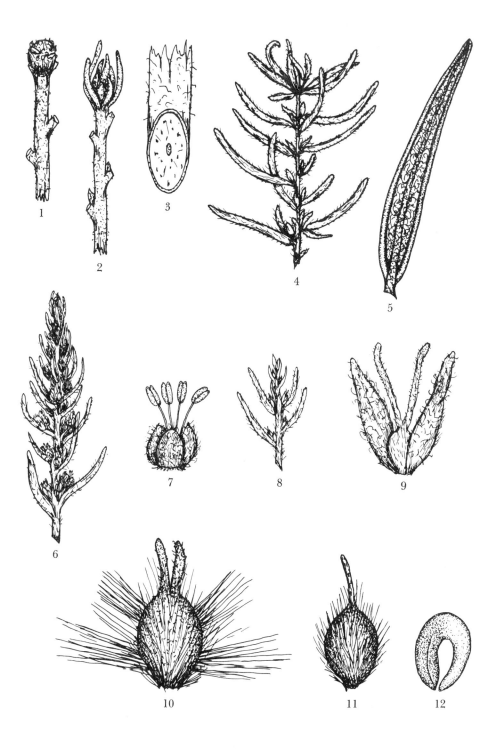

1. Winter twig with naked bud of minute leaves
2. Twig with naked bud of larger leaves
3. Section of twig
4. Vegetative branch
5. Underside of leaf
6. Staminate inflorescence
7. Staminate flower
8. Pistillate inflorescence
9. Pistillate flower with bracts
10. Fruit with bracts
11. Fruit
12. Embryo, pericarp removed

CHENOPODIACEAE

Sarcobatus vermiculatus (Hook.)
Torr.

S. baileyi Cov.; *Fremontia vermicularis*
Torr. & Frem.

Greasewood

A dense shrub to 1.5 m high, with spinescent branches.

LEAVES: Alternate, simple, tardily deciduous. Linear, succulent, nearly terete to elliptical or quadrangular in section; 1-2.5 cm long, 1.5 mm wide, 1 mm thick; acute and gland-tipped, sessile; margin entire; surface occasionally somewhat scurfy near the base; no stipules.

FLOWERS: June-August; monoecious. Staminate flowers in catkin-like spikes 1.5-3 cm long, 4-5 mm thick; peduncle 3-5 mm long, puberulent; the peltate scales of the catkin on stalks 0.5-1.5 mm long, the scale rhombic with the apical side longer and pointed, green, and thin-margined; stamens 2-4, the very short filaments inserted on the catkin axis at the base of the scale stalk; anthers 1-1.5 mm long, greenish-yellow, becoming black after anthesis. Pistillate flowers solitary in the leaf axils below the staminate catkins and set in a deep, pubescent cavity in the stem; perianth turbinate, 1.5 mm long, 1 mm wide at the top, puberulent, green, adnate to the lower half of the ovary wall; ovary compressed ovate, 1 mm across, extending 1 mm beyond the spreading perianth, dark green, puberulent; style short, the 2 stigmas linear, 2 mm long, greenish or reddish, spreading and somewhat papillate.

FRUIT: August-September. Perianth persistent, the upper margin accrescent, forming a lobed disk at right angles to the axis of the ovary and about midway on the ovary; disk 6-10 mm across, scurfy, eventually pale brown; embryo enclosed in a thin membrane, dark green, coiled, circular, flattened, 2 mm across, 1 mm thick, the hypocotyl forming a complete coil with the cotyledons in the center.

TWIGS: 1.4-1.6 mm diameter; bark thick, gray-yellow to whitish, smooth, glabrous, the inner bark reddish or brown, becoming yellowish on older twigs; branches rigid and spinescent, with a ridge extending downward from the center of the leaf scar; wood of the twigs greenish; leaf scars deep U-shaped or V-shaped, with one main bundle scar; pith small. Buds set deeply into the stem, only the tip extended above the leaf scar, 1.5-2 mm long, 0.75 mm wide, ovate, scurfy-puberulent, the 2 outer scales valvate, thick; the leaf scar splits as the bud expands; buds on the lower stem broadly ovate, 2 mm long, 2 mm wide, and less deeply set in the twig, the scales gray-yellow and puberulent on the outside, dark green and glabrous inside.

TRUNK: Bark light yellow-gray on upper branches, light brown and slightly exfoliated on the main branches, and dark yellow-brown and fissured or exfoliated on the trunk; the trunk usually with deep, elliptic pits caused by the deciduous branches. Wood hard, fine-grained, light salmon-colored, with a wide, yellow sapwood.

HABITAT: Usually on flat areas at the base of eroded hills and in alkaline soils of dry regions, often in sandy soil; abundant on the flat flood plains along prairie creek banks.

RANGE: Eastern Washington, south to southern California, east to New Mexico, north through Colorado to the Dakotas and Saskatchewan, and west to British Columbia.

Sarcobatus is a plant of the Great Basin region, but its range extends eastward into the Dakotas and Nebraska where it is quite abundant in some locations. The branches are low to the ground, rigid and spinescent, thus giving excellent protection to small animals. Sheep browse on the upper branches, but during some seasons the plant appears to be poisonous.

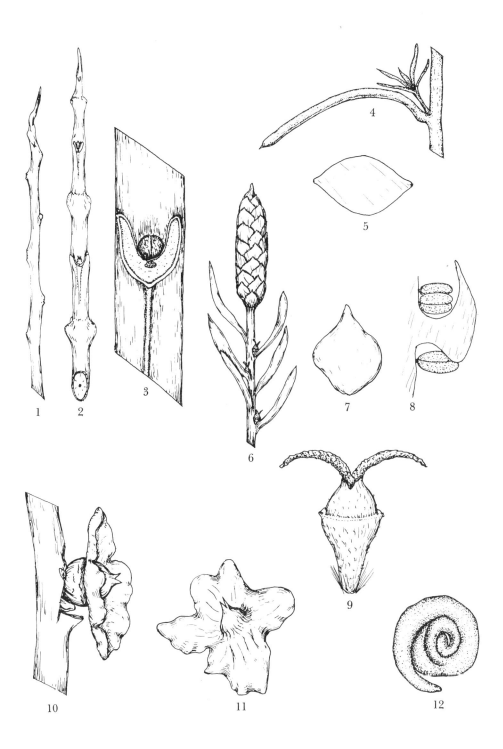

1. Winter twig, without visible buds
2. Twig with buds
3. Detail of twig
4. Leaf
5. Transverse section of leaf
6. Flowering spike
7. Top view of staminate scale
8. Sectional view of scale
9. Pistillate flower
10. Side view of fruit
11. Face view of fruit
12. Embryo

CHENOPODIACEAE
Suaeda intermedia Wats.

Dondia intermedia Heller; often described as *S. fruticosa* (L.) Forsk

Seepweed, seablite

An erect or decumbent, suffrutescent shrub, the stems to 40 cm long.

LEAVES: Alternate, simple, the nodes very short so the leaves appear to radiate from the stem; no visible veins; leaves only slightly reduced upward; tardily deciduous. Linear, nearly terete, 10-15 mm long, 1.5 mm wide, 1.25 mm thick, succulent, glabrous, slightly ridged on the sides, curved upward; tip acute, base sessile or abruptly narrowed to a very short petiole of 1 mm; margin entire.

FLOWERS: Early July; polygamous. Flowers sessile in the axils of the upper leaves or occasionally along the whole length of the stem, giving the appearance of a dense, elongated, leafy spike; perianth segments 5, about 2 mm long, distinct, glabrous, rounded on the dorsal side, green with a somewhat hyaline margin, cucullate and cupped over the stamens in bud; stamens 5, filaments greenish-white, 2 mm long; anthers 0.75 mm long, yellow, extend beyond the perianth; ovary green, 2 mm long, 1.5 mm wide, ovoid to ellipsoid, glabrous; style short or absent; stigmas 2, linear, 0.5 mm long, red-brown, spreading, often extended from the bud before the perianth spreads or the anthers appear.

FRUIT: August-September. Perianth persistent, accrescent, 3-4 mm long, surrounding the one seed. Seeds mostly vertical, discoid, with a projection at the hilum, 1.5-1.7 mm wide, 1 mm thick, black, glossy, minutely pitted (use 30× magnification).

TWIGS: 0.5-1 mm diameter, soft, flexible, yellow-brown, often with a white, scurfy pubescence, roughened by the old, indefinite leaf scars and leaf bases; leaf scar when the leaf is removed, semicircular to nearly round with no obvious bundle scar; pith white, continuous, one-third of stem. Buds naked, consisting of a globular cluster of green leaves in a few of the leaf axils of the upper leaves,

progressively more abundant toward the base of the stem.

TRUNK: Can hardly be called a trunk. Bark yellow-brown, tight, smooth except for the old leaf bases and scars; inner bark thick, green, succulent. Wood soft, greenish, with a small pith.

HABITAT: Barrens and eroded hillsides, usually in rocky or clay soil. Although the habitat is often given as moist soil, in our area the plants occur only on the most exposed, dry, semi-barren hillsides.

RANGE: Washington, Alberta, and North Dakota, south across northwestern South Dakota to Colorado and New Mexico, north and west to Utah, California, and Oregon.

This plant is listed as *S. fruticosa* (L.) Forsk in many manuals and is often considered to be identical to the European form. It is not a true woody plant under most definitions, but the base is definitely woody and often becomes 1 cm in diameter well above the ground level. The branches often die back to within 6-9 cm of the ground during the winter, creating a scrubby crown above ground. Most of the new stems come directly from this crown and have 2-3 lateral branches.

Some manuals give the position of the seed as horizontal, others make no mention of the position. In working with the fruiting material in the field, the author found the number of vertical seeds to be greater than the number of horizontal seeds. So it would appear that the seeds may be in either position.

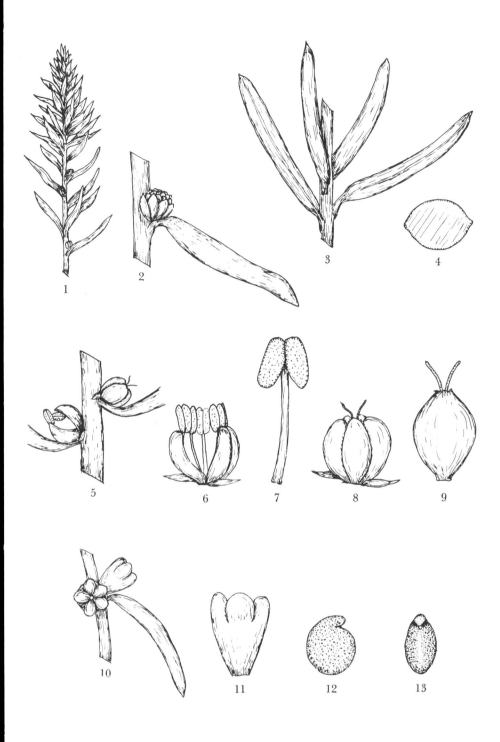

1. Twig in November
2. Naked bud
3. Leaves
4. Transverse section of leaf
5. Staminate and pistillate flowers
6. Staminate flower
7. Stamen

8. Pistillate flower
9. Pistil
10. Fruit
11. Fruit
12. Side view of seed
13. Ventral view of seed

RANUNCULACEAE
Clematis ligusticifolia Nutt.
in T. & G.

C. brevifolia Howell; *C. suksdorfii*
Robins.

Virgin's bower, white clematis

Vines climb by twining and by the tendril-like leaf petiole and rachis to a length of 10 m; freely branched and often sprawling over bushes; tips of branches usually die back several internodes during the winter.

LEAVES: Opposite, pinnately compound, 3-6 cm long, 5-7 leaflets, thin, tough, the ridge on the underside of the petiole decurrent on the stem. Leaflets ovate, 4-7 cm long, 2-4 cm wide, 3-5 veins from the base, anastomosing with the lateral veins from the midrib, basal leaflets largest; tip long acuminate, the base cuneate, rounded or subcordate; margin coarsely toothed or lobed; both surfaces glabrous or sparingly strigose, with no pronounced color difference; petiole 4-7 cm long, ribbed, short pubescent; petiolules 1-3 cm long, both the rachis and the petiolules may curl around objects to aid in climbing; no stipules.

FLOWERS: July-August; dioecious. Inflorescence axillary in corymbiform panicles 5-10 cm long, a pubescent bract at the base of each branch; staminate flowers on pedicels 1-1.5 cm long, sepals 4-5, oblanceolate with acute tip, white, 8-10 mm long, 2-2.5 mm wide, densely puberulent on both sides, spreading or reflexed; stamens numerous on a slightly rounded, pubescent receptacle; filaments stout, 5-8 mm long; anthers pale yellow, 2 mm long. No petals or pistil. Pistillate flowers on pedicels 1.5-4 cm long, sepals 4-5, narrowly ovate to oblanceolate, tip acute, white, reflexed or spreading, finely puberulent on both sides; numerous staminodes; ovary obovate, 0.75 mm long, silky pubescent, green, numerous on a pubescent dome-shaped receptacle, style 4-6 mm long, with long, silky hairs, stigma curved, 0.5-0.75 mm long.

FRUIT: Late August. Fruit obovate with one margin more rounded, flat, short pubescent, 3 mm long, 1.5 mm wide; style elongated, 3-5 cm long, plumose, the hairs white and 5-6 mm long

at the center of the style; stigma persistent.

TWIGS: 1-3 mm diameter at the point of dying back, flexible, tan-brown, somewhat 6-ribbed, glabrous; lenticels small, numerous and not noticcable; old petioles often remain; inner bark green; pith white, continuous, three-fourths of stem. Buds ovoid, 2 mm long, dark brown, the outer scales slightly keeled, apiculate, glabrate, inner scales greenish-brown, pubescent; live buds winter over up to 5 m from the ground.

TRUNK: Bark light gray-brown, shredded into thin strips, inner bark yellow, fibrous. Wood greenish-yellow, porous, soft, lightweight, rays obvious, central pith white or brown.

HABITAT: Wooded or brushy hillsides and creek banks, either in dry, rocky soil or moist, sandy soil.

RANGE: British Columbia, south to California, east to New Mexico, north through Colorado and western Nebraska to North Dakota, and west to Montana and Alberta.

C. ligusticifolia is a common vine in the western and northern part of our range. It forms a dense mass sprawling over low bushes, often weighting them down and smothering them. The five to seven leaflets should distinguish it from *C. virginiana* L. which has a three-foliolate leaf, and is the only plant of our area which might be confused with *C. ligusticifolia*.

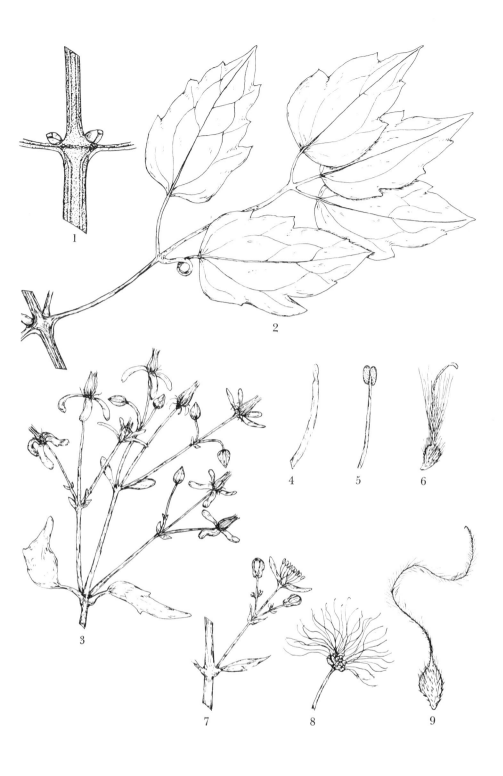

1. Winter twig
2. Leaf
3. Pistillate panicle
4. Sterile stamen of pistillate flower
5. Stamen
6. Pistil
7. Staminate flower and buds
8. Fruit cluster
9. Achene

BERBERIDACEAE
Berberis repens Lindl.

B. nana Greene; *B. brevipes* Greene;
Odostemon repens Cockerell in Daniels;
Mahonia repens G. Don var.
rotundifolia Fedde

Oregon grape, mahonia, barberry

A suffrutescent, erect shrub 15-40 cm high, leaves near the top, single stems from underground stolons.

LEAVES: Alternate, evergreen, pinnately compound, 15-20 cm long, 3-7 leaflets. Leaflets ovate, rarely oblong, 4-6 cm long, 2.5-3.5 cm wide, usually less than twice as long as wide, terminal leaflet largest, tip acute, spinulose, base of terminal leaflet broadly cuneate, truncate or subcordate; lateral leaflets obliquely truncate, with the basal side longer; margin definitely thickened, coarsely spinose-serrate, 1-4 teeth per cm, sinuses rounded, 0.5-4 mm deep; upper surface dark green, semiglossy; lower surface pale green, minutely papillate (visible only with 15× magnification or more); glabrous, coriaceous; petiole 6-10 cm long, broad based, definite articulation line at the base, rachis often reddish; lateral leaflets sessile or nearly so, terminal petiolule 2-3 cm long; stipules narrow, adnate for 5-7 mm, free tip lanceolate to awl-shaped, 3 mm long, brownish.

FLOWERS: Early June; perfect. Racemes dense, 3-6 cm long, 2-2.5 cm wide, erect from the base of new growth, glabrous, a yellow-green bract at the base of each pedicel; pedicels slender, 4-7 mm long, ascending; the 3 bracts at the base of each flower, ovate, 2-2.5 mm long, yellow, spreading or reflexed; sepals 6, distinct, the outer 3 ovate to obovate, 4-5 mm long, spreading, the inner 3 obovate, 6-7 mm long, erect or spreading; petals 6, erect, elliptic to oblong, 5 mm long, cleft at the tip, deeper yellow than the sepals; stamens 6, inserted on the base of petals, filaments 2.5 mm long, yellow, enlarged at the middle and with 2 reflexed prongs; anthers pale yellow; pistil urceolate, 2-3 mm long, 1-1.2 mm wide, pale green; style short with a discoid stigma.

FRUIT: Late August. Fruits in racemes between the upper leaves, the axis glabrous, reddish, 2-4 cm long, 5-20 fruits; pedicels 5-8 mm long, pinkish;
fruits ovoid or top-shaped, 9-12 mm long, 8-11 mm diameter, dark blue with a heavy bloom not easily removed, surface finely granular; 5-7 seeds per fruit. Seeds dark red, shape of an orange section, 3-4 mm long, 2-2.5 mm wide, smooth, glossy, the dorsal side rounded, ventral side with a large hilum and distinct raphe.

TWIGS: 2.5 mm diameter, brown, glabrous, leaves or their bases persistent on the upper stem; leaf scars on the lower stem crescent-shaped, extending nearly around the stem; numerous, indistinct bundle scars; a small, rounded, green bud above the leaf scar; pith white, continuous, one-third of stem. Terminal bud 7-10 mm long, 2-5 mm wide, scales ovate, green, often brown-edged, a linear tooth on each side near the tip and a smaller tooth on the inner scales.

TRUNK: Bark light yellow-brown, lightly fissured, roughened by old leaf scars; inner bark bright yellow. Wood hard when dry, yellow, fine-grained, the rays obvious, pith hard.

HABITAT: Usually on thinly wooded slopes, in rocky, slightly moist soil. Occasionally on open hillsides.

RANGE: Washington to Alberta, southeast to the Black Hills of South Dakota, south to northwestern Nebraska, Colorado, western Texas, west to Arizona, north to Utah, and west to northern California and Oregon.

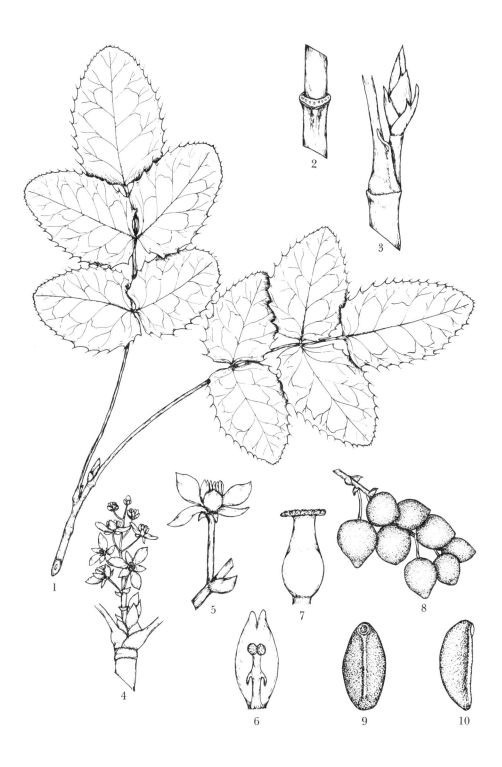

1. Winter twig with leaves
2. Detail of twig, leaf removed
3. Detail of terminal bud
4. Floral cluster
5. One flower
6. Stamen and petal
7. Pistil
8. Fruit cluster
9. Ventral view of seed
10. Lateral view of seed

MENISPERMACEAE
Cocculus carolinus (L.) DC.

Epibaterium carolinum Britt.; *Cebatha carolina* Britt.

Snailseed, coral berry, coral bead

A twining vine to 15 m high.

LEAVES: Alternate, simple, 3-5 main veins. Shape variable, ovate, pandurate, somewhat lobed, or ovate with concave sides, 6-10 cm long, 4.5-8.5 cm wide; margin entire or lobed, thickened, finely ciliate; tip acute, acuminate or obtuse, mucronate; the base obtuse, cordate, or truncate; upper surface dark yellow-green with scattered pubescence; lower surface paler and with soft pubescence; petiole 2.5-5.5 cm long, pubescent, occasionally sharply bent near the blade.

FLOWERS: July-August; dioecious. In the preparation of this work, over 100 plants were examined in the field and none were found to be only staminate. Strong, well-developed pistils were present in all flowers. Staminate flowers (from herbarium specimen) in loosely flowered panicles 12-15 cm long; calyx lobes 6, distinct, 3 of the lobes small, 0.7 mm long, 0.4 mm wide, acute, ciliate, the 3 larger lobes 1.4 mm long, 0.9-1 mm wide, rounded, cupped, toothed or erose; petals 6, yellowish, 1 mm long, 0.6 mm wide, obtuse tip, base lobed and folded around a filament; stamens 6, filaments stout, 0.7 mm long, upper end sharply bent so that end of the anther is toward the center of the flower; anther 0.4 mm long, yellow. Pistillate flowers in axillary, loosely flowered panicles on new growth, 2-18 cm long, axis pubescent, 1-3 small leaves in the panicle; pedicels 2-6 mm long, pubescent; calyx lobes 6, the 3 small lobes 1.4-1.5 mm long, 1 mm wide, ovate to deltoid, the margin translucent and pubescent, the 3 larger lobes 2.4-2.6 mm long, 2.1-2.3 mm wide, broadly ovate, whitish, thick but with a thin, translucent, minutely toothed margin; petals 6, yellow, thick, 1.8 mm long, 1.2 mm wide, lower portion with 2 sharply incurved lobes; tip rounded, minutely toothed; staminodes 6, stubby, greenish with a small yellowish spot; ovaries 6 on a dome-shaped receptacle with 5-6 lobes;

each ovary 1 mm long, 0.4 mm thick, green, glabrous; style and stigma 0.6 mm long, cylindric, recurved, yellow to brown. Petals, staminodes and pistils opposite each other.

FRUIT: Late September-October. In grape-like clusters 3-8 cm long, or in globular clusters 2.5-3 cm diameter; pedicels 3-6 mm long, glabrous. Fruits bright red, glossy, smooth, globose, 5-7 mm diameter; skin thin, flesh gelatinous; stigmatic scar lateral, 1-2 mm from the pedicel. Seeds creamy-white, flat, 5.4-5.8 mm wide, 2.3-2.7 mm thick; rim knobby or minutely ridged; ends of the rim do not meet squarely, but slightly offset, giving a coiled effect, the center concave.

TWIGS: 1.2 mm diameter, pubescent or glabrate, flexible, gray-green or gray-brown; leaf scars half-round with concave upper margin, often deeply depressed in the center; bundle scars indistinct; pith greenish, continuous, one-fourth of stem. Buds concealed under a dense, pale tomentum.

TRUNK: Bark pale gray-green to gray-brown, slightly fissured, the thin edges recurled; old vines with ridges and warty spots of corky bark. Wood soft, whitish, porous with obvious rays.

HABITAT: At the edge of a woods, along fence rows, in roadside thickets, on stream banks or brushy, rocky hillsides.

RANGE: Mexico; Texas east to Florida, north to Virginia, west across southern Indiana to Kansas, and south to Oklahoma.

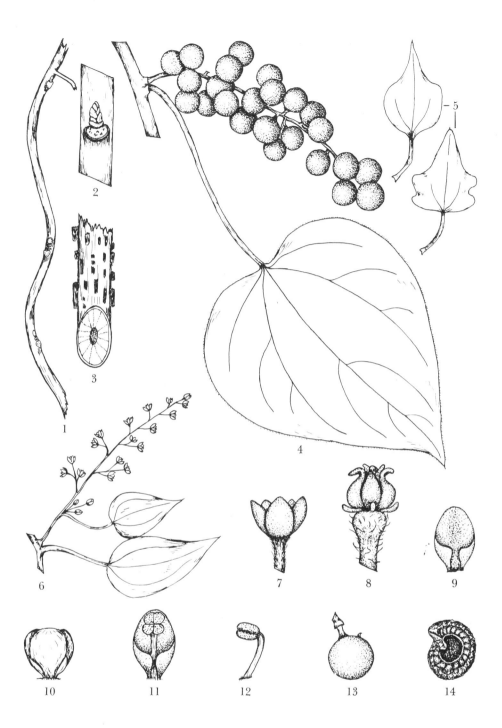

1. Winter twig
2. Detail of bud, tomentum removed
3. Section of old trunk
4. Leaf
5. Leaf variations
6. Pistillate panicle
7. Pistillate flower
8. Pistils
9. Petal of pistillate flower, ventral view
10. Staminate flower
11. Petal and stamen
12. Stamen
13. One fruit
14. Seed

MENISPERMACEAE
Menispermum canadense L.

Moonseed

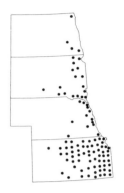

A twining vine to 8 m high, climbing or sprawling.

LEAVES: Alternate, simple. Nearly circular, 6-12 cm long, 8-12 cm wide; peltate with the petiole attached near the basal margin; entire or with 3-7 angles or lobes, the tip and some lobes sharply mucronate; base of leaf cordate or nearly truncate with an indentation near the petiole; upper surface dark green, dull or semiglossy, glabrate, usually some hairs at the petiole attachment; lower surface pale, thinly pubescent; petiole 5-11 cm long, enlarged for 0.5-1 cm at the base, often sharply bent, pubescent; no stipules.

FLOWERS: Late May; dioecious. Staminate panicels 3-5 mm above a leaf, the panicles drooping, 4-4.5 cm long, 2.5 cm wide, 40-50 flowers, a linear pubescent bract at the base of the branches; pedicels slender, 2-5 mm long, pubescent; calyx lobes 5-6, distinct, elliptical, 2.6 mm long, 1.4 mm wide, greenish-white glabrous; petals 5-6, obovate, 1.8 mm long, 1.4 mm wide, white, margins revolute, some petals enclosing a filament; stamens 18-20, filaments white, 2.8 mm long; 2 anther locules embedded in the enlarged end of the filament, opening by a split in the outer surface. Pistillate flowers above the leaf axil, in drooping panicles, 1.5-3.2 cm long, 1.5-2 cm wide, 15-30 flowers, glabrate; pedicels 3-6 mm long, sparingly pubescent; calyx lobes 5-6, distinct, linear to spatulate, 2-2.6 mm long, 0.5-1.2 mm wide, greenish, the margin irregular; petals 5-6, obovate, 1.3-1.5 mm long, white; stamens 15-20, vestigial; pistils 2-4 on a central stalk 0.7-0.8 mm long; ovary 1 mm long, gibbous on the dorsal side, concave on the ventral; stigma sessile, 2-lobed, yellow, recurved.

FRUIT: September. Fruits in a grape-like cluster, peduncle 5-7 cm long, slender, glàbrate, red-brown, minutely ridged; pedicels 7-12 mm long, red-brown, enlarged and jointed near the fruit; fruit globose, 6-9 mm diameter, dark blue with a bloom; stigma scar lateral, 2-3 mm from the pedicel—this distance directly over the notch in the seed; flesh thick, juicy, the skin tough. Seeds yellowish, circular, with a triangular wedge missing, 6.5-8 mm diameter, 2.6-3 mm thick, flat, center depressed and smooth, a ridge near the margin and a marginal rim, each minutely transversely ridged. Fruits somewhat poisonous.

TWIGS: 1.5-2 mm diameter, flexible, glabrous, red-brown or green-brown, lustrous, striate; leaf scars circular, usually split at the top, exposing a bud beneath the scar tissue; 3 groups of bundle scars; pith white, continuous, one-sixth of twig. Buds depressed globose, 1 mm across the top, hardly extended from the circular, hairy crater beneath the leaf scar.

TRUNK: Bark red-brown to greenish-brown, smooth on young stems; brown to gray-brown, scaly or warty with short ridges of corky material near the base of old stems. Wood soft, white, the rays prominent.

HABITAT: Low, moist woods, along streams, ravines, fence rows, roadside thickets, or on rocky bluffs.

RANGE: Quebec to Manitoba, south to Oklahoma, east to Georgia, and north to western New England.

This is an attractive vine but should not be planted in a yard. The fruits closely resemble wild grapes, but are toxic to some people and should be kept from the reach of children.

178

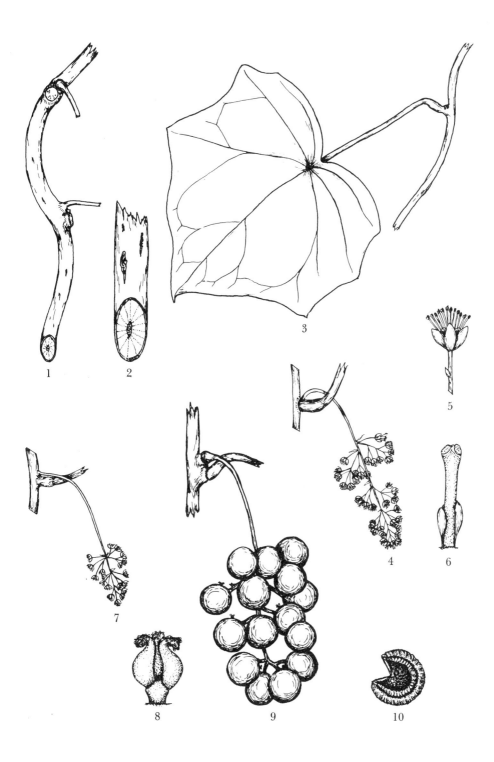

1. Winter twig
2. Section of winter twig
3. Leaf
4. Staminate panicle
5. One staminate flower
6. Stamen
7. Pistillate panicle
8. Pistils
9. Fruit cluster
10. Seed

MENISPERMACEAE
Calycocarpum lyoni (Pursh) Gray

Cupseed

A semiwoody vine to 10 m, usually clambering over low bushes.

LEAVES: Alternate, simple, palmately veined with 5-7 veins. Variable in shape, mostly rounded to broadly ovate, 7-20 cm long, 8-18 cm wide; margin with 3-5 lobes, deep or shallow, the terminal lobe usually ovate, the sinuses rounded; tip of lobes acuminate, the base of the leaf broadly cordate; upper surface dark yellow-green, glabrous; lower surface paler, a few long, straight hairs on the main veins; petiole 6-15 cm long, glabrous, striate.

FLOWERS: Late May; dioecious. Flowers in axillary panicles on new growth, the staminate flowers in loose, many-flowered panicles, 8-14 cm long, the axis, peduncles and pedicels finely pubescent; a minute, tightly appressed bract at the base of each branch; pedicels 0.7-1.2 mm long, green; calyx lobes 6, distinct, spreading, white, the outer 3 lobes elliptical, 2.8-3 mm long, 1.2-1.4 mm wide, the inner lobes elliptical, 3.4-3.7 mm long, 1.8-2 mm wide; corolla none; stamens 12 in 2 rows of 6 (2 opposite each calyx lobe), the outer stamens short-stalked, 1.6 mm long including the anther; filament flattened, white, incurved; inner stamens 2.2 mm long, erect or slightly spreading; pistil absent. Pistillate flowers in 15-20 flowered, pubescent panicles, 8-10 cm long; pedicels 1.5-2.5 mm long, green, with 1-3 minute, tightly appressed bracts at the base of, or on the pedicel; calyx lobes 6, distinct, white, ovate, acute, spreading; the outer 3 lobes 2.8-3 mm long, 1-1.2 mm wide, the inner 3 lobes 2.8-2.9 mm long, 1.6-1.7 mm wide; corolla none; stamens 12, vestigial; ovaries 3, fused toward the base on the ventral side, dorsal side gibbous, 2 mm long, 0.8 mm thick; style absent; stigmas thin, flattened, membranaceous, deeply lobed, 0.8 mm long, reflexed and slightly appressed to the ovary.

FRUIT: October. Grape-like clusters, 7-9 cm long, fruits crowded; pedicels 8-10 mm long, finely pubescent; fruit oval, smooth, 2-2.5 cm long, 1.5-1.8 cm thick, green, becoming slightly yellowish and soft before falling, the fleshy part soon decomposing; 1 seeded. Seed a yellow-brown, hollow bowl, keeled on the dorsal side, 16-19 mm long, 13-14 mm wide, the hollow 6-8 mm deep, shell thin and hard, the edge of the rim with small, close, sharp teeth.

TWIGS: 1.5-2.5 mm diameter, glabrous, flexible, yellow-brown, striate, die back to the ground or near the ground each winter; leaf scars large, oval; bundle scars indistinct; pith brown, continuous, four-fifths of stem. Buds broadly ovate, 1 mm long, reddish-brown, smooth.

TRUNK: Bark greenish to brown, finely ridged. Wood white, soft, can hardly be called wood.

HABITAT: Bases of bluffs, rich flood plains, moist soil, usually near streams.

RANGE: Kentucky, west to southeastern Kansas, south to Louisiana, east to Florida, and north to Tennessee.

Most of the manuals give the fruit color as black, but after following them through several different years, the author has never seen fruits of that color. They fall from the vine while still green or only slightly yellow and are soft. The flesh soon decays, leaving the bowl-shaped seeds on the ground surface. These seeds are one of the most distinctive characters of the plant, and no other seeds of our area resemble them.

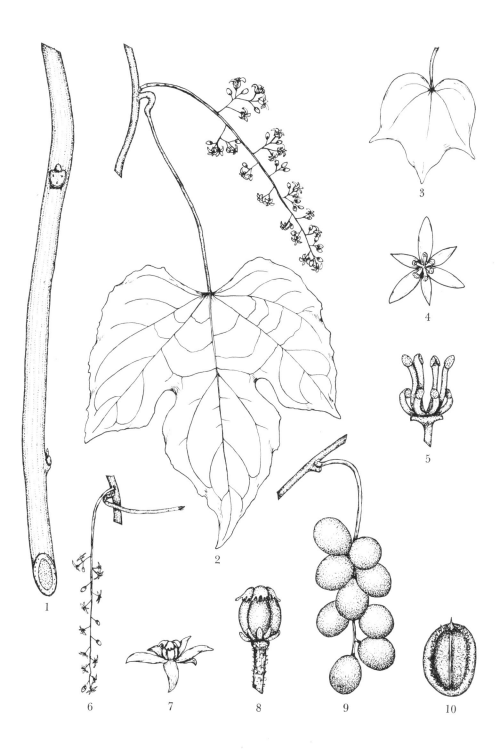

1. Winter twig
2. Leaf with staminate panicle
3. Leaf variation
4. Staminate flower
5. Staminate flower with outer parts removed
6. Pistillate panicle
7. Pistillate flower
8. Pistil
9. Fruit cluster
10. Seed

ANNONACEAE
Asimina triloba (L.) Dunal

Pawpaw, custard apple

A small tree to 7 m high, the trunk single, growing in small colonies.

LEAVES: Alternate, simple. Obovate, 20-30 cm long, 8-15 cm wide; margin entire; tip abruptly acuminate, base gradually tapered, the blade often decurrent on the petiole; upper surface yellow-green with some pubescence on the midrib; lower surface paler, with rusty pubescence at first, becoming glabrate; petiole 8-10 mm long, strongly grooved above, pubescent with red-brown hairs; no stipules.

FLOWERS: April-May, with the leaves; perfect. Solitary, axillary on previous year's growth, pedicels stout, 0.8-1 cm long, recurved, velvety with red-brown hairs; calyx lobes 3, ovate, 0.8-1 cm long, acute, cupped, pubescent with both black and brown hairs; petals 6, deep purple, rugose, leathery, arranged in 2 series, the outer 3 petals broadly ovate, 2.3-2.5 cm long, acute, reflexed, the outer surface pubescent with black hairs, the inner surface glabrous; the inner 3 petals erect or spreading, ovate, 1-1.3 cm long, acute, cupped, yellowish near the base, sparsely black pubescent on the outer side, glabrous inside; stamens numerous, 1.6 mm long, the filaments short and stout, densely packed around a globular receptacle, pale yellow, 4-loculed, the locules adnate along a stout, central connective enlarged at both ends, the locules splitting open for their entire length; pistils 4, tightly appressed, 3.6-5 mm long, white hairy, surrounded by the stamens at the tip of the receptacle; ovary cylindric, 1.9-2.3 mm long, greenish, 1-celled with about 16 ovules arranged in 2 rows; style 1.9-2.3 mm long, slightly flattened; stigma globose.

FRUIT: October. Cylindric, curved, 5-13 cm long, 3-5 cm thick, greenish-yellow, becoming brown; thin-skinned, flesh succulent, yellow, sweet and edible. Seeds 2-10 per fruit, oval, 2.3-2.7 cm long, 1.4-1.6 cm wide, 6-7 mm thick, brown, lustrous.

TWIGS: 2-2.5 mm diameter, light red-brown, pubescent, becoming glabrous; leaf scars large, half-round with concave upper margin; 5-7 bundle scars; pith greenish, becoming brown, continuous but with closely spaced disks 0.25 mm thick, about one-fifth of stem. Terminal bud flattened, acute, 4-7 mm long, covered with a red-brown pubescence; leaf buds ovate, 3-4 mm long, acute, flattened; flower buds with a 3-4 mm stalk, buds globose, 4-5 mm diameter, red-brown, woolly.

TRUNK: Bark thin, dark brown, often with ashen blotches and shallow, irregular fissures; smooth on young trunks. Wood light, soft, weak, pale greenish-yellow, with a narrow, light sapwood.

HABITAT: Deep, rich soils along streams and on flood plains; usually as an undergrowth.

RANGE: New York, west across southern Ontario to Michigan, southwest to Iowa and Nebraska, south to eastern Texas, east to Florida, and north on the west side of the Appalachians to New Jersey.

Pawpaw trees produce very few fruits in comparison to the number of flowers, and it is often difficult to find any quantity. The fruits do not need frost for ripening and may be eaten as soon as they become yellowish or have brownish blotches. Some people develop a skin rash from handling or eating the fruits, so care must be taken.

The stigmas are exserted some time before the stamens are mature. This indicates that the flowering period must be fairly long in order to insure cross-pollination.

182

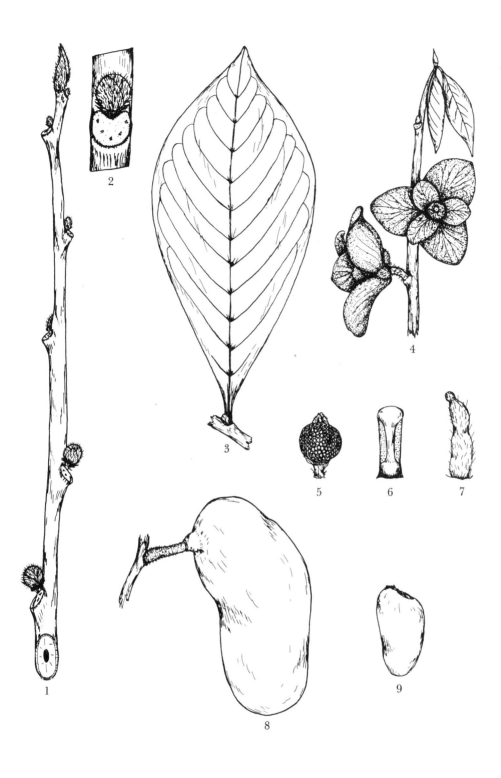

1. Winter twig
2. Detail of twig
3. Leaf
4. Flowers
5. Cluster of stamens and pistils
6. Stamen
7. Pistil
8. Fruit
9. Seed

LAURACEAE
Sassafras albidum (Nutt.) Nees var.
molle (Raf.) Fern.

S. *variifolium* (Salisb.) Ktze.; S. *officinale*
Nees & Eberm.

Sassafras

A tree to 20 m high, but in our area mostly 4-8 m high and occurring in thickets.

LEAVES: Alternate, simple, the shape variable. Ovate, oval, or obovate, 8-15 cm long, 5-10 cm wide; entire or deeply lobed with 1-2 lobes, the sinuses rounded, margin ciliate and a marginal vein on the underside; tip acute or obtuse, base cuneate; upper surface yellow-green, sparingly pubescent, the lower surface paler and puberulous, especially on the midrib and veins; young leaves densely pubescent; petiole slender, 1.5-3 cm long, pubescent; no stipules.

FLOWERS: Mid-April with the leaves; dioecious. Flowers from an end bud at the tip of previous season growth. Staminate flowers in 4-5 corymbs of 4-10 flowers each; peduncle 1.5-1.8 cm long, green, pubescent; pedicels 8 mm long, green, pubescent, a linear, pubescent bract at the base; calyx lobes 6, distinct, elliptical, 4-4.5 mm long, yellow, glabrous; corolla none; stamens 9, 6 in an outer row attached at the base of the calyx lobes and 3 in an inner row and opposite every other stamen of the outer row, a pair of stalked glands at the base of the stamens; filaments 2.5 mm long, yellowish, glabrous; anthers introrse, a small pore at the top and a larger pore on the side. Pistillate flowers in 4-5 corymbs of 4-10 flowers each; peduncle 1-1.5 cm long, green, pubescent, a bract at the base; pedicels 5-7 mm long, green, pubescent, some with a bract; calyx lobes 6, distinct, 3.5-4 mm long, elliptical, yellow, glabrous; corolla none; staminodes 6, short-stalked, deltoid; ovary ovoid to obovoid, 1.3 mm long, green, glabrous; style 2.5 mm long, yellow-green; stigma capitate, lobed, yellowish.

FRUIT: Late August. Usually 4 corymbs from the junction of old and new wood; rays ascending, 2-2.5 cm long, pubescent; 4-5 fruits in each corymb; pedicels 1-1.5 cm long, expanded at the summit to form, along with the accrescent calyx, a cup-like base below the fruit, this base about 5 mm long and 6 mm wide, bright red and glabrous. Fruit blue with a yellowish base, obovoid, 9-11.5 mm long, 7-8 mm thick, glossy or with a slight bloom, the flesh thin. Seeds ovoid to broadly ellipsoid, 7-8 mm long, 5.5-6 mm thick, dull, dark brown, granular surface, ridged on 2 sides.

TWIGS: 2-2.5 mm diameter, flexible, yellow-green, pubescent or glabrous; red-brown on older branches; twigs of staminate trees browner than those of pistillate trees; leaf scars crescent-shaped; 3 bundle scars or all 3 fused into one; pith white, continuous, one-third of stem. Terminal bud 6-8 mm long, 5-6 mm wide, ovoid, tip acute, the scales green with reddish margins, glabrous or pubescent; lateral buds 1-2 mm long and 1 mm thick.

TRUNK: Bark on young trees greenish-brown, smooth or with shallow fissures; on old trees, dark red-brown, thick, fissured, the ridges firm, long, and flat-topped. Wood brittle, durable, soft, weak, aromatic, dull orange, with a light yellow sapwood.

HABITAT: In old fields, along the border of a woods, or in open woods, usually in rocky soil.

RANGE: Maine to Michigan, southwest to Iowa and Kansas, south to Texas, east to Florida, and north to New England.

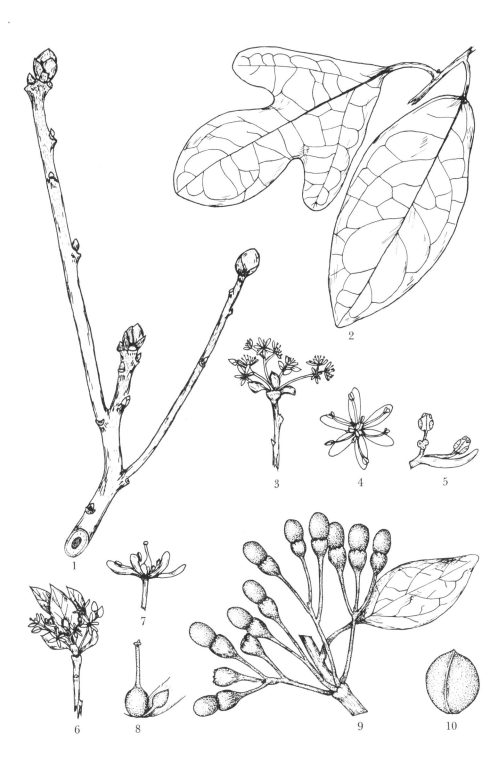

1. Winter twig
2. Leaves
3. Staminate inflorescence
4. Staminate flower
5. Stamens and calyx lobe
6. Pistillate inflorescence
7. Pistillate flower
8. Pistil and staminode
9. Fruit cluster
10. Seed

LAURACEAE
Lindera benzoin (L.) Blume

Benzoin aestivale Nees, not *Laurus aestivalis* L.

Spicebush

An open shrub to 3 m high, with single trunk and usually low branches.

LEAVES: Alternate, simple. Elliptic to obovate, 6-11 cm long, 3-5 cm wide; entire, often ciliate, tip short acuminate, base cuneate; upper surface dark green, glabrous with a few hairs on the midrib; lower surface pale, sparingly pubescent, some hairs in fascicles; petiole 10-12 mm long, pubescent or glabrous; no stipules.

FLOWERS: Early April, before the leaves; dioecious. Staminate flowers in clusters from buds of the previous year; all floral parts glabrous; pedicels 2-2.5 mm long, greenish-yellow; calyx lobes 6, distinct, obovate, 2.5-3 mm long, yellow, the tips erose; corolla none; stamens 9, one opposite each calyx lobe and an inner row of 3 with a pair of yellow, stalked glands at the base of each; filaments greenish, 1 mm long, flattened; anthers yellow, 1 mm long, opening by a pore near the end, the flap valve attached at the upper end of the pore; vestigial pistil in the center. Pistillate flowers clustered on wood of the previous year; floral parts glabrous; pedicels 1-1.5 mm long, green; calyx lobes 6, distinct, oblong to obovate, 2.5 mm long, yellow; corolla none; many staminodes; ovary green, 1-1.2 mm long, ovate, 1-celled, 1-seeded; style 1-1.2 mm long, greenish, often with a slight projection on the side; stigma capitate, slightly lobed, brownish.

FRUIT: September. Single or in clusters of 2-4; pedicel bright green, 5-6 mm long; fruit ellipsoid, 9-12 mm long, 6-8 mm thick, bright red, smooth, glossy, obtuse ends, thin flesh, strongly aromatic of spice; funiculus free or adnate to the inner surface of the flesh, attached at the outer end of the seed. Seed ellipsoid to ovoid, 7-8 mm long, 5-5.8 mm thick, gray-brown to light red-brown, blotched with dark brown; base wide and rounded.

TWIGS: 1 mm diameter, aromatic, flexible, gray-green to brown, glabrous or pubescent, lenticels light, vertical, prominent; leaf scars half-round, 3 bundle scars; pith white or greenish, continuous, one-fourth of stem. Buds single or clustered, ovoid, 1-1.3 mm long; 1 leaf bud near the leaf scar, the flower buds slightly above it; the scales thin, greenish-brown, rounded, the margin brownish and often erose.

TRUNK: Bark light gray-brown, smooth but with raised lenticels; somewhat split on old trunks. Wood soft, greenish-yellow.

HABITAT: Low, moist woodlands and river banks, often in thickets along streams.

RANGE: Maine to Ontario and Michigan, southwest to Iowa and Kansas, south to Texas, east to Florida, and north to New England.

Lindera is common in our area only in the southeast corner of Kansas where it grows as an undershrub along stream banks or near the base of a wooded hillside. The two most outstanding characters are the spicy, brilliant red fruits and the spicy aroma from any part of the plant. Even in midwinter, but especially in early spring, a broken twig gives off a rather strong and pleasant odor.

Var. *pubescens* (Pal. & Steyrm.) Rehd. has been segregated, but the difference between it and var. *benzoin* is purely one of the degree of hairiness, var. *pubescens* being described as with *more* hairy leaves, petioles, and twigs. Since our material shows a wide variation in the number of hairs, even in the same colony, they are both placed here under var. *benzoin*.

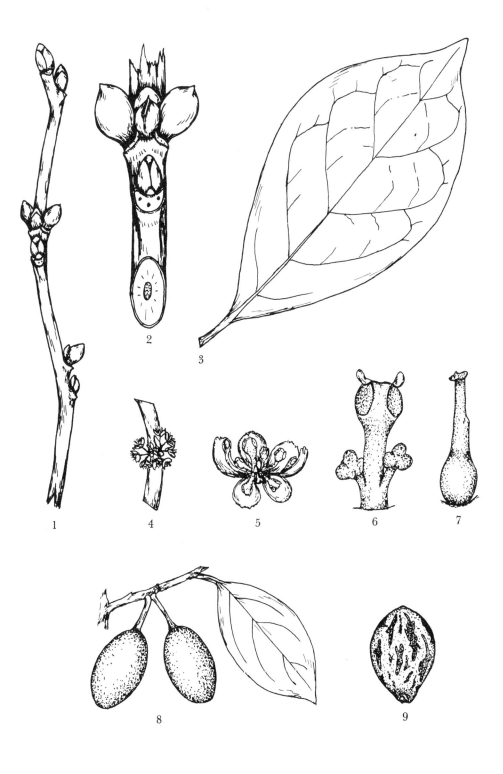

1. Winter twig
2. Detail of twig
3. Leaf
4. Flower cluster
5. One flower

6. Stamen
7. Pistil
8. Fruit
9. Seed

SAXIFRAGACEAE
Hydrangea arborescens L.

Wild hydrangea, seven bark

A shrub to 1.5 m high, sparsely branched; often winterkills to near the ground.

LEAVES: Opposite, simple, 4-6 main lateral veins on each side, forking and joining near the margin. Broadly ovate, 10-15 cm long, 8-13 cm wide; tip abruptly acuminate, the base rounded to subcordate; 2-3 teeth per cm, a veinlet to each tooth; upper surface dark green, sparsely pubescent; lower surface pale green, pubescent, often with tufts in the vein axils and in the tooth sinuses; petiole 2-12 cm long, pubescent, broadened to a semiclasping base; no stipules.

FLOWERS: June-July; perfect or sterile. Inflorescence corymbose, 8-12 cm across, flat or with a round top, branches pubescent; foliaceous bracts beneath the lower branches and reduced to linear bracts or none above; pedicels 1-1.5 mm long; hypanthium hemispheric, 10-ribbed, white, 1.5 mm diameter; calyx lobes 5, white, triangular, 0.3 mm long; petals 5, ovate, white, cucullate, 1.5 mm long, 0.7 mm wide; stamens 10, filaments slender, 2-3 mm long, white; anthers white; ovary embedded in the hypanthium with the dome-shaped top visible, 1-1.5 mm across; 2-3 stout styles 0.7 mm long. Several of the marginal flowers radiate, on pedicels up to 13 mm long, no hypanthium, calyx lobes 3-4, ovate to subrotund, 4-9 mm long, 5-8 mm wide, uneven in size, white, reticulate-veined.

FRUIT: Fruiting head 7-12 cm across, 4-5 cm high, rounded; fruit glabrous, hemispheric, 2.3-2.5 mm across, 1.8 mm high, the 10 ribs joined to a heavy ridge at the top, brown with lighter ribs, opening at the top by a pore in the center between the persistent styles; many-seeded. Seeds flattened ovate to narrowly ovate, often curved, red-brown, striate, 1 mm long, 0.5 mm wide, obtuse or pointed.

TWIGS: 2-2.5 mm diameter, yellow-brown, herbaceous toward the tip, glabrous, the lenticels large and elliptic; leaf scars large, broadly U-shaped, smooth, the ends truncate; 3 bundle scars; pith white, continuous, three-fourths of stem. Buds narrowly ovoid, the scales ovate, acute, brown, glabrous, the outer scales often with a narrow scarious margin, inner scales greenish with brown streaks, ciliate.

TRUNK: Outer layer of bark on branches red-brown, often with small splits; bark of older stems pale tan-brown, peeling into thin sheets; minutely cellular under 15×; inner bark green. Wood soft, greenish-white, open-grained; pith white, one-third to one-half of stem.

HABITAT: Dry or moist hillsides and flood plains, rocky or rich soil, in open or wooded areas.

RANGE: New York, south to Georgia, west to Louisiana, north to eastern Oklahoma and southeastern Kansas, east to Ohio and West Virginia.

Hydrangea arborescens is known in our area from only one general location, along the Spring River in southeastern Kansas. These plants are var. *arborescens*, with the leaf bases rounded or subcordate and with only the marginal flowers radiate. A few miles south of this location, in Oklahoma, the plant grows in abundance. It is an open shrub, with very few branches and large leaves. The seeds are too small to be of much value as food for mammals and birds.

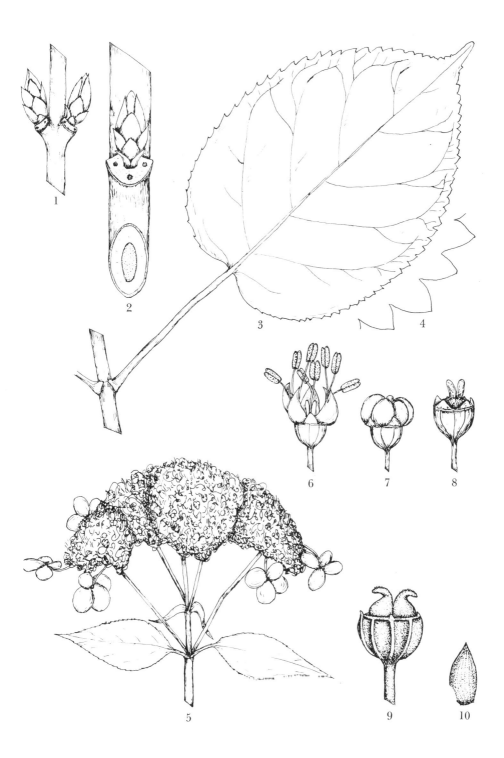

1. Winter twig
2. Detail of twig
3. Leaf
4. Leaf margin
5. Flower cluster
6. One flower
7. Flower bud
8. Flower with petals and stamens removed
9. Fruit
10. Seed

GROSSULARIACEAE
Ribes americanum Mill.

R. *floridum* L'Her.; *Coreosma florida*
Spach

Black currant

A shrub to 1 m high, with few branches.

LEAVES: Alternate, simple, 3-5 lobes, palmately veined. Broadly ovate in outline, 4-7 cm long, 4-9 cm wide, tip of lobes acute; leaf base truncate to subcordate; irregularly serrate, 2-4 teeth per cm, callous-tipped, ciliate, the sinuses acute; upper surface dull yellow-green, glabrous; lower surface paler, sparsely pubescent, with an abundance of sessile, golden-yellow glands; petiole slender, 4-6 cm long, curly pubescent, with a few branched hairs near the broadened base; no stipules.

FLOWERS: Late May; perfect. Racemes drooping, 3-5 cm long, 2-12 flowers, axillary in a cluster of leaves, the axis densely retrorse hairy; pedicels pubescent, 3-7 mm long, jointed at the summit, the basal bracts lanceolate, 5-7 mm long, pubescent and glandular; calyx tube cylindric, 4-5 mm long and about as wide, pale greenish-yellow, glabrate; calyx lobes 5, oblong to oblanceolate, 5-6 mm long, 2 mm wide, pale yellow, spreading or recurved, pubescent on the outside near the tip; petals 5, white, oblong, 2.5-3 mm long, 2 mm wide, inserted on the rim of the calyx tube; stamens 5, opposite the calyx lobes and inserted on the rim, shorter than the petals; filaments yellow-green, broad-based, tapered, the anthers pale yellow; ovary obconic, 2 mm long, green, pubescent along the 5 low ribs; style 7-8 mm long, about even with the end of the petals, divided at the end, the 2 stigmas spreading, capitate, yellow-green.

FRUIT: August-September. Drooping racemes 4-6 cm long, 3-10 fruits; fruits globular, 7-13 mm diameter, deep red-purple to nearly black, smooth, glossy, the floral parts persistent, often a few short hairs around the apex; sweet, edible but not palatable, many seeds. Seeds irregularly ovoid, 2 mm long, 1.4 mm thick, red-purple, the surface unevenly rugose or pitted.

TWIGS: 1.8-2.5 mm diameter, yellow-brown to gray-brown, rigid, puberulent with curly hairs, and dotted with orange, sessile glands; a decurrent ridge from the center and both sides of the leaf scar; leaf scars broadly V-shaped; 3 bundle scars; pith white, continuous, one-third to one-half of stem. Buds with a stalk of 1 mm, ovoid, gray-brown tinged with red, 5-6 mm long, 2-2.5 mm wide, puberulent and with an occasional orange-colored gland; the outer scales apiculate, 3 ribbed, the inner scales densely glandular.

TRUNK: Bark red-brown, smooth, tight, the lenticels small, oval, and light-colored. Wood soft, fine-grained, white, with a brown pith.

HABITAT: Moist, wooded hillsides, margins of bogs, along rock ledges, and occasionally in open, brushy pastures.

RANGE: Alberta, across southern Canada to New Brunswick, south to Delaware, west through Illinois and Nebraska to Colorado, south in the mountains into New Mexico, and north through Wyoming and Montana.

This is the only species of *Ribes* in our area which has the abundance of sessile, golden glands on the lower leaf surface. *R. hudsonianum* Richards also has the glands and has been reported for the region, but no specimens could be located.

The fruits are eaten commonly by birds and small mammals; and since the period of fruit maturation is rather extended, the supply of food lasts for some time. The bushes are too thin to be used much for nesting.

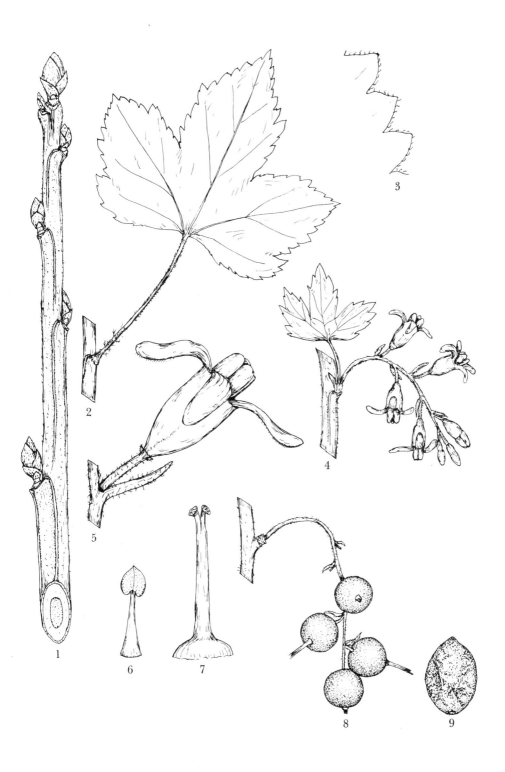

1. Winter twig
2. Leaf
3. Leaf margin
4. Flower raceme
5. Flower
6. Stamen
7. Pistil
8. Fruit cluster
9. Seed

GROSSULARIACEAE
Ribes cereum Dougl. var. *inebrians*
(Lindl.) C. L. Hitchc.

R. *inebrians* Lindl.; *R. reniforme* Nutt.;
R. *cereum* var. *pedicellare* Gray
R. *spathianum* Koehne; *R. pumilum*
Nutt. ex Rydb.; *R. churchii* A. Nels.
& Kenn.

Squaw currant

A shrub to 1 m high, with an abundance of rigid, somewhat crooked branches.

LEAVES: Alternate, simple, often clustered at the end of a short spur, 3-lobed with shallow, acute sinuses, palmately veined. Reniform to broadly rounded, 2-2.5 cm long, 2.5-3 cm wide, lobes rounded, base of leaf subcordate to truncate, occasionally cuneate; margin irregularly lobed and toothed, the teeth rounded, short ciliate and with a few glandular hairs; upper surface darker yellow-green than the lower surface, both surfaces pubescent and glandular hairy; petiole 8-15 mm long, grooved above, pubescent and with glandular hairs; no stipules.

FLOWERS: Late May-June; perfect. The corymbiform racemes, drooping, pubescent and stipitate glandular, 1-6 flowers; peduncle 5-15 mm long, pedicels jointed, 1-3 mm long, the bracts ovate, obovate or flabellate, 3-5 mm long, glandular toothed or incised, pubescent and stipitate glandular; calyx tube cylindric, 5-6.5 mm long, 1 mm wide, pale green, pinkish, purplish, or nearly white, pubescent, glandular, the inside glabrous; calyx lobes 5, triangular, nearly white, about 2 mm long, spreading or reflexed, pubescent and glandular outside and glabrous inside; petals 5, erect, 1-1.5 mm broad, rotund with a short claw, white, inserted just below the rim of the calyx tube; stamens 5, opposite the calyx lobes, inserted lower on the tube than the petals; filaments about 0.25 mm long; anthers pale yellow, 0.75-1 mm long; ovary below the calyx tube ovoid, 1.5-2 mm long, green, pubescent and stipitate glandular; style 6-6.5 mm long, greenish at the base; stigma 2-lobed, often protruding.

FRUIT: July-August. Single or in racemes of 2-5 fruits; fruits globular, 8-12 mm diameter, red, stipitate glandular, the floral parts persistent and the bracts deciduous or persistent; 6-10 seeds. Seeds irregularly ovate, flattened, 2.8 mm long, 1.7 mm wide, 1 mm thick, with a rough, light brown surface.

TWIGS: 1.4-1.6 mm diameter, unarmed, yellow-brown to red-brown, puberulent, the bark splitting by the end of the first winter and showing the reddish inner bark; older twigs purplish, and the bark of the larger branches exfoliating into long, gray, papery strips; many spur branches; leaf scars narrow, extending nearly halfway around the stem; 3 bundle scars; pith greenish-white, continuous, one-fifth to one-fourth of stem. Buds ovoid, 3-5 mm long, 2-2.5 mm wide, the stalk usually less than 0.5 mm; outer scales brown, often 3-ribbed, apiculate, ciliate, glandular ciliate, puberulent, and often with a clear, liquid resin along the margin; inner scales red-brown, soft.

TRUNK: Bark smooth, unarmed, red-brown, often scurfy-pruinose. Wood fairly hard, fine-grained, white, with a small, dark pith.

HABITAT: Dry, rocky hillsides or valleys, prairies or open woods; in South Dakota from the badlands to the tops of the peaks in the Black Hills.

RANGE: British Columbia, south to southern California, east to New Mexico, north to western Nebraska and South Dakota, and west across Montana to Washington.

`This is the most common *Ribes* found in open areas in the northern part of our range. The branches are commonly crooked and gnarled, giving a rugged appearance to the shrub. This and the small rounded leaves are the most outstanding characteristics for field identification.

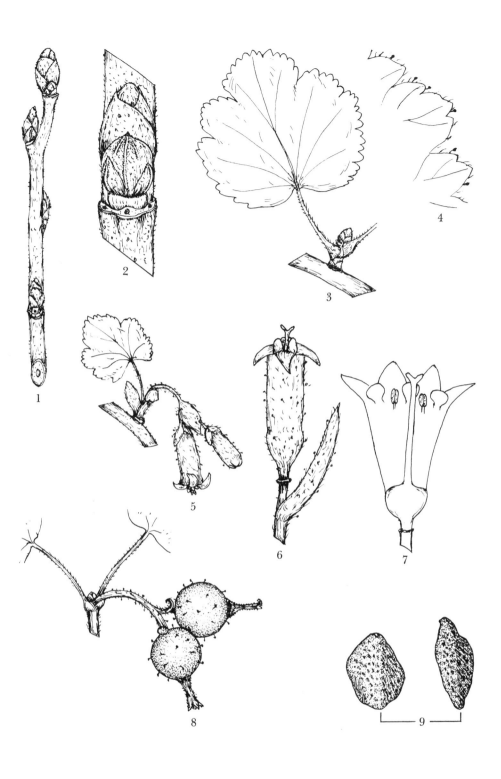

1. Winter twig
2. Detail of twig
3. Leaf
4. Leaf margin
5. Flower cluster
6. Flower
7. Cutaway of flower
8. Fruit
9. Seeds

GROSSULARIACEAE
Ribes cynosbati L.

Grossularia cynosbati (L.) Mill.

Prickly gooseberry, dogberry

A broad, open shrub to 1 m high.

LEAVES: Alternate, simple, palmately 3-5 veined, 3-5 lobes, veins slightly impressed above, prominent below, and often fused at the base. Broadly ovate, 3-6 cm long, and as wide; base truncate to subcordate, occasionally cuneate; both surfaces pubescent and stipitate glandular, the lower more densely so and paler green than above; petiole 2-3 cm long, flat or grooved above, pubescent and glandular, some hairs branched and 2 mm long; no stipules.

FLOWERS: Late May; perfect. Corymbiform racemes, 2-4 flowered, pubescent and glandular pubescent, peduncle 8-14 mm long, pedicels 6-12 mm long, bracts ovate, 1-2 mm long, glandular ciliate; calyx tube cylindric to campanulate, 4 mm long, greenish or pinkish, pubescent outside, glabrous inside, calyx lobes 5, ovate or oblong 2.5-3 mm long, tip rounded, pubescent; petals 5, obovate to rotund, 1 mm across, white to pinkish; stamens 5, inserted opposite the calyx lobes and about 1 mm below the rim, filaments 1.5-2 mm long, white; anthers yellow; ovary below the calyx tube obconic, 2 mm long, green, pubescent, and spiny; style 4.5-5.5 mm long, green, pubescent on the lower half; stigma capitate.

FRUIT: July. Usually 1 or 2 in the raceme, pedicels 10-15 mm long, bracts persistent; fruit globose, 10-14 mm diameter, dark purple, with faint, pale, longitudinal lines, sparingly pubescent, but with many stiff, yellow-brown spines 3-5 mm long, old floral parts persistent; 10-20 seeds per fruit. Seeds irregularly ovoid, with a few facets, dark brown, 2.4-3.2 mm long, 1.5-2.0 mm wide, about 1 mm thick, surface irregular but not rough.

TWIGS: 1.2-1.5 mm diameter, sprouts up to 3 mm; first-year twigs light red-brown, becoming gray and exfoliating by the second year; 2-3 dark red spines 10-14 mm long at each node, few to many retrorse prickles on the internodes, tips of twigs crinkly pubescent with a few glandular, stiff hairs; a slight ridge decurrent from the sides of the leaf scars; leaf scars deep crescent-shaped, red-brown; 3 bundle scars; pith greenish to white, continuous, one-third of stem. Buds narrowly ovoid, acute, 5-6 mm long, 1.2-1.4 mm diameter, pale brown, the scales thin and crinkly ciliate.

TRUNK: Outer bark thin, papery, tan-brown with nodal spines and few to many recurved prickles; inner layer of bark red-brown or purplish, smooth, with numerous light-colored lenticels. Wood hard, fine-grained, white, with a dark pith.

HABITAT: Wooded hillsides and flat areas, usually moist soil.

RANGE: New Brunswick, south through New England to North Carolina, west to Tennessee and southeastern Oklahoma, north through Missouri to eastern South Dakota and North Dakota to Manitoba, east to Quebec.

Ribes cynosbati is our only species of *Ribes* with definitely spiny fruits. *R. lacustre, setosum,* and *cereum* usually have a pubescence of some type but without spines. Also, the ranges of the latter three species do not overlap that of *R. cynosbati* in our area.

The internodes are without spines but may have a few prickles or bristles. Those of the fruiting canes are often without prickles, but the young shoots are densely covered by them. The outer branches are drooping and the fruits are suspended beneath them. The fruiting racemes of our plants are apparently not as long as those of the more eastern plants.

194

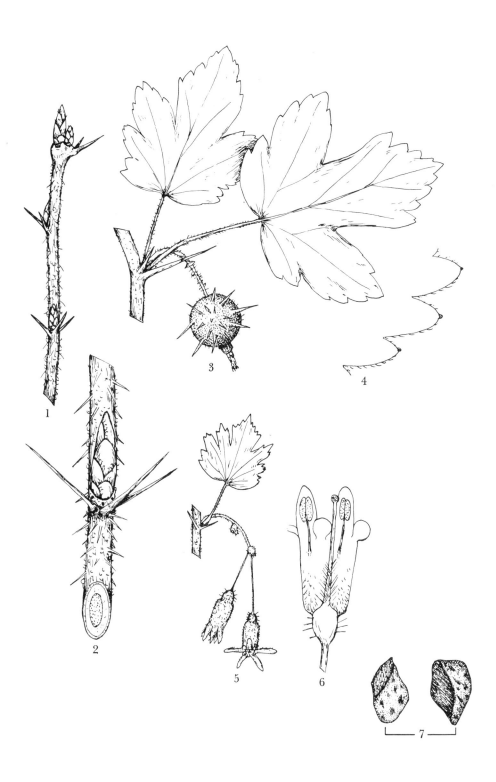

1. Winter twig
2. Detail of twig
3. Leaf and fruit
4. Leaf margin
5. Flowering raceme
6. Diagrammatic cutaway of flower
7. Seeds

GROSSULARIACEAE
Ribes hirtellum Michx.

Gooseberry

A low, often dense, bush to 1 m high, rigid branches.

LEAVES: Alternate, simple, 3-5 deeply cut lobes, palmately veined, impressed at the base above, prominent below, the veins widened in the basal angles. Ovate to rotund in outline, 2-4.5 cm long, 2.5-5 cm wide, base cuneate, truncate or subcordate, lobes often cleft and crenate-serrate; upper surface dark green, glabrous or sparsely pubescent; lower surface paler, glabrate to densely pubescent, especially on the veins; petiole 1.5-3 cm long, slender, flattened or grooved above, base widened and somewhat winged, puberulent, often with long, branched hairs; no stipules.

FLOWERS: Early June; perfect. Drooping corymbiform racemes of 2-4 flowers, peduncles 3-5 mm long, pubescent; pedicels 3-5 mm long, glabrous or slightly pubescent, the bract ovate, sheathing, green, ciliate and often glandular, occasionally two bracts; calyx tube cylindric-campanulate, 2-3 mm long and as wide, glabrous to sparingly pubescent outside and pubescent inside, greenish-white; calyx lobes 5, oblong to obovate, 3.5-4 mm long, longer than the tube, glabrous to slightly pubescent, greenish-white to pinkish, spreading or reflexed; petals 5, obovate, 1.5-2 mm long, erect, thin, white, inserted on the rim of the calyx; stamens 5, opposite the calyx lobes, filaments 3-4 mm long; anthers pale yellow; ovary green, obovoid, glabrous; styles 2, connate and hairy about half of the length, green; stigmas capitate.

FRUIT: July-August. Globular, 8-12 mm diameter, 1-3 on the raceme, smooth, dark red-purple, edible, 4-9 seeds. Seeds ovoid, angular, light red-brown, rugose, 2-2.8 mm long, 1.5-1.8 mm wide, 1.2-1.3 mm thick.

TWIGS: 1.2-2 mm diameter, yellow-gray, puberulent or glabrous, rigid, the bark tight near the tip, exfoliated further back, and red-brown beneath the papery strips; nodal spines 3-5 mm long, few or none, internodal bristles acicular, retrorse, few or none; leaf scars narrow, crescent-shaped, ends pointed, raised and perpendicular to the stem; 3 bundle scars; pith white, continuous, one-half of stem. Buds narrowly ovoid, 5-6 mm long, 2-2.2 mm wide, acute, yellow-brown, axillary at the end of a spur branch, the outer scales broad, apiculate, pubescent, ciliate, the inner scales ciliate.

TRUNK: Outer bark thin, gray, exfoliating into long, papery strips, bark beneath purple-brown, smooth; lenticels prominent, brown. Wood soft, fine-grained, white, with a white pith.

HABITAT: Rocky stream banks, open woods, or occasionally ravine banks in open prairies, dry or moist.

RANGE: According to the manuals, *R. hirtellum* ranges from eastern Saskatchewan to Newfoundland, south to West Virginia, west to Iowa, northwest to the Black Hills and North Dakota. *R. inerme* (Rydb.) Smiley ranges from British Columbia south to California, east to Arizona and New Mexico, north to Wyoming, South Dakota, and North Dakota, and west to Alberta. In brief, *R. hirtellum* is eastern and *R. inerme* is western. The specimens examined in the field and herbaria could be separated only by the cuneate leaf base of *R. hirtellum* and the truncate to subcordate base of *R. inerme,* but even that was not always convincing. Therefore, the above treatment includes the two taxa.

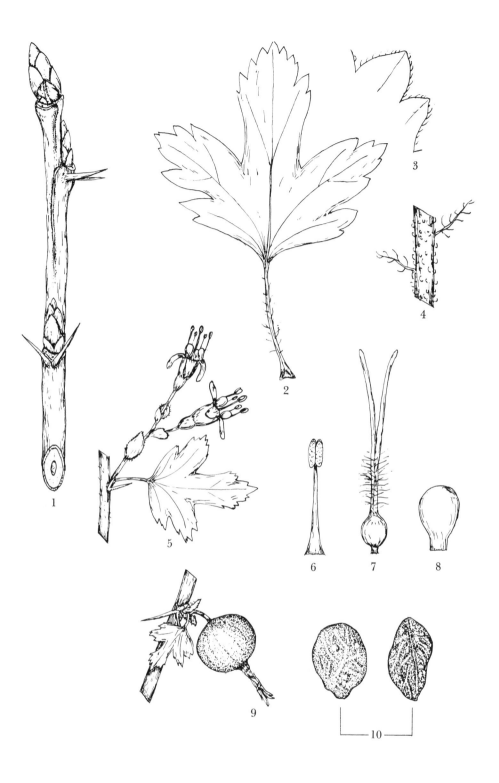

1. Winter twig
2. Leaf
3. Leaf margin
4. Portion of petiole
5. Flowering raceme
6. Stamen
7. Pistil
8. Petal
9. Fruit
10. Seeds

GROSSULARIACEAE
Ribes lacustre (Pers.) Poir. in Lam.

R. parvulum (Gray) Rydb.; *R. oxy-cantthoides* var. *lacustre* Pers.; *R. grossularioides* Michx. in Steud.; *Limnobotrya lacustris* (Pers.) Rydb.

Swamp currant, swamp gooseberry, bristly currant

A thin, spreading, often decumbent shrub to 80 cm high.

LEAVES: Alternate, simple, 3-5 lobed, deeply cleft, the lobes cleft or crenate-dentate, palmately veined. Broadly ovate to circular in outline, 2.5-5 cm long, 2.5-5.5 cm wide, lobes narrowed at the base, the apex acute; upper surface dark green, lower surface lighter, both surfaces with scattered stipitate glands, especially on the veins below, otherwise glabrous; teeth coarse, apiculate, often tipped with a stipitate gland, margin sparingly ciliate and stipitate glandular; petiole slender, 1.5-5 cm long, base widened, slightly winged, pubescent in the shallow groove above, and with stipitate glands up to 2 mm long; no stipules.

FLOWERS: Early June; perfect. Racemes drooping, 3-5 cm long, stipitate glandular, 2-10 flowers, the bracts lanceolate, 2 mm long, peduncles 1-2 cm long, pedicels 3-6 mm long, jointed; calyx tube saucer-shaped, 1 mm long, 1.5-2 mm wide, pinkish or greenish-brown outside, wine-red inside, calyx lobes 5, broadly oblong to obovate, 2 mm long, 1.5 mm wide, erect, pinkish with wine-red lines; petals 5, flabellate, 1.5 mm long, 1 mm wide, pinkish, inserted on the calyx rim; stamens 5, slender, 1-1.5 mm long, the filaments inserted on the pinkish disk and attached at the outer end to the apparently inverted, yellow anthers; ovary obovoid, about 2 mm long and nearly as wide, green, covered with stiff red, glandular hairs; styles 2, united half their length, 1.5 mm long, green, glabrous; stigmas capitate.

FRUIT: August. Drooping racemes of 1-5 fruits. Berry globose, dark purple, 6-9 mm diameter, stipitate glandular, 6-26 seeds. Seeds generally ovoid, irregular, smooth, dull, deep purple-brown, 1.7 mm long, 1.2 mm wide, 0.9 mm thick.

TWIGS: 1-1.5 mm diameter, light brown, fluted, densely covered with soft, acicular prickles 1-2.5 mm long; nodal spines slender, often joined at the base, 3-4.5 mm long, a few short, curled hairs; the bark tight; leaf scars narrowly crescent-shaped, raised with the surface at right angles to the stem; 3 bundle scars; pith greenish, continuous, porous, one-fourth to one-third of stem. Buds narrowly ovoid, yellowish, 3-4 mm long, 0.9-1 mm wide, scales glabrous, thin, tight, finely ciliate, the outer scales keeled and apiculate, the inner scales often pinkish at the base.

TRUNK: Bark red-brown with a purplish cast, young branches smooth, with or without prickles, roughened only with small, brown lenticels; old stems with the outer layer of bark light brown, thin, papery, somewhat loose and scurfy, inner layer of bark purplish-brown. Wood hard, fine-grained, pale brown, with light or dark brown pith.

HABITAT: Moist woods, especially along rock ledges or creek banks, rarely in open areas.

RANGE: Alaska, east across northern Manitoba to Newfoundland and Nova Scotia, south through New England to the mountains of Tennessee, northwest to Ohio, Minnesota, and Manitoba, south through Montana to Colorado, west to Northern California, and north to British Columbia; also in the Black Hills. Apparently is quite spotted within this range.

In our area this species occurs only in the Black Hills and is not very common there. However, it is a low plant, often decumbent, and may have gone unnoticed in many areas of the Hills.

1. Winter twig
2. Leaf
3. Leaf margin
4. Portion of upper side of petiole
5. Leaf and flowering raceme
6. Flower

7. Diagrammatic cutaway of flower
8. Style
9. Petal
10. Stamen
11. Fruit
12. Seeds

GROSSULARIACEAE
Ribes missouriense Nutt.

R. gracile Pursh, not Michx.; *Grossularia missouriensis* (Nutt.) Cov. & Britt.

Wild gooseberry

A prickly shrub to 1 m high, spreading to 2 m wide, with clustered trunks and arching branches.

LEAVES: Alternate, simple. Rotund, 2.5-5 cm long, 3-6 cm wide; 3-5 lobes, often with smaller lobes or rounded teeth, ciliate; tip of lobes rounded, the sinuses narrow; base truncate to subcordate; both sides green; glabrous above and a few hairs on the veins below; petiole 1.5-2.5 mm long, grooved or flattened above, pubescent, and with long, branched hairs near the base.

FLOWERS: April-May; perfect. Solitary or 2-4 in a raceme at a node on twigs of the previous year; peduncle 6-8 mm long, hairy, pedicels 3-6 mm long, glabrous, green; an entire or 3-lobed bract at the base of the flower, margin ciliate and stipitate glandular; calyx tube cylindric, 2-2.5 mm long, white; calyx lobes 5, narrowly oblong, strongly reflexed, 6-7 mm long, white; petals 5, obovate, 2-2.5 mm long, white, tip entire or toothed, tubular around the stamens; stamens 5, filaments 8-9 mm long, flattened, tapered, alternate with the petals; anthers greenish, tinged with red; ovary obovoid, 3 mm long, green, glabrous; styles 2, partly connate, 1 cm long, greenish, hairy toward the base; stigmas green, capitate.

FRUIT: July-August. Globose, 9-12.5 mm diameter, green, eventually red-purple, old style persistent; whitish lines from the base to tip; peduncle 10-12 mm long, pubescent; pedicels 4-7 mm long, glabrous, a bract at the base. Seeds oval, flattened, 2.5-3.5 mm long, 1.4-1.9 mm wide, 1-1.5 mm thick, black, variously angled, granular surface; 8-25 seeds in each fruit.

TWIGS: 1.8 mm diameter, flexible, glabrous, pale to dark brown, often reddish; young shoots light tan, with many dark red-brown, weak prickles and more rigid prickles at the nodes; prickles on fruiting stems dark red-brown, to 1 cm long, nearly straight, semiterete, 1-3 per node below the leaf scar; leaf scars crescent-shaped; 3 bundle scars; pith white, continuous, one-third of stem; pith in old stems brown and alveolate. Buds 4-6 mm long, narrowly ovoid, acuminate, the scales red-brown, long, narrow, tip spreading, keeled.

TRUNK: Bark dark gray or red-brown, exfoliating into thin, papery scales curling laterally. Wood hard, heavy, nearly white.

HABITAT: Dry or moist woods, stream banks, rocky uplands, borders of woods, grazed or cutover pastures.

RANGE: Connecticut, west to North Dakota, south to Kansas, Oklahoma, and Arkansas, and northeast through Tennessee to New York.

The wild gooseberry is a common shrub through most of our area. The fruits are commonly gathered for pies and jellies. The bush furnishes excellent cover for small mammals and birds, some of which eat the fruits. Dead leaves are caught and held by the branches in the autumn, giving good cover for the soil. The thorns are sharp and the bushes dense with many branches.

A variety, *ozarkanum* Fassett, has been named. It has glabrous leaves and simple hairs on the petiole, but has not been reported for our region.

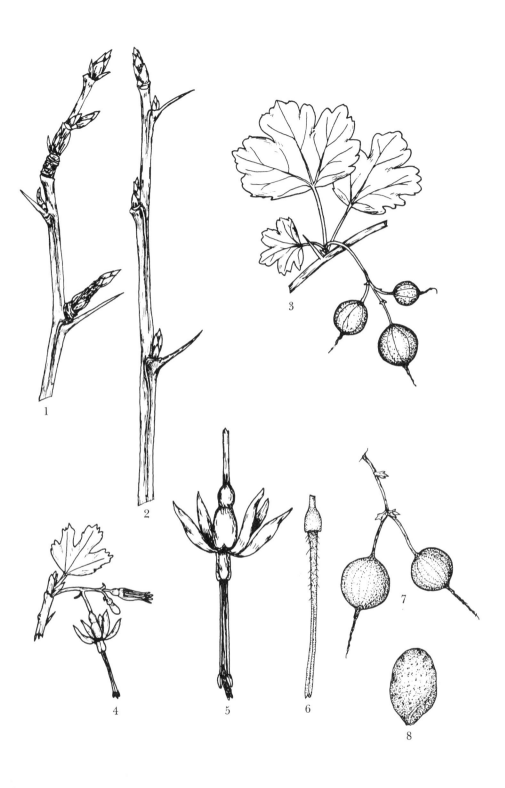

1. Fruiting winter twig
2. Vegetative cane
3. Leaves and fruit
4. Flowers

5. Flower
6. Pistil
7. Fruit
8. Seed

GROSSULARIACEAE
Ribes odoratum Wendl. f.

Ribes aureum of auth., not Pursh;
Chrysobotrya odorata (Wendl.) Rydb.

Golden currant, flowering currant,
Missouri currant

An unarmed, open, colony-forming shrub to 1.5 m high.

LEAVES: Alternate, simple. Broadly ovate, 3-6 cm long, 3.5-7 cm wide; 3-5 palmate veins; 3-5 lobes, tips acute to obtuse, coarsely toothed; base truncate to broadly cuneate; green on both sides, glabrate, pubescent on the veins above and below; margin ciliate; petiole 1.5-5.5 cm long, pubescent.

FLOWERS: Mid-April; perfect. Inflorescence short racemose on old wood, the axis glabrous; pedicels jointed, green, glabrous, 1 mm long, a bract at the base; calyx tube cylindric, 11-12 mm long, yellow; lobes 5, oblong to slightly obovate, 4-5 mm long, spreading or reflexed, golden-yellow; petals 5, ovate, 2.5-3 mm long, pale yellow, often with some red, recurved, apex often toothed; stamens 5, filaments greenish, 2-2.5 mm long, flattened, broad base, alternate with the petals; anthers pink; ovary ovoid, green, 3 mm long; style white, 8-10 mm long, exserted; stigma 2-lobed, capitate, greenish.

FRUIT: June. Pedicels 4-6 mm long, glabrous; fruit globose, 10-12 mm diameter, purple-black, glabrous, veins show as light lines; many seeds. Seeds 2.5-3 mm long, 1.3-2 mm wide, red-brown, irregular shape, pitted.

TWIGS: 1.7-1.9 mm diameter, flexible, reddish-brown, pubescence white, straight or curly on young twigs; fruit spurs gray, rough, densely pubescent; older twigs gray-brown; no spines; leaf scars narrowly crescent-shaped; 3 bundle scars; pith greenish, continuous, one-half of stem. Buds 3-5 mm long, red-brown, narrowly ovate, acute, the outer scales keeled, ciliate.

TRUNK: Bark red-brown to gray, often splits, and the thin sheets curl laterally; lenticels prominent. Wood white, soft.

HABITAT: Borders of open woods, ravines, fence rows, exposed rock ledges, chalk bluffs, or in prairie pastures, often in sandy soil.

RANGE: Minnesota and North Dakota, south on the east side of the mountains in Colorado to Texas, east to Louisiana, north through Arkansas, southern Missouri, and Kansas to South Dakota.

Ribes odoratum is not abundant in any place in our area but is scattered throughout and is more common in the western part. The fruit is full of seeds but makes good jams and jellies and is sometimes used in pies.

In the western part of the prairie states, *R. odoratum* appears to intergrade with *R. aureum* Pursh. *R. aureum* leaves are less deeply lobed, the margin of the lobes are often entire; but if toothed, the teeth are shallow and rounded. The calyx lobes are over half as long as the tube, while in *R. odoratum* the lobes are about one-third the length of the tube. Although *R. aureum* is reported for our area, none of the specimens examined could be definitely placed in that taxon.

Occasionally a colony of *R. odoratum* will have orange-red, or even red, calyx tubes instead of the usual golden-yellow. Nearby may be a colony with the typical yellow color of the calyx.

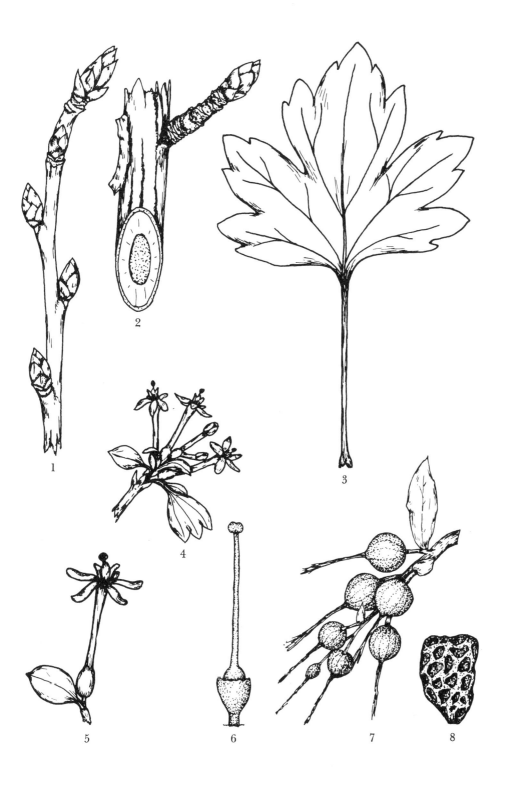

1. Winter twig
2. Detail of twig with spur branch
3. Leaf
4. Flower cluster

5. Flower
6. Pistil
7. Fruit
8. Seed

GROSSULARIACEAE
Ribes oxycanthoides L.

R. saxosum sensu Rydb.; *Grossularia oxycanthoides* (L.) Mill.

Northern gooseberry

A shrub to 1 m high, with several trunks and many branches.

LEAVES: Alternate, simple, 3-5 lobed, usually cleft about halfway to the midrib. Broadly ovate, broader than long, 2.5-4 cm long, 3-5 cm wide, tip of lobes acute, lobes coarsely toothed, ciliate with the hairs bent toward the tip; base of leaf truncate to cordate; sparsely pubescent above, pubescent and glandular below; petiole slender, 1.5-2 cm long, a narrow groove above, long pubescent with occasional branched trichomes and stipitate glands; no stipules.

FLOWERS: Late May; perfect. Flowers in corymbiform clusters of 1-3; bracts broadly obovate, 1-1.5 mm long, encircle the pedicel, the tip spreading; pedicel 2-2.5 mm long, longer than the bract; bracts glandular ciliate, pedicels glabrous; calyx tube campanulate, 2.5-3 mm long, 2.5 mm wide, greenish or pinkish, glabrous outside and pubescent inside; calyx lobes 5, obovate, 4-5 mm long, white, spreading; petals 5, obovate, 2.5 mm long, erect, white; stamens 5, equalling or exceeding the petals, attached to the calyx tube rim and extending downward on the calyx tube; free filaments 2-2.5 mm long, stout, pinkish; anthers greenish; ovary obconic, 1.5 mm long, glabrous; styles 2, 6-6.2 mm long, pinkish, connate and pubescent on the lower half; stigma lobes greenish, capitate.

FRUIT: July-August. Pedicels 6-8 mm long, purplish, glabrous; fruit globular, dark red-purple, smooth, usually single, occasionally 2-3, somewhat juicy, acrid; old calyx tube persistent; 7-26 seeds. Seeds irregularly ellipsoid, dark red-purple, 2.3-3.5 mm long, 1.5-1.7 mm wide, surface irregular and often with flat faces.

TWIGS: 0.9-1.1 mm diameter, gray-brown to purplish, exfoliating into papery flakes, glabrous or with a few curled hairs, internodal prickles weak, 1-2 mm long, nodal spines stiff, 3-4 mm long, and up to 8 mm on lower branches; leaf scars narrowly crescent-shaped, raised; 3 bundle scars; pith greenish, continuous, one-half of stem. Buds narrowly ovoid, 3-5 mm long, 1.5-1.6 mm wide, pointed, light yellow-brown, the scales thin, dry, glabrous, outer scales slightly keeled and apiculate, the margin thin, erose, coarsely ciliate, and turned out, giving a loose effect to the bud, inner scales pinkish at the base. Spur branches 2-2.3 mm diameter, 1-1.3 cm long.

TRUNK: Bark tight on young stems, purple-brown on old stems, large horizontal lenticels; very old stems with wide, thin strips of loose bark. Wood hard, white.

HABITAT: Rocky hillsides, usually in wooded areas, either moist or dry soil.

RANGE: Across southern Canada and extending south into the United States in the Dakotas, Minnesota, and Michigan.

R. oxycanthoides and *R. setosum* Lindl. are often confused; the following chart might aid in the differentiation.

	R. oxycanthoides	*R. setosum*
Nodal spines	1-3, 8-10 mm long	1-5, 10-15 mm long
Internodes	bristly	glabrous or bristly
Leaf	truncate to rounded base, pubescent and glandular	cordate or subcordate base, pubescent, occasionally glandular
Calyx tube	campanulate, 2.5-3 mm long	cylindric, 3-4.5 mm long
Petals	obovate to oblong	obovate to flabellate
Flowers	8-10 mm long	10-12 mm long
Ovary	glabrous	often pubescent

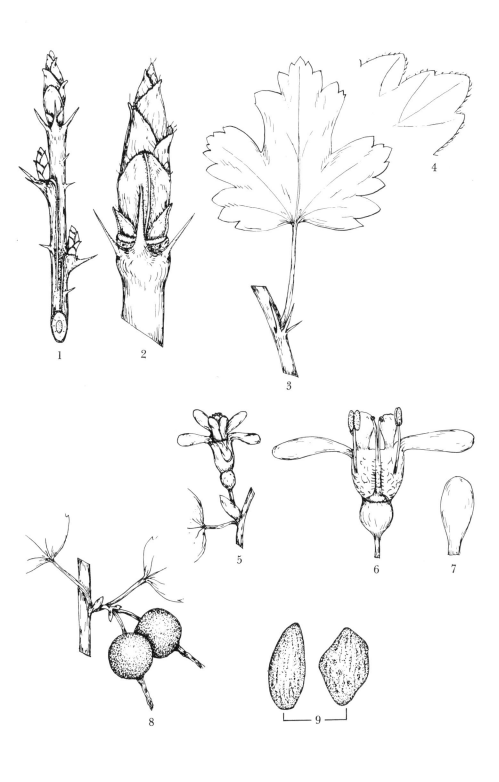

1. Winter twig
2. Detail of bud
3. Leaf
4. Leaf margin
5. Flower

6. Cutaway of flower
7. Petal
8. Fruit
9. Seed

GROSSULARIACEAE
Ribes setosum Lindl.

R. *saximontanum* E. Nels.; *R. camporum*
Blank.; *Grossularia setosa* Cov. & Britt.

Gooseberry

A spreading bush to 1 m high, with arching branches.

LEAVES: Alternate, simple, 3-lobed, deeply cleft, the lobes with rounded, coarse teeth. Broadly ovate, 3-4 cm long, 3.5-4.5 cm wide, lobes and teeth usually with a callous tip, base rounded to cordate; margin short ciliate, upper surface dark green and with scattered hairs, lower surface densely pubescent and occasionally glandular, petiole 1.5-2 cm long, with a shallow groove above, lightly pubescent and often with glandular hairs; no stipules.

FLOWERS: Late May; perfect. Corymbiform clusters of 2-3 flowers, peduncles 2 mm long, with white, spreading pubescence and a few glandular hairs, the bract at the base of the pedicel 1-2 mm long, ovate, toothed and glandular ciliate; pedicels 5-7 mm long, with some spreading pubescence; calyx tube narrow, cylindric, 3-4.5 mm long, greenish-yellow, glabrous or with some pubescence outside, pubescent inside; calyx lobes 5, spreading or reflexed, 3 mm long, 1.5 mm wide, oblong, rounded end, greenish-yellow, glabrous or pubescent; petals 5, broadly oval to flabellate, with a short claw, margin irregular, 1 mm long, 1.5 mm wide, nearly white; stamens 5, inserted opposite a calyx lobe below the rim of the calyx tube; filaments greenish-yellow, 1-1.5 mm long, anthers 1 mm long, yellow-brown, not exceeding the petals; ovary obovoid, green, 2-3 mm long, often with some pubescence; styles 2, green, 5 mm long, connate and pubescent on the lower half, the stigmas capitate, green.

FRUIT: Late July. Fruit globular or slightly depressed, 8-10 mm diameter, deep purple to nearly black, glabrous, old calyx persistent, pulp thick and jelly-like, sweet, edible; 5-12 seeds. Seeds variously shaped, generally oval, with angles and flattened surfaces, dark red-purple,

2.5-3 mm long, 1.3-1.9 mm wide, 1 mm thick, rough surface.

TWIGS: 1.5-2 mm diameter, yellow-brown, densely curly pubescent and usually with a few glands, sprouts with long, straight hairs in addition to the short hairs, nodal spines 1-5, strong, 10-15 mm long, internodes with weak prickles; 2-year-old stems somewhat scurfy and with short spur branches; leaf scars at right angles to the stem, narrowly crescent-shaped, extending halfway around the stem; 3 dark bundle scars; pith greenish, continuous, one-half of stem. Buds narrowly ovoid, yellow-brown, scales densely ciliate at the end, the outer scales broad, keeled, ciliate and somewhat pubescent, inner scales erose ciliate.

TRUNK: Bark purple-brown, slightly split or peeled into thin strips, lenticels large and rough. Wood hard, white with a large, brown, central pith, the rays evident.

HABITAT: Wooded hillsides, rocky soils, either dry or moist, but appears more often in moist conditions.

RANGE: Idaho, southern Saskatchewan, Montana, Wyoming, western Nebraska, and the Dakotas, east in Canada and extending into Michigan.

The range of *R. setosum* is generally more western than that of *R. oxycanthoides;* and in the United States, the two overlap in Michigan and the Dakotas. The two species are often inseparable without the flowers, and even then it may be difficult.

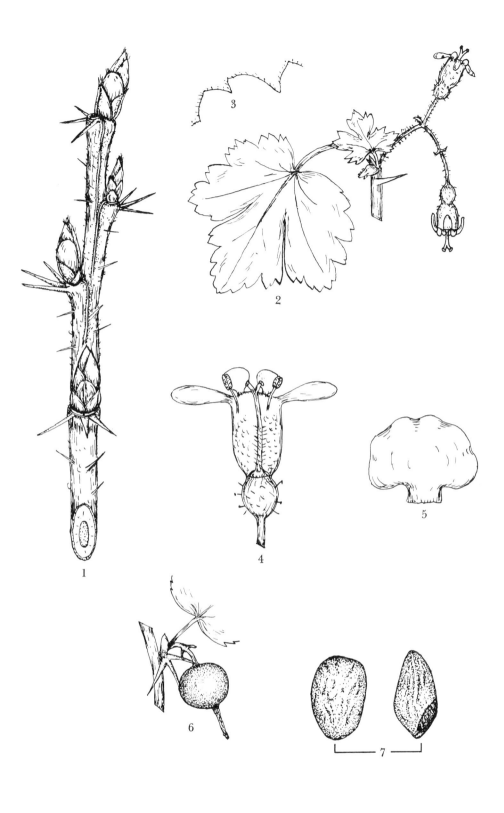

1. Winter twig
2. Leaf and flowers
3. Diagrammatic cutaway of flower

4. Petal
5. Fruit
6. Seeds

GROSSULARIACEAE
Ribes triste Pall.

R. *albinervum* Michx.; R. *ciliosum*
Howell; R. *migratorium* Suksd.;
R. *rubrum* Gray, not L.

Red currant

A low, decumbent bush, with stems creeping above or just below the ground and rooting at the nodes; the erect stem to 60 cm high.

LEAVES: Alternate, simple, 3-lobed, often with an additional, small lobe on the margin of the laterals, palmately veined. Broadly ovate, 5-8 cm long, 7-10 cm wide; lobes shallow, triangular, the margin crenate, teeth callous-tipped, ciliate, 3-4 teeth per cm; base rounded to subcordate; upper surface glabrous or nearly so, lower surface softly pubescent, velvety; petiole 3-5 cm long, grooved above, with a few long hairs on the groove margin, often glandular, sparsely pubescent below, widened toward the base and somewhat winged; no stipules.

FLOWERS: May, when leaves one-third grown; perfect. Racemes drooping, 3-8 cm long, 7-12 flowers, accrescent inner bud scales persistent at the base, axis and pedicels pubescent and glandular, the bract at the base of the pedicel broadly ovate, 1-1.5 mm long, encircles the pedicel, ciliate; pedicels jointed, spreading or reflexed, 3-4 mm long; calyx tube flat, saucer-shaped, 2-3 mm across, reddish-purple; calyx lobes 5, obovate, 2.5-3 mm long and as wide, spreading with reflexed tip, greenish and streaked with red; petals 5, flabellate, 0.5-0.75 mm long and as wide, red-purple, spreading; stamens 5, inserted on the calyx tube, filaments erect, 0.75 mm long, red; anthers yellow; ovary globular, 1-1.5 mm diameter, green, glabrous; the 2 styles 0.75-1 mm long, connate at the base, red with a greenish tip, stigmas capitate, green.

FRUIT: Late August. Drooping racemes 3-5 cm long, pubescent and often glandular, 1-7 fruits; peduncles 10-12 mm long, pedicels 3-5 mm long, bracts persistent; berry globose, bright red, smooth, glossy, 5-8 mm diameter, 1-7 seeds per fruit. Seeds irregularly oval, flattened, 2.3-2.6 mm long, 1.8 mm wide, 1.5 mm thick, yellow-brown, dull, granular surface, often with a slight groove on one side.

TWIGS: 2-2.5 mm diameter, rigid, light red-brown to yellow-brown, sparingly pubescent or glabrous, a few dark glands on the end internodes; bark tight, lenticels not noticeable, a narrow ridge from each end of the leaf scar extends through the internode below; leaf scars crescent-shaped, smooth, tan, extend about halfway around the stem; 3 bundle scars; pith white, continuous, one-half of stem. Buds narrowly ovoid, acute, short appressed pubescent, with a few crinkly hairs toward the end, 3-6 mm long, 2-2.2 mm wide, short stalk; end bud usually narrower, lower scales often spreading at the tip, keeled.

TRUNK: Bark purple-brown, peeling horizontally into thin, curled sheets, lenticels small, light-colored. Wood soft, fine-grained, white with tan pith. Pith often infected with larvae.

HABITAT: Margins of bogs mainly in wooded areas; wet, springy, shaded hillsides and lakeshores.

RANGE: Alaska, across southern Keewatin to Newfoundland, south to West Virginia, northwest to Michigan, Minnesota, and North Dakota; Washington, Oregon, and British Columbia.

This little shrub, growing in wet, brushy areas, is easily overlooked and may be more common than collection records indicate. The flowers are inconspicuous and seldom produce an abundance of fruit.

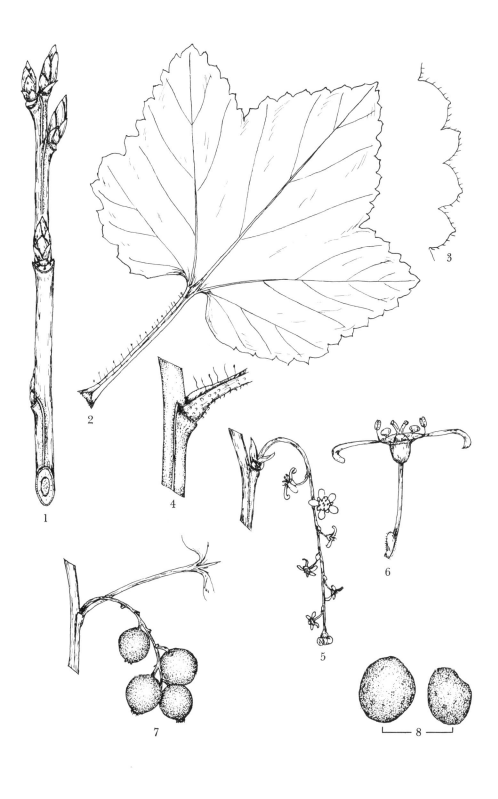

1. Winter twig
2. Leaf
3. Leaf margin
4. Base of petiole
5. Flowering raceme
6. Diagrammatic cutaway of flower
7. Fruit
8. Seeds

PLATANACEAE
Platanus occidentalis L.

Sycamore, plane tree, buttonwood

A tree to 30 m high, with spreading, rounded crown.

LEAVES: Alternate, simple. Broadly ovate to reniform, 8-20 cm long and 8-20 cm wide, with 3-5 coarsely toothed lobes and shallow sinuses; tip long acuminate, the base truncate, subcordate, or attenuate; upper surface bright yellow-green, with a few branched hairs; lower surface pale green, glabrate, with pubescent midrib and veins; stipules attached to the branch above the leaf, the margins fused to form a tube around the stem 5-7 mm long, then becoming foliaceous with acuminate lobes and teeth. Young leaves densely pubescent on both sides.

FLOWERS: April-May; monoecious. Staminate flowers on wood of the previous year, axillary in dense, globose, greenish-yellow heads 8-10 mm diameter; peduncle 6-10 mm long, pubescent; stamens 2.5 mm long, the anther locules 2 mm long and laterally adnate to the central connective which has a disk-like, pubescent apex; filiform staminodes or undeveloped stamens often numerous. Pistillate flowers on old wood, axillary in dense, globose, reddish heads 10-12 mm diameter; peduncle tomentose, 4-5 cm long; ovary green, glabrous, 0.8 mm long, on a pubescent pedicel 0.5 mm long; style 2.5-3 mm long, curved at the end, the red stigmatic surface nearly its full length on one side. Many small, green bracts or staminodes between the flowers.

FRUIT: October. Fruits persistent over winter; peduncles pendent, slender, glabrous, 8-15 cm long, becoming shredded into fibers by spring. Fruit heads brown, globose, 2.5-3.5 cm diameter. Achenes narrowly obconic, 1 cm long, 1.8-2 mm wide at the top, pale brown, angular, the tip truncate or obtuse, the base pointed and with a heavy tuft of buff hairs.

TWIGS: 2-2.5 mm diameter, rigid, brown, tomentose with branched hairs at first, soon glabrous and gray-brown, en-larged at the nodes; leaf scars narrow, surrounding the bud except for a space about 1 mm at the top; 5-10 bundle scars; pith pale brown, continuous, one-fifth of stem. Buds completely surrounded by the petiole base, ovoid, obtuse to acute, 3.5-5.5 mm long, single scale, red-brown, glossy.

TRUNK: Bark of young trees and upper branches of old trees thin, nearly white, the outer brown bark peeling off in short, thin flakes or sheets; bark of old trunks shallowly fissured, forming short, flat, light red-brown or yellow-brown flakes. Wood heavy, tough, hard to split, pale red-brown, with a wide, light sapwood.

HABITAT: Rich bottom lands, flood plains, creek banks. Does well when planted in upland soil.

RANGE: Maine to Ontario and eastern Minnesota, south through eastern Nebraska to Texas, east to Florida, and north to New England.

The sycamore is a common tree along the streams of eastern Kansas, and the white upper branches cause it to stand out sharply among the other trees. It is not uncommon to see a thick stand of young sycamores on a rocky bar in a small stream. The wood is used for butcher blocks, woodenware, boxes, brush handles, furniture, and interior finish. It has a flaky appearance, giving it the trade name of "lacewood."

Forma *attenuata* Sarg. has the blade attenuate on the petiole, but is a questionable form, since it usually occurs only on sprouts or on the lower branches of a tree with otherwise typical leaves. Forma *glabrata* (Fern.) Sarg. is not here recognized as separable.

210

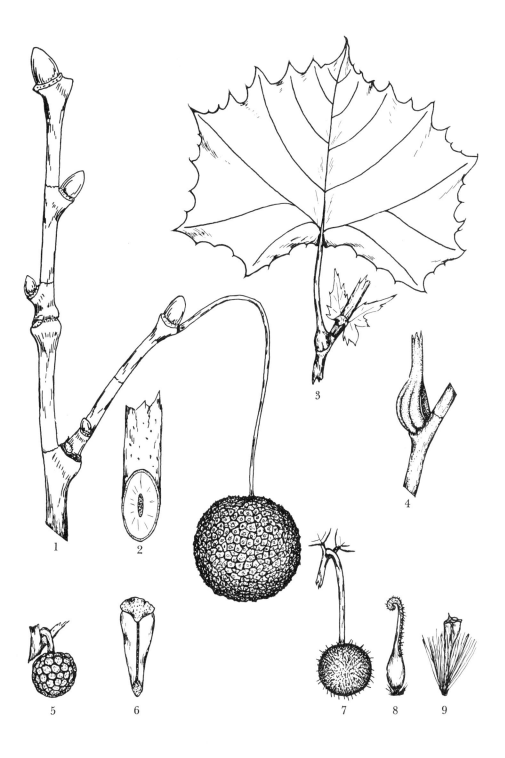

1. Winter twig with fruit
2. Section of twig
3. Leaf
4. Petiole base
5. Staminate flower head

6. Stamen
7. Pistillate flower head
8. Pistil
9. Seed

ROSACEAE
Spiraea alba Du Roi

Meadow-sweet

A shrub to 1 m high, finely branched; often forming small colonies.

LEAVES: Alternate, simple. Elliptic, oblong or oblanceolate, 3-5 cm long, 10-15 mm wide, tip acute, base long tapered, occasionally attenuate on the petiole; sharply serrate except on the lower portion, 4-6 teeth per cm, finely and crinkly ciliate; green on both surfaces and glabrous or puberulent; petiole 3-5 mm long, winged, trough-shaped, somewhat keeled below; no stipules.

FLOWERS: July; perfect. Thyrsoid panicles 7-10 cm long, 3-6 cm wide, open or compact, the lower branches up to 6 cm long, pubescent; the branches subtended by a leaf or leaf-like bract, these decrease in size toward the top of the inflorescence; pedicels puberulent, 3-5 mm long, with a small, linear bract about the middle. Calyx tube turbinate, 1.5 mm long, 1.5-2 mm wide, green, puberulent; calyx lobes 5, ovate, 1.5 mm long, 1 mm wide, obtuse, green, puberulent and ciliate; the disk a glandular ring inside of and just below the rim of the calyx, the disk lobed and bright orange or orange-red; petals 5, ovate, cupped, 3-3.5 mm long, 2.5 mm wide, white, the margin undulate; stamens 30-50, filaments 2-4 mm long, the longest ones on the outside, white; anthers pale yellow; stamens and petals attached outside of the glandular disk; pistils 5, ovary green, about 1 mm long, the shape of an orange section; styles terminal, 1-1.5 mm long, white; stigmas capitate.

FRUIT: September. Follicles 5, pod-shaped, about 2 mm long, partly surrounded by the persistent calyx; follicles red-brown, smooth, glabrous, with a spreading caudate tip; dehiscent on the ventral side for the full length and often about one-fourth of the distance on the dorsal side; 3-10 seeds in each follicle. Seeds narrowly fusiform, slightly rounded at the basal end, 2-2.3 mm long, 0.5 mm wide, golden-brown, minutely papillate.

TWIGS: 0.9 mm diameter, red-brown to gray-brown, somewhat striate and angular, glabrous, the lenticels conspicuous; leaf scars narrow, crescent-shaped; 1 bundle scar; a ridge extending downward from each side and the center of the leaf scar; pith light brown, continuous, one-half of stem; internodes of irregular lengths. Buds ovoid, rounded, 1.5-2 mm long, 1 mm wide, the scales small, red-brown, glabrous and curly ciliate.

TRUNK: Bark thin, red-brown, pruinose, smooth, tight. Wood soft, fine-grained, pale yellow, often slightly greenish, the pith dark brown.

HABITAT: Usually in wet, sandy soil, common in and around shallow potholes and in the low sandhills of the Sheyenne National Grasslands in North Dakota.

RANGE: Vermont, south to North Carolina, north to Ohio, west to northern Missouri, north to South Dakota, North Dakota, and southeastern Saskatchewan, and east to Quebec.

Spiraea alba enters our area from the east and north. Although it extends completely across the north edge of North Dakota, the author knows of only one collection from west of the Missouri River. In our area, at least, it is definitely a plant of sandy soil. It appears as the tallest plant in low spots in the flat prairie of northern North Dakota, or as an undershrub with *Populus tremuloides* in the more hilly areas. In the low, dry sandhills of Ransom County, North Dakota, it grows luxuriously along with *Salix humilis* and *Prunus besseyi*, both of which are dry soil plants. The shrubs are often as wide as they are tall and become quite dense.

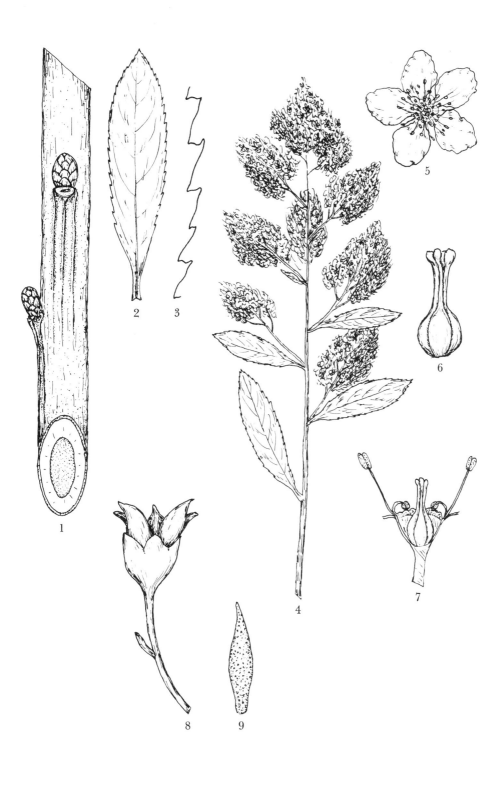

1. Winter twig
2. Leaf
3. Leaf margin
4. Inflorescence
5. Flower
6. Pistils
7. Diagrammatic cutaway of flower
8. Fruit
9. Seed

ROSACEAE
Spiraea betulifolia Pall. var. *lucida*
(Dougl.) C. L. Hitchc.

S. lucida Dougl.; *S. corymbosa* Raf. var.
lucida Zobel

Spiraea

An open shrub to 50 cm high, rhizomatous but not in large colonies.

LEAVES: Alternate, simple. Ovate, ovate-oblong or slightly obovate, 3-5 cm long, 2.5-3.5 cm wide, widest usually near the middle; tip acute to narrowly rounded, base cuneate; coarsely toothed and shallowly lobed, 2-4 teeth per cm, mostly above the middle; upper surface yellow-green, lower surface paler, both sides glabrous; petiole 5-8 mm long, flat or trough-shaped, glabrous; no stipules.

FLOWERS: July-August; perfect. Corymbs glabrous, 5-10 cm across, the branches green, the lower peduncles with foliaceous bracts reduced to linear bracts above, a few hairs at the tip of the bracts. Flowers regular, about 5 mm across; calyx turbinate, glabrous, green; calyx lobes 5, broadly triangular, 0.5 mm long, green, acute, pubescent only at the tip, reflexed; petals 5, obovate to circular, 2 mm across, spreading, often notched at the tip; disk a red-brown, lobed, glandular ring around the inside of the calyx and just below the rim; stamens 15-20, attached outside of the disk; filaments white, 2-4 mm long; anthers yellow; pistils 5, distinct; ovary green, ovoid, 0.5 mm long, glabrous; style terminal, slender, green; stigma capitate.

FRUIT: August-September. Calyx persistent, lobes erect or reflexed; follicles about 2 mm long, smooth, glabrous, red-brown, dehiscent the full length on the ventral suture and about halfway on the dorsal side, the valves shortly apiculate, erect or spreading; 6-10 seeds per follicle. Seeds fusiform, usually curved, 1.8 mm long, 0.5 mm wide, dark red-brown, minutely striate, slightly rounded at the base.

TWIGS: 1-1.5 mm diameter, smooth, glabrous, red-brown, the lenticels small and inconspicuous; leaf scars triangular, dark-colored; one bundle scar; pith white, continuous, over one-half of stem. Buds ovoid, acute, 2.5-3 mm long, 1.5 mm diameter, red-brown, glabrous, often with a few ciliate hairs at the tip of the scales.

TRUNK: Bark smooth, red-brown with a purplish cast, lightly shredded at the base of the stem. Wood soft, greenish-white, with pith about one-half of the stem; often dies back to the ground in winter.

HABITAT: Moist or dry hillsides, usually in open woods or bushy areas; ravine banks or rocky slopes.

RANGE: British Columbia to western Saskatchewan, south to the Black Hills of South Dakota, west to north central Oregon, and north into Washington.

S. betulifolia is somewhat variable but can be recognized by the flat-topped corymbs of white flowers, the only Spiraea species of our area with that characteristic. Our plants are var. *lucida* and they differ from the European var. *betulifolia* by having more coarsely toothed, sublobate leaves, and differ from the eastern var. *corymbosa* in being glabrous, or nearly so, throughout. Var. *corymbosa* usually does not have the sublobate leaves. On one specimen from Oregon, the leaves were entire except for a few teeth at the tip.

The plant is rhizomatous but does not form large or dense colonies. In fact, most of the plants of our area are single, and although it is common, cannot be called abundant in any location. In the Black Hills the seeds may freeze before maturing, and good seeds are often hard to find.

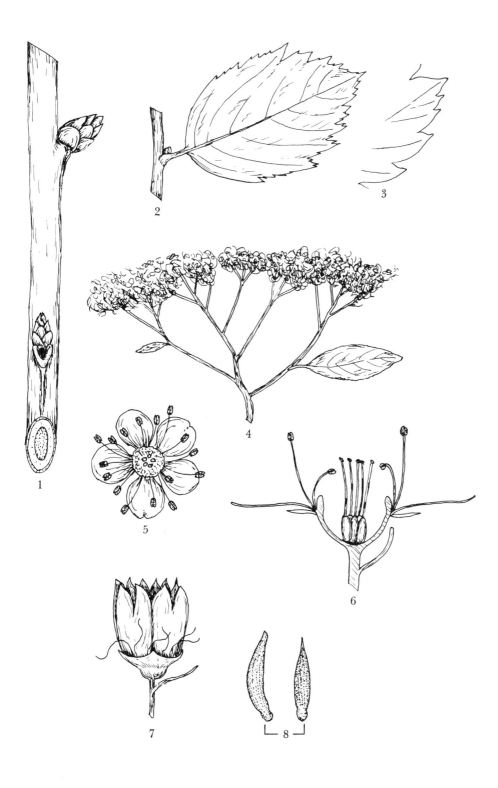

1. Winter twig
2. Leaf
3. Leaf margin
4. Inflorescence
5. Flower
6. Diagrammatic cutaway of flower
7. Fruit
8. Seeds

ROSACEAE
Physocarpus monogynus (Torr.)
Coult.

Opulaster monogynus (Torr.) Kuntze;
P. torreyi Maxim.

Small ninebark

A shrub to 1 m high with many fine branches, most often branching near the ground.

LEAVES: Alternate, simple, palmately veined with 3 veins. Ovate to nearly rotund, 3-5 lobes, sinuses between the lobes acute; the leaves 2-3 cm long, usually under 2.5 cm and about the same width; tip acute to rounded, the base rounded to subcordate; margin of lobes biserrate, often incised, teeth irregular, 6-8 per cm; upper surface dark green, glabrous or nearly so, lower surface paler and sparsely stellate pubescent; petiole 8-10 mm long, sharply grooved above, sparsely pubescent; stipules lanceolate, 2-3 mm long, ciliate.

FLOWERS: June-July; perfect. Corymb 3-4 cm across with rounded top, 15-25 flowers; pedicels slender, 1-1.5 cm long, stellate pubescent, a lanceolate bract 2-3 mm long at the base of each pedicel, it being deciduous above its base, leaving a truncate stub; flowers 10-13 mm across; hypanthium cup-shaped, 2-2.5 mm high, 3.5-4 mm across, yellow-green, stellate pubescent, golden-yellow inside; calyx lobes 5, triangular, 2-2.5 mm long, greenish with nearly white margin and brownish tip, stellate pubescent on both sides; petals 5, rotund, 3-4 mm across, white, slightly cupped, spreading; stamens 30-40, filaments white, slender, 3-4 mm long, spreading or incurved; anthers pinkish to yellow; ovaries 2-4, ovate, 1.5 mm long, connate about half their length, densely stellate pubescent; style 3 mm long, stigma capitate, greenish.

FRUIT: August-September. Calyx lobes persistent; capsules 2-3, occasionally 4, ovoid, 6-8 mm long, 3.5-4 mm wide, inflated, constricted to a stalk-like base 0.5 mm long; capsule apiculate, densely stellate pubescent, greenish-brown, thick margined at the sutures, splitting on both margins, and containing 1-3 seeds. Seeds obovoid, 2-2.4 mm long, 1-1.3 mm diameter, one side nearly straight and with a slight flange, yellow to pale brown, smooth, glossy.

TWIGS: 0.8 mm diameter, red-brown, glabrous to sparingly stellate pubescent; bark starts to shred by the first winter; twigs often winterkill; leaf scars bean-shaped, a ridge on each side decurrent through the internode, 3-5 bundle scars; pith pale brown, continuous, one-third of stem. Buds ovoid, 2.5-3 mm long, 1 mm diameter, obtuse or acute, red-brown, the outer scales loose with a spreading tip, ciliate and often pubescent.

TRUNK: Bark peels into several thin, wide, papery strips, the outer bark striped dark and light gray-brown, the inner bark pale brown and smooth. Wood medium hard, fine-grained, fibrous, pale tan with a large brown pith.

HABITAT: Hillsides in dry, open woods, open slopes, and along rock ledges.

RANGE: South Dakota, southwest to eastern Arizona, east to west Texas, and north to Colorado.

This is a variable species and is quite similar to *P. malvaceus* (Greene) Kuntze of the northwestern United States and to *P. opulifolius* (L.) Maxim. var. *intermedius* (Rydb.) Robins. further south and east. It is a smaller bush with smaller leaves than either of the above and is more commonly found in dry habitats. It usually has 3 carpels, but a flower with 2 or 4 is not uncommon.

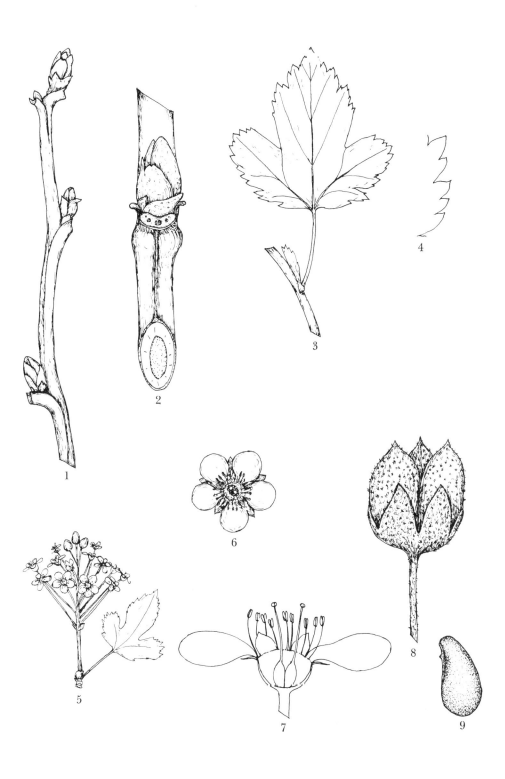

1. Winter twig
2. Detail of twig
3. Leaf
4. Leaf margin
5. Inflorescence
6. Face view of flower
7. Diagram of flower
8. Fruit
9. Seed

ROSACEAE
Physocarpus opulifolius (L.)
Maxim. var. *intermedius* (Rydb.)
Robins.

Physocarpa intermedia (Rydb.) C. K.
Schneider; *Opulaster intermedius* Rydb.

Ninebark

A shrub to 3 m high, usually open and branching above the middle, trunks single or clustered.

LEAVES: Alternate, simple. Ovate, 5-12 cm long, 4.5-9 cm wide; margin with 3-5 small lobes on each side, irregularly crenate-serrate; leaf tip acute, base truncate, cuneate or somewhat rounded; upper surface dark green, glabrous; lower surface paler with some stellate pubescence on the veins and in the axils; petiole 1.5-2.5 cm long, glabrous, grooved above; stipules foliaceous, 7 mm long, ovate, acuminate, toothed, pubescent.

FLOWERS: Early May; perfect. Flowers in terminal corymbs 2-6 cm across on new growth, 10-30 flowers; pedicels 10-15 mm long, glabrous; calyx a shallow cup, 5-lobed, lobes triangular, 4 mm long, pubescent on both sides; petals 5, white, orbicular, 4 mm long, 3.5 mm wide, spreading; stamens 25-35, attached to the rim of the calyx; filaments white, 2-4 mm long; anthers yellow-brown; pistils 4-5, ovaries 3-3.5 mm long, connate for half their length, woolly; style white, filiform, glabrous, 4.5-5.5 mm long; stigma capitate; glandular disk pale yellow. In bud, the anthers are pink and the disk bright orange-yellow.

FRUIT: June-July. Pedicels 2-3 cm long; each fruit with 3-5 follicles, usually 4, 0.8-1.2 cm long, brown, densely stellate pubescent, dehiscent on both sutures. Seeds obovoid, 2-2.4 mm long, 1.3-1.5 mm wide, yellow, glossy, smooth, raphe ridge extending two-thirds the length of the seed, apex rounded, base slightly hooked at the hilum.

TWIGS: 1-1.4 mm diameter, flexible, glabrous, yellow-brown, a narrow decurrent ridge from each stipule scar and often extending along the stem for 2 internodes; leaf scars short, shield-shaped with truncate ends; 1 large and 2 small bundle scars; pith pale salmon to brown, continuous, one-half of stem. Buds ovoid, 2.6-3.8 mm long, obtuse, appressed to the stem, brown, slightly pubescent.

TRUNK: Bark brown, exfoliating into long thin shreds; on larger trunks, light red-brown and peeling into long, papery sheets, exposing the reddish or pale brown inner bark and giving a shaggy appearance. Wood soft, pale brown.

HABITAT: Rocky or sandy stream banks. Moist soil in wooded areas.

RANGE: New York to Minnesota, South Dakota, Nebraska, Colorado, Kansas, Oklahoma, Arkansas, Illinois, and Indiana.

Ninebark is scattered in our area and is not common in any one place. The bush grows to 3 meters high, usually on a stream bank next to the water, with some of the branches arching over the stream. It is a taller, more open shrub and with larger leaves than *P. monogynus* of the Black Hills, the two ranges overlapping in that area. However, the more northern plants of *P. opulifolius* in the plains states have smaller leaves than those of the south, thus tending toward *P. monogynus*.

Our plants are var. *intermedius,* the more western form, and do not extend as far north as var. *opulifolius*. The most obvious difference between the two varieties is that var. *intermedius* has a permanent stellate pubescence on the follicles and var. *opulifolius* does not.

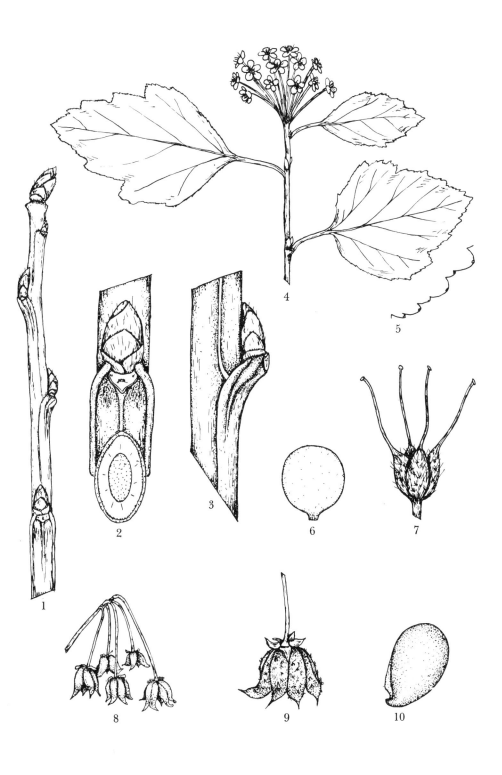

1. Winter twig
2. Detail of twig
3. Detail of twig
4. Leaf and flowering corymb
5. Leaf margin
6. Petal
7. Pistils
8. Fruit cluster
9. One fruit, mature
10. Seed

ROSACEAE
Petrophytum caespitosum (Nutt.)
Rydb.

Spiraea caespitosa Nutt. in T. & G.;
Eriogynia caespitosa Wats.; *Luetkea
caespitosa* Kuntze

Rock spiraea

A creeping, caespitose plant forming mats up to 50 cm across, the stems and leaves no more than 3 cm above the ground.

LEAVES: Alternate-rosulate, evergreen, simple, 1-nerved. Spatulate, 5-12 mm long, 1-2.5 mm wide near the tip; tip rounded, may appear apiculate from the converging appressed hairs, the base tapered, but with the sides nearly parallel on the lower half of the leaf; margin entire; densely sericeous on both sides, light green beneath the hairs; sessile, with a widened base; no stipules.

FLOWERS: Late August; perfect. Peduncles 3-8 cm long, usually slightly reddish, an angular ridge decurrent from each of the reduced, appressed, lanceolate, bract-like leaves; pubescence slightly spreading; raceme 1.5-3 cm long, 10-15 mm thick, flowers crowded; pedicels pubescent, 1-3 mm long with a lanceolate bract below the middle; calyx tube turbinate, 1 mm across, green, pubescent outside and long, hairy at the rim inside; calyx lobes 5, ovate, 1 mm long, acute, pubescent on both sides, green but often red tinged, the tip ascending and somewhat spreading; a glandular ring near the rim of the tube inside; petals 5, narrowly obovate, 2 mm long, 0.5 mm wide, white, erect or slightly incurved, the rounded end often notched; stamens 15-25, erect or incurved, filaments 2-3.5 mm long, white; anthers pale yellow; pistils 5, tightly clustered but distinct, ovary obovoid, 0.75 mm long, pale green, long pubescent; styles terminal, 2 mm long, white, filiform, pilose on the lower half; stigma filiform.

FRUIT: Late September, often freezing before maturing. Calyx persistent; usually 5 follicles, 1.5-2 mm long, yellow-brown, smooth with a few long hairs near the apex, splitting on both sutures, 5-10 seeds per follicle. Seeds fusiform, 1.3-1.6 mm long, 0.4-0.5 mm wide, light yellow-brown with darker ends, rounded at the base, minutely and irregularly striate.

TWIGS: Coarse, the leaves remain greenish over winter, then turn brown but remain attached and persist for several years; growth rate very slow; pith continuous, greenish, most of the stem. Buds naked, terminal, consist of a small group of embryonic leaves in the center of a rosette, no other buds were found on any part of the stem.

TRUNK: Stems usually visible only if the plant is turned over; 1 cm diameter, bark comparatively thick, smooth, dark brown. Wood white, porous, soft, with a pale brown, rubbery pith.

HABITAT: Rock crevices, semiexposed rock surfaces on wooded hillsides and ravine banks under moist condition.

RANGE: Oregon to northern California, east to Montana and the Black Hills, south to Colorado, New Mexico, Texas, and Arizona.

Petrophytum caespitosum is well named since the name means a matted rock plant. It is rigid and grows flat against a rock surface, its strong roots wedged down in a crack or crevice. Although it is found in exposed spots and apparently will not tolerate close competition, the situation must have a fair amount of moisture.

It is closely related to Spiraea, and the flowers and fruits are quite similar. Seeds for this study were difficult to find, for in two consecutive years an early freeze in the Black Hills destroyed them.

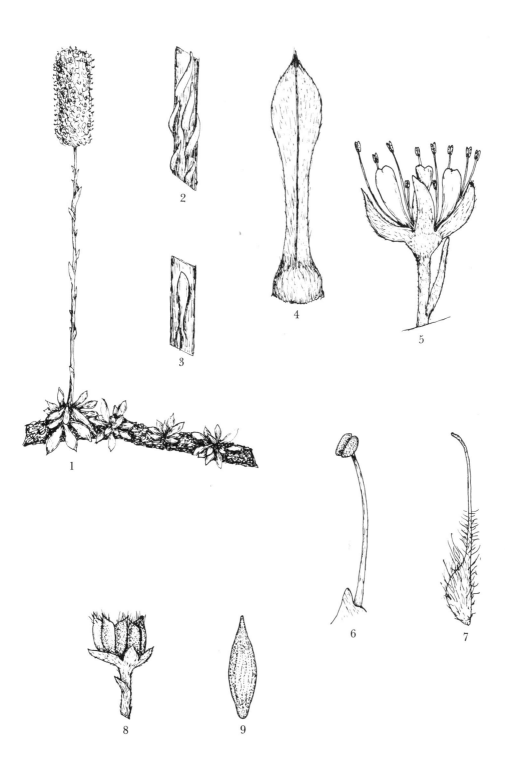

1. Plant with inflorescence
2. Bracts on upper part of peduncle
3. Bract of lower part of peduncle
4. Leaf
5. Flower
6. Stamen
7. Pistil
8. Fruit
9. Seed

ROSACEAE
Pyrus ioensis (Wood) Bailey

Malus ioensis (Wood) Britt.; incl. var. *bushii* Rehd. and var. *palmeri* Rehd.

Wild crabapple

A tree to 6 m high with low, crooked branches; thicket forming from sucker shoots.

LEAVES: Alternate, simple. Ovate to elliptic, 6-9 cm long, 3-6 cm wide; coarsely serrate, 4-6 teeth per cm, or with shallow lobes; tip acute to obtuse, base cuneate to slightly rounded; upper surface dark green, semilustrous, glabrate; lower surface pale, tomentose; young leaves densely tomentose on both sides; petiole 1-2.5 cm long, tomentose; stipules linear, 3-4 mm long, pubescent.

FLOWERS: May, the leaves about half-grown; perfect. Flowers in corymbose clusters; pedicels 8-12 mm long, tomentose; hypanthium obconic, 3-4 mm long, densely tomentose; calyx lobes 5, lanceolate, tomentose on both sides; petals 5, obovate, 1.3-1.6 cm long, 1.2-1.4 cm wide, white to bright pink, margin undulate; stamens many, attached to the rim of the hypanthium; filaments white, 1 cm long; anthers yellow; ovary inferior; styles 5, tomentose at the base, 12.5-13 mm long, gradually widened to a greenish, capitate stigma.

FRUIT: September-October. Pedicels 2-2.5 cm long, pubescent; fruit globose, 2.5-4 cm diameter; apex slightly depressed, base depressed or not; green-yellow with small light spots, bitter, edible if cooked. Seeds ovate, 7-8 mm long, 3.7-4.3 mm wide, 2-2.7 mm thick, chocolate brown, smooth.

TWIGS: 2 mm diameter, rigid, red-brown with pale lenticels; woolly at first, becoming glabrous in the second year; some of the spurs ending in a spine; leaf scars crescent-shaped; 3 bundle scars; pith white, continuous, one-fourth of stem. Buds ovoid, 2.5-4 mm long, end bud largest, the scales red-brown, slightly keeled, pubescent.

TRUNK: Bark thin, red-brown to grayish, shallow fissures and narrow persistent scales. Wood hard, heavy, red-brown, with a yellowish sapwood.

HABITAT: Open woods, rocky hillsides, pastures, creek banks, waste ground, usually in limestone areas.

RANGE: Minnesota south to Texas, east to Louisiana, and north to Indiana and Michigan.

Pyrus ioensis is fairly common in southeastern Nebraska and eastern Kansas. The branches are often thorny, especially on the sprouts which may form a dense, impenetrable thicket around the parent plant. It produces an abundance of pink flowers early in the spring, and at that season, stands out sharply wherever it is. The fruits remain fairly hard and are bitter but are excellent for making jelly. A few birds and mammals eat the fruit and honey bees visit the flowers regularly.

It is occasionally transplanted and used in landscaping but is objectionable because of the sprouts. Since there are many desirable horticultural forms of the crabapple, there is no specific need for using the native ones.

Several specimens in the herbaria of our area have been labeled *P. angustifolia* Ait. or *P. coronaria* L. but all of these examined had the pubescent lower leaf surface and pubescent hypanthium of *P. ioensis*. No specimens of the other two species were located.

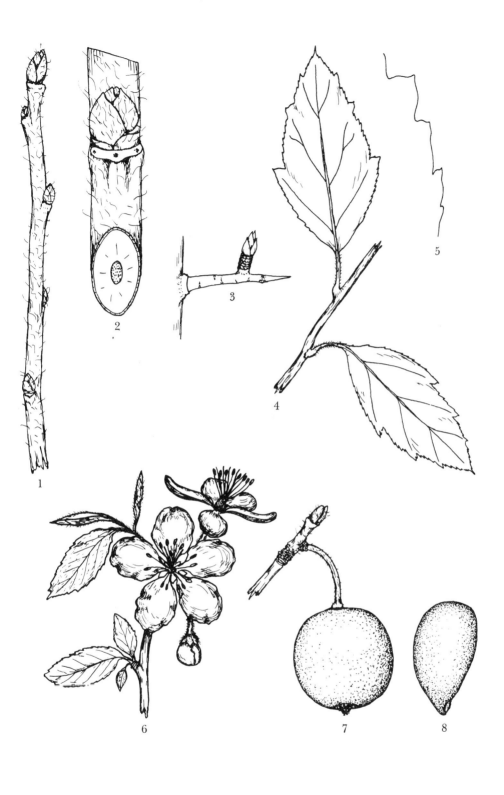

1. Winter twig
2. Detail of twig
3. Spur branch
4. Leaves
5. Leaf margin
6. Flowers
7. Fruit
8. Seed

ROSACEAE
Sorbus scopulina Greene

Pyrus scopulina Longyear; *Sorbus angustifolia* Rydb.

Mountain ash

A small tree or large shrub to 5 m high, a few branches near the top.

LEAVES: Alternate, pinnately compound, 10-18 cm long, 9-13 leaflets, rachis grooved or flattened and pubescent above, rounded and glabrous below, tufts of long, stiff, often fulvous hairs and a few red glands at the leaflet base. Leaflets sessile, elliptic or elliptic-oblong, 3-6 cm long, 1-2.5 cm wide, the longest ones at the middle; tip acute to short acuminate, base cuneate or rounded; margin sharply serrate, 4-5 teeth per cm, ciliate; glabrous or glabrate on both sides; petiole 2.5-3.5 cm long, widened at the base, pubescent along the shallow groove above and glabrous to glabrate below; stipules ovate-lanceolate, long acuminate, green, toothed on one margin, occasionally fimbriate at the tip, either persistent or early deciduous.

FLOWERS: June; perfect. Flowers in terminal, flat-topped corymbs 10-14 cm across, the branches and pedicels green and sparingly pilose with white hairs; pedicels 5-10 mm long, a brown, linear bract 3-4 mm long on the pedicel; hypanthium obconic, yellow-green, 3 mm long and 3 mm wide, pubescent; calyx lobes 5, triangular, 1.5-2 mm long and as wide, acute, green, often red-tipped, pubescent and curly ciliate, ascending between the petals; petals 5, white, 6 mm long, 5 mm wide, ovate with a short claw, slightly reflexed, irregular on the margin, pubescent at the base inside; many stamens, filaments 3-4 mm long, tapered, white, attached on the hypanthium rim; anthers nearly white; exposed top of ovary pubescent; disk around the ovary yellow; 3-5 styles, 3 mm long, pubescent at the base; stigmas capitate, small, yellowish.

FRUIT: September. Corymbs drooping, 10-15 cm broad, the branches reddish and pubescent. Fruit globose, 9-12 mm diameter, orange-red, smooth and semilustrous, slightly juicy; 1-5 seeds.

Seeds flattened ovate, 3.8-4.3 mm long, 2-2.3 mm wide, 1.1-1.5 mm thick, orange-brown, with a yellowish area near the micropyle, smooth, enclosed in a cartilaginous locule.

TWIGS: 2.5-3.5 mm diameter, smooth, green-brown to red-brown, glabrous with a thin waxy coat or pubescent with crinkly or straight, tawny or white hairs, especially dense at the nodes; numerous elliptic to oval, light-colored lenticels; leaf scars crescent-shaped or broadly U-shaped, raised, dark-colored; 5 bundle scars; apparently 2 abscission layers, the petiole breaks off about 1.5 mm above the actual base, later this remaining band of petiole base is deciduous, exposing the true leaf scar; pith white to tawny, continuous, two-thirds of stem. End and upper buds narrowly ovoid, pointed, dark red-brown, 12-17 mm long, 3-4 mm diameter, the two outer scales often keeled, scales ciliate, sparsely pubescent or glabrous, glutinous; buds on the lower part of the stem smaller, obtuse, and appressed to the stem.

TRUNK: Bark mottled gray-brown, tight, roughened with brown, horizontal lenticels, bark sometimes pruinose. Wood soft, fine-grained, pale brown, with a wide, white sapwood.

HABITAT: Rocky, wooded ravine banks or creek banks in moist soil, often at the edge of a stream in wet, mossy soil.

RANGE: Alaska, south to northern California, east to Colorado and New Mexico, north through Wyoming and western Montana to western Alberta; the Black Hills of South Dakota.

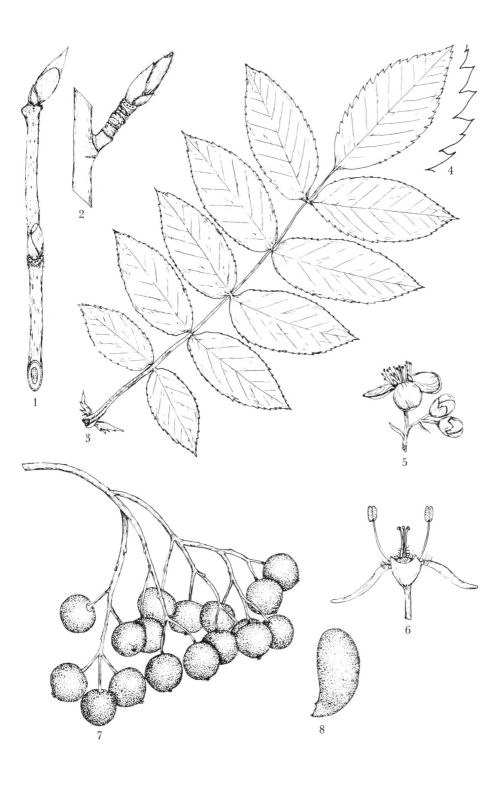

1. Winter twig
2. Spur branch
3. Leaf
4. Leaflet margin
5. Flower
6. Semidiagrammatic flower
7. Fruit cluster
8. Seed

ROSACEAE
Amelanchier alnifolia Nutt.

A. carrii Rydb.; *A. leptodendron* Lunell;
A. macrocarpa Lunell; *A. alnifolia* var.
dakotensis Nielson; *Aronia alnifolia*
Nutt.

Saskatoon service berry, june berry,
shadbush

A shrub with single or clustered trunks up to 4 m high, usually with the branches near the top.

LEAVES: Alternate, simple, the main lateral veins entering the teeth. Oval to obovate or oblong, 3-6 cm long, 2.5-4 cm wide, the tip truncate, rounded or obtuse; the base rounded, subcordate or broadly cuneate; sharply serrate, 2-6 teeth per cm, the teeth often only above the middle; emerging leaves folded lengthwise and with both surfaces tomentose, but becoming glabrous above and glabrate or lightly tomentose below; petiole 12-18 mm long, slender, pubescent or glabrate, a narrow groove above; stipules linear or narrowly lanceolate, 1-3 mm long, pubescent, caducous.

FLOWERS: Late April-May, occasionally into June at higher elevations; perfect. Racemes terminal on new growth, 3-4 cm long, 2-10 flowered, axis and pedicels green and silky pubescent; pedicels 5-10 mm long, ascending, with 1-2 linear, brown, pubescent, caducous bracts; hypanthium campanulate, 3 mm long, 3-4 mm wide, green and short pubescent; calyx lobes 5, triangular, 1.5-2 mm long, reflexed, glabrous on the outer surface and hairy or tomentose inside, densely ciliate; petals 5, white, oblong or obovate, 5-11 mm long, 2-3 mm wide, spreading, a few silky hairs on the upper surface, especially at the rounded tip; stamens 10-15, inserted along the margin of a purplish disk at the top of the hypanthium; filaments white, 1-3 mm long, often connate at the base, the young stamens curved over the ovary; anthers pale yellow, 0.75 mm long; ovary inferior, the summit densely white tomentose; styles 5, distinct, 2 mm long with brownish stigmas.

FRUIT: July-August. Racemes 3-5 cm long, ascending, pedicels 6-10 mm long, glabrous, usually reddish; fruit globose to obovoid, 8-11 mm long, 7-9 mm wide, dark red-purple to nearly black, minutely white-dotted, the slight bloom easily removed, calyx lobes and stamens often persistent; 3-6 seeds per fruit. Seeds red-brown, the shape of an orange slice, 2.5-3.8 mm long, 1.4-2 mm wide, 1 mm thick, smooth, the sides concave or convex.

TWIGS: 1.4-1.6 mm diameter, red-brown, glabrous or densely pubescent on shoots, pruinose; lenticels elliptical, yellow-brown; leaf scars reniform or elliptic, dark with 3 light bundle scars; pith white, continuous, about one-third of stem. Buds ovoid, 4-5 mm long, 1.9-2.1 mm diameter, chestnut-brown or purple-brown, glabrous or with a few hairs, dull, acute, often pruinose; outer scales keeled and apiculate.

TRUNK: Bark gray, tight, finely fissured, which gives a light and dark streaked appearance. Wood hard, heavy, close-grained, brown, with a wide, white sapwood.

HABITAT: Brushy hillsides, open woods, creek banks, usually in well-drained soil, but occasionally around bogs.

RANGE: Alaska to Oregon east of the Cascades, east to Alberta and North Dakota, south to Nebraska and Colorado, and northwest to Idaho.

The fruits of the Amelanchiers are often used for making jam, the one main objection being the numerous seeds. Small mammals and birds eat them regularly, and the long ripening period furnishes food for some time. The flesh is often dry and mealy, which makes them unpalatable when raw.

226

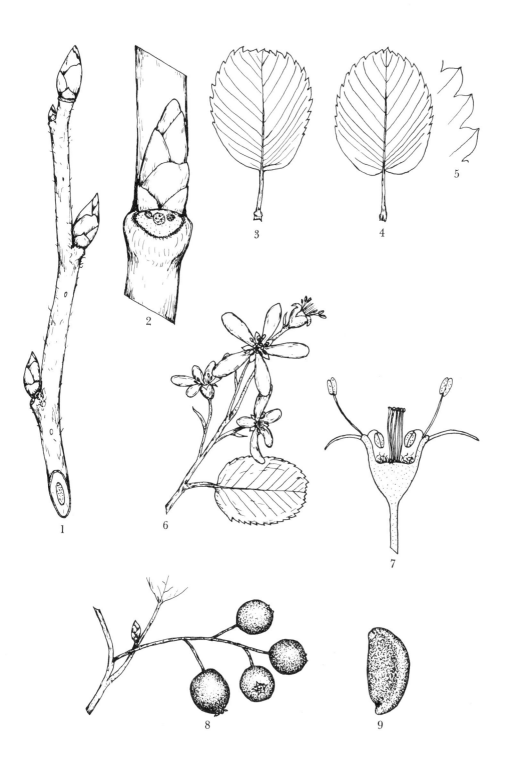

1. Winter twig
2. Detail of twig
3. Typical leaf
4. Leaf variation
5. Leaf margin
6. Raceme of flowers
7. Cutaway of flower
8. Fruit cluster
9. Seed

ROSACEAE
Amelanchier arborea (Michx. f.) Fern.

A. canadensis f. *nuda* Pal. & Steyrm.;
A. canadensis of auth., not (L.) Medic.

June berry, service berry, shadbush

A shrub or small tree to 8 m high, usually with a single trunk, occasionally the trunks clustered.

LEAVES: Alternate, simple. Ovate to oval, 5-9 cm long, 3-5.5 cm wide; finely serrate with the tips pointing definitely forward, 4-8 teeth per cm; leaf tip acuminate, base rounded or cordate; upper surface dark green, glabrous; lower surface pale with some pubescence; emerging leaves densely white tomentose and folded lengthwise; petiole 1-2 cm long, glabrous; stipules minute, truncate, usually reddish.

FLOWERS: Early April, just before the leaves; perfect. Inflorescence a bracted, tomentose raceme on new growth; pedicels 8 mm long, with 1-3 tomentose, red bracts; hypanthium obconic, pubescent; calyx lobes 5, green, deltoid, 2-3 mm long, tomentose, sharply recurved; petals 5, oblanceolate, 10-14 mm long, 4-6 mm wide, white, straight or twisted, tip often emarginate; stamens 20, filaments 4-5 mm long, usually grouped with 3 opposite each petal and 1 opposite each calyx lobe; anthers yellow; ovary 5-celled, partly inferior, 2.5 mm wide, green, glabrous or with a few curly hairs; styles 5, greenish-white, 4 mm long, glabrous; stigmas capitate.

FRUIT: June. Racemes open, pedicels 1.5-2 cm long, glabrous or pubescent; fruit globose, becoming red, then dark purple, 6-10 mm diameter, 5-8 seeds per fruit, calyx persistent. Seeds black, oval, with one side nearly straight, 3.5-4 mm long, 1.5-2 mm wide, 1 mm thick, surface granular.

TWIGS: 1.2-1.7 mm diameter, flexible, glabrous, red-brown; leaf scars narrowly crescent-shaped, enlarged at the center; 3 bundle scars; pith green or brown, continuous, one-fourth of stem. Terminal bud 10-12 mm long, 2 mm diameter, the tip sharply pointed, the scales red-brown with some yellow-green areas, often ciliate; lateral buds 1.1-2.2 mm long, flattened, appressed to stem, blunt.

TRUNK: Bark light gray and smooth on young trees; dark gray with shallow furrows and long ridges on old trees. Wood heavy, hard, dark brown, with a wide, white sapwood.

HABITAT: Open rocky woods, stream banks, bluffs, usually on well-drained slopes.

RANGE: Maine to Minnesota, south to northeastern Texas, east to Florida, and north to New England.

This is the only *Amelanchier* in the southern part of our range although others have been listed for Kansas. Too, it has often been misidentified, probably due to the lack of agreement in the keys in various manuals and other publications. There is no doubt about the Amelanchiers being a difficult group even though they have been studied quite extensively.

Amelanchier arborea usually has juicy fruits but these are seldom used because the birds take them as soon as they ripen. Also, the ripening period is long and the plants are not numerous enough for one to get any quantity of the fruit at a given time.

The wood is of good quality, but the size of the shrub, or tree, is too small for the wood to be used for anything but tool handles and other small articles.

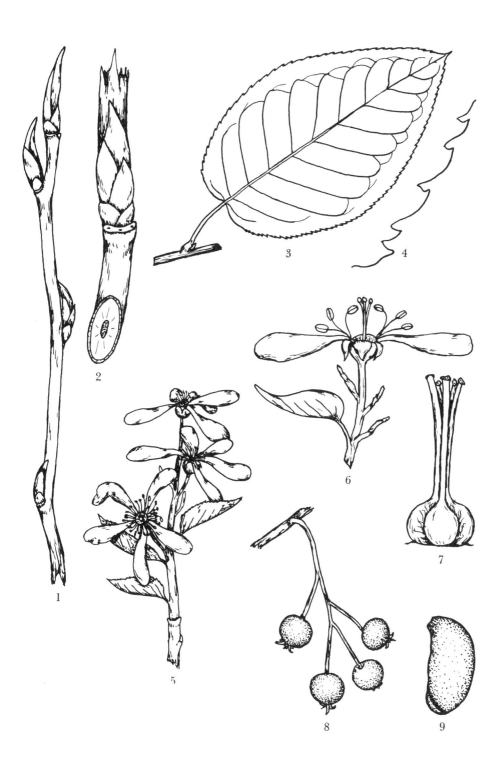

1. Winter twig
2. Detail of twig
3. Leaf
4. Leaf margin
5. Flowering raceme
6. Semidiagram of flower
7. Pistils
8. Fruit cluster
9. Seed

ROSACEAE
Amelanchier sanguinea (Pursh) DC.

A. humilis Wieg.; *A. spicata* in part, not
K. Koch; *A. huronensis* Wieg.

Dwarf june berry, shadbush

A shrub to 3 m high, stoloniferous and forming colonies, usually branched above the middle.

LEAVES: Alternate, simple, the lateral veins entering the teeth directly or branching before entering. Oval to oblong, 3-6 cm long, 2-4 cm wide, acute or obtuse, base rounded or subcordate; margin variable, entire, serrate at the tip or along the upper margin, or serrate to near the base, 3-5 teeth per cm; young leaves densely tomentose and folded lengthwise, glabrate in age but may remain pubescent below; petiole 1-2 cm long, slender, pubescent or glabrous; stipules linear, 8-12 mm long, pubescent, caducous.

FLOWERS: May; perfect. Racemes tomentose, 3-4 cm long, 10-15 flowers, erect or ascending; pedicels 5-12 mm long with 1-2 small, linear, brown, caducous bracts. Sepals 5, broadly triangular, 1.5-2 mm long, reflexed at the middle after anthesis, green, tomentose; petals 5, white, oblanceolate, 7-10 mm long, 3-4 mm wide, flat, attached at the margin of the yellow-green disk; stamens about 20, usually grouped with 3 opposite a petal and 1 opposite a sepal; filaments white, 1.5-2.5 mm long, anthers light yellow to dark brown; hypanthium broadly campanulate, 4-5 mm across, ovary inferior, densely tomentose at the top; styles 5, united below the middle, 3-3.5 mm long; stigmas capitate, yellowish.

FRUIT: July. Racemes erect, ascending or drooping, 4-5 cm long, compact, the axis pubescent, pedicels 5-15 mm long, the lowest ones longer than the upper, glabrous or pubescent; fruit dark purple to nearly black, globose, 9-11 mm diameter, slightly pruinose, calyx lobes persistent; sweet, edible, often mealy; 5-10 seeds per fruit. Seeds red-brown, oval, one end curved or slightly hooked, 2.6-3.5 mm long, 1.7-2.2 mm wide, 0.9-1.4 mm thick, smooth, convex or flat on the sides.

TWIGS: 1.6-2 mm diameter, wine-red or with an orange cast, semirigid, glabrous except at the nodes and the end; lenticels small and not noticeable; leaf scars long and narrow, dark; 3 bundle scars; pith white, continuous, about one-third of stem. Buds broadly ovoid, 3.5-4 mm long, 2.5-2.8 mm wide, the terminal bud longer than the laterals, the scales ovate, wine-red, often with a brownish margin, ciliate, glabrate, the inner scales densely silky pubescent on the outer side and glabrous inside.

TRUNK: Bark red-brown, tight, a few irregular splits expose the inner, light brown bark. Wood white, hard, fine-grained, the heart-sapwood line not obvious.

HABITAT: Open, brushy hillsides, woods or creek banks, usually in well-drained soil and commonly in rocky, clay soil.

RANGE: Quebec, west to Manitoba, south to Nebraska, east to Ohio and New England. One specimen from Kansas.

After examining several hundred plants in the field and with a microscope, the author has been unable to find any consistent characteristic or combination of characteristics that would separate *A. sanguinea* from *A. humilis* Wieg. For this reason they are here placed in synonymy. Perhaps *A. humilis* should be a variety of *A. sanguinea,* but that would require a thorough study of the plants in a far wider range than the north central plains.

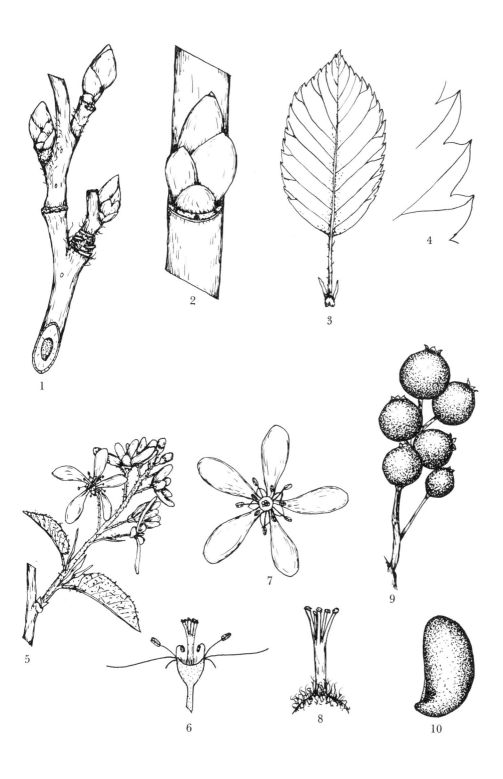

1. Winter twig
2. Detail of twig
3. Leaf
4. Leaf margin
5. Flowering raceme
6. Face view of flower
7. Diagram of flower
8. Styles with top of ovary
9. Fruit cluster
10. Seed

ROSACEAE
Cercocarpus montanus Raf.

Cercocarpus betuloides Nutt. in T. & G.;
C. *betulaefolius* Nutt. ex Hook.;
C. *parvifolius* var. *betuloides* Sarg.;
C. *parvifolius* Nutt.; C. *flabellifolius*
Rydb.

Mountain mahogany, alderleaf

An erect, freely branched shrub to 3 m high.

LEAVES: Alternate, simple, often clustered at the end of a spur branch, the lateral veins prominent beneath. Obovate, rhomboidal, to oval or elliptic, 2-3 cm long, 1-2 cm wide, tip rounded or acute, base cuneate to attenuate; crenate-serrate above the middle, 2-4 teeth per cm, lateral veins nearly straight and ending in a tooth; upper surface dull yellow-green, pubescent, lower surface whitened with a dense tomentum beneath the longer hairs; firm, pliable; petiole 3-5 mm long, flat above, pubescent, blade decurrent on at least half of the petiole; stipules lanceolate, pubescent, early deciduous.

FLOWERS: Early June; perfect. Single in leaf axils at the end of a short spur, the close proximity of the leaves giving an effect of clustered flowers; pedicels 2-4 mm long, pubescent; hypanthium cylindric, slightly larger toward the base, 7-9 mm long, 1 mm thick, the summit turbinate or shallow cup-shaped, 4-5 mm across, 1-2 mm deep, soft pubescent outside, glabrous inside, green; calyx lobes 5, green, triangular, 1-1.5 mm long, densely pubescent on both sides, sharply reflexed after anthesis; no petals; stamens 20-40, inserted along the rim of the hypanthium; filaments 0.5-1 mm long, white, curled outward; anthers pubescent, pale yellow; ovary surrounded by the hypanthium, cylindric, 6-8 mm long, green, densely pubescent with straight, appressed, white hairs; style short, pubescent, stigma 5-6 mm long, reddish, pubescent on one side.

FRUIT: August. Ovary surrounded by the brown hypanthium which splits and falls at fruit maturity, the expanded limb of the calyx and the stamens deciduous shortly after fertilization, falling in one piece; at this time the style elongates to form a plume 4-8 cm long, twisted at the base, the hairs about 3 mm long; fruit body cylindric, 10-13 mm long, with straight pubescence, ridged dorsally and with a faint ridge on the ventral side, 1 seed per fruit. Seed slender, 10-12 mm long, pointed at both ends, brown, rounded on the dorsal side, somewhat flattened on the ventral side with a small ridge along the raphe; hilum about 1 mm from the lower end.

TWIGS: 1.4-1.6 mm diameter, dark red and pubescent, gray-waxy by the second season; many short spur branches; leaf scars narrowly crescent-shaped to triangular; 3 bundle scars; pith white or pale brown, continuous, one-fourth of twig. Buds ovoid, 2-3 mm long, acute, the scales brown, pubescent, ciliate.

TRUNK: Bark gray, roughened with irregular, horizontal, brown lenticels. Wood hard, fine-grained, white with a pinkish cast and indefinite sapwood.

HABITAT: Dry areas, on eroded hillsides or open, rocky woods; often in "badlands" type of soil.

RANGE: Southwest Oregon, south to Baja California, east to Texas, north across the Oklahoma panhandle, Colorado, western Nebraska, and South Dakota, and west across Wyoming to Utah and northern Nevada.

Cercocarpus is a shrub of the western and southwestern states but enters the west edge of our area where the plants are scattered and seldom occur in abundance. The wood is extremely hard but deer browse on the more tender young branches. The one record reported from Kansas was based on some seeds found "in the mud of the roots of a *Scirpus.*"

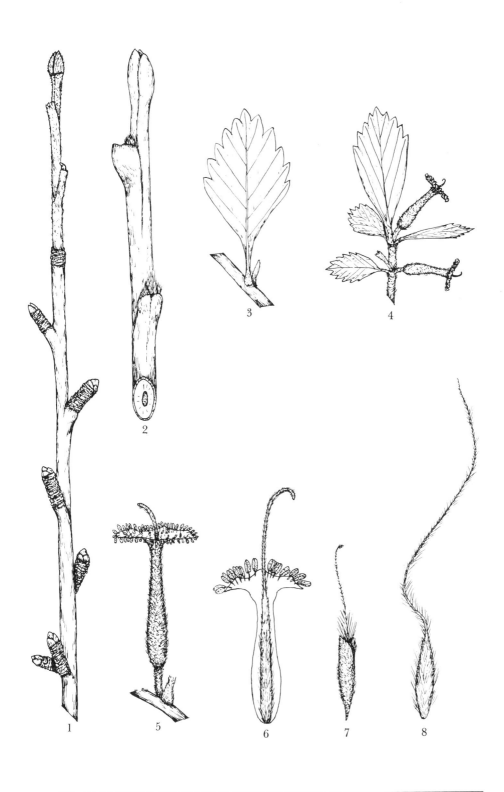

1. Winter twig
2. Detail of twig
3. Leaf
4. Twig with flowers
5. One flower

6. Cutaway of flower
7. Developing fruit, surrounded by hypanthium
8. Achene

ROSACEAE
Crataegus chrysocarpa Ashe

C. rotundifolia Moench var. *pubera* Sarg.

Hawthorn, red haw

A small tree to 4 m high, with stout branches; often in large thickets.

LEAVES: Alternate, simple. Broadly ovate to nearly rotund, broadest below or about the middle, 4-7 cm long, 3.5-6 cm wide; tip acute, base cuneate, the blade often decurrent; margin with 2-4 small lobes or long teeth and biserrate or triserrate, the large teeth up to 1 cm across. with 3-6 small teeth, the sinuses acute and each tooth tipped with an obvious red gland; upper surface semi-glossy, dark yellow-green and with a few hairs, lower surface lighter with a few hairs on the veins; petiole 1-1.5 cm long, flattened and pubescent above, commonly with red glands; stipules linear, 5-6 mm long, glandular, early deciduous.

FLOWERS: Early June; perfect. Corymb 3-4 cm across, 6-10 flowers, peduncles and pedicels green, lightly pubescent and with linear, glandular bracts at the base of the pedicels; hypanthium short cylindric to obconic, 2-2.5 mm long, green, pubescent; calyx lobes 5, broad at the base with long tapered tip, 4 mm long, 2 mm wide, glandular, pubescent on both sides; disk green; petals 5, white, about 5 mm across, nearly circular and cupped, abruptly narrowed to a short claw; stamens 4-8, filaments 3-5 mm long, white; anthers pale yellow; styles 3-5, green, glabrous, 4 mm long; stigmas capitate, yellow-green.

FRUIT: September. Corymbs erect or drooping; fruit globular, 8-11 mm diameter, red, hard, the sepals persistent; the 3-5 seeds usually cohere tightly, forming an ellipsoidal cluster. Seeds the shape of an orange section, 7-8 mm long, 4-5 mm wide, and 3.5-4 mm dorso-ventrally; the sides irregularly concave, the surface rough and yellowish.

TWIGS: 1.4-1.6 mm diameter, red-brown to purplish-red, glabrous, glossy, often pruinose; lenticels small, light-colored; leaf scars narrowly crescent-shaped; 3 bundle scars; pith white, continuous, about one-third of stem; thorns 5-8 cm long, curved or straight, stout, deep red or purplish, pruinose. Buds globular, 1.5-2.1 mm diameter, red, glossy, glabrous; outer scales gibbous, apiculate, and with a narrow scarious margin.

TRUNK: Bark gray-brown, furrowed, the ridges thin, flat, wide, and somewhat scaly. Wood hard, fine-grained, white, with no visible line between heartwood and sapwood.

HABITAT: Stream banks, open hillsides, or at the margin of a woods, usually in rocky ground.

RANGE: Newfoundland across southern Canada to Saskatchewan, south to Colorado and New Mexico; South Dakota east to New York and New England.

C. chrysocarpa and *C. succulenta* Link are the only common red haws of the northern part of our northern plains states. They are often difficult to separate since the distinguishing characters vary or are a matter of degree. *C. chrysocarpa* is a smaller, often shrub-like tree with stout thorns, the petioles and teeth of the leaves with definite red glands, the flowers with 10 or less stamens, the fruits hard, and the seeds usually cohere tightly. *C. succulenta* is a larger, more tree-like plant with slender thorns, the petioles and teeth without red glands, the flowers with 10 or more stamens, the fruits mellow, and the seeds separate easily. Occasionally trees will be found which have a mixture of these characteristics, and the question arises whether they are hybrids or merely variants of one species.

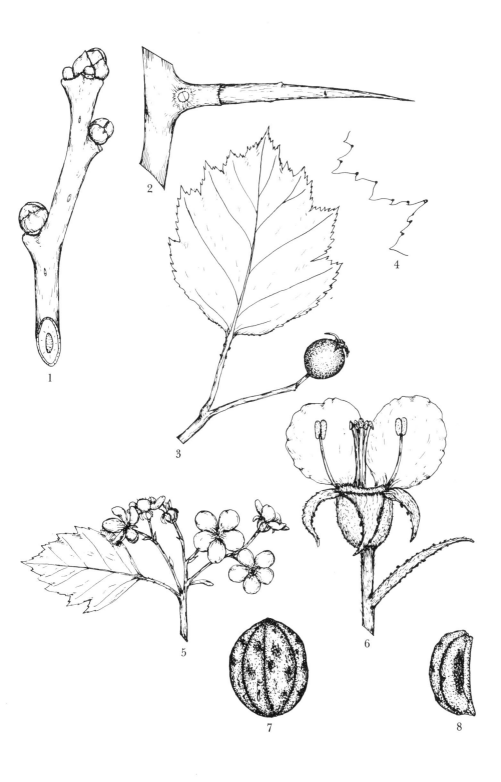

1. Winter twig
2. Thorn
3. Leaf and mature fruit
4. Leaf margin
5. Flower cluster
6. Flower, some petals and stamens removed
7. Seed group
8. One seed

ROSACEAE
Crataegus crus-galli L.

C. regalis Beadle var. *regalis*

Cockspur hawthorn

A small tree to 4 m high, with rigid, nearly horizontal, main branches low to the ground.

LEAVES: Alternate, simple, coriaceous. Obovate or elliptic, 3-5 cm long, 1-2.5 cm wide; sharply serrate except toward the base, 6-7 teeth per cm; tip rounded or acute, base gradually narrowed or cuneate; upper surface dark green, lustrous, glabrous; lower surface pale, glabrous, the veinlets obvious; petiole 5-7 mm long, glabrous; stipules on shoots, 7 mm long, 5 mm wide, stalked, glandular serrate.

FLOWERS: April-May, after the leaves; perfect. Flowers in glabrous corymbs, terminal on short branches; pedicels 7-8 mm long, with a bract near the base; hypanthium obconic, green, glabrous; calyx lobes 5, lance-linear, 4.5-5 mm long, entire or serrate, glabrous, red-tipped; petals 5, broadly oval, 8-8.5 mm long, 7.5-8 mm wide, white, margin often irregular; stamens 10, filaments 3.8-4.2 mm long, white; anthers yellow, turning to rose after the flower opens; ovary inferior; styles 2-4, white, 4.9-5.1 mm long with a small tuft of hairs at the base; stigmas capitate, disk-like.

FRUIT: October, often persistent until late winter; pedicels 1.5-2 cm long, glabrous; fruit a subglobose pome, 9-11 mm diameter, greenish to dull red with a slight bloom, hard, smooth, old calyx lobes persistent. Nutlets 2, yellow-brown, the ventral side straight, the dorsal side rounded, with a broad longitudinal ridge.

TWIGS: 1.4-2 mm diameter, rigid, glabrous, reddish or gray with a waxy covering in the first year, gray-brown later; spines straight or slightly curved, chestnut brown to ashy, 4-7 cm long; leaf scars narrowly crescent-shaped; 3 bundle scars; pith white, continuous, one-fourth of stem. Buds globose, 1.5-3 mm diameter; scales red, glossy, the outer ones apiculate.

TRUNK: Bark dark brown, fissured, the ridges short and scaly, often peeling. Wood salmon-color, hard, fine-grained.

HABITAT: Rich woods, rocky pastures, open areas of grazed pastures or cutover woods.

RANGE: Quebec, west to Minnesota, south to eastern Texas, east to Florida, and north in the Appalachians.

As is the case in most of the species of this genus, *C. crus-galli* varies so much that it has been divided into several varieties and forms. Because of the difficulty in separating them, these are not given here but are left to publications devoted to Crataegus.

C. crus-galli is a short, shrub-like tree from 2 to 4 meters high. It has many branches low to the ground, often extending horizontally from the trunk. These branches are crooked and well-equipped with thorns. The general appearance of the tree in the open is of a tall shrub with gnarled branches and a flat top. When grown in the shade, the tree is more open and the branches more erect.

The fruits are small and often remain green and persist on the tree until early winter. They are hard and very few animals use them for food. After the fruits have fallen, small rodents may cut them open and eat the nutlets.

The following species of *Crataegus*, with synonyms, are listed for Kansas but are not described in this book: *C. berberifolia* T. & G. (*C. engelmanni*); *C. calpodendron* (Ehrh.) Medic. (*C. hispidula*, *C. spinulosa*); *C. coccinioides* Ashe; *C. lanuginosa* Sarg.; *C. palmeri* Sarg.; *C. pruinosa* (Wendl.) K. Koch (*C. disjuncta*, *C. mackenzii*); *C. punctata* Jacq. (*C. collina*); *C. reverchoni* Sarg. (*C. discolor*, *C. stevensiana*).

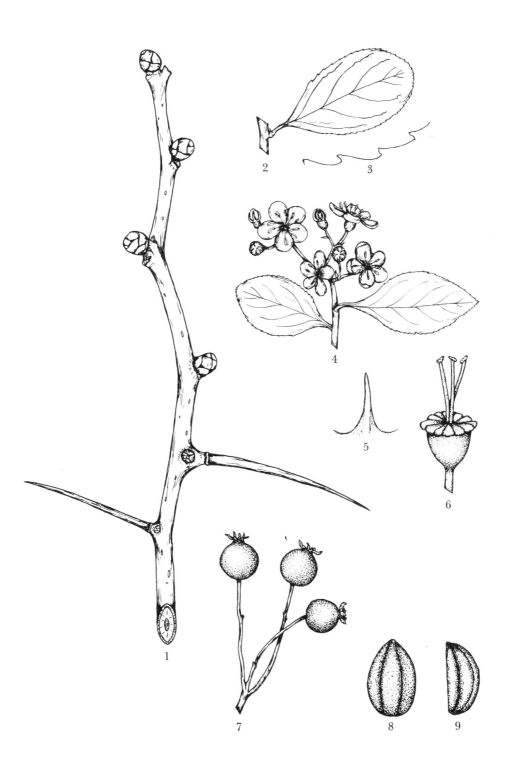

1. Winter twig
2. Leaf
3. Leaf margin
4. Inflorescence
5. Calyx lobe

6. Pistils, disk, and hypanthium of flower, other parts removed
7. Fruit cluster
8. Dorsal view of seed
9. Side view of seed

ROSACEAE
Crataegus mollis (T. & G.) Scheele

C. submollis Sarg.

Red haw, hawthorn, summer haw,
turkey haw

A tree to 6 m high with stout, usually crooked branches.

LEAVES: Alternate, simple. Broadly ovate, 7-10 cm long, 7-8 cm wide; more or less lobed, coarsely and irregularly serrate, 3-5 teeth per cm, ciliate; tip acute, base cordate or rounded; upper surface dark yellow-green, dull, slightly pubescent and rugose; lower surface paler and softly pubescent, the midrib stout; petiole 1.5-4 cm long, stout, pubescent; stipules 10-12 mm long, falcate, toothed, pubescent.

FLOWERS: April-May, the leaves half-grown; perfect. Inflorescence a villous cyme of 5-7 flowers on a short branch; the bracts with red, glandular teeth; pedicels villous, 1-1.3 cm long; hypanthium obconic, villous, calyx lobes 5, broad-based, acuminate, 2-4 mm long, coarsely glandular-serrate, villous on the outer side, tomentose within; petals 5, white, orbicular, 1-1.2 cm across, margin irregular; the disk in the center of flower yellowish and does not turn red with age; stamens 20, usually 3 opposite each petal and 1 opposite each calyx lobe; filaments white, 4-6 mm long; anthers light yellow; ovary inferior; styles 4-5, greenish, 7-8 mm long, finely pubescent at the base; stigma capitate, yellowish.

FRUIT: September. Pedicels stout, 1-1.5 cm long, pubescent, bracted; fruits solitary or in small clusters; globose, 1.4-1.8 cm diameter, scarlet with small yellow dots; old calyx lobes erect to spreading; flesh thick, yellow, mealy when fully ripe, edible. Nutlets 3-5, the shape of an orange section, 7.5-7.7 mm long, 3.7-3.9 mm wide, 3.8-4 mm dorso-ventrally, pale flesh color; ventral edge nearly straight, sides flat or slightly concave; dorsal side ridged.

TWIGS: 2.5-3.5 mm diameter, rigid, white villous during the first season and glabrous in the second, brown to gray beneath the vestiture; leaf scars narrow, wider at the ends and middle; 3 bundle scars; pith white, continuous, one-fourth of stem. Buds broadly ovoid, 3-5 mm long; scales ovate, reddish, thick, pubescent at first; scales of the end bud sharply acute with slightly spreading tips; the outer scales usually gibbous.

TRUNK: Bark red-brown to yellow-brown, with shallow fissures and flat-topped, somewhat blocky and flaky ridges. Wood heavy, hard, light brown, with a wide, pale sapwood.

HABITAT: Low rich woods in open areas; rocky hillsides and hilltops in poor soil.

RANGE: Ontario, west to Minnesota, south to eastern Texas, east to Mississippi, and north to Tennessee and Ohio.

C. mollis is probably the best-known species of the genus in the southern part of our range. This may be due to the fact that it varies less than the other species or because the fruits are large and edible. It is often planted in parks, but is seldom used as a street tree. It has a broad, rounded crown and somewhat crooked branches but is not large enough to be called a good shade tree. Often many sprouts are produced on the main trunk near the base.

The dense pubescence on the under leaf surface, the petiole, and in the inflorescence will distinguish C. mollis from most of the other species. The large, broad leaves and the large fruit are also good characteristics.

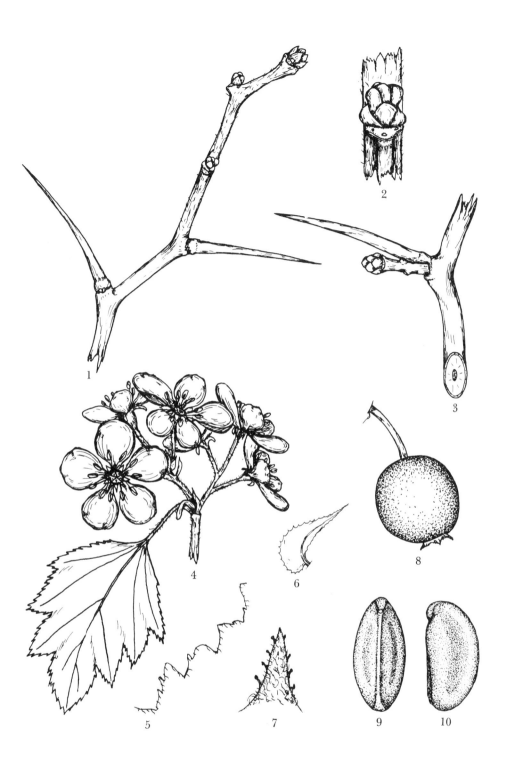

1. Winter twig
2. Detail of bud
3. Thorn and spur branch
4. Leaf and inflorescence
5. Leaf margin

6. Stipule
7. Calyx lobe
8. Fruit
9. Ventral view of seed
10. Side view of seed

ROSACEAE
Crataegus succulenta Link

C. laxiflora Sarg.; incl. var. *pertomentosa* (Ashe) Palmer

Hawthorn, red haw

A tree to 5 m high, either single or in thickets, often with a densely branched crown.

LEAVES: Alternate, simple. Coriaceous, broadly ovate to broadest at or above the middle, 4-6 cm long, 3.5-5 cm wide; tip acute, base broadly cuneate, decurrent on the petiole; margin with small lobes or biserrate with 5-7 teeth per cm, the teeth acute and without obvious glands; upper surface yellow-green, pubescent, the hairs somewhat appressed, midrib impressed toward the base and often densely pubescent; lower surface paler, sparingly pubescent but often quite hairy along the outstanding veins; petiole 1-1.5 cm long, flat and pubescent above, rounded and glabrous below; stipules narrowly linear, 5-9 cm long, glabrous, glandular; not to be confused with the accrescent, obovate bud scales.

FLOWERS: May-June, perfect. Flowers in pubescent, flat-topped corymbs 4-6 cm across with 10-20 flowers; branches of the corymb with linear, glandular, reddish bracts; flowers 15-20 mm across; hypanthium obconic, 3 mm long, villous, green; calyx lobes 5, lanceolate to ovate, 4-5 mm long, reflexed, some pubescence on both sides, margin glandular serrate; petals 5, white, nearly circular, 7-8 mm wide, with or without a claw, margin undulate; disk yellow-green; stamens usually 10, occasionally 20, filaments 6-7 mm long, white, the broad base inserted on the rim of the hypanthium, anthers yellow; styles 3-5, greenish, 5-6 mm long; stigmas capitate, greenish; ovary inferior with only the villous top showing.

FRUITS: September. Corymbs erect or drooping, the pedicels 4-15 mm long, brownish, glabrous or with a few long hairs; fruit globular to slightly oval, 12-15 mm diameter, bright red, mealy, smooth, usually with small light spots, 3-5 seeds per fruit; seeds light brown, the shape of an orange section, 6-6.5 mm long, 3-5 mm wide, 3-3.5 mm dorsoventrally; the surface rough with ridges and grooves, the sides concave; the seeds usually separate readily.

TWIGS: 2-2.2 mm diameter, coarse, rigid, dark red, glabrous, glossy with a few small, light lenticels; thorns curved, usually slender, 5-6 cm long, dark red, becoming gray; leaf scars small, elliptical with rounded ends; 3 bundle scars; pith white, firm, continuous, about one-third of stem. Buds broadly ovoid to nearly spherical, dark red, glossy, 4 mm long, 3.5 mm wide; outer scales gibbous and with scarious margins.

TRUNK: Bark dark gray-brown, soft, with scaly, flat-topped ridges. Wood heavy, hard, tough, fine-grained, pale brown, with a wide, nearly white sapwood.

HABITAT: Creek banks, open woods, margins of woods, or open hillsides. Usually in rocky soil.

RANGE: Quebec, west to Saskatchewan, south to Colorado, east across northern Nebraska, Iowa, and Missouri, and northeast to New England. Steyermark lists it for Kansas, but no specimens were seen from Kansas or southern Nebraska.

The genus *Crataegus* is a most complicated group. Not only do the plants vary within a species, but the separate species hybridize freely. All of this has led to taxonomic confusion. Stevens places all of the *Crataegus* of North Dakota under *C. rotundifolia* (Ehrh.) Borckh. Gleason and Cronquist put *C. chrysocarpa* as synonymous with *C. rotundifolia* Moench, but keep *C. succulenta* as a distinct species.

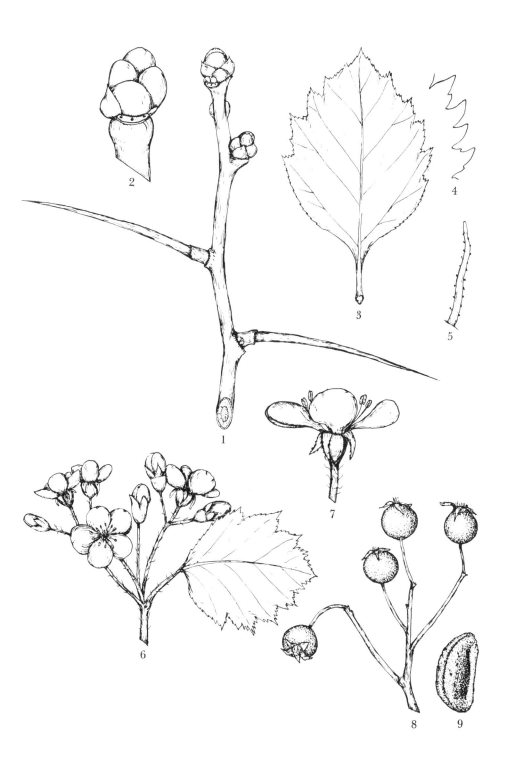

1. Winter twig
2. Detail of bud
3. Leaf
4. Leaf margin
5. Stipule
6. Inflorescence
7. One flower
8. Fruit cluster
9. Side view of seed

ROSACEAE
Crataegus viridis L.

Incl. var. *lutensis* (Sarg.) Palmer

Red haw, hawthorn

A tree to 8 m high with low branches, often at right angles to the trunk.

LEAVES: Alternate, simple. Broadly ovate, 6-9 cm long, 5-9 cm wide; tip acute, base truncate to broadly cuneate; margin with or without small, shallow lobes, sharply serrate, 5-7 teeth per cm, without glands; upper surface dark green, semiglossy, glabrous or sparingly pubescent near the base; lower surface paler, glabrate, usually with some pubescence in the vein axils; petiole 2-3 cm long, slender, flattened or grooved above, glabrous or pubescent; stipules lanceolate, falcate and stalked, 6-8 mm long, 1-1.5 mm wide, glandular serrate.

FLOWERS: April-May; perfect. Corymbs flat-topped, erect, terminal on a leafy branch of the season, 3-5 cm across, 6-12 flowers; peduncle and pedicels green, glabrous, slender, with linear bracts at the base of the corymb branches; pedicels 4-7 mm long; flower 15-17 mm across; hypanthium obconic, green, glabrous, 2.5-3 mm long, 2 mm across; calyx lobes 5, broadly triangular, 2 mm long, 2-2.5 mm wide, green to whitish, recurved, tip acute, finely and sparsely ciliate; petals 5, white, 5-7 mm across, circular with a very short claw, cupped, undulate margin; stamens about 20, attached at the rim of the hypanthium, filaments 2-2.5 mm long, white, tapered; anthers pale yellow, becoming orange; disk yellow at anthesis, becoming bright red after the anthers wither; styles 5, whitish, 4 mm long; stigmas capitate.

FRUITS: September. Corymbs erect, fruits globular, 6-10 mm diameter, dark red, occasionally a few hairs on the surface; flesh somewhat juicy; 4-6 seeds per fruit. Seeds the shape of an orange section, 5-6 mm long, 2.4-3.4 mm broad, 3-4 mm dorso-ventrally, pale brown, fairly smooth, a small groove on the dorsal side; the flat surfaces not pitted or concave.

TWIGS: 1.5-2 mm diameter, glabrous, pruinose, gray-brown to red-brown, rigid; lenticels elliptic, light-colored, obvious; the few thorns 4-5 cm long, straight or nearly so, slender, approximately at right angles to the stem; leaf scars half-round or somewhat elliptic or 3-angled; 3 bundle scars; pith greenish, continuous, one-third of stem. Buds broadly ovoid, obtuse, 1.2-1.5 mm long, 2 mm wide, scales gibbous at the base, red, glabrous, pruinose, the margin of the outer scales scarious, erose, and usually waxy.

TRUNK: Bark reddish-brown, thin, the furrows shallow and the ridges flat, scaly, and somewhat loose. Wood hard, fine-grained, pale brown, with a narrow, nearly white sapwood.

HABITAT: Usually on low ground, around creek banks, flood plains, or low pastures.

RANGE: Virginia to Florida, west to eastern Texas, north to Kansas, and east across southern Illinois to Kentucky.

C. viridis often grows on low ground which is often quite wet, and in this respect it is different from most of the other species. The tree itself is a more delicate-looking tree; the branches are finer and the tree more gracefully proportioned than the others. A few trees are nearly conical, even though the main branches are at right angles to the trunk.

The species has been divided into a number of varieties which may not always be distinguished. If it were necessary to place our plants in a variety, they would have to go into var. *lutescens* (Sarg.) Palmer, even though the fruits are quite often juicy until early winter.

242

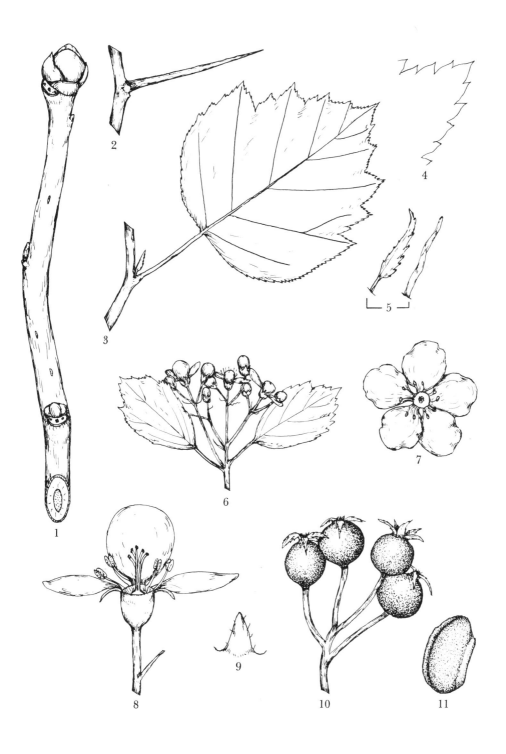

1. Winter twig
2. Thorn
3. Leaf
4. Leaf margin
5. Stipule variation
6. Inflorescence
7. Face view of flower
8. Cutaway of flower
9. Calyx lobe
10. Fruit cluster
11. Side view of seed

ROSACEAE
Potentilla fruticosa L.

Dasiphora fruticosa Rydb.; *Fragaria fruticosa* Crantz; *P. floribunda* Pursh

Shrubby cinquefoil

A spreading shrub to 1 m high, with many fine branches.

LEAVES: Alternate, pinnately compound, 5-7 crowded leaflets, the apical pair attenuate on the rachis, the basal pair of a 7-foliolate leaf usually attached to the rachis at the same point as the pair above; midvein impressed above, prominent below. Leaflets elliptic to narrowly obovate, 10-18 mm long, 3-5 mm wide, acute or rounded, occasionally mucronate, long-tapered at the base; margin entire, somewhat revolute, especially in dried specimens, the pubescence long and silvery, appressed or slightly spreading; pubescence thin above and dense on the lower surface; petiole 8-10 mm long, grooved above, pubescent, jointed at the junction with the stipule; stipules 6-10 mm long, sheathing, adnate to petiole about half of its length, free points acuminate, long pubescent, thin, papery, persistent into winter.

FLOWERS: June-August; perfect. Single in leaf axils or in groups of 2-4 on the end of a new branch; flower 2-2.5 cm across, all green parts pubescent; pedicels 3-5 mm long, hypanthium saucer-shaped, 3.5-4 mm across; bracteoles 5, oblong to elliptic, 4-6 mm long; calyx lobes 5, ovate, 5-6 mm long, acuminate, pale green to yellowish, spreading; petals 5, rotund, 8-10 mm diameter, light yellow, entire, spreading; stamens 20-30, inserted on the rim of the hypanthium; filaments white, 1-1.5 mm long, anthers pale yellow, 1 mm long, flattened; receptacle dome-shaped, densely pubescent; numerous ovaries, 0.4 mm long, bean-shaped, pubescent; style midlateral, stout, 1 mm long; stigma capitate.

FRUIT: July-September. Fruit stalks persistent into winter. Calyx lobes curve inward above the 10-20 achenes on the receptacle. Achenes broad based, acute tipped, 1.5 mm long, 0.7 mm thick, hirsute, the ventral side straight or concave with a small raphe ridge, the dorsal side rounded.

TWIGS: 1.5 mm diameter, red-brown, bark split and shredded, glabrate, lenticels not obvious; pith brown, continuous, one-fourth to one-third of stem; leaves break at the joint on the petiole, scars indistinct, stipules persistent. Buds ovoid, obtuse, 2.5 mm long, concealed beneath the stipules, the scales thin, pale brown, glabrous and ciliate.

TRUNK: Bark red-brown, loose with long, wide, thin shreds. Wood soft, fine-grained, white, with no distinct sapwood.

HABITAT: Mostly in dry, rocky, exposed areas but occasionally in moist soil of ravines and small canyons.

RANGE: Alaska, east across northern Manitoba to Nova Scotia, south to New Jersey, west to Illinois, South Dakota, and Wyoming, south to New Mexico, west to California, and north to British Columbia.

Potentilla fruticosa should not easily be confused with any other plant since it is our only large, woody *Potentilla*. The plant becomes a somewhat globular shrub, occasionally 1 meter high, but usually about 50 centimeters, and is as broad as it is high.

The natural habitat of this plant, from mountains to plains and from arid conditions to moist forests, indicates its adaptability for cultivation. It is often transplanted from the wild state and used as a border plant around homes. Several variants are available in nurseries for landscaping purposes. The flowering period is long but the bushes are seldom heavily covered with flowers at any one time.

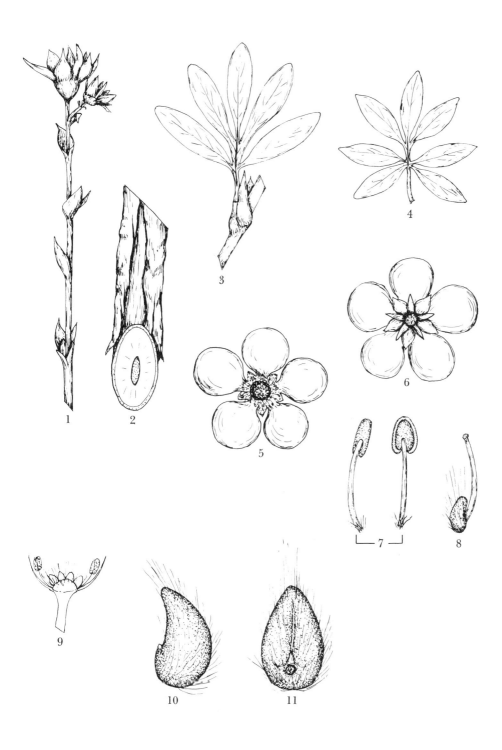

1. Winter twig
2. Detail of bark and stem section
3. Leaf with stipules
4. Leaf variation
5. Face view of flower
6. Dorsal view of flower
7. Stamens
8. Pistil
9. Diagram of section through fruiting head
10. Lateral view of seed
11. Ventral view of seed

ROSACEAE
Potentilla tridentata Ait.

Sibbaldiopsis tridentata (Solander) Rydb.

Three-toothed cinquefoil

A suffrutescent shrub, the erect stems 5-10 cm high, not including the inflorescence; the horizontal stems often covering an area of 40 cm in diameter, rooting at most of the nodes.

LEAVES: Alternate, palmately 3-foliolate, tardily deciduous in our area, clustered at branch tips, the leaflets flat or slightly folded lengthwise, lateral veins hardly larger than the veinlets; leaflets spatulate to narrowly obovate, center leaflet largest, 1.5-3 cm long, 5-12 mm wide, tip truncate with 3 acute teeth 1-2 mm long, tip of teeth often red; leaf sides entire, base long cuneate; young leaflets with white appressed pubescence on both sides and with yellowish papillae-like hairs which later become a brownish scurf; upper surface glabrate, dark green; lower surface densely appressed pubescent, hairs to 1 mm long; petiole 1-2 cm long (up to 5 cm in some more eastern specimens), grooved above, loosely appressed pubescent, reddish, leaflets sessile or nearly so; stipules sheathing, appressed pubescent, adnate for 10 mm, the free ends lanceolate, 5 mm long. Abscission layer at the summit of the petiole, leaving the petioles on the stem.

FLOWERS: Mid-June; perfect. Flowers in cymes of 2-10 flowers, each flower 1.5-2 cm across; pedicels 1-2 cm long, strigose, expanded at the summit to a broad, saucer-shaped hypanthium, glabrous outside and pubescent within; calyx lobes 5, ovate, acute to acuminate, 4-5 mm long, green, strigose, reflexed but often with the tip turned upward, these alternating with 5 oblong bracteoles which are green, strigose, and about as long as, or longer than, the calyx lobes; petals obovate, 7-7.5 mm long, 4 mm across, white, often reflexed; stamens about 20, filaments slender, white, 4-5 mm long, inner filaments pubescent at the base, outer ones glabrous; anthers pale yellow to pinkish; numerous ovaries on a cylindric receptacle; ovary 0.6-0.7

mm long, 0.4 mm wide, oval, green, densely pubescent; style white, 3 mm long, attached laterally near the ovary base; stigma capitate.

FRUIT: August-September. Fruiting stalk 3-7 cm long, erect, slender; calyx lobes erect or incurved, fruiting heads 3-4 mm high, 4-5 mm across, 10-15 achenes, receptacle densely pubescent; achenes irregularly ovoid, 1-1.2 mm long, 0.75-1 mm wide, smooth, red-brown, pubescent with tawny, crinkly hairs as long as the achene.

TWIGS: Erect branches completely covered by old stipules and petioles, about 2 mm diameter; creeping stems 3-5 mm diameter, 5-20 cm long, bark tight, red-brown, shiny, sparingly pubescent; pith greenish or brown, continuous, one-half of stem. End buds of leafy twigs naked, concealed beneath old stipules; lateral buds of creeping stems ovoid, obtuse, 1-2 mm long, delicate, the scales thin, papery, brown or pale green; buds on underground stems broadly ovoid, red-brown.

TRUNK: Bark dark brown, loose, often exfoliating into small sheets. Stems hardly woody, soft, nearly white, with a small, pale brown pith.

HABITAT: In our area, known from only one place in North Dakota, growing in dry, shale-clay soil on northwest exposure of a semibarren prairie hillside.

RANGE: Mackenzie, east to Labrador, south to New York, west to northeastern Iowa and northeastern North Dakota, and northwest to Alberta. Also in Greenland and in the mountains of Georgia.

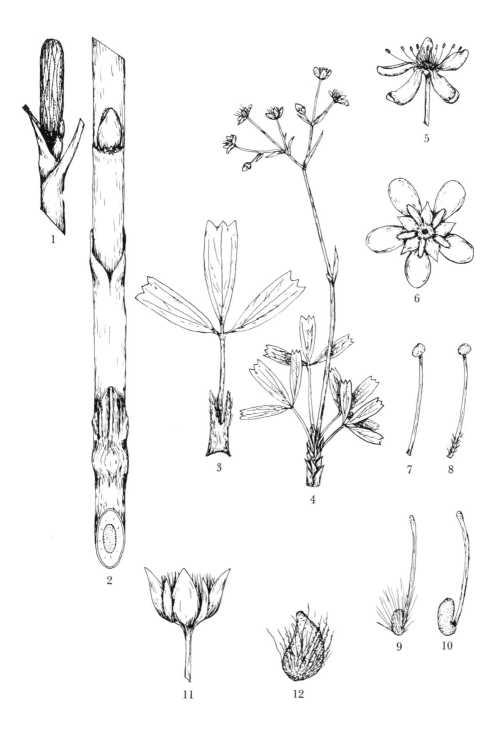

1. Winter naked end bud with old leaves removed

2. Winter twig of runner, sheath removed at bud

3. Leaf

4. Leaves and inflorescence

5. Flower

6. Underside of flower

7. Outer stamen

8. Inner stamen

9. Pistil

10. Pistil with pubescence removed

11. Fruiting head

12. Seed

ROSACEAE
Rubus allegheniensis Porter

R. nigrobaccus Bailey

Highbush blackberry

An erect shrub, the branches occasionally to 3 m and arching high or being supported by surrounding trees or shrubs.

LEAVES: Alternate, 3-5 foliolate. Leaflets ovate to elliptic; terminal leaflet of primocane 6-9 cm long, 3.5-4.5 cm wide, widest at or below the middle; lateral leaflets 4-7 cm long, 2-3.5 cm wide, the lower pair smaller; coarsely sharp-toothed, single or double serrate; teeth on the apical side of the laterals commonly lower and more sharply serrate than those on the basal side; tip acute to short acuminate; base rounded to obtuse; upper surface dark, dull green, glabrous; lower surface paler, pubescent, a few stipitate glands on the midrib; thick and often rugose; petiole 2.5-3 cm long, with hooked prickles; terminal petiolule 1.5-2 cm long, lateral petiolules 1-5 mm long, lower petiolule attached to the base of the middle petiolule; stipules linear, 5-10 mm long, pubescent, stipitate glandular.

FLOWERS: May-June; perfect. Terminal raceme 10-12 cm long, 5-6 cm wide; the lower 2-3 flowers subtended by a leaf, upper flowers with a bract; axis pubescent and with red stipitate glands, raceme prickles short, broad-based, recurved; pedicels green 3-4.5 cm long, stipitate glandular and prickly, a small bract at mid-length; calyx lobes 5, ovate, 5 mm long, 4 mm wide, green with a reddish cast, acuminate, short caudate tip, pubescent on both sides, cupped, reflexed in flower; petals 5, oval, 17-19 mm long, 11-13 mm wide, with a short claw, white, pinkish in bud, tendency to "double flower" up to 15 petals; stamens many, filaments white, 2-4 mm long, slender, not drooping when the petals fall; anthers yellow, turning brown after opening; pistils many, crowded on the rounded receptacle; ovaries green, 0.9-1 mm long, ovate, flattened, glabrous; style 2 mm long, green, curved, clavate; stigmas yellow-green.

FRUIT: Early July. Pedicels 1-2 cm long, pubescent, stipitate glandular; fruit a globose or cylindric cluster of drupelets on the elongated receptacle; each drupelet 4-5 mm diameter, black, glossy, juicy, sweet; receptacle falls with the fruit. Seeds straw-colored, unevenly ovate, 2.8-3 mm long, 1.8-2 mm wide, 1.2-1.4 mm thick, reticulate; the reticulations more coarse than those of *R. ostryifolius*.

TWIGS: 1.2-1.4 mm diameter, greenish to red-brown, flexible; prickles broad-based, recurved, 3-3.5 mm long, red at the base, yellowish tip; pubescent, few stipitate glands; leaf scars indefinite, petiole base persistent; pith greenish to white, continuous, two-thirds of stem. Buds narrowly ovoid, 3-4 mm long, 1.5-1.6 mm wide; scales red-brown, acute pubescent.

TRUNK: Biennial. Primocanes green to reddish, ribbed; prickles numerous, straight or recurved, flat, medium-broad base, reddish, usually darker about the middle; floricanes brown. Wood soft, large pith.

HABITAT: Dry areas, pastures, prairies, roadsides, open woods, margins of woods, rocky hillsides, or rich bottom lands.

RANGE: New Brunswick, west to Minnesota, south to Kansas, east to Tennessee and North Carolina, north along the west side of the Appalachians.

The blackberries are a difficult group to identify. They vary within the species and intergrade with other species, and the description of any one species is difficult. No attempt is made here to determine varieties.

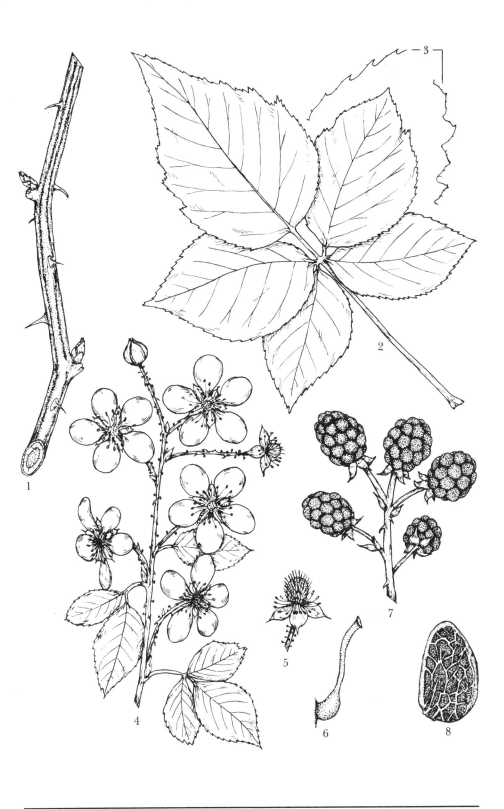

1. Winter twig
2. Leaf
3. Variation in leaf margins
4. Flowering raceme
5. Flower after petals fall
6. Pistil
7. Fruit cluster
8. Seed

ROSACEAE
Rubus flagellaris Willd.

R. baileyanus Britt.; *R. plicatifolius* Blanch.; *R. aboriginum* Rydb.; *R. invisus* (Bailey) Britt.; *R. occidualis* Bailey

Dewberry

A trailing vine-like plant, the branches up to 4 m long.

LEAVES: Alternate, 3-5 leaflets. Terminal leaflet of primocanes 6-8 cm long, 4.5-6 cm wide, broadly ovate, widest below the middle, petiolule 1.5-2 cm long; lateral leaflets 4-6 cm long, 2.5-4 cm wide, lower ones smaller, ovate to obovate, middle leaflet petiolule 2-3 mm long, lower leaflets sessile; margin coarsely serrate to double serrate, 2.5-3 of the larger teeth per cm, ciliate, teeth pointed, often reddish; leaflet tip acute, base rounded; upper surface dark green, slightly scabrous, a few long hairs; lower surface paler with soft pubescence; petiole 4-6 cm long, often with a few prickles; stipules lanceolate, 8-10 mm long, pubescent.

FLOWERS: Mid-May; perfect. Flowers in few-flowered corymbs or single and axillary; pedicels 2-4.5 cm long; calyx lobes 5, divided to near the base, ovate, 5-5.5 mm long, tip subulate, densely woolly inside and out; petals 5 or more, 15-16 mm long, 6-7 mm wide, obovate, cleft at the end, white, often with a pinkish cast; stamens many in 2-3 rows on the rim of the calyx; filaments white, 3-5 mm long, anthers yellow; pistils many on a slightly elongated receptacle; ovary obovate, 1 mm long, 0.6 mm wide, green, glabrous; style 2.4-2.7 mm long, green; stigma capitate, 2-3 lobed, green.

FRUIT: Mid-June. An oval aggregate cluster of many druplets, 1.2-2.2 cm long, 1.4-1.8 cm wide, each druplet 3.5-5.5 mm across; fruits turn red then black at maturity, lustrous; receptacle elongates and drops with the fruit; sweet, juicy, edible. Seeds unevenly ovate, 2.9-3.2 mm long, 2-2.3 mm wide, 1.2-1.4 mm thick, straw-color, strongly reticulate.

TWIGS: 1.5 mm diameter, flexible, green or reddish, angled or terete, glabrous, armed with recurved prickles with wide bases; leaf scars broadly crescent-shaped; 3 bundle scars; pith greenish, continuous, one-half of stem. Buds ovoid, 3-5 mm long, brown, outer scales slightly keeled, appressed pubescent.

TRUNK: Bark of primocanes green, usually brown on the floricanes. Wood soft, white, a large pith; branches often root when covered with dirt.

HABITAT: Rocky, open woods, thickets, roadsides, railroad banks, prairies, pastures.

RANGE: Nova Scotia, west to Minnesota, south to Texas, east to Florida, and north to New England.

The only other dewberry with which this might be confused is *R. enslenii* Tratt, which has obovate terminal leaflets on the primocanes and the lower margins of these leaflets are nearly straight. The lower margins of *R. flagellaris* may also be straight.

If varieties are used our plants are mainly var. *occidualis* Bailey as described above, contrasted to the nearly glabrous lower leaf surface of var. *flagellaris*.

The fruits of *R. flagellaris* are large and of good quality. The flavor is excellent and the berries are often gathered for sauce, pies, or jams or eaten raw with a bit of sugar. They also add a nice flavor to cold cereals for breakfast. Birds and small mammals use them for food; even the fox is quite fond of them.

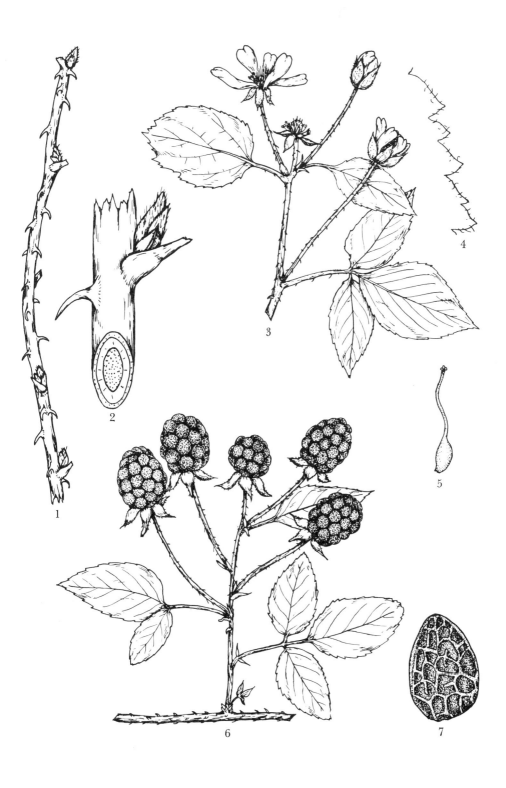

1. Winter twig
2. Detail of twig
3. Leaf and flower
4. Leaflet margin
5. Pistil
6. Fruit cluster
7. Seed

ROSACEAE

Rubus idaeus L. ssp. *sachalinensis* (Levl.) Focke var. *sachalinensis*.

R. strigosus Michx.; *R. idaeus* var. *strigosus* (Michx.) Maxim.; *R. melanolasius* Dieck; *Batidaea strigosa* Greene

Red raspberry

A shrub to 1.5 m high, the canes erect or arching, thicket-forming.

LEAVES: Alternate, 3-5 leaflets. Terminal leaflet ovate to oblong-ovate, 5-7 cm long, 2.5-5 cm wide, tip acuminate, base rounded to cuneate, serrate or biserrate, 3-6 teeth per cm, the apical side of the lateral leaflets cuneate, the basal side rounded and wider than the apical side; upper surface dark green, sparingly pubescent, lower surface densely white tomentose, midvein with a few bristles and glandular hairs; petiole 6-8 cm long, nearly glabrous or densely pubescent, bristly glandular; terminal petiolule 1-2 cm long, laterals sessile; stipules lance-linear, 9-11 mm long, 0.5 mm wide, pubescent and glandular, adnate to the petiole for about 4 mm, abscission layer above the stipules.

FLOWERS: June-July; perfect. Inflorescence racemose, corymbose, or cymose, terminal on short branches of the season, but often with a flower or two in the upper leaf axils, flowering extending over 3-5 weeks, the branches of the inflorescence glabrous or densely pubescent, glandular and bristly. Pedicels 1.5 cm long; calyx saucer-shaped, 6-7 mm broad, 2 mm deep, the 5 lobes narrowly ovate, 7-8 mm long with a short slender tip, glandular, pubescent and often bristly, reflexed or spreading; petals 5, obovate or spatulate, 5-6 mm long, white, often ciliate at the tip, ascending; stamens numerous, inserted on the rim of the calyx cup; filaments 3-5 mm long, white, flattened, attached at the center of the pale yellow anther; receptacle dome-shaped, pubescent; ovaries ovoid, green, 1 mm long, densely puberulent, style 2-3 mm long, slender, whitish, glabrous; stigma capitate.

FRUIT: July-August. The aggregate fruit hemispheric, 13-17 mm diameter, red, separates easily from the receptacle; druplets 3-4 mm across, 1-seeded, finely puberulent; calyx lobes often erect on green fruits, reflexed at maturity. Seeds bean-shaped, 2.4 mm long, 1.4 mm wide, 1 mm thick, ivory-colored, reticulate.

TWIGS: 2-3 mm diameter, pale salmon to purplish, often glaucous, nearly glabrous or covered with soft, slender, yellowish prickles up to 2.5 mm long, some glandular; leaf scars indistinct, covered by the old petiole base which partly surrounds the bud; pith whitish, continuous, three-fourths of stem. Buds 5 mm long, 2 mm wide, ovoid, acute, red-brown, scales puberulent and ciliate.

TRUNK: Bark red-brown, thin, covered with acicular, soft, yellowish prickles; wood soft, consists of a narrow rim around a large brown pith.

HABITAT: Open or wooded hillsides, ravine banks, talus slopes; moist or dry soil.

RANGE: Alaska, east across northern Manitoba to Labrador and Newfoundland, south through New England to the mountains of North Carolina, west to Ohio and northwestern Missouri, north to South Dakota, west to Wyoming, south to New Mexico, west to California, and north to British Columbia.

The plants of our area are var. *sachalinensis,* differing mainly from the more western var. *peramoenus* (Greene) Fern. by having the white, tomentose under leaf surface. Occasionally an eglandular plant appears, but it is considered a mutant form and not the ssp. *idaeus* of Europe.

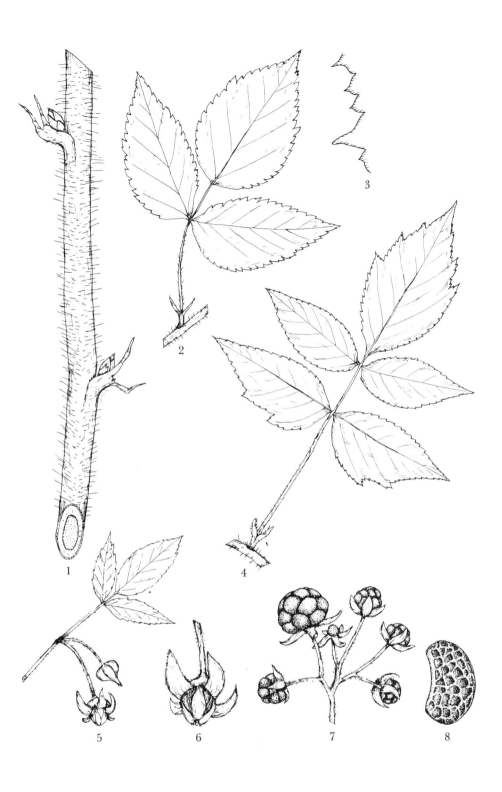

1. Winter twig
2. Leaf of fruiting cane
3. Leaflet margin
4. Leaf of primocane
5. Flower cluster
6. Flower
7. Fruit cluster
8. Seed

ROSACEAE
Rubus occidentalis L.

Black raspberry

An arching shrub, the canes to 2.5 m long, occasionally trailing.

LEAVES: Alternate, 3-5 leaflets. Leaflets ovate to broadly ovate, 6-8 cm long, 4-6 cm wide; margin coarsely double serrate, sometimes with small lobes about 1 cm across, ciliate; tip acuminate, base rounded to truncate and often uneven on the laterals; primocanes usually with 5 leaflets, the flowering canes with 3; upper surface dark green, some pubescence; lower surface densely white tomentose; petiole 5-6 cm long, glabrous, 1-2 prickles; lateral leaflets sessile or with a petiolule to 2 mm long; terminal petiolule 3-3.5 cm long; stipules linear, 5-7 mm long, pubescent.

FLOWERS: Mid-May; perfect. Flowers in an umbelliform cluster, the branches with hooked prickles, the main stalk 5-8 cm long with 1-2 small simple leaves, pubescent; pedicels 5-7 mm long, green, pubescent, a bract at the base; calyx flat, 5-lobed, lobes 6.5-8 mm long, 3-4 mm wide, broad base, acuminate, pubescent on both sides, green, spreading; petals 5, white, obovate, 4-4.5 mm long, 2 mm wide, narrow at the base, early deciduous; stamens many, attached to the calyx rim in one row, erect; filaments white, 2-3.5 mm long, anthers pale yellow, turning dark after the release of the pollen; pistils many on a dome-shaped receptacle; ovary green, 1-1.2 mm long, rounded on one side, straight or concave on the other, pubescent at the base of the style, along the flat side, and at the base; style 2.3 mm long, pale green, curved; stigma 2-3 lobed, capitate.

FRUIT: Mid-June. Pedicels 1.4-1.6 cm long; fruit hemispherical, 12.5-14 mm across, purple-black with many small drupes; falls free from the receptacle at maturity; each drupe 2.6-2.9 mm wide. Seeds 2.3-2.6 mm long, 1.4-1.6 mm wide, 1.2-1.4 mm thick, stramineous, ovate to crescent-shaped, one edge nearly straight, reticulate.

TWIGS: 3 mm diameter, flexible, glabrous, first-year canes glaucous, light red-purple, prickles broad-based, recurved; leaf scars, when visible, narrow; 3 bundle scars; pith brown, continuous, one-half of stem. Buds narrowly ovoid, 3-5 mm long, partly hidden by the old leaf base, red-brown, pubescent.

TRUNK: Bark smooth, red-purple and glaucous, prickles recurved. Wood soft, white, large pith. Tips of branches often root when touching the ground.

HABITAT: Open woods, bluffs, thickets, ravines, pastures; in rocky or good, rich soil.

RANGE: Quebec, west to North Dakota, south to Colorado and New Mexico, east to Georgia, and north to New York.

The black raspberry is common but not abundant in the eastern part of our area from Kansas to the southeastern corner of South Dakota. It is usually found in patches, either open or densely matted. The prickles are stout, recurved, and sharp, and it is an excellent plant to use as cover for song and game birds which inhabit the woods. Mammals also use the cover and eat the fruits, the fox being especially fond of it. The mass of canes catches and holds the leaves in autumn and makes a good soil cover against erosion and water evaporation. The fruits are gathered and eaten raw or made into jam and jelly, or put in pies, muffins, and ice cream.

This is one of the few plants in which the sepals exceed the petals in length, the petals being only about half as long as the sepals.

254

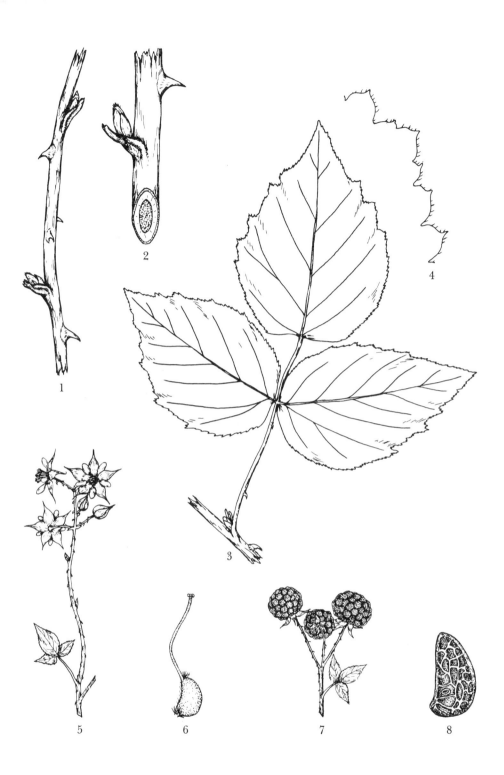

1. Winter twig
2. Detail of twig
3. Leaf
4. Leaflet margin
5. Flower cluster
6. Pistil
7. Fruit
8. Seed

ROSACEAE
Rubus ostryifolius Rydb.

R. argutus Link; *R. kansanus* Bailey;
R. ozarkensis Bailey

Highbush blackberry

Usually a low, thicket-forming shrub to 1 m high, but occasionally to 2.5 m with the canes arching.

LEAVES: Alternate, 3-5 leaflets. Leaflets ovate, broadly elliptic or slightly obovate; terminal leaflet of primocanes 6-10 cm long, 3.5-5 cm wide, widest near the middle; irregular, coarse teeth, single or double, 3-5 teeth per cm; lateral veins to the larger teeth; tip of leaflet short acuminate; base acute to obtuse, often unequal; upper surface dark green, glabrate; lower surface paler, soft pubescent, hooked prickles on the midrib of the terminal leaflet; petiole 3-6 cm long, pubescent, recurved prickles on the lower side; terminal petiolule 1.5-2.5 cm long, middle leaflet petiolule 3-4 mm long, the lower leaflets nearly sessile and attached to the base of the middle petiolule; stipules narrowly ovate, 5-7 mm long, entire or toothed.

FLOWERS: May-June; perfect. Short terminal racemes, 1-7 flowers, the axis pubescent, leaf-like bracts on the 2 lower pedicels, small bracts on the upper pedicels; pedicels 2-3.5 cm long, green, pubescent, an occasional recurved prickle; calyx lobes 5, divided to near the base, 5-6 mm long, ovate, acute, pubescent on both sides; petals 5, white, ovate, 1.8-2 cm long with a short claw; stamens many, inserted at the base of the calyx lobes; filaments white, 4-8 mm long; anthers yellow; pistils many, crowded on the ovoid receptacle; ovary green, 1 mm long, ovate, flat; style 2 mm long, greenish-yellow, slightly clavate; stigmas capitate, yellowish.

FRUIT: July. Short racemes of 5-8 fruits, often several racemes along the side of the cane near the end; pedicels 1.5-3 cm long, the lowest subtended by a leaf; fruit a cylindric or globose aggregate cluster of drupelets on an elongated receptacle; fruit 1.5-2.5 cm long, 1.5 cm thick; drupelets obovoid, nearly terete, 4-5 mm diameter, red when immature, glossy black at maturity; receptacle falls with the fruit. Seeds ovate, flattened, 2.7-3 mm long, 1.9-2.2 mm wide, 1.3-1.6 mm thick, stramineous, reticulate.

TWIGS: 1.5-1.7 mm diameter, semi-rigid, red-brown, or blotched red and green, ridged, pubescent; prickles broad-based, recurved, 3.5-4 mm long, red near the base and with yellowish tips; leaf scars indefinite; pith greenish, continuous, one-half of stem. Buds ovoid, 2.8-3.2 mm long, 1.6-1.7 mm wide; scales red or pinkish, acute, pubescent.

TRUNK: Biennial. Primocanes green, blotched with red-brown, 5-ridged and with a few nearly straight prickles. Floricanes red-brown, slightly ridged, and with straight or hooked prickles. Wood soft with a large, brown pith.

HABITAT: Fence rows, pastures, prairies, open woods, margins of woods, gullies, and roadsides.

RANGE: Maine, west to Michigan and Minnesota, south to Oklahoma, east to Kentucky and Virginia, and north to New England.

The canes of *R. ostryifolius* appear to be weaker than those of *R. alleghen-iensis* and are often more trailing. Also, the inflorescence lacks the abundance of red stipitate glands of *R. allegheniensis*. It is also similar to *R. pensilvanicus* Poir. but lacks the dense, velvety pubescence on the under leaf surface, and the terminal leaflets are widest near the middle rather than at the base as in *R. pensilvanicus*.

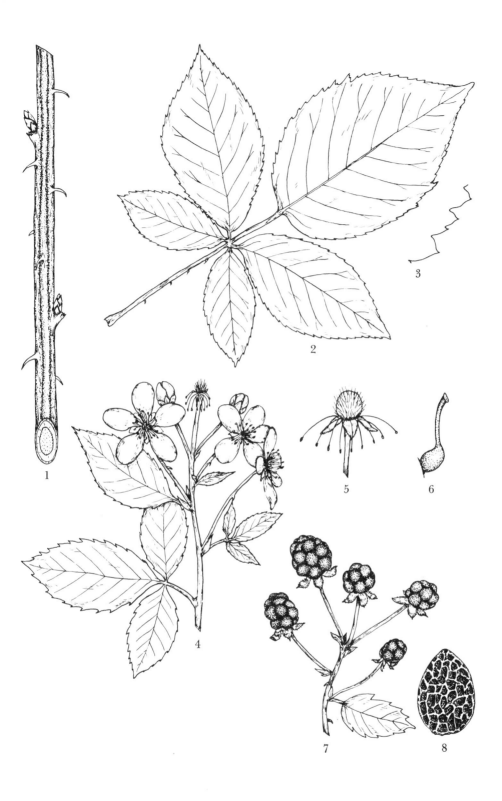

1. Winter primocane
2. Primocane leaf
3. Leaflet margin
4. Flowering raceme
5. Flower after fertilization
6. Pistil
7. Fruit
8. Seed

ROSACEAE
Rubus parviflorus Nutt.

R. nutkanus Moc. ex Ser. in DC. var.
nuttallii T. & G.; *R. velutinus* H. & A.;
R. parviflorus var. *grandiflorus* Farw.;
Rubacer parviflorum Rydb.; *Bossekia
parviflora* Greene

Thimble berry

An open, few-branched shrub to 1 m high, the branches coarse.

LEAVES: Alternate, simple, 5-lobed, palmately veined, the veins below slightly winged in the angles. Broader than long, 10-14 cm long, 11-18 cm wide, lobes ovate, tips acute to short acuminate, base of leaf deeply cordate, the sinus wide or close; dentate-serrate or biserrate, 2-4 teeth per cm, apiculate, sparingly ciliate, and an occasional stipitate gland; dark green, dull, and sparingly pubescent above; paler below, glabrous except for stipitate glands on the larger veins; petiole 6-13 cm long, terete, puberulent and stipitate glandular or glabrous, abscission layer above the stipules, slightly enlarged at this point; stipules ovate to lance-ovate, 5-15 mm long, 2-3 mm wide, pubescent or glabrous, stipitate glandular outside.

FLOWERS: June; perfect. Corymbs 2-6 flowered, puberulent and glandular, peduncle 5-8 cm long, pedicels 1.5-3 cm long, bracts lanceolate, ovate or obovate, 3-12 mm long, 1-7 mm wide, larger bracts deeply toothed at the apex, puberulent, stipitate glandular outside, glandular ciliate and ciliate; calyx saucer-shaped, 6-8 mm across, 2.5-3 mm deep, wrinkled transversely at the base, densely pubescent and glandular; calyx lobes 5, ovate, abruptly narrowed to a caudate tip, total length 10-14 mm, caudate tip about half of the length, pubescent and glandular, densely white ciliate, tomentose inside; petals 5, obovate, 2-2.2 cm long, 13-16 mm wide, white, margins overlap; numerous stamens, filaments 2-5 mm long, white, slender, crowded around the receptacle, anthers yellow; receptacle slightly rounded, 4-5 mm across, pubescent; ovary oblong, 1 mm long, 0.5 mm wide, green, densely pubescent at the top; style greenish, stout, 1-1.2 mm long; stigma capitate.

FRUIT: August. Sepals reflexed or spreading, the aggregate fruit hemispheric to nearly globose, 13-17 mm across, druplets 2-2.5 mm across, covered with deep red, papillae-like hairs, sweet and edible, does not remove easily from the receptacle. Seeds ovoid, 2 mm long, 1.3 mm wide, 1 mm thick, reticulate, the raphe ridged.

TWIGS: 2-4 mm diameter, slightly zigzag, rigid, pale brown, mottled and streaked with dark brown, few stipitate glands, long internodes; leaf scars indefinite, petiole breaks above the stipules; pith yellow, continuous, most of stem. Buds ovoid, obtuse, 8-10 mm long, 3-4 mm wide, the outer scales brown, papery, brittle, nearly glabrous; the inner scales densely covered with white or tawny, straight hairs, glabrous on the inner surface.

TRUNK: Outer layer of bark gray-brown, exfoliating readily, inner layer red-brown, smooth. Wood consists of a thin greenish layer, with obvious rays surrounding the large yellow pith.

HABITAT: Margins of woods, along stream banks, and in the Black Hills, often found in narrow, deeply shaded canyons with moist soil.

RANGE: Discontinuous. Alaska, south to southern California, Arizona, and New Mexico, north to Colorado, Montana, and Alberta; the Black Hills; Bruce Peninsula, Ontario, and Minnesota.

The plants of our area are var. *parviflorus* as described above; var. *velutinus* (H. & S.) Greene, a more hairy form, is apparently found only in southern California.

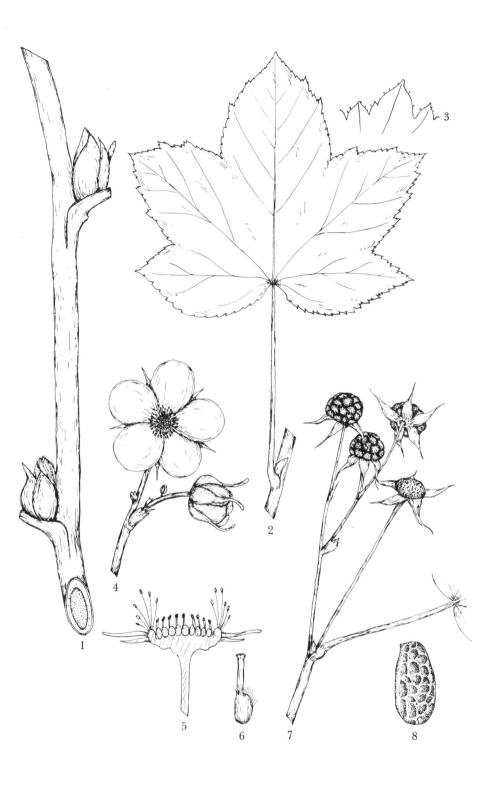

1. Winter twig
2. Leaf
3. Leaf margin
4. Flower
5. Diagrammatic cutaway of flower
6. Pistil
7. Fruit cluster
8. Seed

ROSACEAE
Rubus pensilvanicus Poir.

R. alumnus Bailey; *R. hancinianus* Bailey; *R. frondosus* Bigel.; *R. orarius* Blanch.; *R. laudatus* Berger

Highbush blackberry

A shrub with arching canes to 3 m long, forming open thickets.

LEAVES: Alternate, 5-foliolate on primocanes, 3-foliolate on fruiting canes. Leaflets sparingly pubescent above and rather densely soft pubescent below; terminal leaflet of primocanes ovate, definitely broadest below the middle, 8-10 cm long, 6-8 cm wide, acuminate, base rounded or subcordate; petiolule 3-5 cm long, pubescent with a few prickles; lateral leaflets smaller, middle leaflets with a petiolule 2-3 cm long, basal pair nearly sessile. Floral cane leaflets smaller, more variable in shape, often obovate with cuneate base and acute tip. Irregularly serrate, biserrate, or with large teeth about 1.5 cm across and 4-5 smaller teeth on their margin, usually with a reddish callous tip; petiole 8-14 cm long, sharply grooved above, densely soft pubescent, a few prickles, and an occasional scurfy gland; stipules lanceolate, 6-7 mm long, minutely serrate, ciliate, pubescent.

FLOWERS: Late May; perfect. Corymbiform panicles surpassing the leaves, the terminal flower opening first, lower pedicels usually subtended by small leaves diminishing to bracts at the upper pedicels; axis and pedicels densely soft pubescent; pedicels 2-4 cm long; calyx lobes 5, divided nearly to the base, ovate, apiculate, pubescent and densely tomentose inside and on the margins; spreading or reflexed; petals 5, ovate to obovate, narrowed to a claw of 3-4 mm, total length 17-20 mm, width 10-12 mm, white; stamens numerous, filaments slender, 5-10 mm long, white, attached at the center of the pale yellow anthers; receptacle cylindric; ovaries numerous, ovoid, 1.2 mm long, 0.75 mm wide, green, glabrous; style terminal, curved, gradually widened upward, 2 mm long, green, glabrous; stigma capitate, irregular.

FRUIT: July, time of ripening and the size depend greatly on the temperature and amount of moisture. Racemes 6-10 cm long, 4-6 cm wide; aggregate fruit 1-1.5 cm long, 10-12 mm wide, receptacle falling with the fruit; red before ripening, black at maturity, glossy, smooth, the calyx lobes reflexed; drupelets about 4 mm across, globular to oval, 1-seeded; sweet, delicious. Seed stramineous, unevenly ovoid, reticulate, ridged on both dorsal and ventral sides, 3 mm long, 2 mm wide, 1.2 mm thick.

TWIGS: Primocanes 1.8-2 mm diameter, fluted, red, red-brown or greenish, pubescent, glandular, prickles broad-based, recurved, lenticels not obvious; leaf scars indefinite, irregularly broken at the stipules; pith white, continuous, one-half of stem. Buds ovoid, acute or rounded, 4-7 mm long, 2-3 mm wide, outer scales broadly ovate, red-brown, appressed pubescent, acute to short acuminate, low keel, the inner scales densely white pubescent, the inner surface glabrous and brown; innermost scales densely long pubescent, the inner surface green and glabrous.

TRUNK: Primocanes nearly erect, fruiting canes arched and with longer lateral branches. Canes fluted, bark tight, red-brown or greenish, the broad-based, recurved prickles mainly on the ridges. Wood hard, greenish-white, porous, the rays obvious, consists of a thin rim around the 5-angled, tawny pith.

HABITAT: Open woods, ravine banks, unused prairie lands, usually in rocky, limestone soil, dry or slightly moist.

RANGE: Newfoundland, west to Ontario, south through Minnesota to southeastern Nebraska, eastern Kansas, and Oklahoma, east through Arkansas, Tennessee, and Virginia, and north to New England.

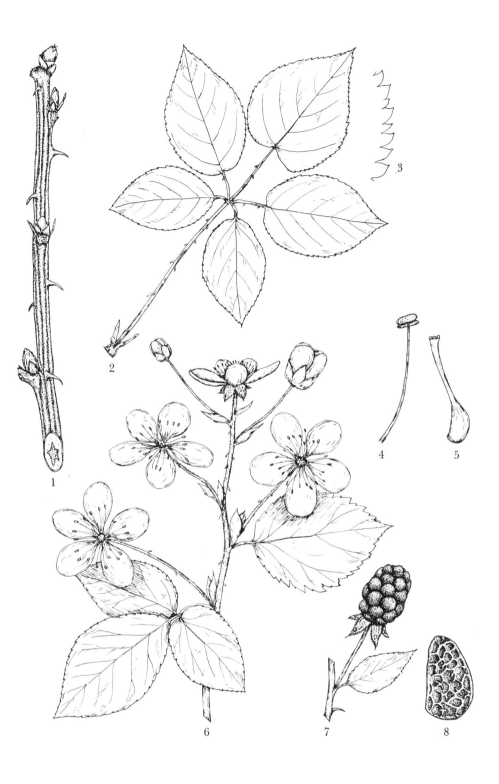

1. Winter primocane
2. Primocane leaf
3. Leaflet margin
4. Stamen
5. Pistil
6. Inflorescence
7. Fruit
8. Seed

ROSACEAE
Rubus pubescens Raf.

R. *saxatilis* L. var. *canadensis* Michx.;
R. *saxatalis* L. var. *americana* Pers.; *R.*
triflorus Richards.; *R. americanus* Britt.

Creeping blackberry

A suffrutescent shrub, the erect branches about 30 cm high, trailing stems to 75 cm long.

LEAVES: Alternate, 3-foliolate, tardily deciduous, and a few may remain green over winter under a mulch of dead leaves. Terminal leaflet rhombic, 4-7 cm long, 2-6 cm wide, sides below the middle nearly straight; laterals symmetrically or obliquely ovate, the basal side rounded and often lobate, the apical side cuneate; serrate or biserrate, 1-2 teeth per cm, stiffly ciliate; upper surface dark green, sparsely appressed pubescent; the lower surface pilose, the veins often winged in the angles; petiole 3-6 cm long, slender, sharply grooved above, pubescent; terminal petiolule 8-10 mm long, laterals sessile or on a petiolule to 3 mm long, stipules obovate to oblanceolate, 6-10 mm long, 3-4 mm wide, entire or toothed, ciliate.

FLOWERS: June-July; perfect. 1-2 flowers at the end of a new branch; pedicels 1-3 cm long, green, finely puberulent, a few stipitate glands; calyx tube saucer-shaped, 5 mm across, 2.5 mm high, green, puberulent and glandular, lobes 5, narrowly ovate to broadly lanceolate, long attenuate, pubescent and glandular, glabrous inside, ciliate, erect or reflexed; petals 5, elliptic to oblong-spatulate, 5-7 mm long, about 2 mm wide, white, erect, and rounded; stamens numerous, filaments 1-4 mm long, flat, white, inserted on the calyx rim; anthers nearly globular; receptacle low, rounded, 15-25 ovaries, bean-shaped, 0.75 mm long, green, glabrous; style terminal, 1 mm long, curved, greenish-white, slightly widened toward the tip; stigma capitate.

FRUIT: July-August. Aggregate fruit 10-13 mm across, 8-9 mm high, hemispheric; drupelets 7-15, globular to oval, 4.5 mm long, 3 mm wide, dark red, often break individually from the receptacle, 2-seeded. Seeds bean-shaped, 2.7 mm long, 1.7 mm wide, 1 mm thick, ivory-white, reticulate with low ridges, the ventral side straight.

TWIGS: Woody stems, below point of dying back, about 1.8 mm diameter, runners filiform, sparingly pubescent or puberulent, red-brown, without prickles; leaf scars indefinite; pith greenish, continuous, most of stem. Terminal buds with 2-3 thin scales about 2 mm long, enclosing a series of densely pubescent, embryonic leaves; lateral buds on woody stems ovate, flattened, 1 mm long, densely pubescent, concealed beneath old stipules.

TRUNK: Hardly a woody plant, dying back to the ground or with a soft, woody stem to 10 cm high, bark smooth, dark red-brown, tight. Wood consists of a rim of greenish, soft material around the large white pith. In late summer the plant sends out filiform runners which take root and leaves appear; the runner continues beyond for about 70 cm, with up to 8 new plants formed. Many of these new plants die over winter.

HABITAT: Wet, shaded ground, usually in moss or leaf mold in boggy wooded areas; rocky, springy, shaded ravine banks.

RANGE: Alaska, east across Canada to Labrador and Newfoundland, south to New Jersey, west to northeastern Iowa, north to North Dakota and Manitoba, west to British Columbia and Washington; the Black Hills and Colorado; rare in Montana.

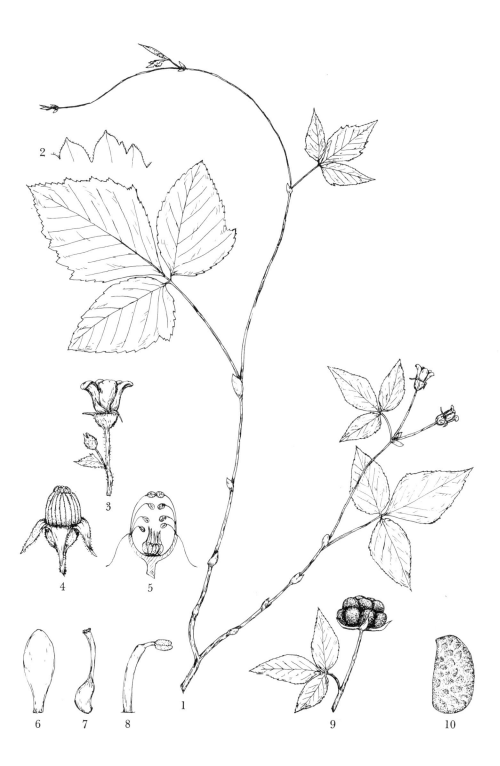

1. Flowering plant
2. Leaflet margin
3. Flower
4. Petals removed to show stamens
5. Diagrammatic cutaway of flower
6. Petal
7. Pistil
8. Stamen
9. Fruit
10. Seed

ROSACEAE
Rosa acicularis Lindl.

R. sayi Schwein. in Keating; *R. acicularis* ssp. *sayi* (Schw.) W. H. Lewis; *R. engelmanni* S. Wats.; *R. pyrifera* Rydb.; *R. butleri* Rydb.; *R. bourgeauiana* Rydb.

Wild rose

An open shrub to 1 m high, usually quite prickly.

LEAVES: Alternate, pinnately compound, 6-11 cm long, 5-9 leaflets, rachis grooved above, downy puberulent and glandular, usually with a few prickles. Leaflets elliptic, rarely obovate, 2-4 cm long, 1-2.5 cm wide, the end pair the largest; tip acute to rounded, base rounded to cuneate, margin sharply serrate, occasionally biserrate, 5-6 teeth per cm, callous-tipped, sinuses acute, ciliate; upper surface sparingly pubescent or glabrous, lower surface softly pubescent, occasionally glandular, especially on the veins; petiole 2-3 cm long, grooved above, short pubescent and glandular; stipules 12-20 mm long, free tips acute to acuminate, 6-8 mm long, glabrous above, pubescent and glandular below, glandular-ciliate.

FLOWERS: Early June; perfect. Single, rarely more, on new shoots from lateral buds; pedicels glabrous, green, 1-2 cm long; hypanthium 5-7 mm long, 4-5 mm wide, obovoid to pear-shaped narrowed to a neck, glabrous, green; calyx lobes 5, 2-2.5 cm long, 3 mm wide at the base, narrowed near the middle and usually expanded at the tip, expanded portion often with low, glandular teeth, lobes puberulent outside, densely soft tomentose inside, densely white ciliate; petals 5, obovate, 2-3 cm long, 1.5-2 cm wide, pink or red; stamens numerous, radiating from a flat collar around the throat of the hypanthium; filaments 4-8 mm long, white, slender; anthers pale yellow, about 1.5 mm long; pistils attached to the sides and bottom of the hypanthium; ovary reniform, 1.5-2 mm long, pale green, densely long hairy on the dorsal side; styles stout, 5-8 mm long, enlarged upward, appressed pubescent at the base; stigmas capitate.

FRUIT: August-September. Globose, ellipsoid, or pear-shaped, 1.5-2 cm long, 1 cm diameter, bright orange-red, smooth; calyx lobes erect; 10-24 achenes; achenes generally elliptic, with a few flattened surfaces, 4-5 mm long, 2 mm thick, light yellow-brown, long white hairs on the dorsal side.

TWIGS: 1-1.2 mm diameter, dark red to red-brown, with internodal, acicular prickles, nodal spines not sharply differentiated; leaf scars long and narrow; 3 bundle scars; pith white, continuous, four-fifths of stem. Buds broadly ovoid, 2-3 mm long, 1.5-2 mm wide, lateral buds often small and rounded, obtuse to acute, brown-red, pubescent at the tip.

TRUNK: Main stems densely covered with stiff bristles or prickles of various sizes, occasionally sparingly so. Bark smooth beneath the prickles, red-brown to purplish-red, tight, the prickles yellowish. Wood, a thin, greenish-white layer around the pale brown pith.

HABITAT: Wooded hillsides, stream banks, rock ledges; usually in moist soil.

RANGE: Alaska, east across northern Manitoba to Quebec and Maine, south to New York, west around the Great Lakes to North Dakota, south through Colorado to New Mexico, and northwest to eastern Idaho and British Columbia.

Our plants are var. *bourgeauiana;* var. *acicularis* of Eurasia has glandular pedicels, the leaves are 5-foliolate and eglandular, and the sepals are less than 3 mm wide at the base.

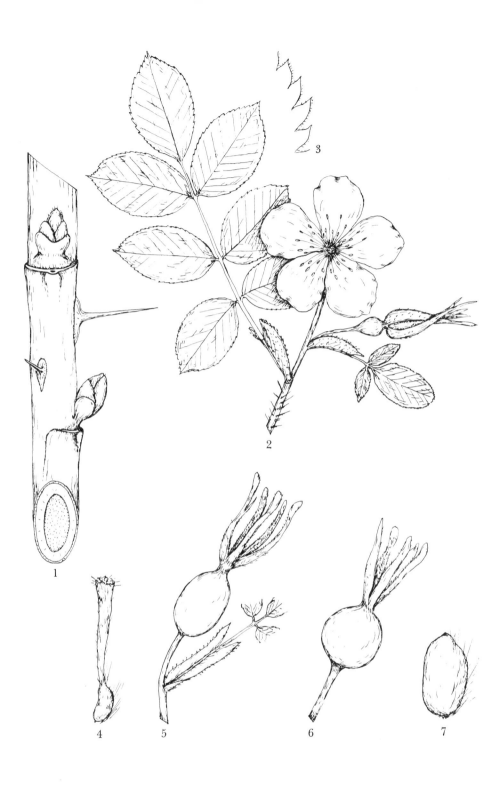

1. Winter twig
2. Leaves and flower
3. Leaflet margin
4. Pistil
5. Fruit
6. Fruit variation
7. Seed

ROSACEAE
Rosa suffulta Greene

R. arkansana Porter var. *suffulta* (Greene) Cockerell; *R. subglauca* Rydb.; *R. pratincola* Greene; *R. lunellii* Greene; *R. alcea* Greene; *R. conjuncta* Rydb.

Prairie rose

A shrub usually under 50 cm high, the main trunk densely armed with yellowish prickles.

LEAVES: Alternate, pinnately compound, 8-10 cm long, 5-9 leaflets. Leaflets oval to obovate, 1.5-2.5 cm long, 8-10 mm wide; sharply serrate nearly to the base, 5-7 teeth per cm; tip of leaf rounded or acute; base cuneate; upper surface dark green, glabrate, lower surface paler, soft pubescent; petiole 1.5-2 cm long, pubescent; rachis pubescent; stipules 1.5-1.8 cm long, free tip 4-5 mm long, acute, margin entire or with small teeth, pubescent, often reddish toward the base.

FLOWERS: Early June; perfect. Flowers terminal, solitary or in clusters of 2-4 on new growth, the new branch lateral or from the ground; pedicels 10-14 mm long, glabrous, an occasional stipitate gland and 2 basal bracts, which are entire, woolly, often with glandular margin; hypanthium 4-5 mm long, green, glabrous, occasionally with a few hairs; calyx lobes 5, lanceolate, 6-8 mm long, with long acuminate tip, entire or with a few lateral projections, pubescent and stipitate glandular, white woolly inside, persistent in fruit; petals 5, broadly obovate, 2-2.5 cm long and as wide, notched at the outer end, pink; stamens many, attached to the hypanthium rim; filaments yellow, 3-6 mm long; anthers yellow; ovaries many, attached to the bottom and inner walls of the hypanthium, half-round, 2 mm long, 1 mm wide, green, hairs on the rounded side; styles several, 1.5 mm long, hairy, slightly exserted, clavate; stigmas broad, yellow-green, often short hairy.

FRUIT: Late August. Solitary, or in clusters of 2-3, central pedicel shorter and stouter than the laterals; pedicels 1.2-2.5 cm long, glabrous, glossy; fruit subglobose, 10-13 mm long, 13-16 mm wide, sometimes longer than broad, bright red, calyx erect or deflexed. Seeds ovoid, 4-5.5 mm long, 2.8-3.5 mm wide, flattened on the sides, straw-colored to light red-brown, smooth, hairy on the apical half of the rounded side and often a tuft of hairs at the rounded base.

TWIGS: 1-1.5 mm diameter, flexible, red to red-brown; prickles numerous, terete, 1-3.5 mm long infrastipular prickles not differentiated; leaf scar a narrow line halfway around the stem; pubescent on the margin; 3 bundle scars; pith white, continuous, one-third of stem. Buds ovoid, 2.3-2.5 mm long, rounded tip, usually about 1.5 mm above the leaf scar.

TRUNK: Stem prickles abundant, straight, small-based, 4-8 mm long, some of which are flexible. Bark green to red-brown, smooth except for the numerous prickles. Wood hard, pale red-brown, large pith.

HABITAT: Prairies, open banks, loess hills, bluffs, thickets, roadsides, railroad banks, margin of woods; usually dry soil.

RANGE: Alberta to Manitoba, southeast through Indiana to the District of Columbia, west to Kansas, southwest to Texas and Mexico, and north to Montana.

Rosa suffulta is the most common wild rose in the southern part of our area, growing along the roadsides and in any unused area where the seeds may chance to fall. It is hardy, even withstanding the mowing of the roadsides. It is a short bush, not forming dense thickets, but grows in loose colonies. In our area it usually does not die to the ground in winter, and most of the flowers are produced on lateral branches.

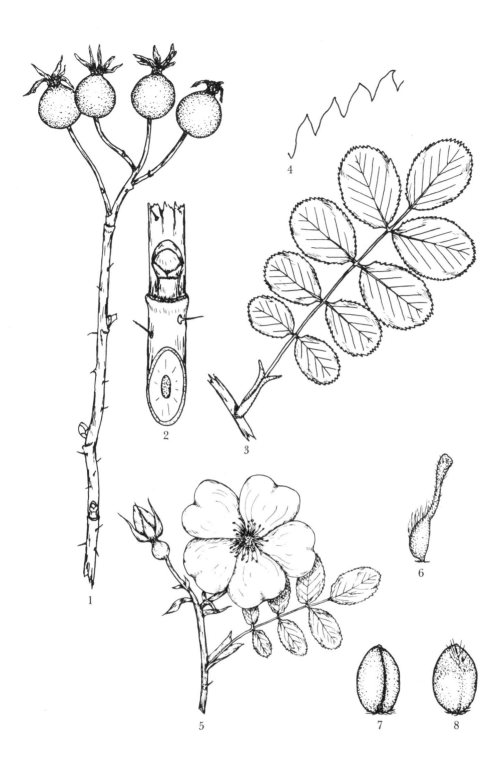

1. Winter twig with fruits
2. Detail of twig
3. Leaf
4. Leaflet margin
5. Flower
6. Pistil
7. Ventral view of seed
8. Dorsal view of seed

ROSACEAE
Rosa blanda Ait.

R. *solanderi* Tratt.; R. *subblanda* Rydb.;
R. *gratiosa* Lunell

Wild rose

An erect, colony-forming shrub about 1 m high, usually unarmed except at the base.

LEAVES: Alternate, pinnately compound, 6-7 cm long, 5-7 leaflets. Leaflets oval to obovate, 1-2.5 cm long, 8-15 mm wide; coarsely serrate, except near the base, 5-7 teeth per cm, ciliate; tip rounded to acute; upper surface dull, yellow-green, glabrate; lower surface paler, pubescent; petiole 1.5-1.8 cm long, pubescent; rachis pubescent; terminal petiolule 8-10 mm long, laterals 1 mm long; stipules adnate for 1.3-1.5 cm, free tips 2-3 mm long, spreading, pubescent on the lower side.

FLOWERS: Early June; perfect. Flowers solitary or in corymbs on new growth from lateral buds. The central flower opens first, its pedicel shorter and stouter than the others, and without bracts; pedicels 2-4 cm long, smooth, 1-2 bracts near the base; hypanthium ovate, 7-10 mm long, glabrous or with a few small hairs, weak prickles, or glandular hairs; calyx lobes 5, long attenuate, 10-20 mm long, often with foliaceous tip and some lateral teeth, pubescent on both sides, some stipitate glands, reflexed or spreading after flowering; petals 5, obovate to obcordate, 2.5-3 cm long, 2.3-2.5 cm wide, pink, often streaked with red, outer end notched; stamens many, usually horizontal against the petals until petals fall, then become erect; filaments white, 4-8 mm long; anthers yellow; ovaries many, 2-2.3 mm long, half-round, short stipe, long hairs on the rounded side; style 5 mm long, densely pubescent, clavate, not exserted; stigma yellow-orange, lobed, capitate.

FRUIT: Late August. Bright red, smooth, globose, 1.2-1.5 cm diameter; calyx lobes spreading or erect. Seeds ovoid, 5-6.5 mm long, 3.5-4 mm wide, yellow to dark brown, pubescent near the tip on the dorsal side, attached to the bottom and side walls of the hypanthium.

TWIGS: 1.2-1.4 mm diameter, flexible, glabrous, often glaucous, red for 1-2 years; prickles few or none; leaf scars narrow, halfway around the stem; 3 bundle scars; pith white, continuous, one-third of stem. Buds ovoid, 1.9-2 mm long, the scales ovate, red, somewhat keeled.

TRUNK: Prickles only at the base of the stem, small and straight, often absent entirely, or occasionally higher and on the upper branches. Bark gray-brown, smooth, and tight. Wood white, soft, a large, white pith.

HABITAT: Dry hillsides, roadsides, fence rows; rocky or sandy soil.

RANGE: Quebec to Ontario, south to Kansas, east to Missouri and Ohio.

Rosa blanda is uncommon in our range, although it may have been overlooked because of its similarity to other species. It is a sturdy rose, the stems are stout, and the seeds are much larger than those of the other prairie roses. It is more thicket-forming than either R. *suffulta* or R. *carolina*, with which it is often confused.

The roses of our area are quite variable, and the species' characteristics usually given do not always hold true. In addition, one plant may have the characteristics of two or three other species. Such characters as leaf tip, number of prickles, pubescence and glandular hairs are not constant in most of our species.

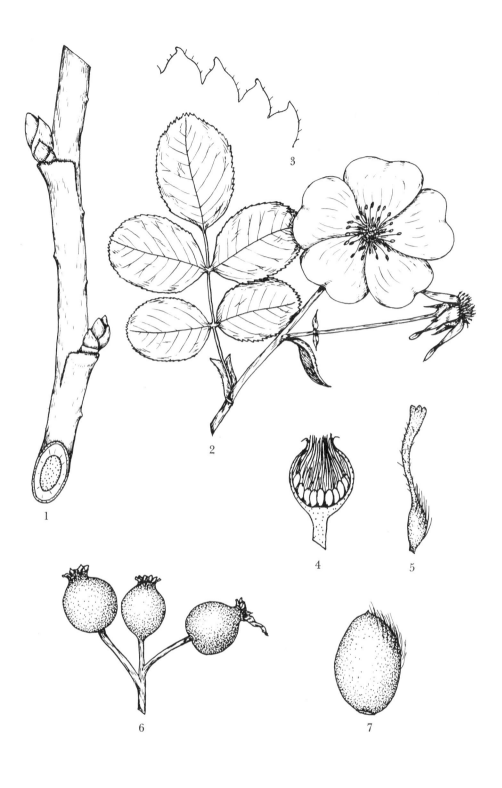

1. Winter twig
2. Leaf and flower
3. Leaflet margin
4. Section of hypanthium
5. Pistil
6. Fruit
7. Seed

ROSACEAE
Rosa carolina L.

R. humilis Marsh.; *R. lyoni* Pursh;
R. serrulata sensu Rydb. not Raf.

Pasture rose, Carolina rose

A shrub 50-80 cm high, the branches near the top, the main trunk armed.

LEAVES: Alternate, pinnately compound, 7-10 cm long, 5-7 leaflets. Leaflets elliptic, 2-4 cm long, 0.8-1.8 cm wide, the lower ones smaller; coarsely serrate, 5-6 teeth per cm, teeth with or without glands, often ciliate; tip acuminate, base acute; upper surface dark green, glabrous; lower surface paler, a few hairs on the veins; rachis with some prickles, with or without stipitate glands; petiole 10-12 mm long; stipules adnate nearly the whole length, 11-13 mm long, free tip 3 mm, acute, margin stipitate glandular and pubescent.

FLOWERS: Early June; perfect. Terminal on new branches, solitary or corymbose; pedicels, hypanthium, bracts, and calyx lobes stipitate glandular; pedicel 1-2 cm long, 2 foliaceous bracts at the base, margin pubescent; hypanthium green, globose, 6-8 mm diameter; calyx lobes 5, 1.5-2 cm long, margin with narrow projections, long attenuate, some with foliaceous tips, coarse pubescence outside, soft pubescence inside; petals 5, obovate, 2-2.5 cm long, 2-2.5 cm wide, pink, notched or a slight tip at the apex; stamens many, attached to the hypanthium rim; filaments white, 2-4 mm long, anthers yellow; pistils many, attached only to the bottom of the hypanthium; ovary pale green, 2-2.3 mm long, straight on one side, rounded on the other, and with long, straight hairs on the upper part; style pubescent, 3 mm long, curved, not exserted; stigma enlarged, indistinctly lobed, yellowish.

FRUIT: Late August. Pedicels 1-2 cm long; the glands or their scars remaining; fruit globose, 9-12 mm diameter, orange-red, surface with scattered stipitate glands or their scars; calyx lobes present or absent; 2-12 viable seeds attached to the bottom of the hypanthium. Seeds ovoid, 3.7-4.9 mm long, 2.4-2.8 mm wide; one margin nearly straight, sides

flattened if crowded, base rounded, minute tip at the apex; straw-colored to dark red-brown; straight, white hairs on the apical half of the rounded side.

TWIGS: 1.7-1.9 mm diameter, flexible, glabrous, green to green-brown, covered with short prickles; infrastipular prickles long, terete; leaf scars narrow, extending halfway around the twig; 3 bundle scars; pith white, continuous, one-third of stem. Buds 2-3 mm long, ovoid; scales red, acute, slightly keeled; end bud pubescent at the tip.

TRUNK: The stem prickles 2-8 mm long, straight, terete, the larger ones with a large, oval base. Bark green to red-brown. Wood soft, white, with a large, white pith.

HABITAT: Upland woods, roadsides, prairies, fence rows; usually rocky soils.

RANGE: Nova Scotia to Minnesota, south to Texas, east to Florida, and north to Maine.

There are four variations of this species in the central Plains. Forma *carolina,* lower surface of leaflets glabrate, tip of teeth without glands; forma *glandulosa* (Crepin) Fern., lower surface of leaflets glabrate, tips of teeth with glands; var. *villosa* (Best) Rehd., lower surface of leaflets soft hairy; var. *grandiflora* (Baker) Rehd.; large flowers, obovate to broadly oval leaflets. These are not always distinct.

Rosa carolina does not appear to be very hardy in our area; drought and cold winters cause the plants to die back, often to near the ground. It does not form dense thickets.

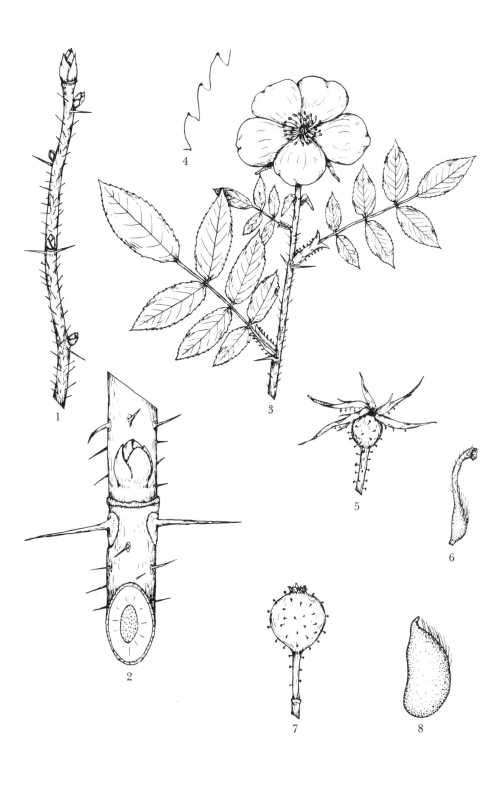

1. Winter twig
2. Detail of twig
3. Leaves and flower
4. Leaflet margin
5. Hypanthium and sepals, petals removed
6. Pistil
7. Fruit
8. Seed

ROSACEAE
Rosa foliolosa Nutt.

Leafy rose, prairie rose

A small, few-branched shrub to 50 cm high.

LEAVES: Alternate, pinnately compound with 7-11 leaflets; leaf shape obovate, 5-6 cm long, 2.5-3.5 cm wide, lower leaflets are smaller. Leaflets oblanceolate to elliptic, 8-20 mm long, 3-7 mm wide, tip acute, base narrowly cuneate, upper surface dark green and glabrous, lower surface paler and glabrate, often pubescent on the midrib and rachis, an occasional prickle on the rachis; margin finely toothed, 6-8 teeth per cm, gland tipped, margin entire near the base; petiole 7-10 mm long, slender, broadly grooved above; terminal petiolule 3-5 mm long, lateral leaflets sessile; stipules narrow, 6-8 mm long, the free portion about one-third of the total length, ciliate and stipitate glandular, lowest leaflets often below the stipule tips.

FLOWERS: June; perfect. Single or 2-3 at the end of seasonal branch; hypanthium globose or elongated, 5-7 mm long, 3-5 mm wide, green, glabrous or with stipitate glands; calyx lobes 5, lanceolate gradually or abruptly contracted to an awl-like tip, 12-20 mm long, 3-4 mm wide, occasionally with long teeth; pubescent or glabrate and densely ciliate, tomentose inside; petals 5, obovate about 2 cm long and as wide, rounded or notched at the tip, pink, spreading; numerous stamens, filaments slender, 5-7 mm long; anthers 1-1.5 mm long, yellow; numerous pistils, ovary obovoid, 1.5 mm long, pubescent on the dorsal side; style 3-4 mm long, pubescent; stigma capitate, 0.5-0.75 mm across.

FRUIT: Late July. Globular, 8-10 mm diameter, red, glabrous or stipitate glandular, old calyx lobes spreading; 4-8 seeds. Seeds brown, irregularly ovoid with nearly straight ventral side, dorsal side rounded and long pubescent, 3-3.5 mm long, 2-2.5 mm wide.

TWIGS: Fine, flexible, 1-1.2 mm diameter, red or greenish, smooth, mi-nutely white-dotted, no prickles or with a few yellow, weak prickles; leaf scars narrow, at right angles to the stem, extend three-fourths of the distance around the stem, scar dark color, often edged with yellow; 3 bundle scars; pith white, continuous but often alveolate, one-half of stem. Buds narrowly ovoid, 2 mm long, 1 mm wide, the scales red, often with a dark, apiculate tip, glabrous, ciliate; lateral buds smaller.

TRUNK: Bark greenish-brown, often slightly split, upper branches smooth, red and minutely white dotted; prickles few or entirely wanting. Wood hard, yellow, with a large pith.

HABITAT: Rocky or sandy soils, in open woods or prairies, occasionally in moist areas.

RANGE: Texas, Oklahoma, Arkansas, and southeastern Kansas.

The most obvious characteristics of *R. foliolosa* are the short, spindly plant and the narrow leaflets. It is most liable to be confused with *R. carolina*, but the length-width ratio of *R. carolina* leaflets is 2-1, while that of *R. foliolosa* is 3-1. The other differentiating characters are a matter of degree—*R. foliolosa* has fewer glandular hairs on the fruit, fewer prickles, shorter stipules, and shorter petioles. These are not good characteristics because the two typical species must be on hand before differentiation can be made.

The geographic range of *R. foliolosa* lies entirely within the range of *R. carolina*, so they cannot be separated in that manner.

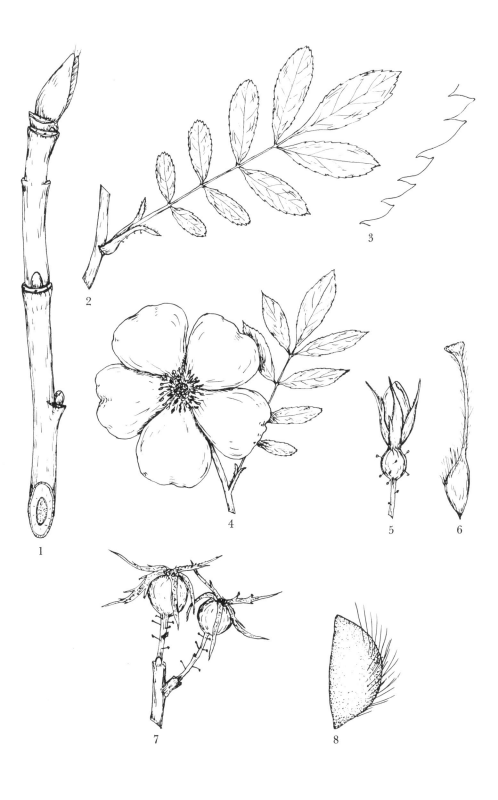

1. Winter twig   5. Flower bud
2. Leaf   6. Pistil
3. Leaflet margin   7. Fruit
4. Flower   8. Seed

ROSACEAE
Rosa multiflora Thunb.

R. polyantha Sieb. & Zucc.

Multiflora rose, Japanese rose

A diffusely branched shrub, the branches to 4 m long and arching or sprawling.

LEAVES: Alternate, pinnately compound, 8-11 cm long, 7-9 leaflets. Leaflets narrow or broad, ovate or elliptic, 2.5-4 cm long, 1.3-2.3 cm wide; serrate with sharp teeth, 5-8 per cm; tip acute to rounded, base cuneate; upper surface dark green, glabrous; lower surface pale, finely pubescent; mature leaves often thick and coriaceous; petiole 1-1.3 cm long, broad based, pubescent; rachis pubescent, often stipitate glandular; terminal petiolule 8-10 mm long, laterals 1-2 mm long; stipules 6-8 mm long, tips acuminate, spreading, margin with linear lobes or teeth, stipitate glandular.

FLOWERS: Late May; perfect. Inflorescence a rounded or pyramidal panicle, 8-15 cm long, with 6-30 flowers, pedicels 1-1.5 cm long, pubescent, green, stipitate glandular, 1-2 lance-linear bracts at the base; hypanthium obconic to oval, 3-4 mm long, 2.5-3 mm thick, green, glabrous to slightly pubescent; calyx lobes 5, 7-8 mm long, variable from oval with entire margin and long acuminate tip to lanceolate, appendaged, and stipitate glandular; lobes pubescent on both sides, reflexed and appressed to the hypanthium; petals 5, pale pink to white, obovate with nearly straight sides, 6-7 mm long, 8.5-9.5 mm wide, truncate and often indented; stamens many, attached to the rim of the hypanthium; filaments white, 2-3.5 mm long, anthers yellow; ovaries several, greenish, 1 mm long, ovoid, long straight hairs on the rounded side; style 4-4.6 mm long, curved, greenish-white, slightly clavate, exserted from a central conical disk; stigmas yellowish, capitate, irregular.

FRUIT: Late September. Drooping or erect clusters; fruits oval to obovoid, 6.8-8.5 mm long, 6-8 mm wide, red, glossy, smooth; calyx deciduous. Seeds yellow, ovoid, 4-4.7 mm long, 2-2.6 mm wide, flattened on two sides, dorsal side rounded with some hairs near the tip, ends acute.

TWIGS: 1.5 mm diameter, flexible, glabrous, red to green, armed with stout recurved prickles; leaf scars long, half around the stem; 3 bundle scars; pith white, continuous, one-half of stem. Buds 2.5-3 mm long, ovoid, acute, often at right angles to the stem; scales red, often toothed.

TRUNK: Bark of young stems red or greenish, prickles 7-8 mm long, sharp, brown, flattened, broad-based; bark of old trunks gray-brown or brown, smooth. Wood hard, nearly white, with a large pith.

HABITAT: Introduced and planted around farmsteads, fields, and pastures, often used in parks; escaped in many places, and long persistent.

RANGE: Introduced from Asia and has escaped from New England, southwest to Texas.

Rosa multiflora has been planted extensively for decoration, as a fence, and for wildlife cover. However, it takes a few years for a row to become dense enough to be a good fence, and a wire fence must be provided until that time. The branches, often quite long, entangled, and prickly, create a barrier so impenetrable that the larger animals will not attempt to pass through a dense row of the bushes. This makes an excellent cover for song birds, game birds, and small mammals, during both summer and winter. The fruits are not eaten to any extent, but a few of the small mammals make use of them during the winter.

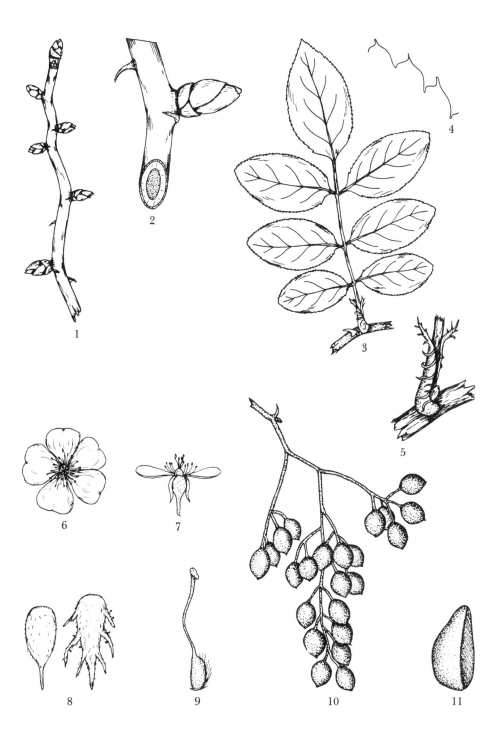

1. Winter twig
2. Detail of twig
3. Leaf
4. Leaflet margin
5. Stipules
6. Flower

7. Diagrammatic flower
8. Sepal variation
9. Pistil
10. Fruit cluster
11. Seed

ROSACEAE
Rosa setigera Michx. var.
tomentosa T. & G.

R. rubifolia R. Br. in Ait.

Prairie rose, climbing rose

A climbing or trailing vine to 6 m long, supported by shrubs or small trees; in the open a dense shrub to 1.5 m high and 2-3 m across.

LEAVES: Alternate, pinnately compound, 3-5 leaflets, usually 3. Terminal leaflet 4-5 cm long, 2.5-3.5 mm wide, narrow to broadly ovate, petiolule 6-8 mm long, red stipitate glandular; lateral leaflets 3.5-4 cm long, 2.3-2.6 cm wide, ovate, nearly sessile; margin of leaflets sharply serrate, 5-6 teeth per cm; tip acuminate or acute, base rounded; upper surface dark green, glabrous; lower surface pale, tomentose; often thick and rugose; petiole 1.5-2.5 cm long, red, stipitate glandular; stipules adnate for 1-1.3 cm, free end 2-3 mm long, acuminate, narrow, slightly pubescent, margin stipitate glandular.

FLOWERS: Early June; perfect. A many-flowered corymb, terminal on new growth; peduncles, pedicels, hypanthium, and calyx lobes stipitate glandular. Peduncles 1-2 cm long, green; pedicels 1-1.8 cm long, green; hypanthium green, obovoid, 5-7 mm long, 5-6 mm thick; calyx lobes 5, ovate, 16-17 mm long, 6-7 mm wide, acuminate, entire or with marginal projections, pubescent, reflexed in flower, appressed tomentose inside; petals 5, pink, obovate to heart-shaped, deeply notched, 3-3.3 cm long, 3-3.2 cm wide; stamens many, attached to the margin of a flat, yellow disk; filaments white, 4-12 mm long; anthers yellow; styles many, united, 1 cm long, exserted for 3-4 mm, green, glabrous; stigmas yellowish, lobed; ovary ellipsoid, 1.7-1.9 mm long, green, long pubescent on one side, attached to side walls and bottom of the hypanthium.

FRUIT: September. Globose to obovoid, 10-11 mm long, 9-10 mm wide, red, a few stipitate glands; peduncles 3-4 cm long, stipitate glandular; calyx lobes deciduous. Seeds variously angled, usually one margin rounded, 3.5-4 mm long, 1.5-2.2 mm thick, yellow, smooth, a few

hairs at the base and on the dorsal side.

TWIGS: 1-1.5 mm diameter, flexible, glabrous, green or reddish-green; prickles straw-colored or pale brown, expanded base, recurved, 6-8 mm long, often in pairs at the nodes; leaf scars narrow, enlarged at the center and the ends; 3 bundle scars; pith white or brown, continuous, two-thirds of stem. Buds ovoid, 1.5-1.8 mm long, red, glabrous.

TRUNK: Bark on young stems green to red-brown, smooth, glossy, under magnification showing rows of minute white dots; prickles scattered along the whole length; bark of old stems gray-brown, split with narrow, long slits. Wood soft, pale brown, with a large, brown pith.

HABITAT: Moist or dry ground, rock ledges along streams, open woods, fence rows, pastures, prairie thickets.

RANGE: Texas to western Georgia, north to Ohio, west to Kansas, south to Oklahoma. Naturalized in New England.

This is the only climbing rose native to our area and the only one commonly with just 3 leaflets. In the woods it extends up into tall bushes and low trees, making passage difficult. In the open it forms a dense mass of arching, trailing canes. This type of growth gives excellent cover to birds and mammals and holds the moisture and winter snow.

No specimens of *R. setigera* var. *setigera,* if separable, were found for our range, but it probably grows along the eastern edge. In this variety, the under surface of the leaflets is nearly glabrous.

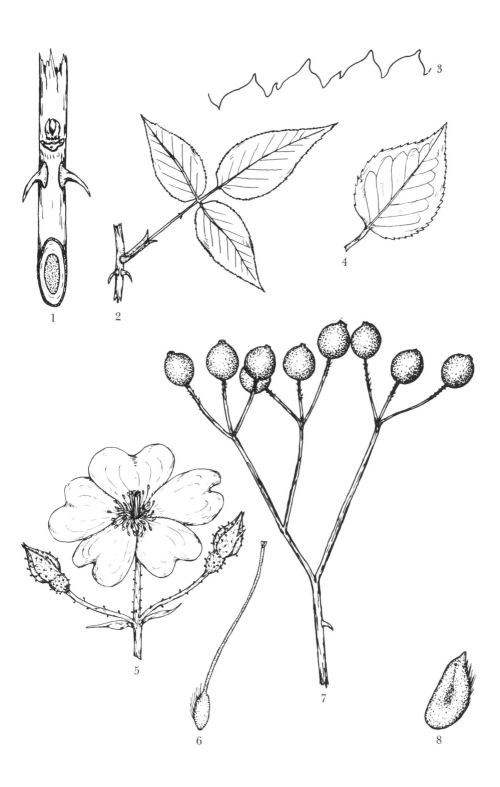

1. Winter twig
2. Leaf
3. Leaflet margin
4. Leaflet variation
5. Flower
6. Pistil
7. Fruit cluster
8. Seed

ROSACEAE
Rosa woodsii Lindl.

R. fendleri Crepin; *R. macounii* Greene;
R. fimbriatula Greene; *R. sandbergii*
Greene; *R. woodsii* f. *hispida* W. H.
Lewis; *R. woodsii* (var.) *fendleri* (Crepin)
Rydb.

Wild rose

A shrub to 1.5 m high, often forming dense thickets.

LEAVES: Alternate, pinnately compound, 6.5-8 cm long, 7-9 leaflets. Leaflets oval to obovate, 2-2.5 cm long, 10-13 mm wide; coarsely serrate, often double serrate, 3-7 teeth per cm, teeth sharp, narrow, glandular; tip of leaflet obtuse or rounded, base cuneate; upper surface dark green, mostly glabrous; lower surface paler, pubescent; rachis with short prickles; petiole 1.5 cm long; stipules 12 mm long, the free tips short, spreading, pubescent, and with glandular teeth.

FLOWERS: Late May; perfect. Pedicels 1-1.3 cm long, green, glabrous; hypanthium ovoid, 6.1-6.3 mm long, 4.9-5.1 mm wide, green, glabrous; calyx lobes 5, lance-linear, 3.7-3.9 mm long, 1.7-1.9 mm wide, some with lateral projections, often with foliaceous tip, pubescent outside, cobwebby hairs inside; petals 5, obovate, 2.2 cm long and as wide, pink, notched at the apex; stamens many, filaments white, 5-7 mm long; anthers yellow; ovaries oval, green, flattened on one side, 1.8 mm long, covered with long, slightly curly hairs; style pubescent, not exserted, often sharply bent near the ovary, 4 mm long, tapered toward the tip; stigma capitate, lobed.

FRUIT: Late August. Pedicels glabrous, 2.5-3.5 cm long, often with prickles; fruits subglobose, often compressed slightly, orange-red, glabrous; calyx spreading or erect; seeds on the sides and bottom of the hypanthium. Seeds light tan, 3.7-4.3 mm long, 2.6-3 mm thick; shape variable, generally one margin straight and the other rounded, ridged or with 1-2 flat surfaces, straight hairs on the upper half.

TWIGS: 1.2-1.4 mm diameter, flexible, glabrous, red, often with a purple cast; 2 infrastipular prickles 5-7 mm long, straight or slightly recurved, the base broad; leaf scars narrow, long, three-fourths around stem; pith greenish, continuous, one-half of stem. Buds ovoid, 2-2.6 mm long, red, glabrous, glossy.

TRUNK: Bark red-brown, irregularly split; some straight or slightly recurved prickles with obovate base. Wood white, soft, and a large pith.

HABITAT: Sandy or light clay soils, in rocky ravines in open prairies or at the margin of a woods.

RANGE: Minnesota, west to British Columbia, south to Arizona, northern Mexico, and western Texas, and north to western Kansas and North Dakota.

Our plants are var. *woodsii,* as described above. To the west and mainly in mountainous regions is var. *ultramontana* (Wats.) Jeps. The leaflets may be 5 cm long, and are without glands on the teeth.

R. woodsii is a shrub of dry regions. It is often found at the upper end of a steep ravine and forms an impenetrable mass of upright stems. In such locations it is quite useful in the prevention of further soil erosion and also furnishes good cover for birds and mammals. It is also common in open woods on the sandy flood plain of the Arkansas River.

The stout infrastipular prickles and the lack of internodal bristles should differentiate *R. woodsii* from *R. acicularis* and *R. suffulta.* The other differences are rather subtle or variable. The leaflets of *R. woodsii* are smaller and there is a gland tip on the teeth, the bushes are stiffer, and the branches have a purplish cast.

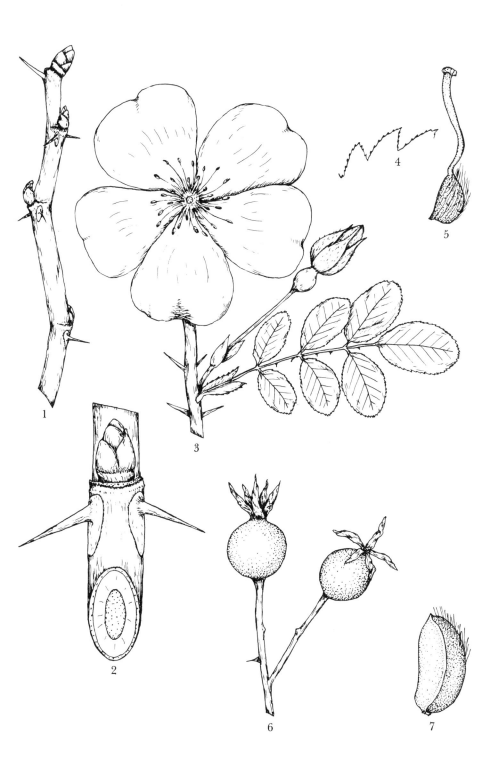

1. Winter twig
2. Detail of twig
3. Leaf and flower
4. Leaflet margin
5. Pistil
6. Fruit
7. Seed

ROSACEAE
Prunus americana Marsh.

Wild plum

A small tree to 5 m high, forming thickets by sprouts from the roots.

LEAVES: Alternate, simple. Oval to obovate, 6-10 cm long, 3-5 cm wide; margin sharply serrate or double serrate, 3-6 teeth per cm, glandless, often acuminate; leaf tip acuminate, base acute; upper surface dark green, dull, often rugose, glabrous; lower surface paler with some pubescence, especially on the veins; petiole slender, 9-16 mm long, pubescent on the upper side only, occasionally with glands at the summit; stipules linear, 6-12 mm long, toothed, pubescent, divided into 2-3 linear segments, caducous.

FLOWERS: Early April, with or before the leaves; perfect. Flowers in clusters of 2-5 on the end of a spur or from lateral buds; pedicels 8-14 mm long, green, glabrous; hypanthium obconic, 3 mm long, glabrous, green or reddish; calyx lobes 5, oblong, 3 mm long, entire or with few small teeth, obtuse, ciliate, green, glabrate on the outer surface and pubescent inside; petals 5, white, ovate, 9-11 mm long, tip rounded, abruptly narrowed at the base; stamens 20-30, filaments 5-7 mm long, white or with pinkish base; anthers yellow; ovary 15-20 mm long, ovoid, green, glabrous; style slender, 12-15 mm long; stigma capitate, brownish.

FRUIT: August. Pedicels 8-15 mm long, slender, glabrous; fruit globose to oval, 2.2-2.7 cm long, 2-2.5 cm thick, purple or red with a slight bloom; thick skin, the flesh sweet, juicy, and edible. Stone cream-colored, oval, 1.6-1.8 cm long, 1.1-1.2 cm wide, 7-8 mm thick, surface somewhat rugose, ridged on one edge and narrowly grooved on the other, obtuse at both ends.

TWIGS: 1.5 mm diameter, rigid, glabrous to glabrate, light orange-brown becoming purplish, often spine-tipped; leaf scars half-round, with a fringe of hairs on the upper margin; 3 bundle scars; pith brown, continuous, one-fourth of stem. Buds ovoid, acute, 3-5 mm long, the scales chestnut brown, usually with light margins.

TRUNK: Bark dark gray-brown, tinged with red or purple, the lenticels horizontal and prominent; older trunks with persistent large, thin, scaly plates. Wood hard, heavy, strong, dark red-brown, with a narrow, light sapwood.

HABITAT: Stream banks, woodlands, upland pastures, thickets, fence rows; in rocky, sandy, or rich loam soils.

RANGE: Massachusetts, west to Manitoba and Montana, south to Utah, Colorado, and Oklahoma, east to Florida, and north to New York.

Prunus americana is a common plum throughout our area and is often found in dense thickets. It is most commonly confused with *P. mexicana* S. Wats. and is not always easily distinguished. The twigs of *P. americana* are glabrous or nearly so and the petioles are pubescent only on the upper side. The pubescence on the underside of the leaf is sparse, while that of *P. mexicana* is often rather dense. The fruit of *P. americana* is red or purple and seldom over 2.5 cm long, while that of *P. mexicana* is purple with a heavy bloom and usually over 2.5 cm long.

The confusion has not been helped by manuals which give *lanata* as a variety, as has been done in a number of cases. *Lanata* as used in most manuals should be synonymous with *P. mexicana*.

The fruits of *P. americana* are used greatly in pies, jams, or jellies and often as a sauce. In good years, a thicket will produce enough plums to make many jars of jelly or jam, providing a person can get into the center of the thicket to gather them. The branches are rigid and often somewhat spiny. Birds and small mammals eat the fruits; the blue jay especially appears to be quite fond of them.

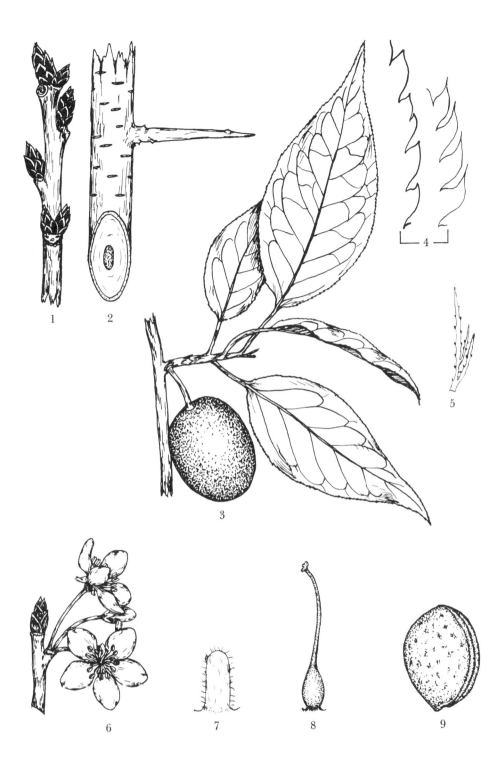

1. Winter twig
2. Detail of twig
3. Leaves and fruit
4. Variation in leaf margin
5. Stipule

6. Flower cluster
7. Sepal
8. Pistil
9. Stone

ROSACEAE
Prunus angustifolia Marsh.

Chickasaw plum, sandhill plum

A colonial shrub to 2 m high, forming thickets up to 15 m across.

LEAVES: Alternate, simple. Elliptic to lanceolate, 2-6 cm long, 1-2 cm wide; finely glandular serrate, 10-14 teeth per cm, the glands often turn outward; leaf tip acute to acuminate, base rounded to cuneate; upper surface bright yellow-green, glabrous, lustrous; lower surface paler, glabrous; blade usually slightly folded lengthwise and the tip curled down; petiole 7-14 mm long, often reddish, pubescent when young; 2 glands at the summit; stipules linear, often incised, glandular serrate, green, ciliate, early deciduous, and usually seen only on fast growing branches.

FLOWERS: Mid-April, with or before the leaves; perfect. Flowers 1 cm across, in clusters of 2-4 from buds of the previous year; pedicels 7-10 mm long, slender, glabrous; hypanthium obconic, 1-2 mm long, glabrous; calyx lobes 5, ovate, 1-1.5 mm long, ciliate, a few hairs at the base; petals 5, white, obovate, 3.5-4 mm long, rounded at the apex and abruptly narrowed at the base; stamens 20, filaments white, 4-5 mm long, glabrous; anthers yellow-orange; ovary 1 mm long, ovoid; style yellow-green, 4-6 mm long; stigma capitate, irregular, dark.

FRUIT: June-July. Pedicels 8-11 mm long; fruit a globose drupe, 2-2.2 cm long, 1.7-1.9 cm thick, red or yellowish with a slight bloom, a crease on one side, thin-skinned, the flesh juicy, tangy, and edible. Stone ovoid, 12 mm long and nearly as wide, slightly pointed at both ends, surface somewhat rough, lightly pitted.

TWIGS: 1.8 mm diameter, red-brown, becoming dark and dull brown to gray-brown, glabrous, rigid, often ending in a spine, lenticels inconspicuous; leaf scars small, elliptical, often ciliate on the upper margin; 3 bundle scars; pith pinkish, continuous, one-fifth of stem. Buds acute, 1.5-2.5 mm long, mostly grouped at the ends of twigs or spurs, the scales red-brown, acute.

TRUNK: Bark thin, dark red-brown to gray-brown, old trunks with shallow furrows, or the outer bark splitting and curling. Wood hard, heavy, red-brown, with a wide, light sapwood.

HABITAT: Fence rows, pastures, fields, prairie stream banks, sand dunes, waste ground, or rocky banks; most common in sandy soil.

RANGE: Texas to Florida, north to New Jersey, west to Indiana and Nebraska, south to Kansas and Oklahoma.

Two hybrids have been named and apparently occur in our area: X *P. orthosepala* Koehne, a hybrid with *P. americana;* and X *P. slavini* Palmer, a hybrid of *P. angustifolia* var. *varians* and *P. gracilis.*

Also two varieties have been named but are here placed with the species, since they are not separable in our material: Var. *varians* Wight & Hedrick, a larger plant with longer pedicels, more pointed stone, and yellowish fruits; var. *watsonii* (Sarg.) Waugh, with smaller leaves and a small fruit with a red, thin skin.

In the western part of Kansas *P. angustifolia* is fairly constant in its characteristics, but in the eastern part it is quite variable and approaches *P. munsoniana* Wight & Hedrick in appearance. It is the common plum in sandy areas and is appropriately known as the sandhill plum. The trees grow to a height of 2 meters and form a dense, thorny thicket. This is excellent cover for small birds and also produces a quantity of fruit for human consumption. Rodents often open the hard stone by gnawing and eat the kernel from the inside.

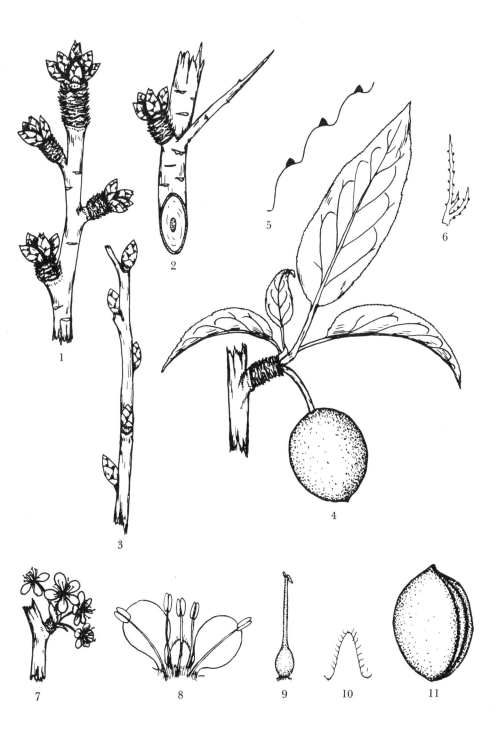

1. Winter twig
2. Twig with spur branch and spine
3. Twig with leaf buds
4. Leaves and fruit
5. Leaf margin
6. Stipule
7. Flower cluster
8. Diagram of floral parts
9. Pistil
10. Calyx lobe
11. Stone

ROSACEAE
Prunus besseyi Bailey

P. pumila L. var. *besseyi* (Bailey) Gl.;
P. prunella Daniels

Sand cherry, dwarf cherry

An open shrub to 7 dm high, but on rocky, exposed hillsides often prostrate and gnarled.

LEAVES: Alternate, simple, often clustered near the end of a short branch. Elliptical, obovate or oblanceolate, 4.5-6.5 cm long, 1-2.5 cm wide, narrow on fruiting branches; finely serrate, 6-8 teeth per cm, longer teeth toward the base; leaf tip acute to obtuse, base attenuate; upper surface dark green, lustrous, glabrous; lower surface slightly glaucous, glabrous; margin thickened; petiole 6-12 mm long, grooved, green, glabrous, occasional glands at the summit; stipules vary from narrowly lanceolate to a narrow base with a foliaceous outer end, both forms with narrow lobes; early deciduous.

FLOWERS: Late April, with the leaves; perfect. Flowers 13-15 mm across, usually 2 from a bud on last year's growth; pedicels 8-12 mm long, green, glabrous; hypanthium campanulate, 3-3.5 mm long, glabrous, green, often reddish; calyx lobes 5, oblong, 1.2-1.5 mm long, obtuse, toothed and occasionally glandular, green or somewhat reddish, glabrous inside and out; petals 5, oval, 6.5-7.5 mm long, 3-4 mm wide, white, base abruptly narrowed, tip rounded, often undulate; stamens 25-30, attached on the inner surface of the hypanthium; filaments white, 3-5 mm long; anthers yellow; ovary green, 1.2 mm long, ovoid, glabrous, slightly gibbous on one side; style 5-6 mm long; stigma capitate, obliquely lobed, yellowish.

FRUIT: Mid-July. Clustered in leaf axils or on the side of a spur; pedicel 10-15 mm long, glabrous; fruit a globose to ovoid drupe, 13-14 mm long, 13.5-15 mm thick, deep purple, glossy, indented at the pedicel, apex with a minute tip; flesh purple, acrid, but edible. Stone oval, flattened, 7.8-8 mm long, 6.5-7 mm wide, 5.8-6.2 mm thick, acute tip and rounded base, flesh-colored,

granular with irregular, shallow pits, ridged on the dorsal side and with a small groove on the ventral side.

TWIGS: 1.7-1.9 mm diameter, flexible, glabrous, red in the first year becoming grayish-pruinose, the waxy coating often striate; lenticels small and orange-colored; leaf scars broadly crescent-shaped to short shield-shaped; 3 bundle scars; pith salmon, continuous, one-fifth of stem. Leaf buds 1 mm long, obtuse, the scales red-brown, glabrous, glossy, slightly keeled; flower buds 1.8-2 mm long, obtuse, red-brown.

TRUNK: Bark reddish-gray, smooth, with horizontal, orange lenticels; latent buds often occur completely to the base of the plant. Wood hard, nearly white.

HABITAT: Sandy, prairie hillsides, often hidden in tall grass; rocky ledges or in gravel soil.

RANGE: Manitoba to Wyoming and Colorado, east to Kansas, and north to Minnesota.

Prunus besseyi does not form thickets, and the shrubs vary from a few scraggly branches to a dense, globose shrub 70 cm in diameter. In rocky, hilly pastures it is difficult to locate because of the small size and the fact that it grows scattered among the tall grasses. For this reason it is probably overlooked and not recorded for many areas in which it grows. The flowering period is short, only a few days, but usually an abundance of flowers is produced and the plant stands out clearly. The fruits are tart but have a pleasant flavor when fully ripe.

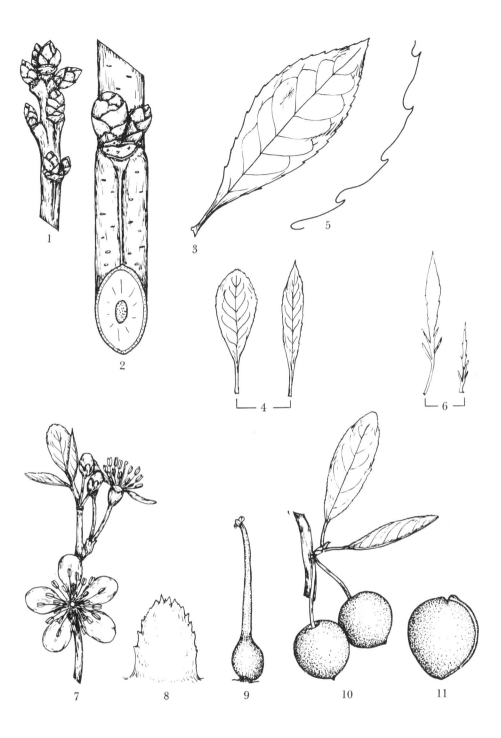

1. Winter twig
2. Detail of twig
3. Leaf
4. Leaf variation
5. Leaf margin
6. Stipules
7. Flowers
8. Sepal
9. Pistil
10. Fruit
11. Stone

ROSACEAE
Prunus gracilis Engelm. & Gray

Prunus normalis (T. & G.) Small

Sand plum, Oklahoma plum

A shrub to 1.5 m high, single or in open colonies.

LEAVES: Alternate, simple, flat (not folded), thick, 4-6 main lateral veins. Broadly elliptic, 3-5 cm long, 1.5-2 cm wide, tip acute to obtuse, base cuneate; margin finely serrate, 8-11 teeth per cm, glandular tipped when young, often biserrate, sinuses acute; upper surface dull, dark yellow-green, sparsely pubescent, the lower surface slightly paler, densely pubescent, the veins prominent; petiole 5-15 mm long, often stout, densely pubescent; stipules lanceolate, 3-4 mm long, glandular serrate, occasionally incised, pubescent, more common on fast-growing branches.

FLOWERS: Early April, before or with the leaves; perfect. Axillary in clusters of 2-8, on wood of previous year or on short spur branches; pedicels 7-10 mm long, pubescent; flower 12-16 mm across; hypanthium obconic, 2.5-3 mm long, 2-2.5 mm wide, pubescent, green; calyx lobes 5, ovate, 1.5 mm long, 1 mm wide, obtuse or acute, pubescent, ciliate at the tip, sparsely glandular-toothed, cupped, spreading; petals 5, white, nearly circular with a claw about equalling the calyx lobes in length, total 5.5-6.5 mm long, 4.5-5 mm wide, spreading; stamens 20, filaments slender, 6-9 mm long, white; anthers yellow; ovary dark green, ovoid, 2 mm long, 1 mm wide; style slender 8-9 mm long, stigma capitate, yellowish.

FRUIT: July. Pedicel 10-12 mm long, pubescent; fruit nearly globular, 17-18 mm long, yellow-red, smooth, semi-glossy, juicy, tart. Stone ovoid, compressed, 10-12 mm long, 8-9 mm wide, 6-7 mm thick, yellowish, with rough surface, somewhat ridged on one margin.

TWIGS: 1-1.3 mm diameter, red-brown to purplish, often pruinose, pubescent, the lenticels large, broadly oval and orange-colored; leaf scars pale, bean-shaped; 3 main bundle scars; a small ridge from each end of the scar, extending downward on the stem; small stipule scars; pith white to pale brown, continuous, one-fourth to one-third of stem. Buds ovoid, obtuse, 2.5-3 mm long, scales purplish-brown, waxy, ciliate near the tip, the basal scales often erose-margined; flower buds, when present, lateral to the leaf bud.

TRUNK: Bark brown, roughened by large, horizontal, pale brown lenticels. Wood hard, fine-grained, white, no obvious sapwood.

HABITAT: Dry, sandy soil, open creek banks, hillsides, or fence rows, often in open woods.

RANGE: Kansas, Oklahoma, Texas, and New Mexico.

From the few specimens available from central Kansas it would appear that the northernmost plants have the lower leaf surface with less pubescence and the veins are less prominent. The plant is often confused with *P. angustifolia* and *P. rivularis*.

	P. rivularis	*P. gracilis*	*P. angustifolia*
Sepals	glandular margin	few glands	entire
Pedicels	glabrous	pubescent	glabrous
Hypanthium	glabrous or pubescent at summit	pubescent	glabrous
Leaves	folded, glabrous or pubescent below, teeth 9-11 per cm	flat, pubescent below, teeth 8-11 per cm	folded, glabrous below, teeth 15-18 per cm
Flowers	over 1 cm wide	over 1 cm wide	under 1 cm wide
Twigs	glabrous	pubescent	glabrous

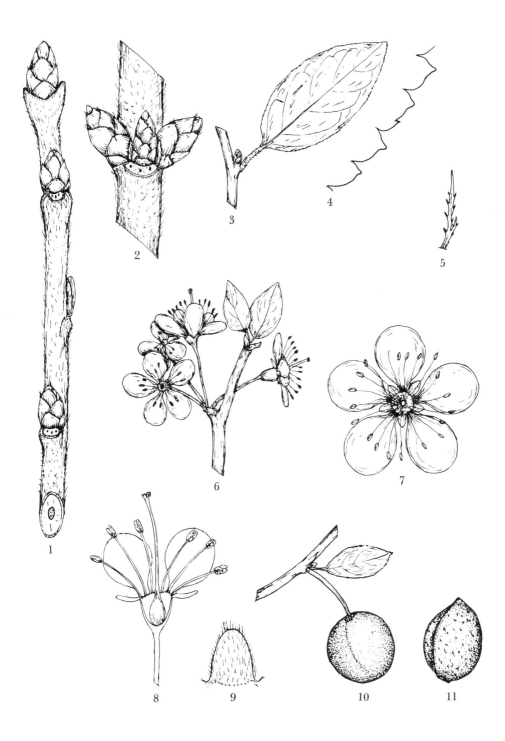

1. Winter twig
2. Detail of buds, leaf bud in center
3. Leaf
4. Leaf margin
5. Stipule
6. Flower cluster
7. Flower
8. Diagrammatic cutaway of flower
9. Sepal
10. Fruit
11. Stone

ROSACEAE
Prunus hortulana Bailey

P. palmeri Sarg.

Wild goose plum

A small tree to 4 m high, single or in open thickets.

LEAVES: Alternate, simple, conduplicate in bud. Ovate to ovate-lanceolate, 6-10 cm long, 3-4.5 cm wide; margin finely serrate, 7-9 teeth per cm, most of the teeth pointing outward, gland-tipped; leaf tip acuminate, base rounded to cuneate; upper surface dark green, lustrous, glabrous; lower surface paler, pubescent especially on the midrib and veins; blade flat, thin, and firm; petioles 1-1.5 cm long, often reddish, grooved and slightly pubescent above, usually 2 glands at the apex; stipules linear, 5-9 mm long, glandular, and occasionally with linear lobes, caducous.

FLOWERS: Mid-April, when the leaves about a third grown; perfect. Flowers clustered on short spurs; pedicels 11-18 mm long, glabrous; hypanthium obconic, 4-5 mm long, green, glabrous; calyx lobes 5, ovate, 3-3.5 mm long, acute or rounded, glandular serrate, pubescent at the base on the inner side, glabrate outside; petals 5, ovate to oval, 6-9 mm long, abruptly narrowed to a short claw, apex rounded or slightly notched; stamens about 30, filaments white, 1 cm long, attached at the rim or on the inner wall of the hypanthium; anthers yellow, ovary green, ovoid, 1 mm long, glabrous; style slender, 7-8 mm long, nearly white, glabrous; stigma capitate, yellowish, lobed.

FRUIT: August-September. Pedicels 13-20 mm long, glabrous; fruit a subglobose drupe, 1.9-2.8 cm thick, slightly wider than long, bright red, little or no bloom, minutely yellow-dotted, thin skin, the flesh firm, sweet, juicy. Stone ovoid, 1-1.1 cm long, 9.4-9.8 mm wide, 7.2-7.8 mm thick, gibbous, apex rounded, base rounded to obtuse, pinkish, the surface with shallow pits and several indistinct, irregular ridges on one margin, a very narrow ridge and groove on the other.

TWIGS: 2.5 mm diameter, rigid, glabrous, red-brown with a purplish cast, often gray-waxy; lenticels not obvious; leaf scars oval, small, 3 bundle scars; pith brown, continuous, one-fifth of stem. Buds ovoid, 3-4 mm long, the scales chestnut brown, slightly ciliate or erose.

TRUNK: Bark thin, dark red-brown; on old trunks splitting into large, thin, persistent plates, often curling laterally, light brown inner bark. Wood hard, heavy, reddish-brown, with a wide, white sapwood.

HABITAT: Rich bottom lands, open woodlands, borders of woods, common in the strip-pit area of southeast Kansas.

RANGE: Texas, east to Alabama, north to Kentucky, northwest to Iowa, and southwest to Oklahoma.

P. hortulana usually grows in open thickets and does not form sucker shoots, but creates the thicket from seedlings. Several horticultural varieties have been developed from it.

It is often confused with *P. munsoniana,* and although both species will vary, the main contrasting characteristics of *P. hortulana* are: leaf blades flat, nonsuckering, flowers usually after the leaves, teeth glands pointing outward, and the fruit not elongated.

The flesh of the fruit is hard and firm until fully ripe, then becomes soft, juicy, and edible. It may be eaten raw or cooked in pies or made into jams or jellies.

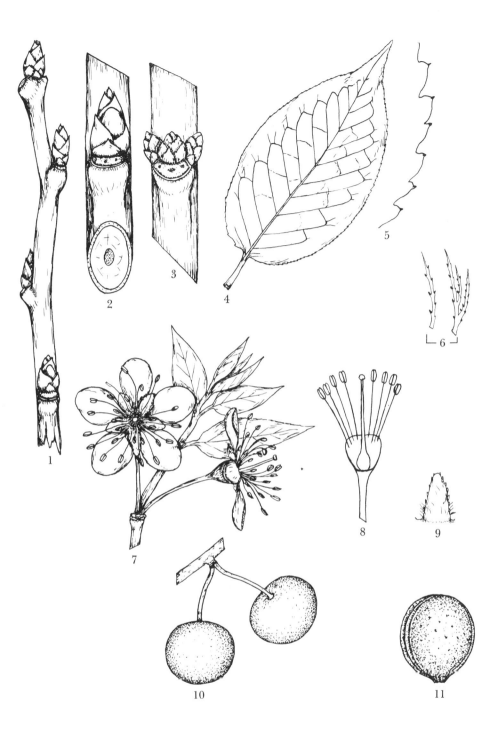

1. Winter twig
2. Detail of twig
3. Detail of buds, leaf bud in center
4. Leaf
5. Leaf margin
6. Stipules

7. Flowers
8. Diagram of flower cutaway
9. Sepal
10. Fruit
11. Stone

ROSACEAE
Prunus mahaleb L.

Mahaleb cherry, perfumed cherry

A tree to 7 m high, with low branches, often forming small thickets around the parent tree. Trunks often clustered.

LEAVES: Alternate, simple. Ovate to oval, 3-5 cm long, 2-3.5 cm wide; finely crenate-serrate, 6-8 teeth per cm, a gland on the incurved side; leaf tip acute to abruptly acuminate; base rounded to subcordate; upper surface dark green, lustrous, glabrous; lower surface paler, with a few hairs on the midrib; petiole 6.3-10 mm long, glabrous, flattened or grooved on the upper side, often 1-2 greenish or red glands at the summit; stipules narrowly ovate, 2 mm long, red-glandular, caducous.

FLOWERS: Late April, after the leaves partly grown; perfect. Flowers 1.3-1.5 cm broad; 6-8 flowers in a corymbose raceme terminating a new leafy shoot; pedicels 10-13 mm long, green, glabrous, the bract at the base glabrous, 1.5-2 mm long, with glandular margin; hypanthium obconic, green, glabrous, 2-3 mm long; calyx lobes 5, ovate, 1.5-2 mm long, entire, obtuse, glabrous on both sides, reflexed sharply when flower is fully open; petals 5, white, oval to obovate, 5-8 mm long, gradually narrowed to the base, rounded at the apex; stamens 20, usually 3 opposite each petal and 1 opposite each calyx lobe; filaments white, 3.5-4 mm long, attached at the rim of the hypanthium; anthers orange-yellow; ovary 1-1.2 mm long, ovoid with one side more rounded, glabrous; style 3.5 mm long, green, glabrous; stigma disk-like.

FRUIT: Mid-June. Pedicels 14-15 mm long; fruit an ovoid drupe, 8.2 mm long, 7.4 mm wide, dark red-purple, smooth, glossy, bitter. Stone long ovoid, 6.5-7 mm long, 4.3-5.5 mm wide, 3.6-5 mm thick, pinkish, smooth, tip acute, a ridge on the dorsal margin and a small groove on the ventral side.

TWIGS: 1.4-1.6 mm diameter, flexible, red-brown to gray-brown, pubescent at first, later glabrous and pruinose, lenticels prominent; leaf scars narrow, half-round; 3 bundle scars; pith greenish, becoming brown, continuous, one-sixth of stem. Buds acute, 2.5-3 mm long, scales red-brown, with pale gray edges and some pubescence at the tip.

TRUNK: Bark on young trees not fissured, gray-brown, with large transverse lenticels; on old trees, dark gray and scaly, with thin, recurved scales; branches gray, smooth. Wood hard, light red-brown, with a light sapwood.

HABITAT: Naturalized, escaped. Around old farmsteads, in rocky woods, hillsides, creek banks.

RANGE: Introduced from Europe, becoming naturalized from New England, Delaware, and Ontario, southwest to Indiana, Missouri, and Kansas.

Most of our trees of *P. mahaleb* which have attained any size are found around old farmsteads where they have persisted for several years. It has escaped and established itself in a number of areas along creeks, brushy hillsides, or ravine banks. The small, nearly round leaves make it fairly easy to identify, and the young branches have a gray color instead of the reddishness of most cherries. The fruits are small and bitter but can be used in making jelly if a sufficient quantity can be found. Even though they are bitter, birds seem to like them and will keep the ripe fruits stripped from the branches.

The species is horticulturally important in that the seedlings are used as grafting stock for cherries.

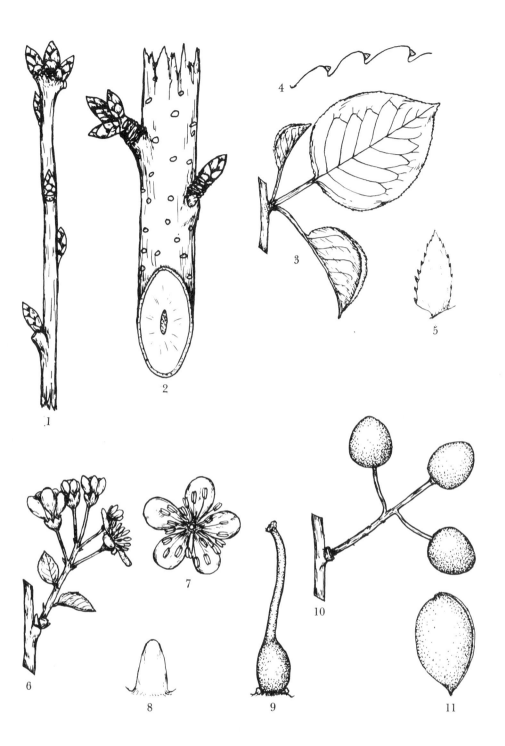

1. Winter twig
2. Twig with spur branch
3. Leaves
4. Leaf margin
5. Stipule
6. Flower cluster
7. Flower
8. Sepal
9. Pistil
10. Fruit
11. Stone

ROSACEAE
Prunus mexicana S. Wats.

P. americana, var. *lanata* of auth., not
Sudw.; *P. arkansana* Sarg.; *P. reticulata*
Sarg.; *P. tenuifolia* Sarg.

Big tree plum

A tree to 4 m high, single or in thickets.

LEAVES: Alternate, simple. Ovate to elliptic, 5-10 cm long, 3-5 cm wide; single or double serrate, 4-6 teeth per cm, teeth often acuminate, leaf tip abruptly short acuminate, base rounded; upper surface dark yellow-green, often rugose, glabrate; lower surface paler, pubescent; petiole stout, 1-2 cm long, usually glandular at the summit, pubescent on all sides; stipules lanceolate, toothed or lobed, pubescent, 5 mm long, caducous.

FLOWERS: Late April, with the leaves; perfect. Flowers 1.8-2.5 cm across, in clusters of 3-4 from buds of the previous year. Pedicels green, 1 cm long, finely pubescent; hypanthium narrowly obconic, 3-3.5 mm long, finely pubescent, often reddish; calyx lobes 5, oblong, 3.4-3.6 mm long, rounded or toothed at the apex, pubescent on both sides, reflexed after anthesis, ciliate margin; petals 5, white, ovate to orbicular, 6.5-7.5 mm, puberulous at the base, margin often undulate; stamens about 30, filaments white, 6-7.5 mm long; anthers yellow; ovary 0.5 mm long, ovate, yellow-green; style greenish, 7.5-8 mm long; stigma dark, disk-like.

FRUIT: Late July. Single or clustered; pedicels 1.2-1.8 cm long, pubescent; fruit a subglobose drupe, slightly elongated, 2.5-3 cm long, 1.9-2.5 cm thick, red before ripening, finally purple with a glaucous bloom; flesh thick, juicy, and has a tendency to stick to the stone. Stone ellipsoid to nearly circular, asymmetric, 1.3-1.8 cm long, 1.1-1.5 cm wide, 7-8.5 mm thick, smooth, tip rounded, a broad ridge on one margin and a groove on the other.

TWIGS: 1.5-1.9 mm diameter, rigid, puberulent, light red-brown, becoming gray in the second year; leaf scars oval, often pubescent on the upper margin; 3 bundle scars; pith white, brown in older stems, continuous, one-fifth of stem. Buds ovoid, 3.5-4 mm long, acute, the scales finely pubescent, red-brown, with a pale margin.

TRUNK: Bark nearly black, exfoliating into thin, plate-like scales on young branches; older trunks rough, deeply furrowed, the outer bark thin and curling; bark of branches tight and with large transverse lenticels. Wood hard, heavy, brown, with a narrow, light sapwood.

HABITAT: Open woods, rich woodlands, bottom lands, or prairie hillsides in rocky soil.

RANGE: Indiana west to South Dakota, south to Texas and Mexico, east to Arkansas, and northeast to Tennessee.

P. mexicana is often confused with *P. americana* but may be distinguished by its pubescent pedicels, petioles, and hypanthium, and the slightly larger purple fruit with a glaucous bloom. Normally it is not thicket-forming, but may be found in open colonies, especially at the head of a ravine or as individual plants scattered loosely over a small area. The range does not extend as far north as that of *P. americana* but overlaps with it in Nebraska and Kansas.

The fruit is sweet and juicy, but, too often, is assumed to be ripe when it turns red. At that time it is still strongly acrid and not really edible.

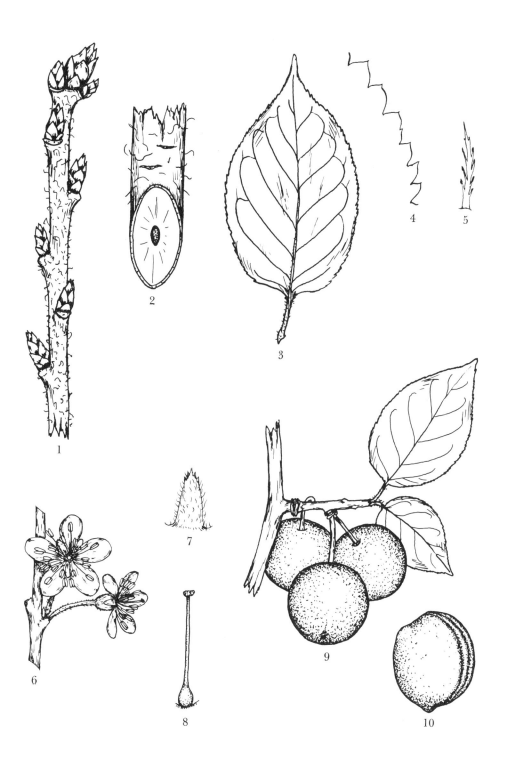

1. Winter twig
2. Section of twig
3. Leaf
4. Leaf margin
5. Stipule

6. Flower cluster
7. Sepal
8. Pistil
9. Fruit
10. Stone

ROSACEAE
Prunus munsoniana Wight & Hedrick

Wild goose plum

A tree to 5 m high, forming thickets by sucker shoots.

LEAVES: Alternate, simple, convolute in bud. Elliptic, 6-10 cm long, 2-4 cm wide; margin finely crenate-serrate, 7-9 teeth per cm, teeth rounded with a gland on the incurved face; leaf tip acuminate, base cuneate, upper surface yellow-green, lustrous, glabrous; lower surface paler, with sparse rusty or white hairs, densely villous when unfolding; blades have the tendency to fold lengthwise; petiole slender, 1-1.8 cm long, usually reddish, and often 2 glands at the summit, grooved and pubescent above; stipules 5-7 mm long, linear with linear lobes, glandular, early deciduous.

FLOWERS: Early April, as the leaves unfold; perfect. Flowers in clusters of 2-4 from lateral buds or on short spurs; pedicels slender, 10-13 mm long, glabrous, often a small, toothed bract at the base; hypanthium obconic, 2-4 mm long, glabrate; calyx lobes 5, acute to obtuse, 2-3 mm long, minutely glandular serrate with some pubescence; petals 5, white, 4-7 mm long, obovate to oblong-obovate, margin entire or with slightly erose tip; stamens many, filaments 7-8 mm long, in 2 rows, attached on the rim and inner wall of the hypanthium; ovary green, ovoid, 0.8 mm long, glabrous; style 6-7 mm long, slender, greenish, stigma capitate, yellow, lobed.

FRUIT: August. Pedicels 1.4-3 cm long, glabrous; drupe 1.8-2.2 cm long, 1.7-2 cm thick, bright red with minute, pale dots and a slight bloom; flesh firm, juicy, edible, slightly fibrous, tends to cling to the stone even after being cooked. Stone ovoid, compressed, 1.3-1.5 cm long, 8.6-10.5 mm wide, 6-6.7 mm thick; surface pebbled, apex acute, base acute to truncate; a wide ridge on the dorsal margin and a narrow groove on the ventral.

TWIGS: 1.5-2 mm diameter, flexible, glabrous, red-brown, becoming gray-

ish in the second year; a few pale lenticels; leaf scars half-round, ciliate upper margin; 3 bundle scars; pith white, continuous, one-fourth of stem. Buds ovoid, 3 mm long, obtuse; scales chestnut brown, glabrous, often with an erose margin.

TRUNK: Bark thin, reddish-brown, usually with horizontal light patches; the lenticels horizontal; old trunks gray to gray-brown, flaky. Wood hard, heavy, pale red-brown, with a light sapwood.

HABITAT: Rich bottom lands or rocky hillsides, prairies, borders of woods, along streams or in fence rows.

RANGE: Ohio, west to Kansas, south to Texas and Louisiana, and northeast to Kentucky.

P. munsoniana is the most common plum in the southeast corner of our area, where it occurs in thickets up to 15 meters across. The main tree may be 5 meters high and the sucker shoots nearly as high. The original tree may be entirely gone. It apparently spreads rapidly. The fruit is strongly acrid and firm even after falling from the tree, but because of its abundance is often gathered in quantities to be made into jelly. Most of the wild plums have a tendency to be infected by insects and should be carefully examined before using. During years of abundant production there are plenty of good, clean fruits for human use.

The two most constant characteristics distinguishing it from *P. hortulana* are the tendency of the leaves to fold lengthwise and the definitely rounded, incurved teeth, with the gland on the infacing side.

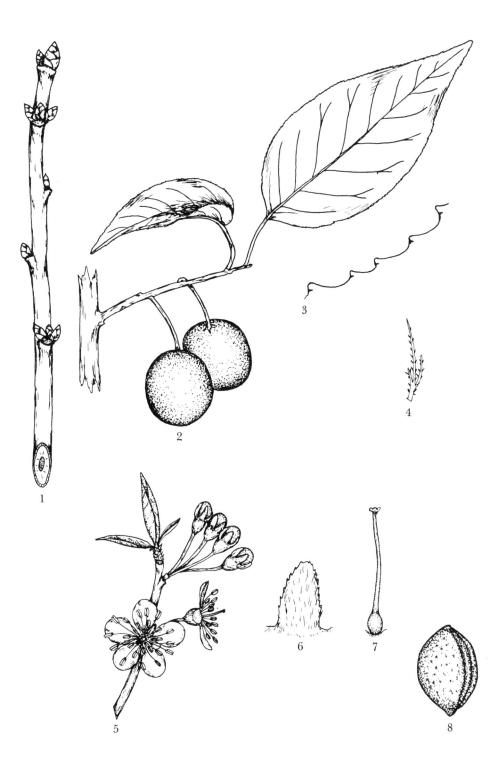

1. Winter twig
2. Leaves and fruit
3. Leaf margin
4. Stipule
5. Flower cluster
6. Sepal
7. Pistil
8. Stone

ROSACEAE
Prunus pensylvanica L.f.

Cerasus pensylvanica (L.) Lois.

Bird cherry, pin cherry

A small tree to 8 m high, the branches high; often in small colonies.

LEAVES: Alternate, simple. Elliptic, lanceolate to oblanceolate, oblong or narrowly ovate, 4-7 cm long, 2-3.5 cm wide, tip gradually or abruptly acuminate, base rounded or acute; margin crenate-serrate, 10-13 teeth per cm, the teeth rounded and irregular, with a large, triangular gland on the incurved (apical) side and occasionally overlapping the next tooth; both surfaces yellow-green and glabrous; petiole 1-1.5 cm long, deeply grooved above, glabrous, glandular at the summit; stipules lanceolate, 2-3 mm long, glandular serrate, early deciduous.

FLOWERS: May-June; perfect. Single, clustered or in short corymbose racemes from buds on wood of the previous year or on short spurs; pedicels 10-15 mm long, glabrous; flowers regular, 11-13 mm across; calyx tube campanulate, 1.5 mm long, pale green, glabrous; calyx lobes 5, triangular, obtuse, thickened at the base, spreading, glabrous, 1.5 mm long, deciduous from the flared summit of the calyx tube; petals 5, obovate, 5-6 mm long, 3-4 mm wide, white, undulate margin, narrowed to a short claw; stamens 10-20, inserted on the flared rim of the calyx; filaments 4-5 mm long, white; anthers yellow; ovary green, ovoid, 1.5-2 mm long, 1-1.2 mm wide, glabrous; style 3.5-4 mm long, greenish-white; stigma capitate, brownish.

FRUIT: July-August. Pedicels 10-20 mm long, green, glabrous, slender; fruit globose or ovoid, 5-8 mm diameter, bright red, glossy, smooth; stone ovoid, about 6 mm long, 4.5 mm wide, 3.5 mm thick, yellow-brown, obtuse base, acute tip, pebbled surface, a broad ridge on the ventral side.

TWIGS: 1.2-1.4 mm diameter, flexible, red-brown, glabrous, often pruinose; lenticels orange-colored, circular, inconspicuous on wood of the season, obvious on older wood; a few short, spur branches; leaf scars broadly elliptic or nearly half-round; 3 bundle scars; pith white, continuous, one-third of stem. Buds broadly ovoid, obtuse, 2.5 mm long, 1.5 mm wide, dark red-brown, glossy, glabrous, margin of scales minutely lacerate; outer scales gibbous at the base.

TRUNK: Bark of young trunks purple-brown, semiglossy, old trunks dark brown, irregularly split, the edges curling outward; the lenticels obvious, horizontal, orange-brown, and raised. Wood lightweight, fairly soft, close-grained, light brown, with a wide, yellowish sapwood.

HABITAT: Moist woods, hillsides, ravine banks, or fence rows.

RANGE: Mackenzie, east to Newfoundland, south to North Carolina and Tennessee, northwest to Iowa and North Dakota, south to Colorado, north to central Montana, and northwest to British Columbia.

During fruiting time the pin cherry should not be confused with any other plant, for the small cherries are quite distinctive. They are acrid but birds eat them regularly. Although the trees may flower heavily, they do not produce a large amount of fruit. The trees are not large and grow in rather crowded conditions, either with their own species or with totally unrelated species, and appear to be quite tolerant of shade.

The range of *P. pensylvanica* overlaps with only one other cherry, *P. virginiana* L., and the two species are entirely different in appearance.

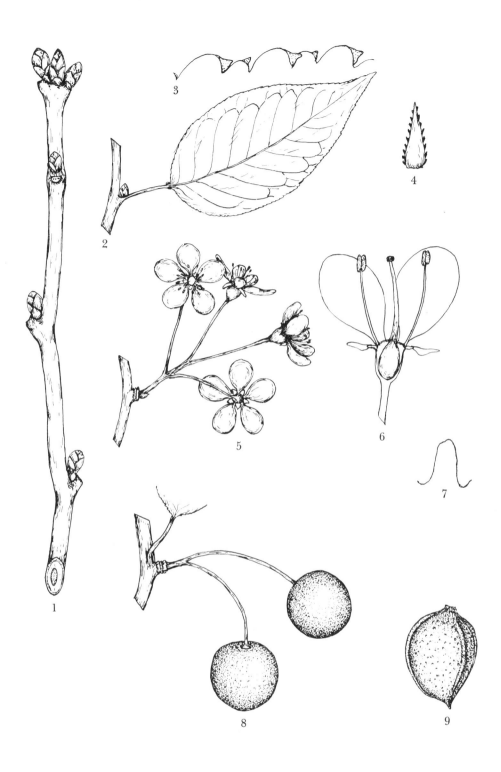

1. Winter twig
2. Leaf
3. Leaf margin
4. Stipule
5. Flower cluster

6. Diagrammatic cutaway of flower
7. Sepal
8. Fruit
9. Stone

ROSACEAE
Prunus rivularis Scheele

P. reverchonii Sarg.; *P. texana* Scheele

Hog plum, creek plum

A large shrub to 3 m high; thicket-forming, the branches low.

LEAVES: Alternate, simple, 6-10 main lateral veins; nearly flat or folded longitudinally, dull to semiglossy. Elliptic, 5-6 cm long, 2-3 cm wide; tip acute to acuminate, base cuneate; margin serrate, 10-11 teeth per cm, gland-tipped; upper surface yellow-green and glabrous, the lower surface slightly paler, glabrous or pubescent along the veins with tufts in the vein axils; petiole 10-12 mm long, a narrow groove above, pubescent along the groove and often on the lower side, occasionally glandular near the summit; stipules lanceolate, glandular, 5-7 mm long, puberulent or glabrous.

FLOWERS: Mid-April, the leaves just starting; perfect. Flowers 12-16 mm across, in clusters of 2-8, in leaf axils or on short spurs; pedicels 10-15 mm long, green, glabrous, enlarged just below the hypanthium; hypanthium cylindric to obconic, 2-2.5 mm long, about 2 mm wide, glabrous; calyx lobes 5, oblong, 2 mm long, the tip rounded to truncate, often with a large gland on each side of the tip, cupped, pubescent on both sides, margins glandular and ciliate; petals 5, nearly circular with a claw of 1 mm, total length 5.5-6 mm, cupped, margins slightly undulate, white; stamens 20, filaments slender, 4-6 mm long, inserted on the rim of the hypanthium; anthers yellow; ovary dark green, ovoid, 1-1.5 mm long; style slender, 5-6 mm long, white; stigma capitate, yellowish.

FRUIT: Late July. Single or 2-3 in a cluster; pedicels glabrous, slender, 13-16 mm long; fruit globular to slightly elongated, orange with a red cheek, or bright red with minute, yellowish dots, smooth, 17-22 mm long and nearly as wide; sweet, juicy. Stone ovoid, flattened, 13-15 mm long, 9-11 mm wide, 7-8 mm thick, strongly ribbed on one margin, bluntly acute at both ends, yellowish, smooth.

TWIGS: 1.4-1.6 mm diameter, red-purple to grayish red-brown, glabrous, lightly covered with a waxy layer which scurfs off easily; the lenticels brownish, vertical and elliptic, horizontal on 2-year-old stems; many short fruit spurs on older branches; leaf scars irregularly oval, dark, with a white pubescence along the upper edge; 3 bundle scars; pith pale tan, continuous, one-fourth of stem. Buds ovoid, red-brown, acute, slightly pruinose, few marginal hairs on the scales; buds usually in three's, the central leaf bud 2 mm long, 1.25 mm wide, the lateral flower buds 3 mm long, 1.5 mm wide.

TRUNK: Bark gray, roughened with large, horizontal lenticels; older trunks with shallow furrows or cracks, the thin edges of bark turned outward. Wood hard, fine-grained, pale brown, lighter toward the outside, but no definite sapwood line.

HABITAT: Creek banks, in open woods or prairies, moist areas along rocky ledges, calcareous soils.

RANGE: Kansas, Oklahoma, Texas.

P. rivularis forms colonies by its numerous, stout rhizomes, these colonies often dense. When growing in a pasture, cattle often form narrow paths through the thicket and use the stiff branches as a scratcher to rid them of flies.

The gland-tipped teeth of the leaves should separate it from *P. americana* and *P. mexicana*; the large flowers and the glandular sepals from *P. angustifolia*; and the glabrous twigs, pedicels, and hypanthium from *P. gracilis*.

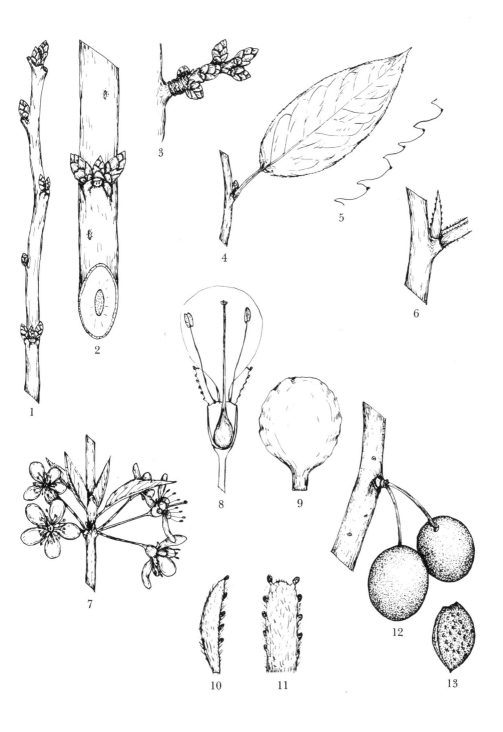

1. Winter twig
2. Detail of twig
3. Spur branch
4. Leaf
5. Leaf margin
6. Stipule
7. Flower cluster
8. Diagrammatic cutaway of flower
9. Petal
10. Side view of sepal
11. Dorsal view of sepal
12. Fruit
13. Stone

ROSACEAE
Prunus serotina Ehrh.

P. virginiana L. (in Rydberg); *Padus serotina* (Ehrh.) Agardh.

Black cherry, rum cherry

A tree to 15 m high, tall and straight when crowded, branching low in the open.

LEAVES: Alternate, simple. Narrowly ovate to oblong-lanceolate, 4-10 cm long, 2-4 cm wide; finely serrate with incurved, callous-tipped teeth, 5-6 teeth per cm; leaf tip abruptly or gradually acuminate, base cuneate to rounded; upper surface dark green, lustrous, glabrous; lower surface paler, glabrous; thick, firm, subcoriaceous; petiole slender, 1-1.5 cm long, glabrous, often red on the upper side, 2 glands at the summit; stipules lanceolate, 4-6 mm long, glandular serrate, caducous.

FLOWERS: April-May, after the leaves half-grown; perfect. Many-flowered, loose racemes, 5-10 cm long, terminal on a short new branch; axis and pedicels glabrous; pedicels 4-6 mm long; flowers 7 mm across; hypanthium campanulate, 1.3-1.5 mm long, glabrous; calyx lobes 5, ovate to narrowly oblong, 0.6 mm long, acute, glabrous, green, with a pale margin, and an occasional gland; petals 5, broadly ovate, 3 mm long, 2.8 mm wide, white; stamens 15-20, filaments white, 1-2.8 mm long, attached to the hypanthium rim; anthers yellow; ovary dark green, 1.5-1.7 mm long, ovoid, glabrous; style 1.6 mm long, stout, green, glabrous; stigma capitate, lobed.

FRUIT: August. Drooping racemes, 12-14 cm long, terminal on a branch with 2-3 leaves; pedicels 5-7 mm long; calyx persistent; fruit a subglobose drupe, 8-12 mm diameter, dark purple, glossy, apex often slightly indented; flesh juicy, slightly bitter, edible. Stone subglobose, 5-7 mm long, 4-5 mm wide, 3.5-4 mm thick, wall thin, smooth, apex acute, broadly ridged on one margin.

TWIGS: 2 mm diameter, flexible, glabrous, red-brown, with minute, pale lenticels; leaf scars elliptical, with the upper edge flattened; 3 bundle scars; pith pale salmon, continuous, one-sixth of stem. Buds ovoid, 3-5 mm long, acute, the scales brown, keeled and short apiculate.

TRUNK: Bark dark red-brown, smooth, with prominent, transverse lenticels; old trunks blackish, the bark broken into thin, irregular, blocky plates; inner bark bitter, aromatic. Wood strong, hard, light red-brown, with a narrow, yellow sapwood.

HABITAT: Rich bottom lands, rocky hillsides, fence rows, woods, and borders of wooded areas.

RANGE: Nova Scotia to Ontario, south to Minnesota, Nebraska, Texas, and Mexico, east to Florida, and north to New England.

Prunus serotina becomes a large tree to 15 meters high, with a trunk circumference of 3 meters. The wood resembles mahogany and is used for furniture, cabinets, interior finish, panels, show cases, scientific instruments, musical instruments, and caskets. The fruit is made into drinks or as a flavoring in candy. It is eaten by many species of birds, especially the robin, blue jay, brown thrasher, and waxwing. Small mammals gnaw the stones open to get the seed.

The leaves contain hydrocyanic and prussic acid and are considered poisonous to cattle. However, deer eat them without harm and will browse on the young twigs during the winter.

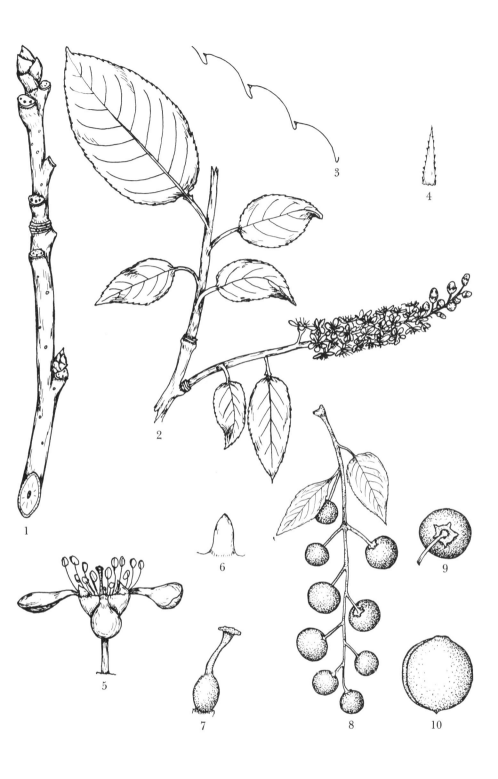

1. Winter twig
2. Leaf and inflorescence
3. Leaf margin
4. Stipule
5. Flower
6. Sepal
7. Pistil
8. Fruit
9. Basal view of fruit
10. Stone

ROSACEAE
Prunus virginiana L.

P. virginiana L. var. *melanocarpa* Sarg.;
P. demissa Walp. var. *melanocarpa* A.
Nels.; *Padus melanocarpa* Shafer in Britt.
& Shafer; *Cerasus demissa* Nutt. var.
melanocarpa A. Nels.

Choke cherry

A tree to 6 m high, but more often found in loose thickets with the tallest tree about 3 m high.

LEAVES: Alternate, simple. Obovate to oval, 5-10 cm long, 3-5 cm wide; finely and sharply serrate, 5-7 teeth per cm, teeth pointing outward or toward the apex, without glands when full grown; leaf tip acute, base cuneate to rounded; upper surface dark green, glossy, glabrous; lower surface pale, sparsely pubescent, the midrib often heavily pubescent; petiole slender, 1.4-2 cm long, grooved above, glabrous, glandular at the summit; stipules linear, 2-3 mm long, incised, promptly deciduous.

FLOWERS: Late April, with the leaves; perfect. Drooping racemes terminating a short new branch with 2-3 leaves; racemes dense, 5-7 cm long, 2 cm wide, pubescent; pedicels 4-5 mm long, pubescent; hypanthium campanulate, 1.3-1.5 mm long, glabrous; calyx lobes 5, obtuse, 1-1.3 mm long, glandular toothed, not persistent in fruit; petals 5, oval to obovate, 3 mm long, white; stamens 20-30, filaments white, 1.4-2 mm long; anthers yellow; ovary green, ovoid, 1.7 mm long, glabrous; style curved, 1.7 mm long, stout, green, glabrous; stigma disk-like, lobed, 1 mm across.

FRUIT: Early July. Racemes 10-13 cm long, axis reddish, glabrous or puberulent; pedicels 5-8 mm long, red, essentially glabrous; fruit a globose drupe, 8-11 mm diameter, dark red-purple, glossy; flesh juicy, bitter, edible. Stone oval, 6.9-7.3 mm long, 5.3-5.5 mm wide, 4.6-4.8 mm thick, cream-colored, pointed at the apex, granular, ridged on one margin and a minute groove on the other.

TWIGS: 2 mm diameter, flexible, glabrous, shiny red-brown to dark brown, with prominent lenticels; leaf scars small, half-round; 3 bundle scars; pith white to tan, continuous, one-fourth of stem. Buds conical, acute, 4-8 mm long; scales rounded, light brown with grayish edges, glabrous.

TRUNK: Bark thin, red-brown, with prominent transverse lenticels on young trees; fissured and scaly on old trees. Wood heavy, hard, red-brown, with a wide, light sapwood.

HABITAT: Rich soils, thickets, fence rows, roadsides, borders of woods, and in sandy or rocky soil on hillsides and in ravine banks.

RANGE: Newfoundland to British Columbia, south to California, east to Texas, northeast to Arkansas, Tennessee, and North Carolina, and north to New England.

The plants of *P. virginiana* are often confused with the young plants of *P. serotina* Ehrh. The most usable characteristics for identification of *P. virginiana* are: the obovate leaves, the teeth pointing outward, the nonpersistent calyx lobes, the thicket-forming habit, and the small size of the tree. The fruits fall from the tree more readily than those of *P. serotina*, and the fruiting season is much shorter. The range of *P. virginiana* overlaps that of *P. serotina* in our area only in southeastern Nebraska and eastern Kansas.

Forma *deamii* G. N. Jones has been described as having more pubescence on the underside of the leaf, but this is purely a matter of degree, and the form is inseparable in our material.

1. Twig
2. Leaf
3. Leaf margin
4. Stipule
5. Leaf variation
6. Flowering raceme
7. Flower
8. Sepal
9. Pistil
10. Fruit
11. Basal view of fruit
12. Stone

LEGUMINOSAE
Acacia angustissima (Mill.) Ktze.
var. *hirta* (Nutt.) Robinson

A. hirta Nutt. in T. & G.; *A. suffrutescens*
Rose; *A. filicoides* (Cav.) Trelease
(misapplied); *Acaciella hirta* (Nutt.)
Britt. & Rose

Prairie acacia, fern acacia

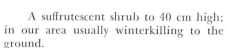

A suffrutescent shrub to 40 cm high; in our area usually winterkilling to the ground.

LEAVES: Alternate, bipinnately compound, 5-12 cm long; pinnae 1-2.5 cm long, 10-14 pairs, each pinna with 10-35 pairs of leaflets. Leaflets sessile or nearly so, oblong, 3-4.2 mm long, 0.9-1 mm wide, entire, ciliate, acute with a rounded or truncate base; both surfaces pale yellow-green, glabrous or with a few hairs; rachis appressed pubescent; petiole 1-4 cm long, with 2 lateral ridges forming a flat or grooved upper surface, appressed pubescent; stipules linear to narrowly lanceolate, 3-3.5 mm long, minutely toothed.

FLOWERS: June-July; perfect. Flowers in short axillary, nearly globose racemes 1.3-1.5 cm diameter with 6-15 flowers; pedicels 1 mm long, glabrous; calyx 0.7 mm long, 1.3 mm wide with 5 low, broad lobes, green, glabrous, loose around the corolla; corolla cylindric; greenish-white, the 5 lobes oblong, 1.5-2.5 mm long, 0.6-0.8 mm wide, acute, thickened and incurved; disk doughnut-like, 0.8 mm diameter, yellow; stamens many, long exserted, attached below the disk at the base of the ovary; filaments white or pinkish, 5 mm long, filiform with minute, pale yellow anthers; ovary cylindric, 1-1.3 mm long, 0.4 mm thick, green, glabrous, slightly curved; style filiform, 4-4.5 mm long, white, glabrous, the stigma not sharply differentiated.

FRUIT: August-September. Pods solitary or clustered, dark brown, flat, 4-6 cm long, 6-8 mm wide, constricted between the seeds, pendent, glabrous, a ridge on both margins, the tip subulate and curved; 3-6 seeds per pod. Seeds brown, often mottled, broadly ovate, 4.4 mm long, 3.8 mm wide, 2 mm thick, tip emarginate, base obtuse, the flattened surfaces with a narrow, oblong marking.

TWIGS: 1-1.3 mm diameter, brown, ridged, often die back to near the ground in winter, glabrate or pubescent with the hairs about 2 mm long and spreading; leaf scars half-round to reniform; 3 groups of bundle scars; pith red-brown, continuous, three-fourths of stem. Buds globular, 1 mm diameter with 2-3 red-green scales, often a small bud at the base of the larger one.

TRUNK: Bark greenish-brown, ribbed, tight. Wood soft, white, with a large brown pith; can hardly be called wood.

HABITAT: Rocky prairie hillsides, exposed limestone ledges, and in sandy, gravel soil.

RANGE: Mexico, New Mexico, Texas, Louisiana, Oklahoma, Arkansas, Missouri, and Kansas.

This plant is mainly herbaceous and is heavily grazed by cattle. In several locations examined where the plants were abundant and healthy on the roadside right of way, there were no plants, or only a few scraggly ones, on the other side of the fence where the cattle could graze. The seeds are apparently nutritious because mice climb the stems, cut the seed pod open and eat the mature seeds.

In the vegetative condition, this plant may be confused with *Desmanthus illinoensis* (Michx.) MacM. The stipules of *Desmanthus* are filiform and 10-14 mm long and the stem hairs are about one-fourth millimeter long and are curled or appressed.

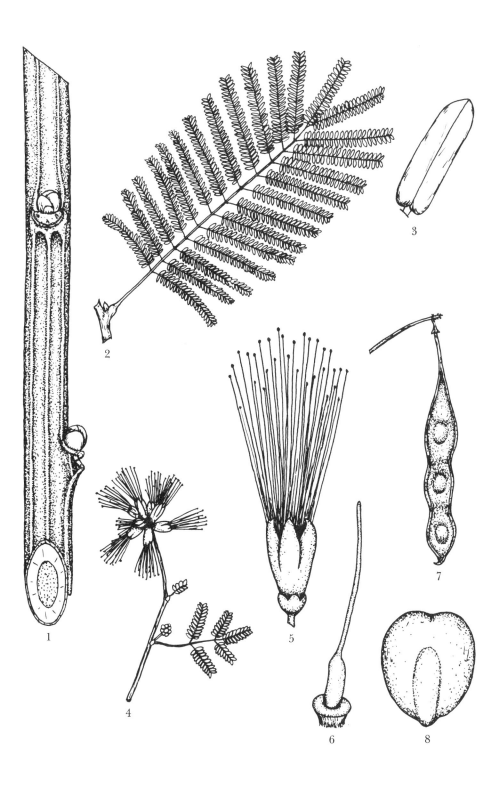

1. Winter twig
2. Leaf
3. One leaflet
4. Cluster of flowers
5. One flower
6. Pistil
7. Fruit
8. Seed

LEGUMINOSAE
Mimosa borealis Gray

M. fragrans Gray; *M. texana* (Gray) Small

Cat claw, fragrant mimosa

An open shrub to 1 m high, the branches crooked and with sharp prickles.

LEAVES: Alternate, bipinnately compound, 2-3.5 cm long; 1-3 pairs of pinnae, 5-15 mm long; 4-6 pairs of leaflets per pinna. Leaflets oblong or ovate, 3-5 mm long, 1-2 mm wide; margin entire, tip rounded or obtuse and mucronate, the base rounded, often obliquely truncate; green on both sides, punctate-dotted, glabrous, 1-3 obscure veins from the base; leaflets fold together during dry, hot weather; petioles 5-15 mm long, green, light dotted, flattened or grooved above; petiolules 0.5 mm long; base of petiole and of the pinnae usually pinkish or translucent; the leaf rachis extending beyond the last pair of pinnae or leaflets; stipules broad-based, subulate, 1.5 mm long.

FLOWERS: Late May; perfect. Axillary on old wood; peduncle 1.8-2 cm long, green, glabrous, no prickles, terminated by the globose or dome-shaped flower cluster; base of calyx stipe-like, 0.5 mm long; upper part of calyx saucer-shaped, 0.6-0.8 mm deep, the 5 lobes small, acute, purplish; petals 5, rose-colored, obovate, 2-2.1 mm long, 0.9 mm wide; stamens 10, attached to the base of the petals; filaments purple, 7-7.1 mm long, exserted; anthers yellow; ovary on a green, glabrous stalk, 0.5 mm long; ovary ellipsoid 1.2-1.3 mm long, 0.3 mm wide, green, glabrous; style filiform 5.5-5.8 mm long, purple; stigma filiform.

FRUIT: Late July. Single or in clusters of 2-8; base of seed pod stipe-like; pod 3-6 cm long, 7-8 mm wide, 5-6 mm thick, brown, curved, with or without a prickle-like projection on the margin; pod constricted between the seeds, tip sharply acuminate, a flat rim extending completely around the pod; dehiscent by the rim breaking off in one piece and allowing the sides of the pod to separate; 2-6 seeds per pod. Seeds flattened ovoid, 5-6 mm long, 4-5 mm wide, 3-4 mm thick, a slight ridge on the edges, smooth, light brown, with a dark brown, U-shaped marking on each surface.

TWIGS: 1.1-1.4 mm diameter, rigid, glabrous, yellow-brown to gray-brown, ridged; prickles numerous, flat, broad-based, recurved at the tip, gray to red or yellow-brown; stipules often persistent and rigid; leaf scars small, oval, raised to the extent that the surface is toward the end of the twig; pith whitish, continuous, one-tenth of stem. Buds 1-1.2 mm long, 1.8-2.2 mm wide, the scales acute, brown, densely white woolly.

TRUNK: Bark on young limbs brown, striate, and with large, nearly white lenticels; old bark gray-brown, flaky or fibrous. Wood hard, red, with a narrow, white sapwood.

HABITAT: Dry, sandy or gravelly hilltops and hillsides, open prairies, and pastures.

RANGE: Oklahoma, Kansas, Texas, and New Mexico.

The shrubs of *Mimosa borealis* growing in southwestern Kansas are much smaller and with fewer branches than those of the same species in Oklahoma and Texas. It is a prickly plant and the foliage is thin; therefore, it is not browsed by the larger mammals.

Often the complete cluster of seed pods drops as a unit, lying on the ground for some time before the outer rim breaks off permitting the two sides of the pod to separate.

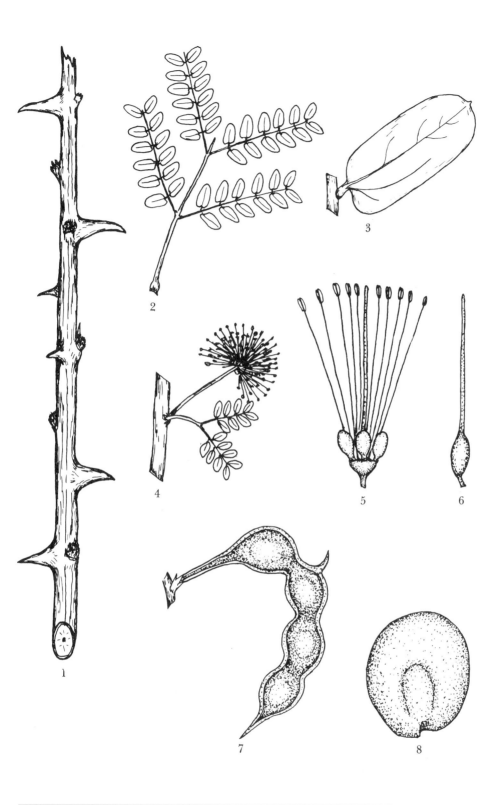

1. Winter twig
2. Leaf
3. Leaflet
4. Flower cluster
5. One flower
6. Pistil
7. Fruit
8. Seed

LEGUMINOSAE
Prosopis glandulosa Torr. var.
glandulosa

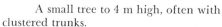

P. juliflora (Swartz) DC. var. *glandulosa*
(Torr.) Cockerell; *P. glandulosa* Torr.;
Neltuma glandulosa (Torr.) Britt. & Rose

Honey mesquite

A small tree to 4 m high, often with clustered trunks.

LEAVES: Alternate, compound with 2 pinnae, commonly clustered, each pinna 6-15 cm long, with 6-10 pairs of leaflets. Leaflets linear, 2-3.3 cm long, 2-3 mm wide, the end pair usually curved; margin entire; tip acute, mucronate, the base rounded; both sides yellow-green and glabrous, the lateral veins indistinct; leaflets sessile or nearly so; petiole 5-8 cm long, grooved above, glabrous, the lower side extending as a point beyond the base of the pinnae, a circular gland on the upper side at the base of the pinnae; rachis of the pinnae angular, the summit similar to that of the petiole; base of the pinnae enlarged, wrinkled, and usually reddish-brown; stipules narrowly lanceolate, 2-3 mm long.

FLOWERS: Late May; perfect. Flowers in pendulous spikes, 7-9 cm long, from buds on old wood; peduncle glabrous, 1-2 cm long, with 1-3 caducous bracts; flowers sessile or with a pedicel about 0.5 mm long; calyx campanulate, 1 mm long, green with a reddish edge; lobes 5, triangular, 0.3-0.4 mm long, finely pubescent at the tip; petals 5, elliptic to obovate, 3 mm long, yellow-green, sometimes reddish, erect, incurved, pubescent at the tip and inside; stamens 10, attached on the inside wall of the calyx; filaments 5-5.5 mm long, white, exserted, becoming yellow as the anthers open; anthers yellow, with a minute gland at the tip between the locules; pistil on a stalk 0.8 mm long, green, glabrous; ovary cylindric, 1.8-2 mm long, 0.7-1 mm thick, green, densely white woolly; style white, 2 mm long with some pubescence at the base; style exserted before the flower opens; stigmas white, 0.5 mm long, hollow cylindric about the same diameter as the style.

FRUIT: September. Pods solitary or in clusters of 2-3; pod 10-15 cm long, 1 cm wide, 6 mm thick; indehiscent, constricted between the seeds, yellowish with short red streaks, abruptly constricted to a point 1-2 cm long; 2-10 seeds per pod, embedded in a pithy, glutinous material. Seeds surrounded by a hard, cream-colored, wrinkled coat, unevenly ellipsoid, 9.5-13 mm long, 5.5-6.5 mm wide, 3-3.3 mm thick; seeds ovoid, flattened, tan-brown, 5.8-6.6 mm long, 4.5-4.6 mm wide, 2.3 mm thick, rounded at the apex, an obovate, indistinct marking on the side.

TWIGS: 1.5-2.7 mm diameter, rigid, glabrous, red-brown, later gray-brown; leaf scars oval, but usually indistinct because of the persistent petiole base; pith white, continuous, one-eighth of stem. Buds ovoid, 4 mm long, rough, knotty-looking, the scales brown with cuspidate, spreading tips.

TRUNK: Bark dark gray-brown, furrowed, the ridges short and flat-topped. Wood heavy, dark brown, with a wide, light sapwood.

HABITAT: Dry, sandy or gravelly hills in open prairie.

RANGE: Mexico, New Mexico, Texas, Oklahoma, Kansas.

Honey mesquite is rare in our area, being found in only a few isolated spots in southwestern Kansas and usually only two or three plants at any one location. It has often been destroyed by the rancher for fear it would take over the pastures. This is not probable at the extreme northern limit of the species.

The compound leaf, with just two pinnae and the linear leaflets, should be sufficient to differentiate it from any other tree or shrub in our area.

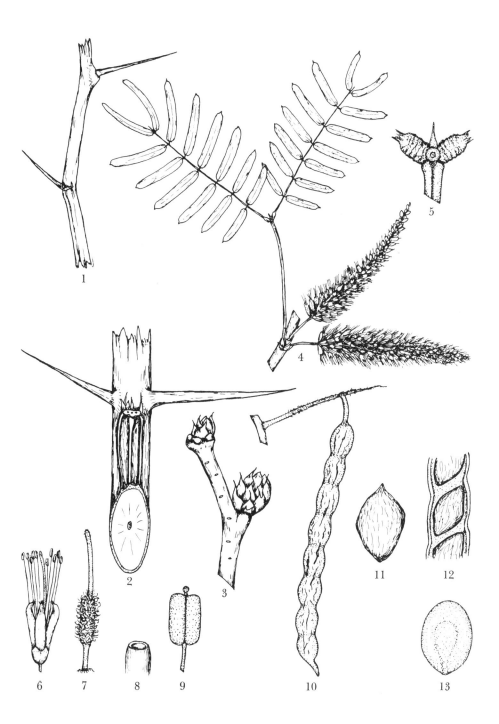

1. Winter twig
2. Detail of twig
3. Detail of bud
4. Leaf and inflorescence
5. Summit of petiole with base of pinnae
6. Flower
7. Pistil
8. Cylindric stigma
9. Anther
10. Fruit
11. Seed with outer coat
12. Portion of pod, one-half removed
13. Seed with outer coat removed

LEGUMINOSAE
Gymnocladus dioica (L.) K. Koch

Kentucky coffee tree

A tree to 23 m high. Trees in a bottom-land woods have long straight trunks, those growing in the open have a short bole and large branches.

LEAVES: Alternate, bipinnately compound, 50-80 cm long with 3-5 pairs of pinnae 2-4 dm long, or a few pinnae replaced by a single, large leaflet 6-8 cm long. Leaflets 10-14 on each pinna, ovate, 4-5 cm long, 3-4 cm wide, acute to acuminate, often apiculate, the base cuneate or rounded and uneven; margin entire; light green on both sides, the upper surface often pubescent near the margin and on the midrib; lower surface sparsely pubescent; young leaves densely pubescent on both sides; petiole 5-6 cm long, stout, enlarged at the base, glabrous; petiolules 2-3 mm long; stipules, if present, lanceolate, glandular serrate at the tip, minute.

FLOWERS: Mid-May, after the leaves; dioecious or perfect. Loose terminal panicles on new growth, 25-30 cm long, 18-20 flowers, the axis glabrous; pedicels 1-3 cm long, longest ones toward the base of the inflorescence, glabrate; hypanthium 8 mm long, 5 mm wide, green urceolate, pubescent, a small ridge extending from the base of the hypanthium to the base of each petal and calyx lobe; ridged inside, with hairs on the ridges; calyx lobes 5, lanceolate, 4-5 mm long, pubescent outside, tomentose inside; petals 5, obovate, 4-5 mm long, whitish, pubescent outside, tomentose inside; stamens 10, filaments slightly tapered, white, pubescent, those opposite a calyx lobe 3.9-4 mm long, those opposite a petal 2.9-3 mm long; anthers yellow; ovary long ovate to ellipsoid, 7-8 mm long, 3 mm wide, flattened, green with white hairs; style 3-3.5 mm long, curved near the end; stigma capitate disk-like.

FRUIT: October. A stalked pod; stalk 2-4 cm long; pod heavy-walled, 10-15 cm long, 4-5 cm wide, 8-12 mm thick, purple-brown to red-brown; remains closed most of the winter; 1-8 seeds surrounded by a green, glutinous material. Seeds nearly circular, 17.5-20.3 mm long, 15.8-17.4 mm wide, 10-12 mm thick, flattened, hard, smooth, dark olive-brown.

TWIGS: 4-8 mm diameter, stout, rigid, reddish pubescent at first, becoming glabrous, light brown, often mottled; the lenticels orange; leaf scars large, somewhat heart-shaped; 3-5 bundle scars; pith salmon to brown, continuous, three-fourths of stem. Buds hardly protruding from a raised, silky-lined cavity 1.5 mm across and about 3.5 mm above the leaf scar; bud scales brownish.

TRUNK: Bark on young trees light brown, smooth; on medium-aged trees thick, light gray, with rigid scales curling laterally and persisting; on old trees thick, tight, with shallow furrows and flat ridges. Wood soft, coarse-grained, durable, brownish-red, with a narrow, light sapwood.

HABITAT: Rich bottom lands, or on rocky hillsides; often planted around farmsteads.

RANGE: New York to Ontario and South Dakota, south to northern Texas, east to Arkansas, Tennessee, West Virginia, and Pennsylvania, mostly west of the Appalachians.

Kentucky coffee tree was formerly planted a great deal around farm homes, but this practice seems to have stopped. It was seldom that young trees were found beneath the old, fruiting trees, probably because of the heavy seed coat which is nearly impermeable to water. By the time the coat is softened, the embryo may have been destroyed in some manner.

The wood is resistant to rot and was used for fence posts, but the wood is soft and the staples would pull out easily.

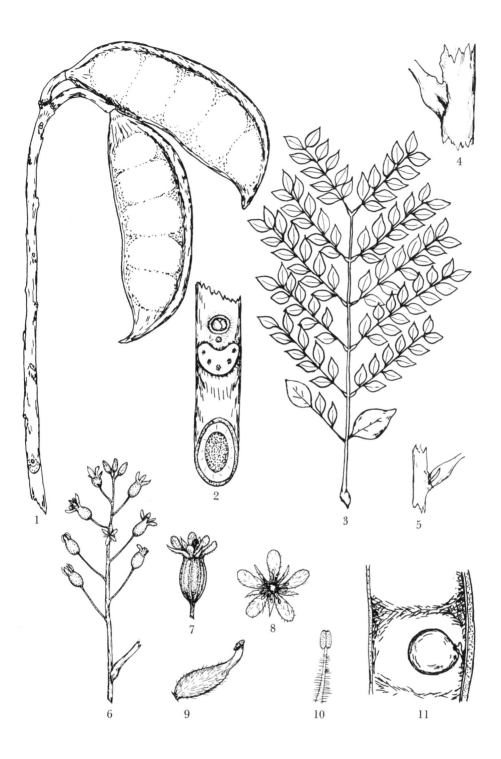

1. Winter twig with fruits
2. Detail of twig
3. Leaf, with occasional variation of basal pinnae
4. Base of petiole
5. Stipule
6. Floral raceme
7. Flower, side view
8. Flower, face view
9. Pistil
10. Stamen
11. Portion of fruit with one seed

LEGUMINOSAE
Gleditsia triacanthos L.

Honey locust

A tree to 15 m high, with short bole and low branches.

LEAVES: Alternate, pinnately or bipinnately compound, leaves often clustered. Pinnate leaves 15-27 cm long, 26-32 leaflets; leaflets 2.5-4 cm long, 9-12 mm wide, lanceolate to ovate, obscurely toothed or entire; tip obtuse or rounded, base acute; green on both sides, glabrous above, pubescent on the midrib below; petiole 3-5 cm long, abruptly enlarged at the base, grooved above, pubescent with erect hairs; leaflets sessile or with a petiolule of 1 mm. Bipinnate leaves 20-30 cm long, 5-8 pairs of pinnae, 18-24 leaflets per pinna; leaflets 1-2 cm long, 4-6 mm wide, lanceolate to narrowly ovate, otherwise similar to those of the pinnate leaves; no stipules.

FLOWERS: Mid-May; polygamo-dioecious. Staminate flowers in dense, many-flowered, paniculate racemes from buds of the previous year. Raceme 4.5-5.5 cm long, 1.5 cm thick, the flowers pedunculate in 3's on the axis, center flower opening first; pedicels 1 mm long; hypanthium obconic, 2-2.3 mm long, finely pubescent; calyx lobes 4, linear, 1.8-2 mm long, 0.5 mm wide, pubescent, green with a red-brown streak in the center; petals 4, obovate, 2-2.2 mm long, 1.4 mm wide, pubescent, green with some red-brown markings; stamens 4, opposite the calyx lobes; filaments 3 mm long, incurved, greenish-white, pubescent at the base; anthers yellow-brown. Pistillate flowers on growth of the previous year, the racemes open, 3-5 cm long, with 8-15 flowers; pedicels 2.6 mm long; hypanthium obconic, pubescent, 3.6 mm long; calyx lobes 4, subulate to linear, 3.5-3.8 mm long, pubescent, partly enclosing an abortive stamen; petals 4, oblong, 4-5 mm long, obtuse to acute, pubescent, yellowish; ovary linear, 5-8 mm long, pubescent, green; style short, 1.5 mm long, sharply curved at the outer end; stigma lobes 2, appressed to the outside curve of the style.

FRUIT: September, falling in early winter. Pedicels 2-3 mm long; pod a pendent, twisted legume, 20-35 cm long, 2.5-3 cm wide, flat, base abruptly narrowed, tip acute to acuminate; walls thin, tough, brown, velvety pubescent or glabrous; 6-27 seeds. Seeds oval, 10-11.4 mm long, 5.1-6.8 mm wide, 3.5-4.6 mm thick, brown, hard, dull, smooth, but with a subsurface showing transverse fracture lines.

TWIGS: 3 mm diameter, rigid, glabrous, lustrous, red-brown to gray-brown, often zigzag, nodes enlarged; thorns simple or branched, stout, red-brown, often 20 cm long; leaf scars large, deeply U-shaped; 3 bundle scars; pith pinkish, continuous, one-third of stem. Buds globose, brown, 1.5-1.7 mm diameter, superposed, the lower one somewhat covered by the edge of the leaf scar, the promeristem of the thorn about 5 mm above the bud.

TRUNK: Bark thick, gray to black, fissured, the ridges long, flat, plate-like, with the sides or ends free and curved outward. Wood hard, durable, brown-red, with a wide, white sapwood.

HABITAT: Rich bottom lands or rocky hillsides, open or wooded pastures, fence rows; commonly planted.

RANGE: New York to South Dakota, south to Texas, east to Florida, north on the west side of the Appalachians to Pennsylvania, becoming established east of the mountains.

The trunk of this tree may be densely covered with long, branched thorns, or it may be totally barren. The latter is forma *inermis* (Pursh) Schneid. The tree was often planted around farmsteads but, possibly because of the thorns, the practice has been discontinued.

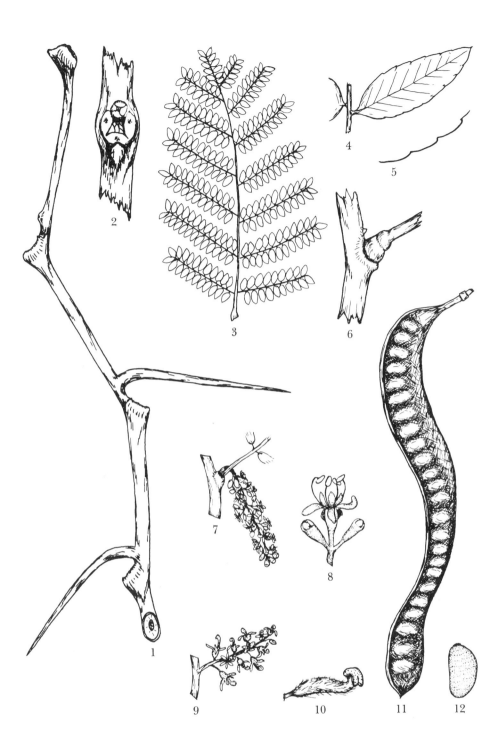

1. Winter twig
2. Detail of twig
3. Leaf
4. Leaflet
5. Leaflet margin
6. Base of petiole
7. Staminate raceme
8. Staminate flower
9. Pistillate raceme
10. Pistil
11. Fruit
12. Seed

LEGUMINOSAE
Cercis canadensis L.

Redbud, Judas tree

A small tree to 8 m high, with a broad, open crown.

LEAVES: Alternate, simple, palmately veined, usually 5 principal veins. Broadly ovate, 8-14 cm long, 5-12 cm wide; entire, tip abruptly short-acuminate or acute, base truncate or cordate; upper surface dark green, glabrous, lower surface paler with axillary tufts and a few hairs on the veins; petiole 4-6 cm long, glabrous, enlarged at the base and summit; stipules minute.

FLOWERS: Mid-April, before the leaves; perfect. Flowers in clusters of 4-8 from lateral buds on old wood. Pedicels 12-17 mm long, red, glabrous; flowers irregular, imperfectly papilionaceous; calyx broadly campanulate, 2.4-2.6 mm long, 3.2-3.6 mm wide; calyx lobes 5, broad, rounded, irregular, purplish, ciliate at the tip; petals 5, rose-purple to pinkish, abruptly narrowed at the base, rounded at the apex; standard 6.9-7.4 mm long, 3-3.2 mm wide, extending upward between the wings; wings 7-7.5 mm long, 3-3.3 mm wide, strongly reflexed; keel petals distinct, 8-9 mm long, 3.8-4 mm wide, surrounding the stamens and pistil; stamens 10, filaments 7-8 mm long, white, surrounding the pistil, a few pinkish hairs at the base; anthers rose-colored; ovary 3.5-4 mm long, 0.5 mm wide, glabrous, green, stalked; style 3.5-4 mm long, red; stigma red, capitate.

FRUIT: August, clustered and remaining on the tree during early winter; pedicels 14-17 mm long; pod 5-7 cm long, 1.2-1.4 cm wide, flat, brown, glabrous, pointed at both ends, a small ridge on both sides of the upper suture; 8-12 seeds. Seeds oval, 4.8-5 mm long, 3.9-4.5 mm wide, 1.7-2.2 mm thick, light brown to chocolate brown, smooth, semilustrous.

TWIGS: 1.5-2 mm diameter, flexible, glabrous, dark gray-brown or blackish, often zigzag; leaf scars broadly crescent-shaped, pubescent along the margin; 3 bundle scars; pith white, continuous, one-fifth of stem. Buds 3 mm long, obovoid, nearly black, somewhat stalked, often 2 buds at a node; scales ciliate, outer scales slightly keeled.

TRUNK: Bark red-brown; on old trees fissured, with short, thin, blocky plates. Wood heavy, hard, brown, with a thin, white sapwood.

HABITAT: Stream banks, wooded areas, rocky hillsides, usually along the edge of a woods or in open areas on the stream bank; does well as a street tree.

RANGE: New York and southern Ontario to Michigan, Iowa, and Nebraska, south to Texas and Mexico, east to Florida, and north to Connecticut.

Forma *glabrifolia* Fern. occurs in our area but most of the plants have at least a few hairs on the lower leaf surface. Forma *alba* Rehd., with white flowers, has not been reported for our range.

Redbud is one of the earliest trees to flower and it produces an abundance of purplish-pink flowers. Since it grows in somewhat open areas, the pinkish color stands out sharply against the brown, barren branches of other trees on a hillside or stream bank. Occasionally a cluster of wild plum grows along with it and the contrast of the white and the purplish-pink flowers creates a most pleasant sight. The tree quite commonly has a flat-topped crown and the height seldom exceeds 6 meters although the bole may become 4 feet in circumference.

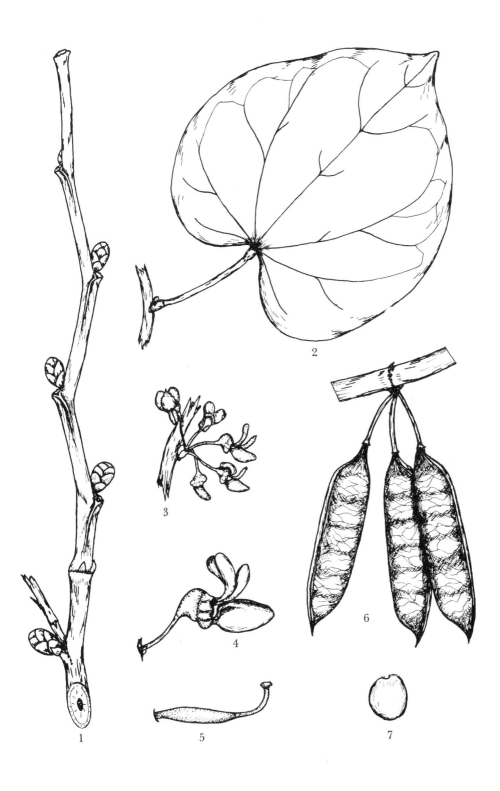

1. Winter twig
2. Leaf
3. Flower cluster
4. One flower

5. Pistil
6. Fruit
7. Seed

LEGUMINOSAE
Amorpha canescens Pursh

Incl. forma *glabrata* (Gray) Fassett

Lead plant

A shrub to 1 m, with single or clustered stems, often densely branched.

LEAVES: Alternate, pinnately compound, 4-8 cm long, 17-20 pairs of leaflets, the terminal leaflet usually small. Leaflets oblong, 7-15 mm long, 3-6 mm wide; entire, acute and mucronate, rounded at the base, densely tomentose on both sides, the lateral veins obscured; petiole, rachis, and petiolules tomentose; petiole 1-1.5 mm long, petiolules 0.6 mm long; stipules and stipels subulate, red-brown, often early deciduous.

FLOWERS: Early June; perfect. Flowers in several terminal, spike-like racemes on new growth, the central raceme usually longest and first to flower; racemes 6-10 cm long, tomentose, densely flowered, the basal flowers opening first; pedicels tomentose, 1-1.3 mm long, subtended by a caducous, subulate, woolly bract; calyx tube obconic, woolly, 2.5-2.7 mm long, purple; calyx lobes 5, triangular, acute, 1-1.6 mm long, purple, woolly; 1 petal, broadly obovate, 4.2-4.4 mm long, 3 mm wide, purple, abruptly narrowed at the lower end, attached to the upper part of the calyx and folded lengthwise over the stamens and pistil; stamens 10, inserted on the lower part of the calyx; filaments red-purple, 6 mm long, connate for 1 mm; anthers yellow; ovary obovate, 0.8 mm long, green, densely white pubescent on the apical half; style 4-4.2 mm long, purple, pubescent; stigmas knob-like, yellowish.

FRUIT: September. Racemes 6-10 cm long, often no pods at the tip; pods 4.1-4.6 mm long, 1.7-1.9 mm wide, densely covered with gray wool, and with reddish pustules showing through the wool, calyx persistent; 1 seed in each pod. Seeds oval, 2.2-2.7 mm long, 1-1.3 mm wide, slightly beaked at the hilum end, olive-brown, glabrous, smooth.

TWIGS: 1.5-2 mm diameter, flexible, pubescent, gray-brown; leaf scars irregularly half-round; 1 large bundle scar;
pith white, continuous, one-half of stem. Buds ovoid, 1.4-2.4 mm long, rounded tip, scales gray or brown, pubescent.

TRUNK: Bark gray, smooth, exfoliating to some extent on very old stems. Wood white, hard, no differentiation of heartwood and sapwood.

HABITAT: Well-drained prairies, open woods, pastures, fence rows, and rocky or sandy hills.

RANGE: Michigan, west to Saskatchewan, south to New Mexico, east to Arkansas, and north to Indiana.

Amorpha canescens is an attractive shrub with its erect branches and canescent leaves, all topped by the spike-like racemes of purple flowers. It is common in all but the western edge of our area and, strangely enough, is quite common in the Black Hills. Several erect stems arise from one base. The plant is palatable to cattle and is less often found in heavily pastured areas than in ungrazed territory.

During the summer months it should not be confused with any other shrub. The purple flowers and the gray wool of the foliage should make identification easy. The more glabrous forms might be confused with *A. nana*, but that species has a solitary inflorescence at the end of the branches and the glands on the leaves stand out prominently. Too, *A. nana* is a much shorter plant and usually has only one main stem.

316

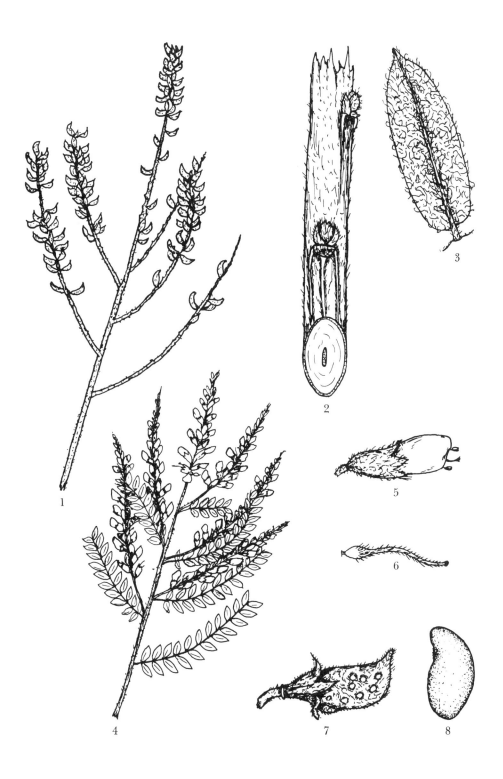

1. Winter twig with fruiting racemes
2. Detail of twig
3. Leaflet
4. Flowering racemes
5. One flower
6. Pistil
7. Fruit
8. Seed

LEGUMINOSAE
Amorpha fruticosa L.

A. fragrans Sweet = var. *angustifolia*
Pursh

False indigo

A shrub with clustered trunks up to 2.5 m high, the branches near the top.

LEAVES: Alternate, pinnately compound, 12-22 cm long, 15-25 leaflets. Leaflets elliptic, ovate or obovate, 2-4.5 cm long, 0.8-1.8 cm wide; entire, the tip rounded and mucronate, the base acute to obtuse; upper surface dark, dull green and appressed pubescent; lower surface paler, appressed pubescent; petiole 1.5-2 cm long, appressed pubescent, lower end abruptly enlarged; stipules linear to subulate, 2 mm long, pubescent; stipels subulate.

FLOWERS: Mid-May; perfect. Flowers in terminal, densely flowered, spike-like racemes which are usually in 3's; racemes 10-15 cm long, the center one longest, basal flowers opening first, the axis pubescent and ridged, a minute bract at the base of each pedicel; pedicels green, pubescent, 2-2.2 mm long and parallel to the axis for one-third of their length; calyx tube obconic, 3.5-4 mm long, glabrate; calyx lobes 5, acute, 0.5-1.2 mm long, pubescent, green, with fine red streaks and circular glands, upper lobes the shortest; 1 petal, broadly obovate, 5 mm long, 4 mm wide, tip rounded and undulate, deep purple, folded around the stamens and pistil; stamens 10, attached at the bottom of the calyx tube; filaments 6 mm long, connate on the lower half, white at the base, and purplish toward the tip; anthers yellow; ovary green, ovoid, 0.8-1 mm long, glabrous; style 4.5-5 mm long, pubescent, green at the base, purplish toward the tip; stigma knob-like, yellowish.

FRUIT: September. Pods 6.6-7.2 mm long, 1.7-2 mm wide, brown, pustuled, curved, the 1 seed surrounded by a sticky fluid. Seed 3.1-3.9 mm long, 1.2-1.5 mm wide, tan-brown, long oval, a slight beak just above the hilum, smooth, glossy.

TWIGS: 2-2.5 mm diameter, rigid, glabrous, red-brown or gray, often with an insect-caused, long swelling near the tip. Leaf scars narrow, crescent-shaped with ends and center enlarged; 3 bundle scars and small stipule scars; pith white, continuous, one-half of stem. Buds ovoid, 1.5-2.5 mm long, obtuse, red-brown, pubescent, usually a small bud adjacent to the leaf scar and a larger bud 2.5 mm above, the upper bud stalked or with the stalk fused to the stem.

TRUNK: Bark of young trunks smooth, brown-gray, with prominent transverse lenticels; old trunks slightly fissured. Wood hard, porous, yellowish, with prominent rays.

HABITAT: Moist stream banks, low, open woods, pasture gullies, and along lakeshores; upland in dry ground, fence rows, and at the border of a woods.

RANGE: New Hampshire, west to Saskatchewan, south to Texas, New Mexico, Arizona, and California, east to Florida, and north to New England.

This is a variable species and has been divided into a number of varieties, four of which may be found in our area: var. *fruticosa,* with 13-25 leaflets, ovate or oblong, acute or obtuse base, about twice as long as broad, pubescence of young stems not appressed; var. *tennesseensis* (Shuttlew.) Palmer, with 21-35 leaflets, 1.2 cm long, 2-3 times as long as broad, pubescent, fruits nearly straight; var. *oblongifolia* Palmer, with 21-35 leaflets, 2-4.5 cm long, 2-3 times as long as broad, sparsely pubescent, fruit curved; var. *angustifolia* Pursh, with 15-25 leaflets, elliptic or obovate, base acute or obtuse, appressed pubescent.

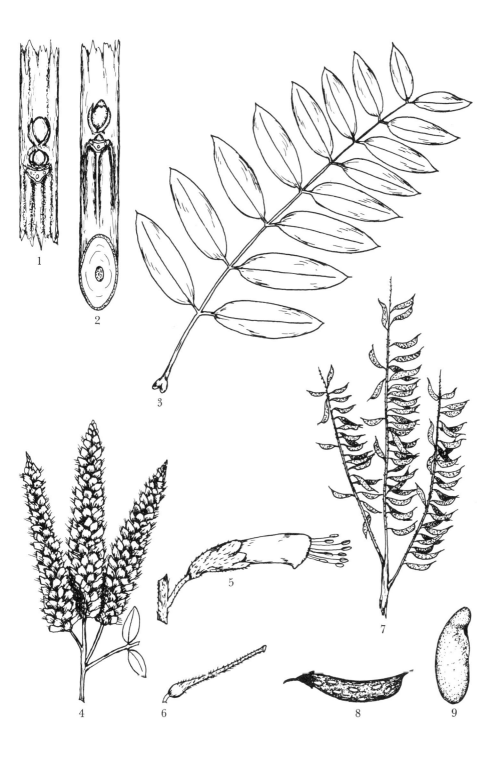

1. Detail of winter twig
2. Detail of winter twig
3. Leaf
4. Flowering racemes
5. One flower

6. Pistil
7. Fruiting cluster
8. One fruit
9. Seed

LEGUMINOSAE
Amorpha nana Nutt.

A. microphylla Pursh

Fragrant false indigo

A shrub to 40 cm high, usually with one trunk and numerous branches.

LEAVES: Alternate, pinnately compound, 5-8 cm long, 25-31 leaflets. Leaflets oblong, 7-11 mm long, 2-4 mm wide, the terminal leaflet often obovate; margin shallowly crenate, often ciliate; tip rounded, emarginate or mucronate, the base rounded or cuneate; upper surface dark yellow-green and glabrous, the lower surface paler, sparingly pubescent and glandular; petiole 5 mm long, pubescent; rachis pubescent, flat or grooved above; petiolules 1 mm long, pubescent; stipules subulate, 3-5 mm long, red-brown, pubescent and with a tuft of hairs at the tip; one stipel subulate and red-brown, the opposing one is a red-brown gland.

FLOWERS: Late May; perfect. Terminal, solitary, spike-like racemes, 3.5-4.5 cm long, 1-2 cm wide; pedicels short; calyx tube obconic, 2 mm long, green with a reddish cast, tube and lobes pustulate; calyx lobes 5, acuminate, 1.2 mm long, upper ones shortest, ciliate, reddish; 1 petal, obovate, 4.6 mm long, 4 mm broad, tip rounded and emarginate, purple with darker veins, attached to the calyx above the ovary and folded over the stamens and pistil; stamens 10, filaments purple, 5 mm long, connate for 1.5 mm; anthers purple; ovary 0.8-1 mm long, ovoid, green, glabrous; style purple, 3 mm long, white pubescent; stigma linear, 0.7 mm long, purple, glabrous.

FRUIT: August. Racemes 5-7 cm long, calyx lobes persistent; pods red-brown, glandular, nearly straight on the upper edge, rounded on the lower, 4-5 mm long, 2 mm wide, a ridge around the margin and pubescent at the tip; 1 seed per pod. Seeds narrowly oval, 2.5 mm long, 1.4-1.5 mm wide, 1.2 mm thick, olive, flattened, the hilum near one end.

TWIGS: 1.2-1.8 mm diameter, flexible, glabrous, gray-brown, a slight ridge on each side of the buds; leaf scars small, crescent-shaped; 1 bundle scar; pith white, continuous, one-half of stem. Buds 1.9-2.3 mm long, 1.8-2.1 mm broad, conical, slightly flattened, obtuse; outer scales brown, glabrate, ciliate; inner scales densely brown tomentose.

TRUNK: Bark gray, split longitudinally, and slightly rough. Wood white, hard.

HABITAT: Dry prairies and rocky or sandy hillsides in the open.

RANGE: Manitoba and Saskatchewan, south to Colorado, northern Nebraska, and east to Iowa and Minnesota. The one station in Kansas is along the southern border and in the western half of the state. This is completely outside of the normal range of *A. nana*, but it is fairly abundant over an area at least one mile across. Those plants are larger than the plants in the northern part of the range, often one meter tall. They are also quite freely branched.

Amorpha nana is often overlooked in tall grass prairies, but is quite noticeable in short grass areas. In dry regions the plant usually grows about 25 cm high, even then, it may be browsed down to 15 cm. The upright, single, spike-like racemes of flowers at the ends of branches are one of the best characteristics for identification.

The distribution map would indicate that it is a common plant in the Dakotas. This is true in some sections, but usually the plants are only common in a small area. Perhaps that one area is the only location in the county.

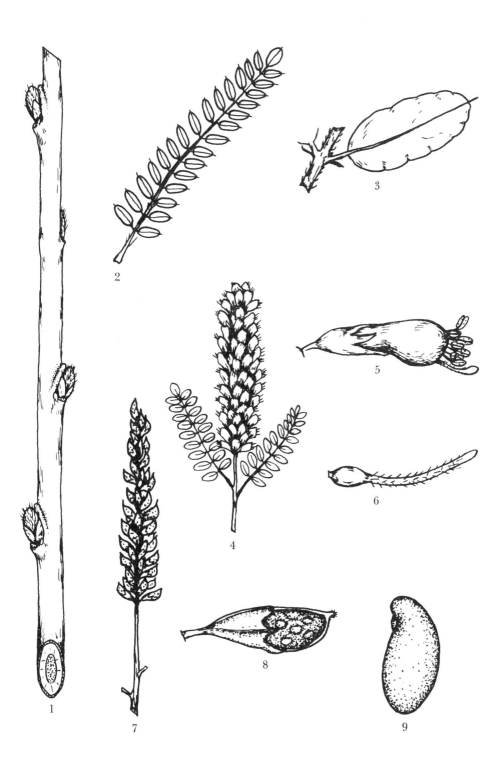

1. Winter twig
2. Leaf
3. One leaflet
4. Flowering raceme
5. One flower

6. Pistil
7. Fruiting raceme
8. One fruit
9. Seed

LEGUMINOSAE
Robinia pseudoacacia L.

Black locust, false acacia, yellow locust,
post locust

A tree to 15 m high with rounded or oblong crown.

LEAVES: Alternate, pinnately compound, 10-18 cm long, 9-19 leaflets. Leaflets 2-3.5 cm long, 1-1.5 cm wide, elliptic to oval; margin entire, tip mucronate, base rounded; upper surface light green, glabrous; lower surface paler, slightly pubescent; petiole 4-5 cm long, slender, glabrous or pubescent, the base enlarged; petiolules short; the tiny stipules at the leaf base becoming replaced by spines. Leaves yellow in the autumn.

FLOWERS: May-June; perfect. Sweet scented. Drooping racemes 10-15 cm long, 10-30 papilionaceous flowers; pedicels 5-6 mm long, purplish, pubescent; calyx campanulate, 7-8 mm long, purplish, pubescent; calyx lobes 5, green or purplish, 2-2.5 mm long, pubescent, the lower 3 lobes acuminate, upper 2 lobes fused, merely notched at the apex; petals 5, white, the standard suborbicular, 15-17 mm long, reflexed, slightly yellow in the center, 3 mm claw; the wings obovate, 20-22 mm long, lobed on one side above the claw, claw 6-7 mm long; the keels 19-20 mm long, semicircular, acute tip, lobed above the claw; stamens 10, diadelphous; the 9 lower filaments 19-20 mm long, connate for 14 mm, the 1 stamen above, connate with the others for 5-7 mm; pistil 1, stalk 3 mm long, glabrous; ovary pod-shaped, 10-11 mm long, green, pubescent; style 7-8 mm long, sharply curved upward, the outer end pubescent; stigma capitate.

FRUIT: August, often persisting through winter. Racemes of 2-12 pods; pedicels 1 cm long, glabrous or finely pubescent, old calyx persistent; pods 5-10 cm long, 1-1.5 cm wide, brown, flat, straight, and ridged on both sides; tip apiculate; dehiscent along both margins, satin-white inside; 4-8 seeds. Seeds 5 mm long, 3 mm wide, 2-2.2 mm thick; hard, smooth, brown mottled with darker brown; hilum near one end, slightly beaked at the hilum.

TWIGS: 1.8-2 mm diameter, rigid, angular, somewhat brittle, glabrous, gray to red-brown, often rough-dotted, lenticels small and scattered; short, stout spines in pairs at the nodes and extending at right angles on either side of the bud; bark of older twigs often splitting, showing inner, light-colored bark; leaf scars broadly triangular; 3 bundle scars; pith brown, continuous, one-fourth of stem. Buds 3 mm long, often nearly hidden by rusty hairs, the scales heavy, brown outside, covered with white wool inside.

TRUNK: Bark gray to red-brown, thick, deeply furrowed, the ridges long, flat-topped. Wood heavy, strong, hard, durable, light brown, with a narrow, yellowish sapwood. One of the strongest woods and shrinks less than other woods in drying.

HABITAT: Moist, well-drained soils; grows well in sandy or rocky soils. Mainly an open-area tree, does not grow well in deep shade.

RANGE: Georgia to Louisiana, north to Oklahoma, Kansas, and Nebraska, east through southern Illinois, Indiana, to West Virginia, north to Pennsylvania and New York, and south in the Appalachians.

Black locust is fast growing but short-lived, and is often used in shelterbelts or for reforestation of poor soil or mine dumps, but is subject to attack by borers, leaf miners, and fungus diseases.

The roots are sweet and licorice-like, but may be poisonous if eaten. There are also reports of cattle being poisoned from eating the young shoots. The flowers may be eaten fried or cooked as a vegetable; and song birds and game birds, including the pheasant, eat the seeds.

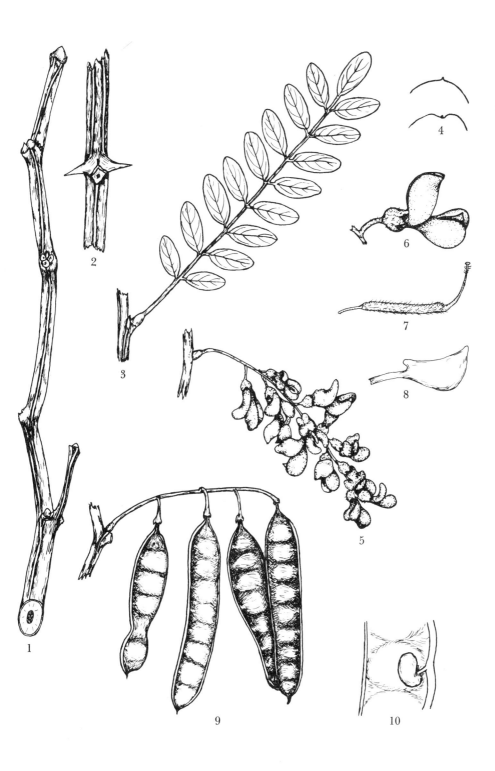

1. Winter twig
2. Detail of node
3. Leaf
4. Variation of leaflet tip
5. Flowering raceme
6. Flower
7. Pistil
8. Keel petal
9. Fruits
10. Seed and portion of pod

RUTACEAE
Zanthoxylum americanum Mill.

Prickly ash, toothache tree

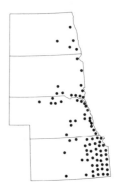

A thicket-forming shrub to 2 m high, often densely branched above the middle.

LEAVES: Alternate, pinnately compound, 15-20 cm long, 5-11 leaflets, aromatic. Leaflets ovate, 3-6 cm long, 1.5-3 cm wide; margin entire or finely crenate; tip acute, base rounded or obtuse; upper surface dark green, glandular dotted, slightly rugose, glabrate; lower surface paler, erect pubescence on the veins, heavily pubescent when young; petiole 3 cm long, rusty pubescent; leaflets nearly sessile.

FLOWERS: April, before the leaves; dioecious. Staminate flowers in clusters of 2-10 on short spurs, or from buds of the previous year. Pedicels 3-4 mm long, glabrous; calyx none; petals 5-6, obovate, 2 mm long, green with some red, a red fringe at the incurved tip, the margin undulate, wider than the petals of the pistillate flowers; stamens 5-6, alternate with the petals; filaments 3-3.5 mm long, white, attached at the base of the vestigial, yellowish pistil; anthers yellow. Pistillate flowers in umbellate clusters of 2-10 on short spurs or from lateral buds; pedicels 3-4 mm long, hairy; calyx none; petals 5-6, ovate, 2 mm long, green and reddish, a red fringe at the acute or obtuse tip, glabrous, undulate margin; pistils 2-5, ovaries distinct, 1 mm long, ovoid, on short stalks from a cylindric base, green, pubescent; styles 2 mm long, green, glabrous, connate above; stigmas globose, green.

FRUIT: Late July. Dense clusters of follicles; peduncle 2.5 mm long, glabrous; pedicels 2.5 mm long, pustuled; follicles globose, 4.5-5 mm diameter, reddish-brown, the surface pitted, strongly aromatic, 2 valves. Seeds oval, 4-4.4 mm long, 3.2-3.6 mm wide, 2.5-2.7 mm thick, finely pitted, glossy black; seed coat oily, aromatic, easily removed exposing the inner, dark brown embryo.

TWIGS: 2 mm diameter, rigid, glabrous, dark brown; 2 prickles at each node, prickles flat, broad-based, recurved; leaf scars crescent-shaped or half-round; 3 bundle scars; pith white, continuous, one-third of stem. Buds globose, 1-1.5 mm diameter, the scales indistinct, covered with red, woolly hairs.

TRUNK: Bark dark gray-brown with light blotches, smooth; slightly fissured on old trunks. Wood soft, light brown.

HABITAT: Upland rocky hillsides in open woods or on rich, lowland soils in wooded areas; often on rocky prairie hillsides or in fence rows.

RANGE: Quebec, west to North Dakota, south to Oklahoma, east to Georgia, and north to Virginia.

Prickly ash forms thickets by sending up shoots from the underground, creeping stems. The single trunk is usually branched above the middle but may have branches low to the ground. The only other plant with which it might be confused would be the young saplings of *Robinia pseudoacacia*, which is also found in thickets and has two flat prickles at each node. The red, woolly buds, clusters of globular aromatic fruits, and the inconspicuous flowers should distinguish *Zanthoxylum*.

The spineless form, f. *impuniens* Fassett, has not been reported for our area and evidently is not common.

The fruits when first mature and before drying have a pleasant, spicy aroma. The oil from the fruit and bark is used in medicine as a tonic as well as a home remedy for the toothache.

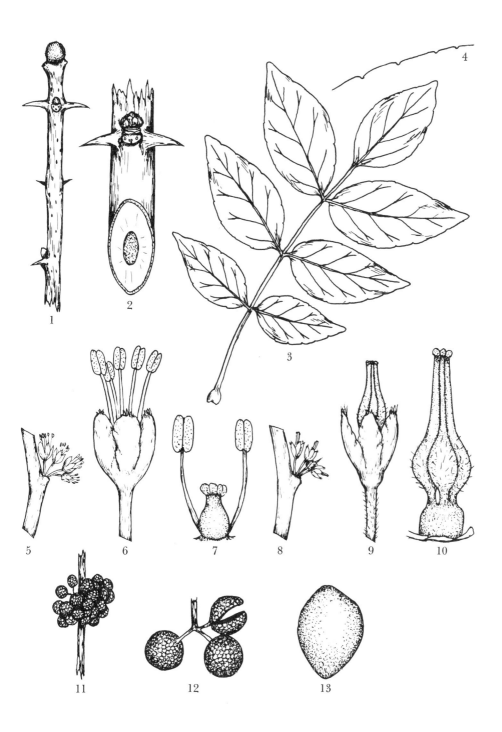

1. Winter twig
2. Detail of twig
3. Leaf
4. Leaflet margin
5. Staminate flower cluster
6. Staminate flower
7. Stamens and vestigial pistil
8. Pistillate flower cluster
9. Pistillate flower
10. Pistil
11. Fruit cluster
12. Fruit
13. Seed

RUTACEAE
Ptelea trifoliata L.

Hop tree, shrubby trefoil, stinking ash, wafer ash

A shrub to 3 m high, with single trunk, usually branching above the middle.

LEAVES: Alternate, 3-foliolate, 10-15 cm long. Leaflets ovate, obovate or broadly elliptic, 6-9 cm long, 3-5 cm wide, the terminal one largest; margin entire or finely serrate with low teeth, 1-3 per cm; tip abruptly acuminate, the base of the terminal leaflet acuminate-decurrent; laterals obtuse, the basal side more rounded; upper surface dark green, semilustrous, glabrate; lower surface paler, sparingly pubescent, densely pubescent at first; minutely glandular punctate on both sides; petiole 6-8 cm long, glabrate, thickened at the base; leaflets sessile; no stipules.

FLOWERS: Mid-May; dioecious. Staminate flowers in terminal cymes of 45-60 flowers on new growth; cyme 2.5-3 cm high, 5-6 cm wide, branches glabrous; pedicels 6-8 mm long; calyx lobes 4, distinct nearly to the base, narrowly ovate, 1.7-1.8 mm long, thick at the base, glabrous except for the ciliate margin, pale yellow; petals 4, pale yellow, elliptical, 5-5.8 mm long, finely pubescent inside and out, spreading or recurved, slightly folded lengthwise; stamens 4, opposite the calyx lobes; filaments 2.5-2.6 mm long, white, tapered, woolly at the base; anthers 1.9 mm long, rich yellow; vestigial pistil in the center. Pistillate flowers in terminal cymes of 20-25 flowers; cyme 2-2.5 cm high, 3-3.5 cm wide, branches glabrous; pedicels 6-8 mm long; calyx lobes 4, distinct to near the base, 1.4 mm long, narrowly ovate, thick at the base, erect between the petals, yellow-green, glabrous, ciliate; petals 4, pale yellow, elliptical, 4-5 mm long, 1.8 mm wide, finely pubescent inside and out, slightly folded lengthwise, not spreading flat; vestigial stamens 4; ovary with enlarged stalk; ovary obovate, 3 mm long, 2 mm wide, glabrous, pitted on the central por-

tion, winged laterally; style 0.9-1 mm long; stigma capitate, 2-lobed, greenish.

FRUIT: Late June. Drooping clusters of samaras, winged all around; pedicels 1-1.5 cm long; samara oval to subcordate, flat, 1.7-2.1 cm long, 1.6-1.9 cm wide, glandular punctate, glabrous, the tip mucronate; seed portion 9 mm long, 6 mm wide, 2-celled, but usually only 1 seed matures. Seeds ovate, 4-6 mm long, angular, nearly black, granular, glossy.

TWIGS: 1.4-1.7 mm diameter, brittle, pubescent at first, becoming glabrous, red-brown, and lustrous; leaf scars shield-shaped to U-shaped; 3 bundle scars; pith white, continuous, one-fourth of stem. Buds depressed globose, 0.5 mm across, pale brown, partly hidden beneath the upper side of the leaf scar, white woolly.

TRUNK: Bark red-brown, smooth except for the prominent, raised, transverse lenticels; older bark slightly cracked or fissured. Wood heavy, hard, yellow-brown, with a narrow, light sapwood.

HABITAT: Rocky hillsides, ravines at the border of a woods or along fence rows; in shaded or open areas.

RANGE: New York, west to Ontario and Minnesota, south to Nebraska, Colorado, and Texas, east to Florida, and north to Pennsylvania. Also in Mexico.

Ptelea trifoliata is common but scattered in southeastern Kansas. It is a bush or low tree up to 3 meters high and is most noticeable after the leaves have fallen and the clusters of fruits remain. All parts of the plant have a faint, disagreeable odor when crushed. The bark of the roots was formerly used as a substitute for quinine, and the fruits were used in place of hops.

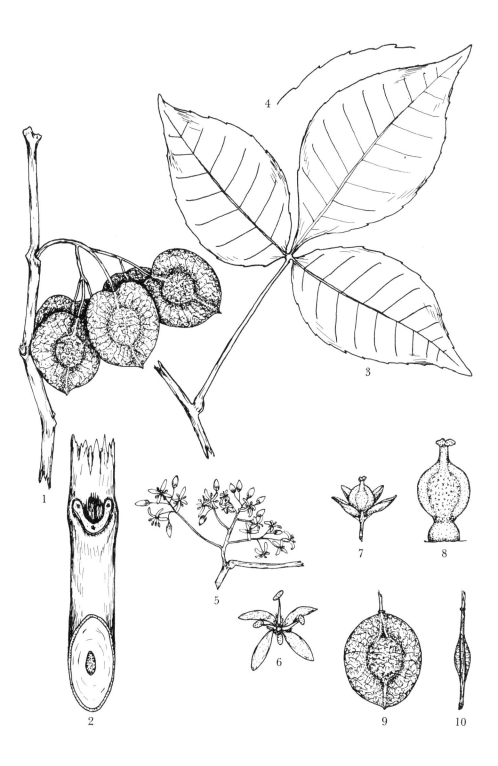

1. Winter twig with fruit
2. Detail of twig
3. Leaf
4. Leaf margin
5. Flowering cyme

6. Staminate flower
7. Pistillate flower
8. Pistil
9. Fruit
10. Edge view of fruit

SIMAROUBACEAE
Ailanthus altissima (Mill.) Swingle

A. glandulosa Desf.

Tree of heaven, tree of paradise,
smoke tree

A tree to 15 m high, with low, spreading branches in open areas.

LEAVES: Alternate, pinnately compound, 25-50 cm long, 11-41 leaflets, ill-scented. Leaflets oblong or lanceolate, 9-13 cm long, 2.5-5 cm wide; tip acuminate, base rounded, truncate or oblique; margin ciliate, entire except for 1-3 coarse, glandular teeth at the base; upper surface dark yellow-green, semilustrous, glabrate; lower surface pale and pubescent; petiole 8-10 cm long, the base greatly enlarged; petiole and rachis velvety pubescent.

FLOWERS: Mid-May; dioecious. Staminate flowers crowded in terminal panicles 15-20 cm long on new growth; pedicels 4-6 mm long, green, glabrous; calyx broadly campanulate, shallow, the lobes triangular, 0.9-1.2 mm long, green, acute, glabrous, ciliate; disk purplish-brown, in 2 sections, the lower portion with 5 large lobes adjacent to the calyx, the upper portion forming a flat-topped structure filling the center of the flower and consisting of many globose lobes; petals 5, valvate in bud, narrowly ovate, 4-5 mm long, strongly cupped and partly enfolding a stamen, yellow-green, pubescent at the base, inserted between the 2 portions of the disk; stamens 10, inserted between the 2 parts of the disk, 1 opposite each petal and each calyx lobe, filaments greenish, 4-5 mm long, pubescent toward the base; anthers yellow. Pistillate flowers in terminal panicles on new growth; panicle 15-20 cm long, open and few-flowered, the branches finely puberulent; pedicels green, 4-7 mm long, puberulent or glabrous; calyx campanulate, shallow, the 5 lobes triangular, acute, 0.8-1 mm long, green, ciliate; petals 5, narrowly ovate, margins folded inward, 3.4-3.6 mm long, green-yellow, tip acute to truncate, heavily pubescent on the lower margin and at the base inside, glabrous outside; vestigial stamens 10, filaments hairy, inserted below the upper

disk; ovaries 5, winged on each side, 1.8-2 mm long, 0.8 mm diameter, green or reddish, glabrous, adnate to a central axis; styles 5, green, 1.4 mm long, connate, twisted, glabrous; stigmas linear, 0.8 mm long, spreading.

FRUIT: August-September. Pendent in dense panicles, frequently persistent through the winter; pedicels glabrous, 8-10 mm long; the samara 2.4-4.9 cm long, 9.3-11.4 mm wide, the central seed area 6.2-6.9 mm diameter, straw-colored, the outer end of the wing usually twisted.

TWIGS: 6-10 mm diameter, coarse, rigid, orange at first, becoming yellow to red-brown, thickened at the nodes, velvety pubescent or glabrous, lenticels conspicuous; leaf scars large, heart-shaped, 6-12 mm wide, a row of bundle scars around the edge; pith white to tan, continuous, one-half of stem. Buds dome-shaped, 3-5 mm across, 1.5-2 mm high, brown, downy.

TRUNK: Bark of young trees smooth, tan-brown, with large, prominent lenticels; bark of old trees grayish, with shallow fissures and tight, flat ridges. Wood soft, weak, pale yellow, with a wide, light sapwood.

HABITAT: Introduced and escaping in many places; uplands, borders of woods, old farmsteads, and fence rows.

RANGE: Native of Asia, occasionally escaping in many parts of the United States.

Ailanthus gets the name of smoke tree from its tolerance of smoke and smog in large cities, where it is often planted in crowded areas. The branches are coarse, allowing the sunlight to pass through in the winter; and the leaves are large, providing good shade in the summer.

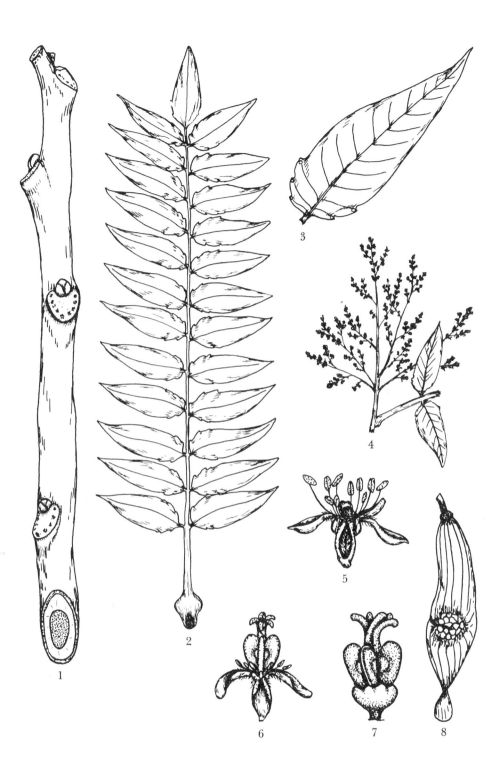

1. Winter twig
2. Leaf
3. One leaflet
4. Staminate panicle
5. Staminate flower

6. Pistillate flower
7. Pistillate flower with outer parts removed
8. Fruit

ANACARDIACEAE
Rhus aromatica Ait.

R. canadensis Marsh.; *R. crenata* (Mill.) Rydb.; *R. nortonii* (Greene) Rydb.; *R. osterhoutii* Rydb.; *Schmaltzia lasiocarpa* Greene; and including the varieties of these species.

Aromatic sumac, fragrant sumac, polecat sumac, lemon sumac

A dense shrub to 2 m high with several trunks from the base.

LEAVES: Alternate, 3-foliolate. Leaflets sessile, ovate to broadly ovate, lobed or coarsely toothed, the teeth pointed or rounded; upper surface dark yellow-green, dull or lustrous, with or without hairs; lower surface pale, glabrate or densely pubescent; terminal leaflet 5-6 cm long, 3-4 cm wide, tip acute to rounded, base acuminate-decurrent; lateral leaflets 3.5-4 cm long, 2-2.5 cm wide; tip acute to rounded, base acute to obtuse; petiole 1-1.5 cm long; no stipules.

FLOWERS: March-April, before, with, or after the leaves; polygamodioecious or perfect. Flowers in dense terminal clusters from catkins formed in the previous season, each flower subtended by a catkin scale. Pedicels 3 mm long, yellow-green, glabrate; calyx lobes 5, united at the base, 1 mm long, obtuse, glabrous to glabrate, yellow-green, pubescent inside, attached below and between the lobes of the yellow, 5-lobed disk, each lobe notched at the tip; stamens 5, usually incurved, attached to the edge of the disk between the lobes; filaments white, 1 mm long; anthers yellow; ovary green, ovate, 0.6 mm long, short pubescent; style green, 0.5 mm long; stigma 3-lobed, capitate, yellow.

FRUIT: Late July. Dense clusters 3-6 cm long, 2-4 cm wide. Drupes 4-7 mm diameter, globose, bright red or orange-red, densely or sparingly hairy with stiff, straight, erect hairs 1-1.3 mm long; 1 seed per fruit. Stone light red-brown or yellow-brown, somewhat circular, 4-5.5 mm long, 3.5-5.5 mm wide, 2-2.5 mm thick, smooth, the reticulate coat clinging tightly; occasional stones are ellipsoidal and without the reticulate coat.

TWIGS: 1.4-1.8 mm diameter, flexible, glabrous or pubescent, gray-brown to red-brown, young twigs often with red glands; winter catkins 4-8 mm long; leaf scars elongated, raised, U-shaped; 8-10 bundle scars; pith white to salmon, continuous, one-third of stem. Buds completely covered by the old leaf base, the scales tan pubescent.

TRUNK: Bark dark brown, smooth on young stems, the transverse lenticels prominent; cracked and slightly fissured on old stems. Wood brown with a light sapwood.

HABITAT: Rocky pastures and prairies, bluffs, rock ledges, and fence rows; usually in open areas, but also grows in the woods where it becomes a more open shrub.

RANGE: Vermont and Quebec to Michigan and North Dakota, south to Nebraska and Texas, east to Florida, and north to New England.

Because of the lack of agreement as to the proper name for this plant and for the various subspecies or varieties, they are here treated as a highly variable entity. Careful studies must be made in both field and laboratory before the problem can be cleared. One of the most common key characters to varieties is the time of flowering in relation to the leaf emergence. During the field work for this book, it was found that the flowering period of any one bush is often over two weeks long. This points out clearly that if the first flowers appear just before the leaves emerge, the same plant will still be in flower when the leaves are well developed. At least in our area, the leaflet shape, the leaf pubescence, the shape of the terminal leaflet tip, and the density of pubescence on the twigs and catkin scales are all extremely variable. Additional work needs to be done on the consistency of correlation of these characters and as to where *Rhus trilobata* fits into the picture.

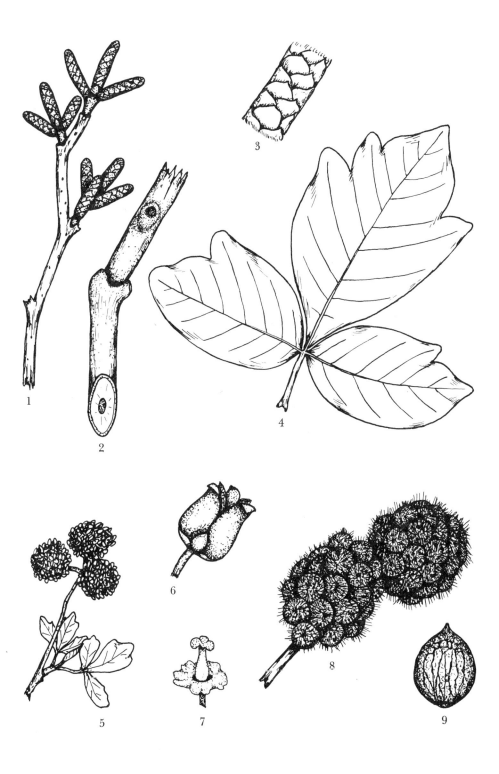

1. Winter twig
2. Detail of twig
3. Detail of winter catkin
4. Leaf
5. Flower cluster
6. Flower
7. Pistil and disk
8. Fruit
9. Seed

ANACARDIACEAE
Rhus copallina L. var. *latifolia*
Engler

Winged sumac, dwarf sumac

A colonial shrub to 2 m high, the branches toward the top.

LEAVES: Alternate, pinnately compound, 15-25 cm long, 9-15 leaflets, rachis winged. Leaflets elliptical to ovate, 5-8 cm long, 2.5-4 cm wide; margin entire or remotely low-toothed, ciliate; tip acute to acuminate, base acute to obtuse; upper surface dark green, lustrous glabrate, lower surface paler and pubescent; petiole 6-8 cm long, pubescent, the rachis often red, each section of the wing along the rachis between the leaflets is wider toward its apex; no stipules.

FLOWERS: Early July; dioecious. Staminate flowers in terminal, pyramidal panicles on new growth, 15-20 cm long, 12-18 cm wide, loosely flowered, axis pubescent; pedicels pubescent, 0.5-1 cm long, a bract at the base of each; calyx lobes 5, distinct, ovate, acute, 1 mm long; petals 5, ovate to elliptic, 3 mm long, 1 mm wide, green-yellow, acute, recurved, often ciliate; stamens 5, opposite the calyx lobes, filaments 2 mm long, pale green; anthers 1 mm long, yellow. Pistillate flowers in terminal, pyramidal panicles on new growth, 8-15 cm long, 6-10 cm wide, densely flowered, branches pubescent; pedicels pubescent, 0.5 mm long, a bract at the base; calyx lobes 5, distinct, deltoid 1 mm long, pubescent, ciliate; petals 5, oval, 2 mm long, 1 mm wide, green-yellow, recurved, ciliate; stamens 5, vestigial; disk 5-lobed, yellow at first, becoming orange-brown, perianth attached at the base of the disk; ovary orange, globular, 0.8 mm diameter, pubescent; styles 3, 0.4 mm long, divergent; stigmas disk-shaped.

FRUIT: Late September. Drooping conical clusters 8-15 cm long, 5-9 cm wide, dense, the axis and branches with crinkly, rusty hairs. Drupes dark red, 4-5 mm long, 3.5-4 mm wide, 2.5-3 mm thick, pubescent with fine, straight, white hairs. Seed oval to bean-shaped, 3.2-3.4

mm long, 2.3-2.4 mm wide, 1.8 mm thick, olive-brown, smooth.

TWIGS: 3.5-5 mm diameter, brownish, brittle, covered with curled, tan hairs; lenticels red; leaf scars U-shaped, often reaching to near the top of the bud; the bundles indistinct; pith white to brown, continuous, three-fourths of stem. Buds short ovoid, rounded, 2.4-2.5 mm long, 3 mm wide, pale pubescent.

TRUNK: Bark gray, with a few shallow fissures; the lenticels red and prominent. Wood soft, brittle, brown, with a white sapwood.

HABITAT: Abandoned fields, pastures, roadsides, fence rows, edge of woods, and in gullies, often in rocky sandstone or limestone soils.

RANGE: Maine to Michigan, south to Texas, east to Florida, north to New England.

Rhus copallina is a smaller shrub than *R. glabra,* usually 1.5 meters high. It is more branched, the branches are finer, and the fruiting heads droop instead of being erect as in *R. glabra.* It forms thickets by sending up shoots from the roots, spreading over wide areas.

The leaves turn a brilliant red in the autumn, but fall early. The thickets are not dense enough to afford protection to birds and small mammals, but are excellent for soil erosion control. The drupes are eaten by a few birds, but would not be called an essential part of their diet. The drupes fall earlier than those of *R. glabra.*

Our plants are var. *latifolia,* with 9-15 leaflets 2.5-4 cm wide. Var. *copallina* has 11-23 leaflets only 1-2 cm wide, and the range is along the eastern edge of the United States.

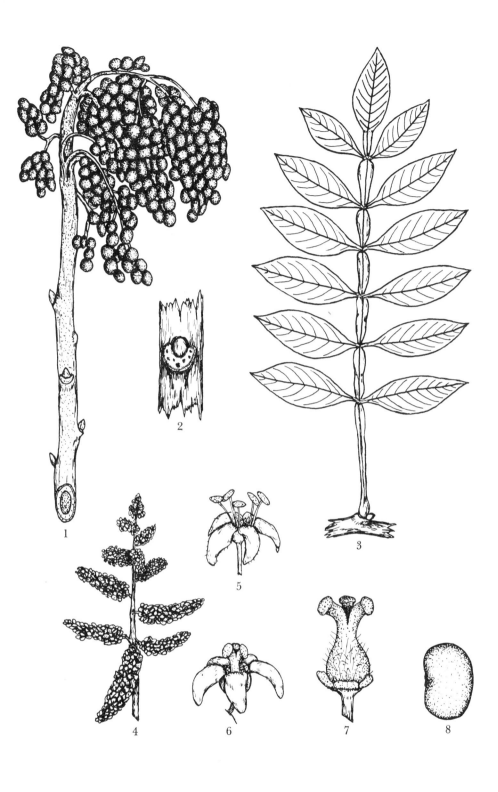

1. Winter twig with fruit
2. Detail of bud
3. Leaf
4. Inflorescence
5. Staminate flower
6. Pistillate flower
7. Pistil and disk
8. Seed

ANACARDIACEAE
Rhus glabra L.

R. angustiarum Lunell; *R. hapemanii* Lunell; *R. sambucina* Greene; *R. cismontana* Greene; *Toxicodendron glabrum* Kuntze; *Schmaltzia glabra* Small

Smooth sumac

A colonial shrub to 3 m high, the branches near the top.

LEAVES: Alternate, pinnately compound, 30-40 cm long, 15-23 leaflets. Leaflets 7-9 cm long, 2-3 cm wide, longest leaflets near the middle of the leaf; elliptic to narrowly ovate, coarsely serrate, 2-3 teeth per cm; tip acuminate, the base obliquely rounded with the apical side wider; upper surface dark green, lustrous, pubescent only at the base and along the midvein; lower surface whitened, glabrous; petiole 6-9 cm long, enlarged at the base, often pubescent and red above; lateral petiolules about 1 mm long, glabrous or pubescent; rachis glabrous or with a narrow line of hairs on the upper surface; no stipules.

FLOWERS: Late May-June; dioecious. Staminate flowers in loose, terminal panicles on new growth; panicle pyramidal, 15-25 cm long, 12-18 cm wide, axis pubescent; lateral branches 10-12 cm long, a lance-linear, ciliate bract 5-8 mm long at the base; smaller bracts at the base of each flower cluster and 1-3 bracts on each pedicel; pedicel 1-1.7 mm long, green, pubescent; calyx lobes 5, distinct, 1.9-2 mm long, narrowly triangular, glabrate, ciliate, acute; petals 5, ovate, 3.3-3.5 mm long, green-yellow, attached below the 5-lobed, golden-yellow disk and opposite the lobes of the disk; tip of petals acute, cupped, and upturned, a line of hairs lengthwise through the center on the inside; stamens 5, attached between the lobes of the disk and opposite the calyx lobes; filaments white, 1 mm long, slightly flattened, tapered; anthers yellow, 1.2 mm long; in bud, the edges of the petal are securely caught between the locules of the anther; vestigial pistil in the center. Pistillate flowers in dense, terminal panicles, 10-15 cm long, 4-6 cm wide, pyramidal; lateral branches 1-4 cm long, a lance-linear bract at the base; lower bracts of the inflorescence 2.5-3 cm long, 2.5-3 mm wide, upper bracts 3 mm long, 0.4 mm wide, early deciduous, entire and glabrous; pedicels 1.5-2 mm long, green glabrous, 1-3 bracts at the base; calyx lobes 5, distinct, narrowly triangular, 9-10 mm long, 2 mm wide, glabrate, acuminate, incurved between the petals; stamens none or with fragments of the filaments; petals 5, ovate, 2.4-2.6 mm long, 1 mm wide, yellow-green, attached at the base of the disk, tip sharply incurved, a line of hairs along the center inside; disk pale yellow in the bud, becoming brilliant orange at anthesis; ovary depressed globose about 1 mm diameter, silvery-white glandular; styles 3, green, 0.4 mm long, glabrous; stigma capitate, yellowish.

FRUIT: August-September. Fruits in a dense pyramidal cluster, 10-15 cm long, erect, persistent over winter; drupes compressed globose, dark red, coarsely pubescent, 3.5-4.5 mm wide. Seed straw-colored, oval, 3.2-3.6 mm long, 2.4-2.8 mm wide, 2-2.1 mm thick, smooth, with a small indentation at the hilum.

TWIGS: 4-6 mm diameter, rigid, glabrous, glaucous, red-brown to purplish, with prominent, gray lenticels; leaf scars narrow, completely around the bud; bundles indistinct; pith brown, continuous, three-fourths of stem. Buds conical, 3-5 mm long, buff or gray pubescent; buds develop inside of the base of the petiole.

TRUNK: Bark gray-brown, roughened with raised lenticels; old trunks with shallow fissures. Wood soft, yellow-brown, with a white sapwood.

HABITAT: Upland prairies, thickets, pastures, fence rows, borders of woods, and abandoned fields; in rocky or rich soil.

RANGE: Throughout most of the United States, southern Canada, and northern Mexico.

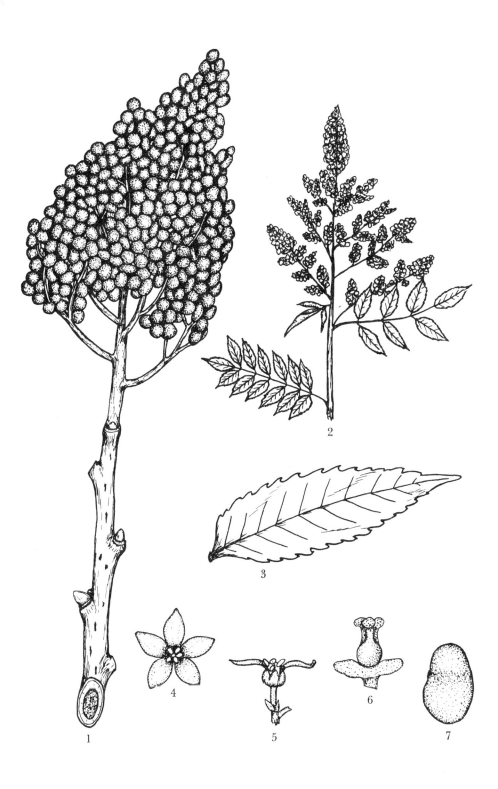

1. Winter twig with fruit
2. Leaves and inflorescence
3. Leaflet
4. Face view of staminate flower
5. Side view of staminate flower
6. Pistil and disk
7. Seed

ANACARDIACEAE
Toxicodendron radicans (L.) Kuntze ssp. *negundo* (Greene) Gillis

Toxicodendron negundo Greene; *Rhus toxicodendron* L. (ssp.) *negundo* (Greene) Gates; *Rhus radicans* L. var. *vulgaris* (Michx.) DC. forma *negundo* (Greene) Fern.; *Toxicodendron aboriginum* Greene

Poison ivy

A vine to 20 m high, climbing by aerial roots; or a low, colonial shrub.

LEAVES: Alternate, 3-foliolate. Leaflets ovate, the terminal leaflet 12-17 cm long, 7-10 cm wide, petiolule 1.5-2 cm long; laterals 10-14 cm long, 5-7 cm wide, petiolules 2-4 mm long; margin coarsely toothed or with small lobes; tip of leaflet acute or acuminate; base obtuse, the basal side of the laterals wider and more rounded; upper surface yellow-green, dull, glabrous; lower surface paler, pubescent with straight hairs; petiole 6.5-20 cm long, green, reddish at the junction with the leaflets, curly pubescent.

FLOWERS: Late May; dioecious. Staminate flowers in axillary panicles on new growth, 5-10 cm long, axis green, pubescent, the flowers crowded; pedicels green, 2.5 mm long, pubescent, a brown, deciduous bract 0.8 mm long at the base; calyx lobes 5, united at the base, 1.5 mm long, ovate, acute, green, with light edges, glabrous, usually recurved; petals 5, ovate, 3.8 mm long, acute, yellow-green, strongly recurved, the margins often revolute; stamens 5, filaments 1.5 mm long, erect and close together, sharply tapered from the base, pinkish, opposite the calyx lobes; anthers 1.5 mm long, yellow; usually a vestigial pistil. Pistillate flowers in axillary panicles on new growth, 4-5 cm long, few-flowered, the axis green and pubescent; pedicels 1.5-2 mm long, green, pubescent, the brown bract at the base 0.5 mm long, deciduous; calyx lobes 5, united at the base, ovate, 1 mm long, obtuse to acute, green, glabrous; petals 5, oval to ovate, 1.9-2 mm long, yellow-green, slightly recurved, margins revolute; stamens 5, vestigial; ovary globose, 0.6-0.7 mm across, green, glabrous; style 0.4-0.5 mm long, stigma capitate, 3-lobed.

FRUIT: September. Grape-like clusters 4-6 cm long, 2.5-3 cm wide, axis and pedicels pubescent; pedicels 2 mm long, fruits globose, 4-5 mm diameter, creamy-white, smooth, semiglossy, indented slightly at the apex, and with shallow, longitudinal grooves, the short style persistent; exocarp dries and cracks off; inner portion subglobose, 3-3.4 mm long, 3.9-4.1 mm wide, 3.1-3.2 mm thick, covered with a fibrous layer which is easily removed, exposing the hard, bony, yellowish, 2-lobed cotyledons, 3 mm long, 4 mm wide, 2.3 mm thick.

TWIGS: 2.8-3.2 mm diameter, flexible, pubescent, light brown with raised lenticels; leaf scars crescent-shaped to shallow shield-shaped; 5-10 bundle scars; pith white, continuous, three-fourths of stem. Buds flattened, acute or obtuse, terminal bud 4-5 mm long, laterals 3.5-4 mm long, tawny pubescent.

TRUNK: Bark gray, smooth except for pinpoint, raised lenticels; bark of large vines flaky; aerial roots bright red at first, becoming brown. Wood soft, porous, yellow-brown, with a wide, light sapwood.

HABITAT: Woods, open pastures, waste ground, creek banks, and fence rows; rocky, sandy, or rich loam soils.

RANGE: Including all the subspecies and varieties, throughout the United States and Mexico, southern Canada, and the West Indies.

Another closely related species of our area is *T. rydbergii* (Small ex Rydb.) Greene. In the key, Gillis (Rhodora 73: 163) describes it as having suborbicular or broadly ovate leaflets with the petiole completely glabrous and as being a subshrub or shrub. Thus, it is differentiated from *T. radicans* ssp. *negundo* which has ovate to lanceolate leaflets, puberulent to densely pubescent petiole, and is a shrub or vine. Synonyms of *T. rydbergii* are: *T. macrocarpum* Greene; *T. desertorum* Lunell; and *T. fothergilloides* Lunell.

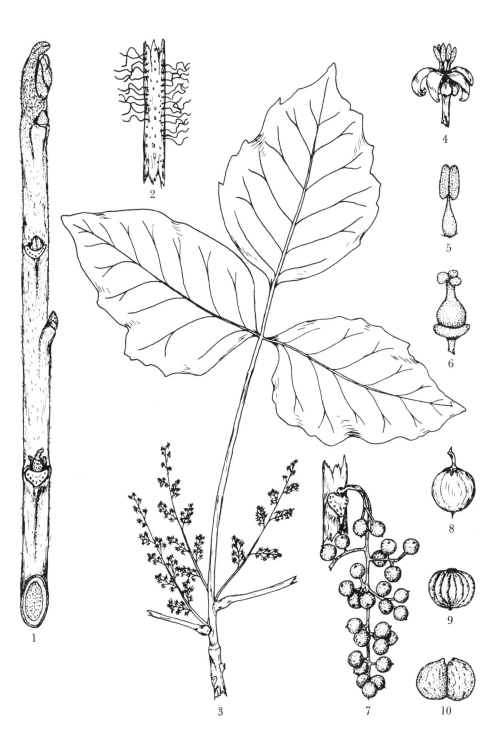

1. Winter twig
2. Aerial roots
3. Leaf and staminate inflorescence
4. Staminate flower
5. Stamen
6. Pistil
7. Fruit cluster
8. Fruit
9. Seed
10. Cotyledons, seed coat removed

AQUIFOLIACEAE
Ilex decidua Walt.

Possum haw, deciduous holly, winter berry

A shrub or small tree to 5 m high, with single or clustered trunks; the shrub often dense.

LEAVES: Alternate, simple. Obovate to spatulate, 4-7.5 cm long, 1-3 cm wide; margin crenulate-serrate, a gland in the sinus or on the apical side of the rounded tooth, 3-4 teeth per cm; leaf tip rounded to obtuse, base cuneate or attenuate; upper surface dark green, glabrous, the lower surface paler and pubescent on the midrib and larger veins; thin, membranaceous when young, becoming thicker; petiole slender, 7 mm long, pubescent; stipules minute, triangular, acute, deciduous or persistent.

FLOWERS: Early May, when leaves are about half-grown; polygamodioecious. Flowers in clusters of 6-11 on the end of a short spur; pedicels 4-10 mm long, glabrous; calyx lobes 4, ovate, acute, 1 mm long, green with a red tip, margin often toothed or ciliate; petals 4, ovate to obovate, 3-4 mm long, pale yellow to white, connate at the base, the end rounded and somewhat recurved; stamens 4, alternate with the petals and attached to their base; filaments white, stout, 1.5 mm long; anthers yellow; ovary 0.8-0.9 mm long, ovoid, green, glabrous; style 0.5-0.6 mm long; stigma capitate and with minute, erect lobes.

FRUIT: October, remaining to late winter. Single or in clusters of 2-5, calyx persistent, the stigma persistent as a flat knob; fruit depressed globose, 5.7-6.8 mm long, 6.2-7.5 mm wide, orange to orange-scarlet, surface smooth or granular; 4 seeds per fruit. Nutlets 3.9-4.9 mm long; 2.5-2.9 mm wide, the shape of an orange section, with 2 flat surfaces and a rounded dorsal side, pale yellow, ridged on all surfaces.

TWIGS: 1.4-1.8 mm diameter, flexible, glabrous, pale gray; greenish, or pale brown on sprouts; short spur branches numerous; leaf scars half-round, stipules often persistent, one bundle scar; pith greenish, continuous, one-third of stem. Buds ovoid, 1.2-1.6 mm long, gray-brown, the outer scales acuminate.

TRUNK: Bark thin, light brown to gray, generally smooth, but with small, warty lenticels. Wood heavy, hard, pale yellow, with a light sapwood.

RANGE: Maryland, west to southern Illinois and southwest to Kansas, south to Texas, east to Florida, and north to Virginia.

Ilex decidua branches freely, with many fine branches forming a dense, rounded shrub. Cattle often browse on the young branches and leaves, causing an isolated plant to become contorted to a flat-topped, crooked-branched shrub. The larger animals such as the deer will also browse on the branches.

The shrub is a heavy producer of fruits, most of which hang on until late winter. This would indicate that birds do not use them as food to any great extent. However, when sleet or ice covers the regular food of the birds, they often flock to these shrubs and clean the fruits from the branches in a short period.

It is an attractive shrub, and along with other species or horticultural varieties is commonly used in landscaping. The long-persistent fruits add a bit of color after most of the other leaves and fruits are gone.

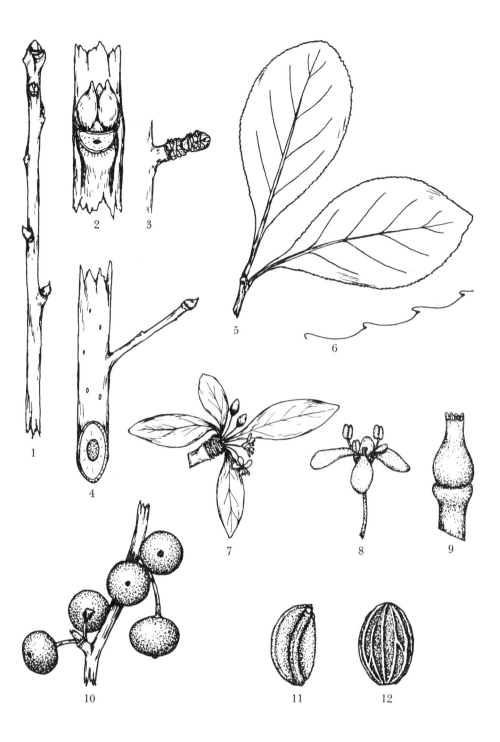

1. Winter twig
2. Detail of twig
3. Spur branch
4. Young branch
5. Leaves
6. Leaf margin
7. Flower cluster
8. One flower
9. Pistil
10. Fruit cluster
11. Side view of seed
12. Dorsal view of seed

CELASTRACEAE
Euonymus atropurpureus Jacq.

Wahoo, burning bush

A shrub to 2 m high, with one trunk, either single or in colonies.

LEAVES: Opposite, simple. Ovate to elliptic, 5-9 cm long, 2.5-5 cm wide; finely crenate-serrate, 7-8 teeth per cm, with glandular tips; leaf tip acuminate, base cuneate; dark green on both sides, glabrous above, pubescent on the veins below; petiole 1-1.5 cm long, glabrous; stipules linear, minute, promptly deciduous.

FLOWERS: Late May; perfect. Flowers in glabrous, axillary cymes on new growth, 7-15 flowers, a minute bract or scar at the base of each branch and pedicel; peduncles 3-4 cm long, secondary branches 1-1.3 cm long; pedicels 2-3.5 mm long, often with minute, red streaks; calyx lobes 4, oval, 1-1.4 mm long, 1.6-1.9 mm wide, reddish-purple with a bright red base, 2 of the lobes usually slightly smaller and somewhat green; petals 4, broadly ovate, 3.4-3.7 mm long, 3.5-3.6 mm wide, obtuse, rich red-purple, spreading flat; disk 4-sided, purple; stamens 4, sessile or nearly so in the corners of the disk, anthers yellow, the locules globose; ovary mostly imbedded in the disk, only the dome-shaped end protruding, 1 mm wide, green, slightly 4-angled; no style; stigma green, 4-lobed.

FRUIT: October, persistent until midwinter. Pendent, solitary or in clusters; peduncles 4.5-6 cm long; pedicels 7-8 mm long, red; fruit red, deeply 4-lobed but 1-2 lobes often undeveloped, 8-10 mm high, 1.8-2 cm across, smooth, calyx usually persistent; lobes splitting along a median line in late autumn, exposing the scarlet-red aril, each lobe of which contains 1-2 seeds. Seeds cream-colored at first, drying to yellow-brown, ovoid, 7.9-8.5 mm long, 4-4.6 mm thick, slightly gibbous on one side, smooth.

TWIGS: 2.5 mm diameter, rigid, glabrous, purplish-brown at first, becoming green, 4-angled, nodes slightly enlarged; leaf scars half-round; 1 large bundle scar; pith white, continuous, one-fourth of stem. Terminal bud 3-5 mm long, narrow, acute, purplish, scales apiculate and slightly glaucous; lateral buds 2.5-3 mm long, acute, scales apiculate, green.

TRUNK: Bark thin, ashy-gray, the old trunks with small thin scales. Wood hard, heavy, white or yellowish, with a narrow, white sapwood.

HABITAT: Rich soils of wooded stream banks, rocky wooded hillsides, and ravines.

RANGE: New York, west to Montana, south to Kansas and Texas, east to Florida, and north to Pennsylvania.

Wahoo has a wide range but is not abundant in any part of our area. The distribution is quite spotted, although a county dot map might indicate an abundance of the species. It is not a dense bush, and has only a few branches from the single trunk. This is probably the reason for it not being used in decorative planting. However, several horticultural species and varieties of *Euonymus* have been developed and are more desirable for landscape purposes.

Under low magnification, the flower is beautiful, the structure is unique, and the colors are rich and contrasting. Pollination or fertilization appears not to be very efficiently accomplished, since only about 10 percent of the flowers produce fruits. This, however, is enough to give a bright red color to the shrubs in the autumn.

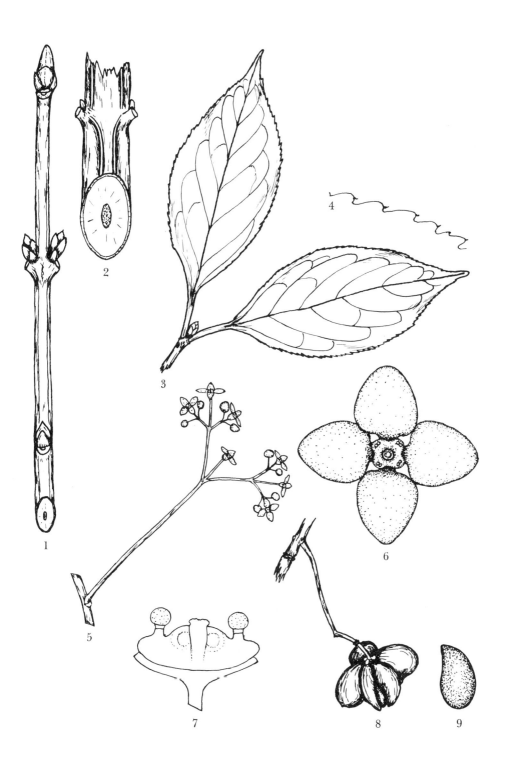

1. Winter twig
2. Cutaway of twig
3. Leaves
4. Leaf margin
5. Inflorescence
6. Face view of flower
7. Diagram of central portion of flower
8. Fruit
9. Seed

CELASTRACEAE
Celastrus scandens L.

Bittersweet

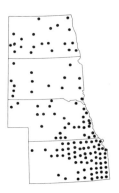

A twining vine to 12 m long, most commonly sprawling over bushes or fences.

LEAVES: Alternate, simple. Oval to elliptic, 7-10 cm long, 4-6 cm wide; finely serrate, 3-6 teeth per cm; leaf tip acuminate, base acute; dark yellow-green above, paler beneath, glabrous on both sides; petiole 1-1.5 cm long, glabrous; stipules minute.

FLOWERS: Mid-May; dioecious. Staminate flowers in terminal or axillary panicles on new growth; panicles 5-6 cm long, many-flowered; pedicels 2.5-3.5 mm long; calyx cup-shaped, 2-3 mm long, the 5 lobes ovate, 1.5-2 mm long, acute to obtuse, the margin irregular, thin, and nearly colorless; calyx cup lined with a bright yellow, fleshy, lobed disk; petals 5, obovate 4 mm long, 1.5-1.7 mm wide, yellowish; stamens 5, filaments white, 1.7 mm long, tapered; anthers 1 mm long, yellow; rudimentary pistil usually present. Pistillate flowers in glabrous, terminal racemes or panicles on new growth, 2.5-4 cm long with 5-25 flowers; pedicels 3-5 mm long, a linear bract at the base, bract of lowest pedicel 4-5 mm long; calyx broad, the 5 lobes oblong, 1.5 mm long, green, truncate or erose, and nearly transparent at the tip; petals 6, obovate to oblong, 3.4-3.5 mm long, 1.7 mm wide, yellowish, truncate to obtuse tip, usually reflexed, margin often toothed; vestigial stamens 5; ovary broadly ovoid, 1.3 mm long, 1.5 mm wide, green, glabrous; style stout, 1.5 mm long, green, glabrous; stigma 3-lobed, 1.4 mm across, yellowish.

FRUIT: Late September. Pendent clusters 6-10 cm long, 6-20 fruits; pedicels 9-12 mm long, jointed, glabrous; fruit a loculicidal, globular capsule, 8-12 mm diameter, orange, the surface granular and slightly wrinkled transversely; the 3 locules with thick, leathery walls and wide-spreading after splitting; each locule with 1-2 seeds. Aril crimson, glo-

bose, 8-10 mm diameter, slightly creased, giving the appearance of 6 sections. Seeds white at first, becoming cream-colored and drying to brown, oval, 5.2-5.8 mm long, 2.3-2.8 mm thick; one side ridged and nearly straight, apex rounded, base obtuse.

TWIGS: 1.5 mm diameter, flexible, twining, glabrous, brown to gray-brown; leaf scars indented or raised, half-round; 1 elongated bundle scar; pith white, continuous, one-third of stem. Buds 1.5-2 mm long, ovoid, acute, the outer scales apiculate, red-brown; buds often at right angles to the stem.

TRUNK: Bark light brown, smooth, with prominent lenticels; bark of old stems peeling into thin flakes and small sheets. Wood soft, porous, white.

HABITAT: Woods, bluffs, creek banks, thickets, fence rows, and rocky slopes.

RANGE: Quebec to Manitoba, south to Wyoming and New Mexico, east to Georgia, and north to southern New England.

Bittersweet is a common vine in the Eastern Deciduous Forest area and extends sporadically through most of the prairie states. It may be a high-climbing vine or it may sprawl over bushes and fences, often forming a dense mass of twisted vines. The brilliant red fruits are attractive and hold their color throughout the winter. For this reason the branches are often cut and used as winter bouquets. The vine may twine tightly around a sapling tree; and, as the tree diameter increases, the vine, unable to expand, will cut into the tree. Larger trees are seldom harmed by the vine.

Birds eat the fruits only to a limited extent, but do use the mass of vines as cover and for nesting sites.

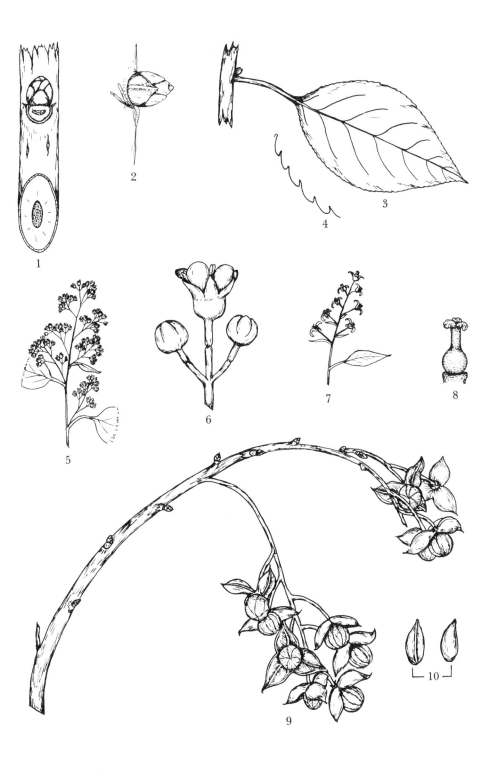

1. Detail of winter twig
2. Detail of bud
3. Leaf
4. Leaf margin
5. Staminate inflorescence
6. Staminate flower and buds
7. Pistillate raceme
8. Pistil
9. Winter twig with fruit
10. Seeds

STAPHYLEACEAE
Staphylea trifolia L.

Bladdernut

A thicket-forming shrub to 3 m high, the branches near the top.

LEAVES: Opposite, 3-foliolate. Leaflets ovate to obovate, 6-8 cm long, 3-4.5 cm wide; margin finely and sharply toothed, 6-8 teeth per cm; tip of leaflet acuminate, base obtuse; upper surface dark green, glabrous; lower surface paler and with fine white pubescence; petiole 6-10 cm long, glabrate; terminal petiolule 3-3.5 cm long, pubescent, grooved above, lateral petiolules 2-3 mm long, pubescent; petiole 6-10 cm long, pubescent; stipules lance-linear, pubescent, caducous.

FLOWERS: Late April; perfect. Flowers in drooping panicles 5-6 cm long from buds of the previous year; pedicels 7-10 mm long, yellow-green, glabrous; calyx campanulate, lobes 5, oblong, rounded, 7.5-8.5 mm long, pale green or white, glabrous except for a few marginal hairs; petals 5, white, 8-9 mm long, narrowly obovate, narrowed at the base, erose or toothed at the tip, ciliate, glabrous outside and pubescent on the lower two-thirds inside; disk green, 5-lobed; stamens 5, filaments 9-9.5 mm long, white, pubescent on the lower part; anthers yellow; ovary 3-lobed, 3.3-3.5 mm long, ovate, pubescent, the base surrounded by the disk; styles 3, greenish, more or less connate, 4.4 mm long; stigmas flat, capitate.

FRUIT: August, persistent until midwinter. Pedicels 1-2 cm long; fruits solitary or in drooping clusters of 2-5; pod 3-5 cm long, 3-4 cm wide, papery, inflated, 3 sections, each section with a free, pointed tip; sparsely pubescent; 1-4 seeds per section. Seeds oval, 5.7-6 mm long, 5.2-5.3 mm wide, 4 mm thick, gray-brown or brown, smooth, glossy.

TWIGS: 2.5-3 mm diameter, flexible, glabrous, red-brown to greenish-brown; often a definite offset at the beginning of a year's growth; bark of the last internode of the previous year splitting and exposing the tan inner bark, giving the effect of stripes; leaf scars half-round; 3-8 bundle scars; pith white, continuous, one-half of stem. Buds ovoid, 2.5-3.5 mm long, obtuse, red-brown, the scales ciliate, 2 short, thick scales at the base.

TRUNK: Bark gray-brown, smooth on young shrubs and slightly fissured and flaky on older trunks. Wood hard, nearly white, with no definite line of sapwood.

HABITAT: Woods and stream banks in shaded areas; rich loam or rocky soils.

RANGE: Massachusetts, west to Quebec and Minnesota, south to Oklahoma, east to Georgia, and north to Connecticut.

Staphylea is a thicket-forming shrub, but the thickets do not become large or dense. It grows as an undergrowth in wooded areas along streams in eastern Kansas and along the Missouri River to northern Nebraska. The bladder-like, clustered pods make it easy to identify in early winter, for there are no other shrubs in the area with that type of fruit. The only other shrub with which it might be confused in a purely vegetative form is *Ptelea trifoliata*. *Ptelea* is a more dense shrub, with branches close to the ground, and the terminal leaflet is sessile, the base of the blade being long attenuate; whereas, *Staphylea* shrubs are open, the branches are high, and the terminal leaflet has a definite petiolule.

The flowers often fall readily from the pedicels when collected and should be pressed as soon as possible.

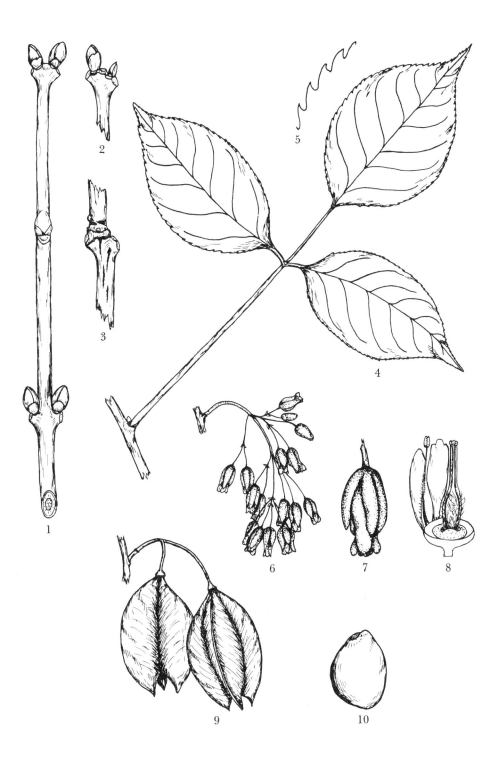

1. Winter twig
2. Variation of end buds
3. Typical junction between two years' growth
4. Leaf
5. Leaf margin
6. Inflorescence
7. Flower
8. Cutaway of flower
9. Fruit
10. Seed

ACERACEAE
Acer glabrum Torr.

A. tripartitum Nutt. ex Torr. & Gray;
A. diffusum Greene; *A. neo-mexicanum*
Greene; *A. torreyi* Greene

Mountain maple, dwarf maple, Sierra
maple

A shrub or small tree to 7 m high, trunks usually clustered and the branches usually high.

LEAVES: Opposite, simple or trifoliolate, palmately veined; broadly ovate, 2.5-6 cm long, and often wider than long, 3-5 lobed, sinuses shallow or divided to the midrib, tip of lobes acute; base truncate or cordate; serrate or biserrate, 3-5 teeth per cm, entire along the basal margin; glabrous on both sides with a few red glands on the veins and at the junction with the petiole; yellow-green above, slightly paler beneath; petiole 3-6 cm long, glabrous, slender, broadly grooved above, usually with a few red glands; no stipules. Leaflets of a trifoliolate leaf widest near the middle, the terminal leaflet attenuate on the short petiolule, lateral leaflets obtuse and sessile or nearly so. Young leaves usually glandular pubescent.

FLOWERS: May-June with the leaves; dioecious or monoecious. Racemes ascending, terminal on a short new branch from a lateral bud, peduncles 5-8 mm long, pedicels 7-9 mm long, a linear bract 2 mm long at the base of the pedicels; calyx lobes 5, distinct, yellow-green, oblong, 5-7 mm long, 1.5 mm wide, end irregular; petals 5, distinct, oblanceolate, 3-5 mm long, 1 mm wide, rounded. Staminate flower with a dark green, lobed disk in the center, with 5-8 stamens from between the lobes; filaments white, 3 mm long, the anthers yellow. Pistillate flowers with a small, dark green, lobed disk and rudimentary stamens; ovary green, broad, flattened, 1.5 mm wide, wings on either side to 1 mm long; styles 2, yellow-green, with linear stigmas, style and stigma 1.5 mm long.

FRUIT: August. Racemes 4-6 cm long, peduncle 1.5-2 cm long, red, glabrous, pedicels 1-1.5 cm long, red or green, glabrous; fruits paired, divergent at 90° or less, each 2-3 cm long, seed portion ovate, wrinkled, 5-7 mm long, 3-4 mm thick; wing terminal, 1.5-2 cm long, 7-10 mm wide, tan-brown to reddish, dorsal side ridged, ventral margin thin, entire or irregular.

TWIGS: 1.5-1.8 mm diameter, rigid, red-gray to wine-red, glossy, smooth, glabrous, lenticels not prominent, many spur branches on the lower twigs; leaf scars narrowly crescent-shaped, nearly halfway around the twig; 3 bundle scars; pith white, often with brown dots, continuous, one-third of stem. Buds 4.5-5 mm long, 2-2.3 mm wide, sessile or with short stalks; 4-5 pairs of valvate scales; outer scales keeled, glabrous outside and pubescent inside, often break at the base and slip off of the bud in one piece; inner scales ovate, acute, wine-red, glossy, and glabrous outside, densely villous inside; scales accrescent and may persist until midsummer.

TRUNK: Bark gray to red-brown, smooth or finely fissured and showing long, light-colored streaks of inner bark. Wood hard, heavy, fine-grained, light brown, with a wide, white sapwood.

HABITAT: Moist, wooded hillsides or ravine banks, usually in rocky soil. Occasionally upland in the foothills and around large boulders.

RANGE: Alaska to California, east to New Mexico, and north to Colorado, western Nebraska, Montana, and Idaho.

Our plants are var. *glabrum,* including the plants with simple leaves, trifoliolate leaves, or both.

346

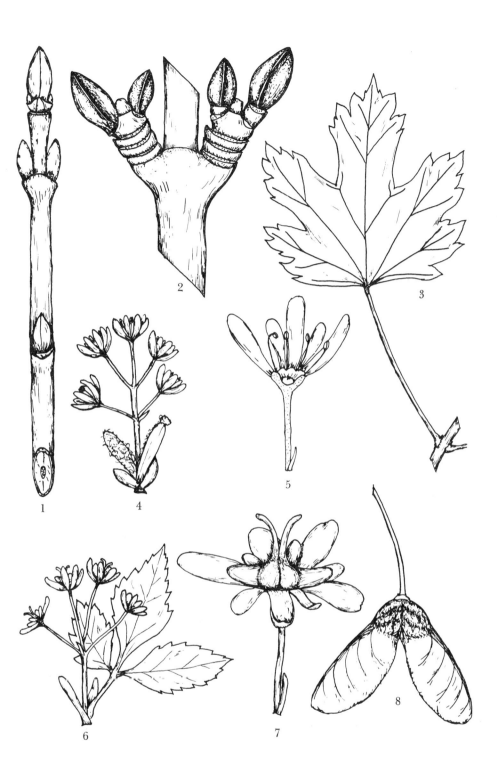

1. Winter twig
2. Spur branch with flower buds
3. Leaf
4. Staminate raceme
5. Cutaway of staminate flower
6. Pistillate raceme
7. Pistillate flower, spread to show pistil
8. Fruit

ACERACEAE
Acer negundo L.

Negundo fraxinifolium Nutt.; *N. nuttallii*
(Nieuwl.) Rydb.; *N. interius* (Britt.)
Rydb.; *Rulac negundo* (L.) A. S. Hitchc.

Box elder, ash-leaved maple

A tree to 15 m high, the trunk short and knotty on trees grown in the open; occasionally tall and straight when crowded.

LEAVES: Opposite, pinnately compound, 20-30 cm long, 3-7 leaflets. Leaflets ovate, 8-12 cm long, 4-7 cm wide, coarsely and irregularly toothed above the middle; tip abruptly acuminate, base obtuse or rounded, laterals with the basal side wider and more rounded; upper surface light yellow-green, glabrous, dull; lower surface paler, pubescent on the veins and with axillary tufts; petiole slender, 6-8 cm long, often pubescent, base enlarged; terminal petiolule 2-3 cm long, lateral petiolules 2-10 mm long, glabrous or pubescent; no stipules.

FLOWERS: Mid-April, before or with the leaves; dioecious. Staminate flowers in drooping clusters of 8-15 flowers from winter buds; pedicels 2-5 cm long, green, pubescent; calyx campanulate, 1-1.5 mm wide and as long, pubescent, yellow-green, 5 minute, acute lobes; petals none; stamens 5, long exserted, filaments white, 0.5-1 mm long, anthers 3-4 mm long, reddish. Pistillate flowers in a 5-9 flowered raceme from winter buds; pedicels 1 cm long, pubescent, green or reddish, a pubescent bract at the base; calyx lobes 5, distinct, elliptic, yellowish, 1-2 mm long, pubescent; ovary 2-lobed, green, 2 mm long, glabrous at the base, long hairy or glabrous on the upper part; style green, 1 mm long; stigmas 2, yellowish, 7 mm long, filiform, divergent.

FRUIT: July-August. Drooping clusters 15-20 cm long; pedicels 1.5-2 cm long; fruit a paired samara with terminal wings; 3-4 cm long, 9-11 mm wide, seed portion 1.2-1.4 cm long, 3.5-4 mm wide; fruits often persistent over winter.

TWIGS: 2.2-3 mm diameter, rigid, glabrous or pubescent, green to purplish, lustrous or glaucous. Leaf scars narrowly crescent-shaped and extending around the twig; 3 bundle scars; pith white, continuous, one-half of stem. Buds ovoid, 2.5-3.5 mm long, obtuse, pubescent, nearly as broad as long.

TRUNK: Bark yellow-brown, fissured, the ridges broad, rounded, and often scaly. Wood light, weak, creamy-white, with a wide, white sapwood.

HABITAT: Stream banks, flood plains, moist soils. Planted upland around farmsteads.

RANGE: Vermont, west to Ontario and Montana, south to Texas, New Mexico, and Arizona, east to Florida, and north to New England.

Four varieties of *A. negundo* have been described and may be found in our area; however, they are not always distinct, and identification of variety may be difficult: var. *negundo,* with green, glabrous twigs; var. *violaceum* (Kirsch.) Jaeg., with gray-purple, glaucous twigs; var. *texanum* Pax, with pubescent twigs, minutely hairy fruits and glabrous petioles; var. *interius* (Britt.) Sarg., with pubescent twigs, glabrous fruits, and pubescent petioles.

When grown in a flood-plain woods, box elder produces a long, straight trunk with a high, open crown. But alone in upland soil it is usually about 8 meters high with low, crooked branches, often with many sprouts near the base of the trunk. The branches are weak and easily broken in a wind.

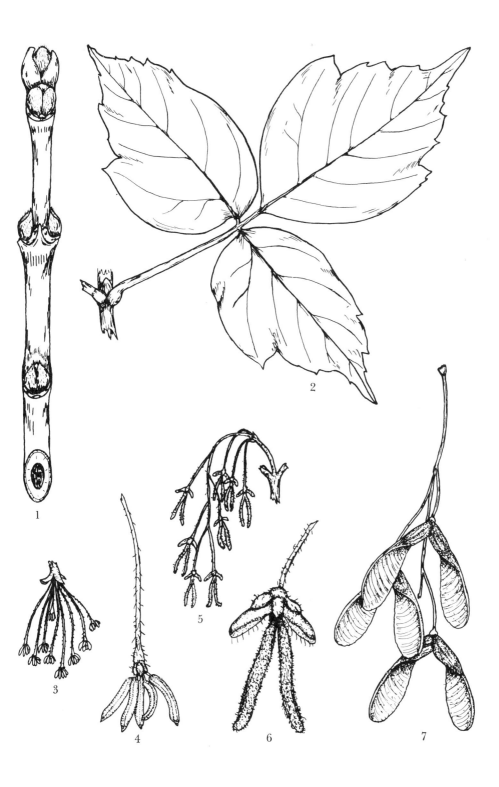

1. Winter twig
2. Leaf
3. Staminate flower cluster
4. Staminate flower
5. Pistillate raceme
6. Pistillate flower
7. Fruit

ACERACEAE
Acer nigrum Michx. f.

A. saccharum var. *nigrum* (Michx. f.)
Britt.; *A. barbatum* Michx. var. *nigrum*
(Michx. f.) Sarg.; *Saccharodendron*
nigrum (Michx. f.) Small

Black maple, hard maple, sugar maple,
rock maple

A tree to 20 m high, with straight trunk and spreading branches.

LEAVES: Opposite, simple, palmately veined. Broadly ovate to broader than long, 7-12 cm long, 7-13 cm wide, 3-5 lobed, the sinuses rounded, usually with an angle of more than 90°, basal lobes often absent or as mere bumps; tip of lobes acuminate, base of leaf rounded or cordate; margin usually drooping, irregular, with coarse teeth or rounded humps; upper surface dark green, dull, glabrate; lower surface yellow-green, glabrous or pubescent and with tufts of hair in the vein axils; young leaves ciliate and red glandular; petiole 4-12 cm long, glabrous or pubescent, grooved above, the base broadened and often with bract-like growths; no stipules.

FLOWERS: April-May, leaves partly grown; monoecious, the staminate and pistillate flowers in the same cluster on a new leafy shoot; racemes drooping, 10-20 flowers, pedicels 3-4 cm long, filiform, pubescent, green; calyx short cylindric to campanulate, the tube 3-4 mm long, 3-4 mm wide, green, glabrous, with 5 low, rounded, long ciliate lobes. Staminate flowers with 5-10 stamens inserted at the base of the calyx tube, filaments yellow-green, 4-5 mm long; anthers yellow, the filament attached near the base. Pistillate flowers with abortive stamens; ovary green, globose, 1.5 mm diameter, long pubescent toward the apex, 2-winged; style 2-3 mm long, green, glabrous; stigmas 2, 5-6 mm long, curled outward, papillose on the inner surface.

FRUIT: August. Paired terminal-winged samaras; body ovoid, 8-10 mm long, 5-7 mm wide, veiny, glabrous or with a few long hairs; wings 1.5-3 cm long, 7-10 mm wide, widest toward the outer end, divergent at less than a 90° angle and often nearly parallel.

TWIGS: 2-2.5 mm diameter, brown, glabrous, semiglossy, the lenticels light-colored, elliptical, noticeable; outer internodes slightly flattened; leaf scars U-shaped, ciliate on the upper margin; 3 bundle scars; pith white, continuous, one-third of stem. Terminal bud ovoid, acute, 4-7 mm long, 2-2.5 mm wide, red-brown, the scales ovate, glossy, long hairy in the center, densely ciliate, the outer scales apiculate; lateral buds similar but smaller, less pubescent and usually with a stalk of 1 mm.

TRUNK: Bark dark brown to black, with shallow furrows and narrow, flat-topped, tight ridges. Wood hard, heavy, fine-grained, light brown, with a wide, white sapwood.

HABITAT: Moist hillsides and stream banks in rich, often rocky soil.

RANGE: New Hampshire, across southern Canada to Minnesota and eastern South Dakota, south to Missouri, east to Tennessee and North Carolina and north to New York.

This species is similar to *A. saccharum* and in our area is hardly distinguishable from it. The two species may grow together in the woods of a hillside or stream bank, and a great many of the trees in such an area have characteristics of both. It is not uncommon to find good trees of *A. saccharum* where the lobe sinuses of the leaves are narrow, and directly beneath that tree will be small saplings with wide, open sinuses. Occasionally one can find both types of leaf sinuses on the same tree. This is also true of the other characteristics which differentiate the two species.

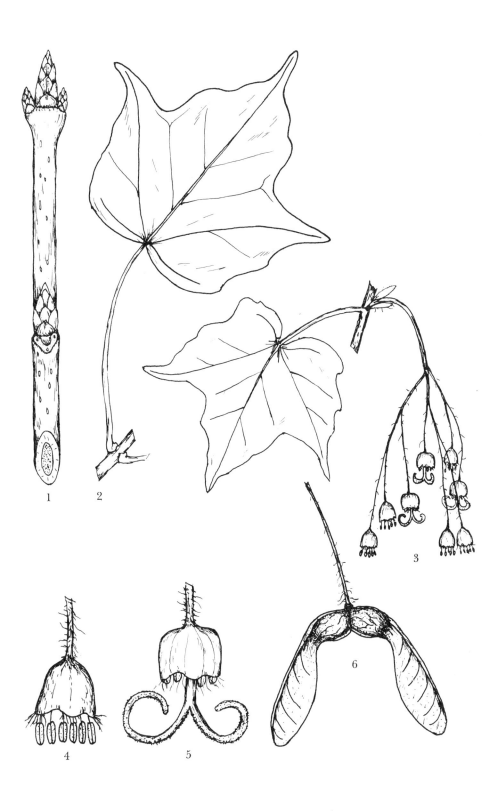

1. Winter twig
2. Leaf
3. Flower raceme
4. Staminate flower
5. Pistillate flower
6. Fruit

ACERACEAE
Acer saccharinum L.

A. dasycarpum Ehrh.

Silver maple, soft maple

A spreading tree to 25 m high, with short, massive bole and large branches.

LEAVES: Opposite, simple, palmately veined. Suborbicular to deltoid-ovate, 10-15 cm long, 9-14 cm wide; deeply 5-lobed, sinuses narrow, acute; lobes irregularly dentate and acuminate, leaf base truncate or cordate; light green above, silvery-white beneath, glabrous; petiole 8-10 cm long, slender, glabrous, often reddish; no stipules.

FLOWERS: March, long before the leaves; monoecious or dioecious. Staminate flowers from a cluster of winter buds, 4-6 flowers per bud, flowers enclosed between 2 broadly rounded scales; receptacle densely pubescent with red hairs; pedicel 1 mm long, glabrous, greenish; calyx campanulate, glabrous, 2-2.5 mm long; lobes 5, rounded or notched, 0.6 mm long, green-yellow; petals none; stamens 5, inserted at the base of the calyx; filaments white, 5-5.5 mm long, anthers yellow, 1.5-1.7 mm long; pistil vestigial. Pistillate flowers from buds of the previous year, enclosed between 2 rounded scales; flowers sessile, with a dense, red pubescence around the base; calyx campanulate, 0.6 mm long; lobes 5, acute or rounded, 0.3 mm long, pubescent; stamens 5, vestigial; ovary 0.8-1 mm long, 1.8-1.9 mm wide, ovate, winged, densely pubescent; style 1-1.1 mm long, green, glabrous; stigmas 2, linear, 3.5-3.7 mm long, red, papillose.

FRUIT: May. Paired samaras with terminal wings, total length 5-5.8 cm, the seed portion 1.3-1.8 cm long and 7-9 mm wide; wings divergent, prominently veined. Seed chestnut brown, 1 cm long.

TWIGS: 2.5-3.5 mm diameter, flexible, glabrous, reddish-brown to gray-brown, lustrous; leaf scars narrowly crescent-shaped, the ends enlarged and with an abrupt point; 3 bundle scars; pith white to brown, continuous, one-half of stem. Flower buds clustered near the end of twigs, with large clusters near the base of the current growth and globular clusters on the end of the short spurs. Buds globose, 3-4 mm diameter; outer scales red, ciliate, slightly keeled. Leaf buds solitary, terminal or lateral, ovoid, acute.

TRUNK: Bark light gray-brown, furrowed, exfoliating into long, thin plates fastened at one end and curling outward. Wood soft, easily worked, brittle, pale brown, with a wide, white sapwood.

HABITAT: Low, moist, rich woods, stream banks, moist hillsides. Planted around farmsteads and homes.

RANGE: New Brunswick, west to Ontario, Minnesota, and South Dakota, south to Oklahoma and Louisiana, east to Florida, and north to New England.

Silver maple is a large, spreading tree with low branches, seldom with a long, straight bole. It was formerly planted extensively around farmsteads, probably because of its rapid growth. It has a number of disadvantages and is no longer used to any extent. The branches are brittle and easily broken, and the wood is soft and susceptible to decay.

Woodpeckers and squirrels take advantage of the soft wood and decayed knot holes and often build their nests in the trees. If the early flowers are not frozen, an abundance of seed is produced, but they appear not to be a favorite food of small mammals. These seeds germinate readily, and seedlings are common beneath the parent tree.

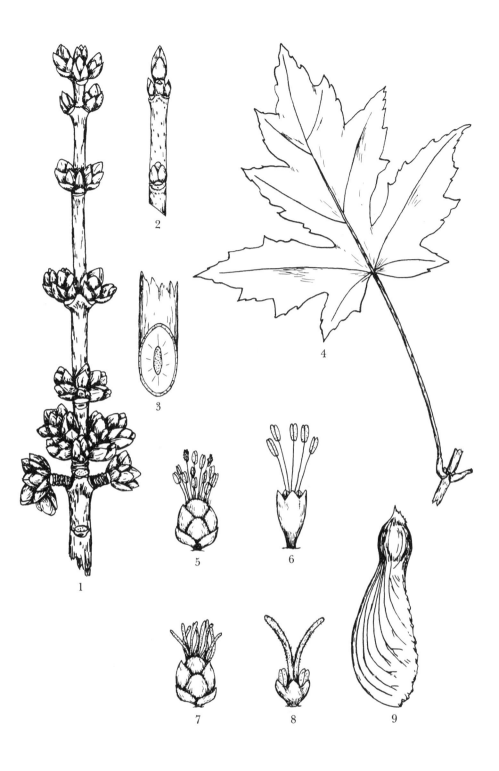

1. Winter twig with flower buds
2. Twig with leaf buds only
3. Cutaway of twig
4. Leaf
5. Cluster of staminate flowers
6. Staminate flower
7. Cluster of pistillate flowers
8. Pistillate flower
9. Fruit

ACERACEAE
Acer saccharum Marsh.

A. barbatum Sarg., not Michx.; *A. saccharinum* Wang., not L.; *A. saccharophorum* K. Koch; *Saccharodendron barbatum* sensu Nieuwl.

Sugar maple, hard maple, rock maple

A tree to 25 m high, usually with straight trunk and small branches.

LEAVES: Opposite, simple, palmately veined. Suborbicular, 8-13 cm long and as wide; 3-5 lobed, each lobe with 1-2 large teeth, the sinuses between lobes narrow and rounded; lobes acuminate, but not sharp pointed; base truncate or subcordate; upper surface dark green, dull, glabrous; lower surface blue-green or somewhat whitened, glabrous except for axillary tufts; flat, firm, thick, becoming yellow to bright red in the autumn; petiole slender, 4-7 cm long, glabrous; no stipules.

FLOWERS: Mid-April; monoecious or dioecious. Flowers from winter buds, in short racemes of 8-14 flowers, the axis 1-1.5 cm long; pedicels 4-5 cm long, filiform, pubescent, often branched, occasionally a staminate and pistillate flower on the same pedicel. Staminate flowers at the base of the raceme; calyx widely campanulate, 5-6 mm long, 4-5 mm wide, lobes 5, obtuse, pale green, glabrous except for the ciliate margin; disk lobed, yellow; stamens 5-8, long exserted, attached at the hairy center of the disk; filaments 6-7 mm long, yellowish; anthers yellow; pistil vestigial. Pistillate flowers at the end of the raceme; calyx narrowly campanulate, 4.7-4.9 mm long, 4.2-4.3 mm wide, lobes 5, obtuse, pale green, incurved, glabrous with a ciliate margin; vestigial stamens concealed by the calyx; ovary ovoid, 1.7-1.8 mm long, 3-3.5 mm wide including the lateral wings, a few straight hairs on the ovary; style short, stigmas 2, linear, 4-4.5 mm long, greenish-brown papillose on the inner surface.

FRUIT: August. Paired samaras with terminal divergent wings, total length 2.5-2.7 cm; seed portion 7.6-8.5 mm long, 5-5.8 mm wide. Seeds oval, red-brown, smooth, 7 mm long.

TWIGS: 1.5-2.5 mm diameter, flexible, glabrous, brown, smooth, lustrous, the lenticels large; leaf scars broadly U-shaped, narrow, enlarged at the center and extending halfway around the stem; 3 bundle scars; pith white to light brown, continuous, one-fourth of stem. Buds solitary, ovoid, acute, 5-7 mm long, the several pairs of scales red-brown and glabrous except for some pubescence at the tip.

TRUNK: Bark dark gray, furrowed, the thick plates tight or cleaving along one edge. Wood tough, durable, hard, heavy, pale brown with a narrow, light sapwood.

HABITAT: Moist rich valleys and stream banks or on moist, rocky hillsides in limestone soils; occasionally on drier hillsides.

RANGE: Nova Scotia, west to Manitoba, south to Minnesota, eastern Kansas, and northeast Texas, east to Georgia, north along the Appalachians to Delaware and New England. Also in the Wichita Mountains of southwestern Oklahoma.

Four forms have been described but are not completely separable in our material: forma *saccharum,* the leaves with 5 lobes, lower surface glabrous and not glaucous; f. *glaucum* (Schmidt) Pax, leaves with 5 lobes, lower surface glabrous and glaucous; f. *schneckii,* leaves with 5 lobes, lower surface densely hairy; f. *rugellii* (Pax) Pal. & Steyerm., leaves with three entire lobes.

Acer saccharum is an Eastern Deciduous Forest tree, and instead of following the streams out into the prairie as many of the trees do, it forms a continuous, narrow strip along the eastern edge of Kansas. In other words, its western limit in our area is a rather sharp, straight line.

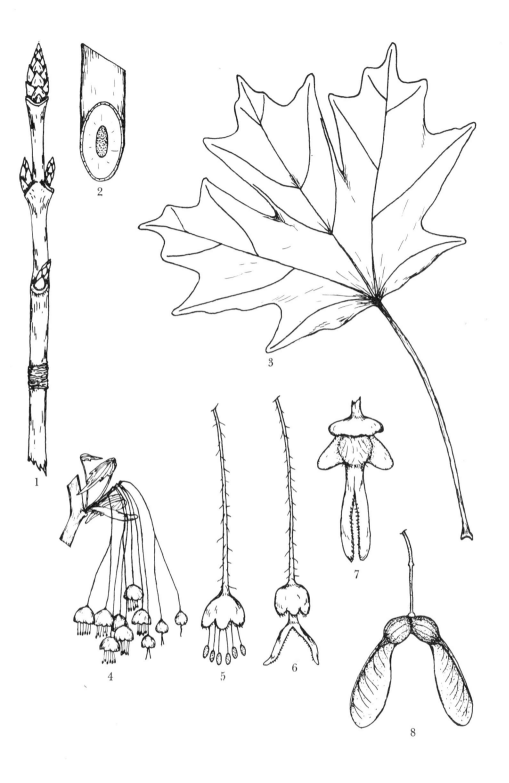

1. Winter twig
2. Cutaway of twig
3. Leaf
4. Flower raceme

5. Staminate flower
6. Pistillate flower
7. Pistillate flower with calyx removed
8. Fruit

HIPPOCASTANACEAE
Aesculus glabra Willd. var. *arguta*
(Buckl.) Robinson

A. glabra var. *sargentii* Rehd.; *A. arguta*
Buckl.; *A. buckleyi* (Sarg.) Bush

Western buckeye, white buckeye

Usually a shrub to 3 m high, but may become a tree to 10 m high.

LEAVES: Opposite, palmately compound, usually 7 leaflets. Leaflets sessile, elliptic, 7-15 cm long, 1.5-3.5 cm wide, outer ones the smallest; tip and base long acuminate; serrate, often double toothed, 4-6 teeth per cm, tip of teeth blunt, sparingly ciliate with curled hairs; upper surface dark green, semilustrous, glabrous or a few hairs on the veins; lower surface paler and with curled hairs on the veins; petiole 10-16 cm long, the base enlarged; no stipules.

FLOWERS: April-May; perfect. Terminal panicles 10-12 cm long, with 50-60 flowers; pedicels 1-2 cm long, puberulent; calyx campanulate, 7-9 mm long, greenish-yellow, short pubescent, lobes 5, obtuse, pink-tipped, 2-3 mm long, glabrate; petals 4, the 2 upper ones spatulate, 2-2.2 cm long, 4-5 mm wide, abruptly narrowed to a 5 mm claw, yellow with 2 orange spots inside, often strongly recurved, the outside and margin heavily pubescent with crinkly hairs; the lateral 2 petals ovate, 13-15 mm long, 6-7 mm wide, abruptly narrowed to a claw 7 mm long, yellow with an orange or red streak in the center, crinkly hairs and glandular hairs on the outside and a line of dense hairs on each side at the upper end of the claw; stamens 7, filaments 1.8-2.2 cm long, exserted, white, curved, pubescent near the tip; anthers orange; ovary stalked, white, 4.5-5 cm long including the stalk, pubescent; style 1.5 mm long, pubescent, elongating after anthesis; stigma red, indistinctly 2 parted.

FRUIT: September. Capsule globose, 3-4.5 cm diameter; hull 3-4 mm thick, leathery, rusty brown, scurfy and spiny, the spines 1-5 mm long with a broad, round base; the 1-3 seeds irregularly globose, often with one or more flattened surfaces, 2-2.5 cm diameter, hard, smooth, glossy, dark red-brown, with a large, pale scar.

TWIGS: 4.5-5 mm diameter, rigid, light red-brown to gray-yellow, glabrous; leaf scars triangular with concave sides; bundle scars in 3 groups, or a line of several scars; pith white, continuous, two-thirds of stem. Terminal bud 1-1.2 cm long, conical, acute, the scales red-brown, glaucous, keeled, often toothed; lateral buds 7-9 mm long, conical, the scales red-brown and keeled.

TRUNK: Bark on young trees pale yellow-brown, smooth but with large lenticels; bark of old trees dark brown, with shallow fissures and short, flat ridges. Wood soft, white, the sapwood hardly distinguishable.

HABITAT: Creek banks, woods, bottom lands and rocky upland hillsides. It is usually an undershrub in the woods along streams and on hillsides bordering a flood plain.

RANGE: Ohio to Iowa, south to Kansas and Texas, east to Mississippi, and north to Kentucky.

Our plants are var. *arguta*, but a few from northeastern Kansas have only 5 leaflets which are wider than those of typical var. *arguta*, tending toward var. *glabra*.

The buckeye seeds are poisonous, and care should be taken that children do not eat them. However, roasting appears to destroy the poison, and Indians roasted and ate them as a source of starch.

The wood is lightweight and soft, and pioneers hollowed out the logs and used them for a number of purposes. Our trees are not large enough for this.

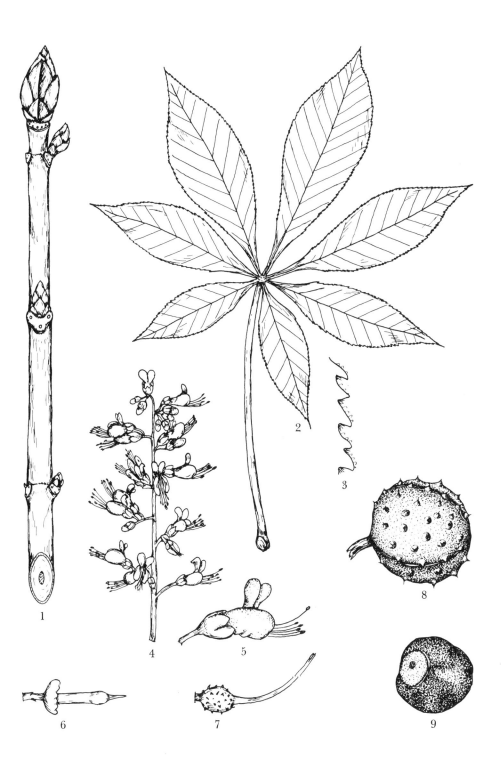

1. Winter twig
2. Leaf
3. Leaf margin
4. Flower panicle
5. One flower

6. Pistil
7. Developing ovary
8. Fruit
9. Seed

SAPINDACEAE
Sapindus drummondii Hook. & Arn.

Sapindus saponaria L. var. *drummondii*
(H. & A.) L. Benson

Soapberry, chinaberry

A tree to 15 m high, solitary or in thickets.

LEAVES: Alternate, pinnately compound, 20-35 cm long, 6-10 pairs of leaflets, the leaflets often alternate on the rachis. Leaflets ovate, 4-6 cm long, 2-2.5 cm wide; margin entire; tip acuminate, often falcate, the base oblique; the basal side of the leaf narrow with a cuneate base, the apical side wider with a rounded base; light green on both sides, glabrous above, pubescent on the veins below and on the petiolules; petiole 4-5 cm long, petiolules 3-4 mm long.

FLOWERS: Mid-June; dioecious. Staminate flowers in a pyramidal panicle on new growth, 15-20 cm long, 15-20 cm wide, loosely flowered, the axis yellow-green, ridged and densely pubescent; the bract at the base of each inflorescence branch subulate, brown, pubescent and caducous; pedicels 0.5-1 mm long, pubescent; calyx lobes 5, distinct, irregularly oval to ovate, 2 mm long, cupped, pale yellow-green, glabrate, ciliate; petals 5, white, ovate, 4 mm long, abruptly narrowed toward the base to a 1 mm pubescent claw with a dense tuft of hairs at the base, the margin ciliate; petals attached below the disk; stamens 7-8, filaments 4-5 mm long, white, slightly tapered, pubescent on the lower half, inserted in the center of the disk; anthers yellow; pistil vestigial. Pistillate inflorescence similar to the staminate; calyx lobes 4-6, broadly oval, 1 mm long, 1.2 mm wide, yellow-green, ciliate; petals 5, white, 3 mm long, 1.4 mm wide, ciliate, the claw pubescent and 1 mm long; stamens 7-8, vestigial; ovary 3-loculed, 1.4 mm long, 1.5 mm wide, dark green, glabrous; style yellow-green, glabrous, 1.5 mm long, exserted before the flower opens; stigma yellowish, capitate, indistinctly lobed; central disk green, 2 mm wide, irregularly lobed.

FRUIT: October. Terminal panicles with 10-30 fruits. Fruit globose, golden, somewhat wrinkled, 12-14 mm diameter, flesh thick, gummy, translucent; persistent most of the winter, turning brown and shriveling toward spring; 1-seeded. Seeds black, obovoid, 8.5-9 mm long, 6.8-7.5 mm diameter, semiglossy, smooth but minutely pitted, a tuft of short, white hairs on the basal end.

TWIGS: 2.3-2.8 mm diameter, brittle, gray-brown; pubescence short, curled, tan; leaf scars shield-shaped or heart-shaped; 3 groups of bundle scars; pith creamy-tan, continuous, one-third of stem. Buds 2-2.6 mm long, the scales brown and thick at the base; buds often superposed, the upper bud about 2 mm diameter and dome-shaped, the lower one much smaller.

TRUNK: Bark gray-brown, with shallow fissures and short ridges or thin plates. Wood hard, strong, heavy, yellow-brown, with a light sapwood.

HABITAT: Along streams, at the margin of a woods, in pasture ravines, on waste ground, and on rocky hillsides. Often planted.

RANGE: Mexico, Arizona, New Mexico, Texas, Louisiana, Arkansas, Missouri, Kansas, and Oklahoma.

Soapberry often occurs in an open woods with other trees and grows to a height of 15 meters, the trunk tall and straight. It is more often found in our area as a small tree no more than 6 meters high and growing in thickets. It is an attractive tree and is often used for yard planting.

The flesh of the fruit may be ground and used as a soap substitute and the hard black seeds are often drilled and strung for beads.

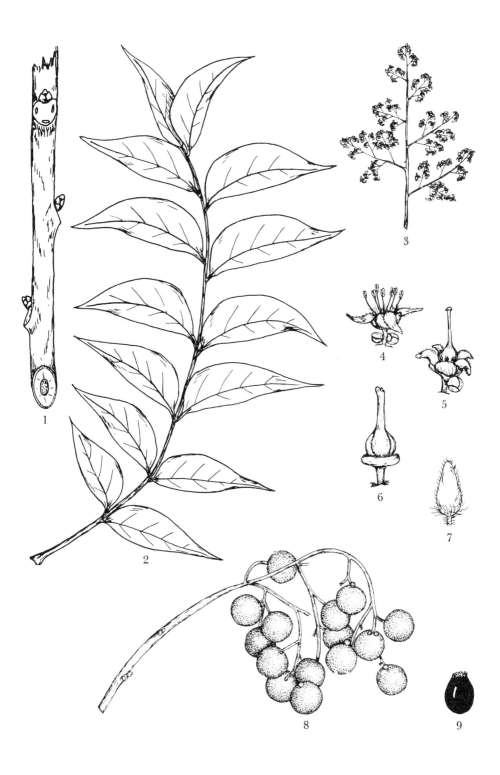

1. Winter twig
2. Leaf
3. Staminate inflorescence
4. Staminate flower
5. Pistillate flower
6. Pistil
7. Petal
8. Fruit cluster
9. Seed

RHAMNACEAE
Rhamnus alnifolia L'Her.

Swamp buckthorn

A small shrub to 7 dm high, 1-2 main stems, each with a few short branches, erect or decumbent.

LEAVES: Alternate, simple. Broadly elliptic or oblong-elliptic, 9-12 cm long, 5-6 cm wide, acute or short acuminate, the base rounded or cuneate; 5-7 lateral veins; serrate, 3-6 teeth per cm, teeth rounded with a small gland on the infacing margin, young leaves ciliate; older leaves glabrate above, sparingly pubescent below, but with short pubescence on the main veins; petiole 1-1.5 cm long, V-grooved above, puberulent; stipules ovate to lanceolate, rounded or acute, 5-6 mm long, 1.5-2 mm wide, finely puberulent and short ciliate, the marginal glands early deciduous; stipules remain until after flowering time.

FLOWERS: Late May, when the leaves half-grown; dioecious. Flowers axillary on growth of the current season, 1-3 in each axil; pedicels green, 5-6 mm long, finely puberulent. Staminate flowers with a broad, saucer-shaped hypanthium, 2.5-2.75 mm across, 0.5-1 mm deep, green, glabrous; calyx lobes 5, ovate, 1.5-1.75 mm across, 1.5-2 mm long, green, glabrous, acute, somewhat folded lengthwise; disk green; no petals; 5 stamens alternate with the calyx lobes, filaments incurved, 1 mm long, white; anthers yellow. Pistillate flowers similar to the staminate; pedicels 6-8 mm long; ovary about half enclosed in the central disk, upper part green and glabrous; 2-3 united styles, 0.5 mm long, green, with greenish stigmas.

FRUIT: August-September. Usually a single, axillary fruit; pedicel ascending or drooping, 7-9 mm long, finely puberulent, green or brownish, widened to a disk beneath the fruit; drupe globular, black or nearly so, 7-9 mm diameter, smooth, glossy, 3-4 seeds; pulp somewhat juicy, and stains a dark purple. Seeds ovate, flattened, 3.5-4 mm long, 2.7-3 mm wide, 1.4-1.6 mm thick, dark gray-brown,

with a mealy surface; raphe ridge sharp, the surface on either side of the raphe nearly flat, dorsal side ridged, with a wide groove on each side of the ridge.

TWIGS: 1-1.5 mm diameter, often slightly flattened, red-brown, finely puberulent, flexible, scurfy-waxy; lenticels inconspicuous on growth of current season, but small and numerous on older branches; leaf scars semicircular, with a concave upper side, light-colored; bundle scars indistinct but appear to be a curved line of small bundles; pith white to pale brown, continuous, one-half to three-fourths of the stem. Buds ovoid, 2.5-4.5 mm long, 1.5-2.5 mm wide, the scales scurfy pubescent and ciliate.

TRUNK: Bark red-brown, slightly scaly or fissured. Wood soft, nearly white, with a large brown pith.

HABITAT: Moist woods, around bogs but not in the bog, often in brushy areas on moist, springy hillsides.

RANGE: Newfoundland, west to British Columbia, south on the east side of the Cascades to the Sierra Nevada in California, northeast to Idaho and Montana, east to North Dakota, southeast to Illinois, and east to Pennsylvania and New England.

Rhamnus alnifolia is an inconspicuous little shrub growing as an undershrub along with other bushes and herbaceous weeds in wet or moist soil. It usually does not occur in the bog proper, but around the margin and often in low, flat, wooded areas. It is more common in sandy soils.

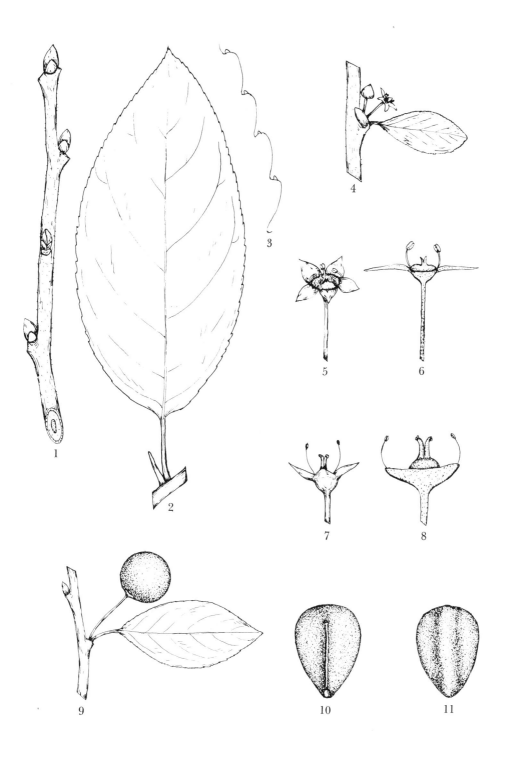

1. Winter twig
2. Leaf
3. Leaf margin
4. Flower cluster
5. Staminate flower
6. Diagram of staminate flower
7. Pistillate flower
8. Diagram of pistillate flower
9. Fruit
10. Ventral view of seed
11. Dorsal view of seed

RHAMNACEAE
Rhamnus cathartica L.

Common buckthorn

A profusely branched shrub to 2 m high and often as wide.

LEAVES: Mostly opposite, simple; broadly elliptic, oval or obovate, 3-7 cm long, 2.5-5.5 cm wide, 2-3 main lateral veins on each side; tip rounded to abruptly acuminate; base rounded to cuneate; serrate with 6-8 teeth per cm, the teeth irregular with the tip usually turned in and glandular; dark green above, paler below, both surfaces glabrous; petiole 1-3 cm long, slightly grooved and puberulent above, rounded and glabrous below; stipules linear, 3-5 mm long, glandular toothed and often persistent.

FLOWERS: Late May; dioecious. Staminate flowers axillary on a short branch of the season, clusters of 2-6, all parts glabrous; pedicels 6-7 mm long, green, expanded to a green, campanulate hypanthium 1.75-2.5 mm long, 1.5-2 mm wide; calyx lobes 4, narrowly ovate, 2.5-3 mm long, yellow-green, spreading or recurved, 3-nerved; petals 4, lanceolate or oblong, 1.25 mm long, yellow-brown, curved around a stamen; stamens 4, attached below the rim of the hypanthium, filaments tapered, 2 mm long, incurved; anthers yellow; usually an abortive pistil. Pistillate flowers 2-15 on a short spur branch of the season; pedicels 8-12 mm long, green; hypanthium hemispheric, green, 1.25-1.75 mm across, 1 mm deep; calyx lobes 4, narrowly triangular, ridged lengthwise on the inner surface, deciduous after fertilization; usually no petals, but if present, 4, linear, yellow-brown, 0.75 mm long; 4 abortive stamens; ovary depressed globose, attached to the hypanthium only at the base, 1.25 mm across, 1 mm high, green, glabrous; 3-4 styles, united at the base, 2-2.5 mm long, divergent above; stigmas capitate, yellow-green.

FRUIT: September, often remains until December. Single or clustered at the nodes or on the end of short spur branches, the spur often extends after the fruits are formed leaving the fruits at the base of the new growth; pedicels slender, glabrous, green or brown, 8-11 mm long, ending in a flat disk 3-4 mm across. Fruit black, depressed globose, 6-8 mm long, 7-10 mm wide, smooth, semiglossy, juicy, 4 seeds per fruit, but usually only 1-2 mature. Seeds variable, generally ovoid, 4-5 mm long, 2.5-3.8 mm diameter, slate gray, smooth, dull, keeled on the raphe side and rounded or with a shallow groove on the dorsal side.

TWIGS: 1.5-2 mm diameter, slightly flattened, gray to yellowish-brown, dull, smooth, glabrous, semirigid; the lenticels are narrow vertical slits, often with the red-brown inner bark showing at the margin, numerous; twigs often ending in a spine; leaf scars narrowly reniform to nearly triangular or half-round; 3 bundle scars near the center; stipule scars small; pith pale brown, continuous, one-half of stem. Buds ovoid, acute, 3-6 mm long, 2-2.5 mm wide, either dark or light red-brown, tip definitely turned in toward the stem so the inner side of the bud is either straight or concave, outer side definitely convex; the outer scales ciliate and often with a double keel.

TRUNK: Bark of young trunks gray, smooth; older trunks with somewhat scaly bark and with long, horizontal lenticel marks resembling *Prunus*, trunk usually blotched with light and darker gray. Wood hard, fine-grained, heartwood small, light brown, with a wide, white sapwood which appears flaky.

HABITAT: Escaping occasionally in most of the northern part of eastern United States, growing in nearly any habitat except arid regions.

RANGE: Introduced from Europe and escaping.

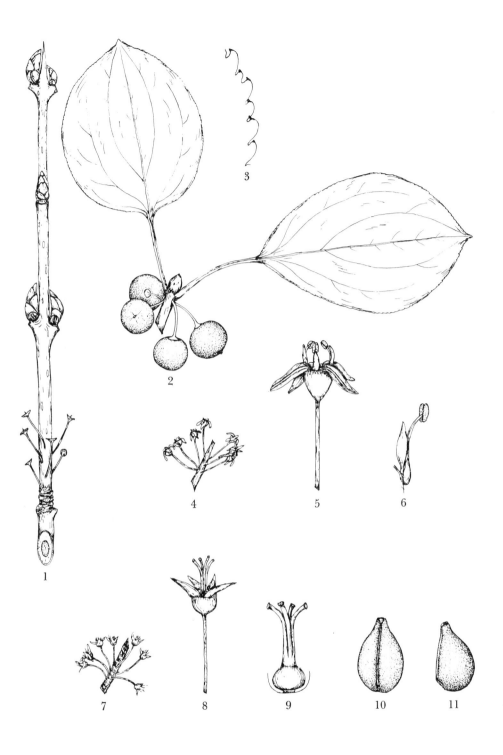

1. Winter twig
2. Leaves and fruit
3. Leaf margin
4. Staminate flower cluster
5. Staminate flower
6. Stamen and petal

7. Pistillate flower cluster
8. Pistillate flower
9. Pistil
10. Ventral view of seed
11. Side view of seed

RHAMNACEAE
Rhamnus dahurica Pall.

Buckthorn

An open shrub to 2.5 m high, with single or clustered trunks.

LEAVES: Opposite, simple. Elliptic to oval or oblong, 7-10 cm long, 2-3 cm wide; crenate serrate with a gland on the incurved side, 6-10 teeth per cm, 3-5 lateral veins on each side, usually 4; acute or acuminate, the base cuneate; glabrous and yellow-green on both sides, slightly paler below; petiole 1.5-1.75 cm long, grooved above, glabrous; stipules linear, 7-8 mm long, the margin glandular.

FLOWERS: June; dioecious. Flowers in clusters of 1-4, axillary on new growth, pedicels 6-8 mm long, slender, glabrous. Staminate flowers with a campanulate hypanthium 1.5-1.75 mm long, 1.3 mm wide, green, glabrous; calyx lobes 4, narrowly ovate, 3-3.5 mm long, 1 mm wide, glabrous, spreading; petals 4, lanceolate, 1 mm long, green or brownish; stamens 4, opposite the petals, filaments 1.5 mm long, white, dilated toward the base; anthers yellow; vestigial pistil usually present. Pistillate flower hypanthium broadly campanulate, green, glabrous, 1 mm long, 1.5-1.75 mm wide; calyx lobes 4, narrowly ovate, somewhat keeled, 2.5-2.7 mm long, 1 mm wide; stamens and petals undeveloped; ovary inferior, only the dome-shaped top exposed; styles 2, connate except at the tip, 1.25-2.5 mm long; stigmas linear, brownish, recurved.

FRUIT: September. Axillary or clustered on a short spur branch; pedicels 10-12 mm long, slender, ascending or drooping; fruit obovoid, dark purple to black, 7-9 mm long, 6-7 mm diameter, smooth, semiglossy, juicy; 2 seeds per fruit. Seeds ovoid, 5-6 mm long, 3.6-3.7 mm wide, 2.9-3 mm thick, gray-brown, smooth, dull, the raphe ridge low and a narrow groove on the dorsal side.

TWIGS: 2-2.5 mm diameter, flattened, gray-brown to red-brown, rigid, smooth, glabrous, often pruinose, occasionally ending in a short spine; lenticels small and inconspicuous; leaf scars broadly crescent-shaped; 3 bundle scars; pith white, continuous, one-fourth of stem. Buds ovoid, 5-7 mm long, 2-2.2 mm wide, pointed, slightly flattened on the twig side and, in general, parallel to the twig; scales red-brown, smooth, the margin thin and scarious or ciliate.

TRUNK: Bark gray-brown, often with a silvery cast, lenticels abundant and rough; the bark often peels back from around an old limb scar. Wood medium hard, red-brown, with a wide, yellowish sapwood.

HABITAT: Cultivated and often escaping in the northern part of our range; appears to grow well in any type of soil, especially when moist.

RANGE: Introduced from Siberia, north China, and Korea, and commonly planted in the windbreaks of our area.

Rhamnus dahurica could easily be confused with *R. lanceolata* Pursh, but the larger and opposite leaves along with the clustered trunks should distinguish it. *R. cathartica* L. is also quite similar but its leaves are usually obovate and the proportion of the width to the length is much greater, 2.5 to 3 as compared to 1 to 3 in *R. dahurica*. These are general proportions, and leaves can be found which will not conform to them. Also the branches of *R. cathartica* are much finer and the bushes more densely branched.

R. dahurica seldom escapes, and in our range has been known to do so only in North Dakota and one location in South Dakota. One reason for this may be that it is less planted than *R. cathartica* and thus has fewer chances of escaping.

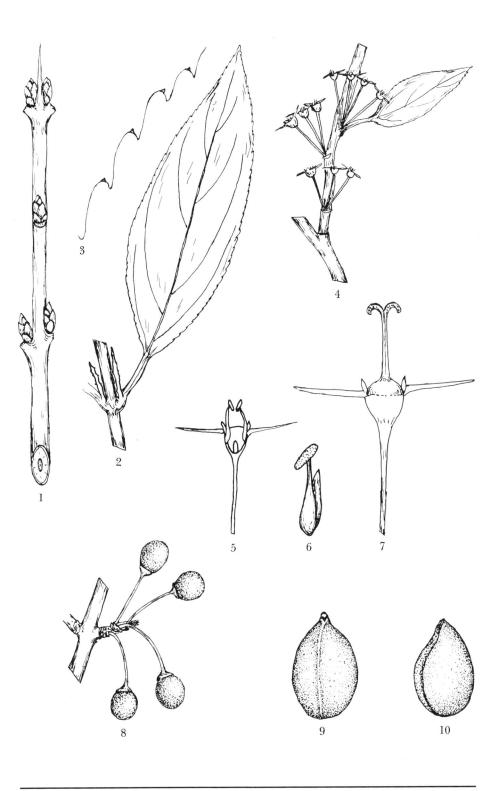

1. Winter twig
2. Leaf
3. Leaf margin
4. Staminate flowers
5. Diagram of staminate flower
6. Stamen and petal
7. Cutaway diagram of pistillate flower
8. Fruit
9. Ventral view of seed
10. Dorsal view of seed

RHAMNACEAE
Rhamnus lanceolata Pursh, var.
glabrata Gleason

Buckthorn

A shrub to 3 m high, often dense with many fine branches.

LEAVES: Alternate, simple. Ovate to oblong-elliptic, 4-7 cm long, 2-3 cm wide; 6-8 lateral veins on each side; finely serrate with incurved, often callous-tipped teeth, 10-13 per cm; leaf tip short acuminate, base broadly cuneate; upper surface dark green, semiglossy, glabrous; lower surface paler with some pubescence on the midrib; petiole 8-10 mm long, often reddish, glabrous except for 2 lines of hairs along the groove of the upper surface; stipules oblong, 2.5-3 mm long, caducous.

FLOWERS: Early May, with the leaves; dioecious. Staminate flowers on short branches from wood of the previous year, 1-3 flowers per leaf axil; pedicels green, glabrous, 1.5 mm long; calyx tube campanulate, glabrous, 2.5 mm long; lobes 4, acute, 2.1 mm long, glabrous, greenish-yellow; petals 4, obovate, 1.3 mm long, each folded around a stamen, yellow with greenish base and brown tip; stamens 4, opposite the petals, filaments green, 1.4 mm long; anthers yellow. Pistillate flowers axillary on short, new branches, some of which continue to grow during the summer, others lose their leaves at fruiting and remain short; 1-3 flowers per axil, smaller and not as showy as the staminate flowers; calyx campanulate, 1.8 mm long; lobes 4, acute, 1.7 mm long, glabrous; petals 4, obovate with the outer end cleft, 0.9-1 mm long, greenish; stamens vestigial; ovary compressed globose, 1 mm long and as wide, 0.6 mm thick, yellow-green, glabrous; styles 2, green, 1.8 mm long, connate most of their length, the ends divergent; stigmas capitate.

FRUIT: Early July. Clustered along the twigs; pedicels 3-4 mm long, green, glabrous; fruit globose, 7-8 mm diameter, dark blue or nearly black, with a thin bloom which does not rub off easily; portions of the calyx persistent; 2 seeds per fruit. Seeds obovate, 4.3-4.5 mm long, 3-3.2 mm wide, 2.4-2.7 mm thick; deeply grooved on the ventral surface, with a ridge around the groove and a small tip on the broad end; light brown, smooth, glossy.

TWIGS: 1-1.5 mm diameter, flexible, glabrous, gray-brown; leaf scars crescent-shaped; 1 central bundle scar; stipule scars prominent; pith white, continuous, one-fourth of stem. Buds ovoid, obtuse, red-brown, the scales often ciliate.

TRUNK: Bark smooth, gray, often with light blotches, the lenticels horizontal and dark; occasionally the bark of an old trunk peels and curls laterally. Wood medium-hard, light red-brown, with a narrow, pale sapwood.

HABITAT: Open, wooded slopes; rocky, limestone hillsides; ravine banks in open prairies.

RANGE: Southern Ohio to South Dakota, south to Kansas and Oklahoma, and east to Arkansas, Tennessee, and Kentucky.

Buckthorn is a solitary plant, only occasionally forming thickets. The branches are low to the ground and the shrub has many small branchlets. It is excellent for song- and game-bird protection, some of them eating the fruits.

The typical variety *lanceolata* has not been reported for our area. It has pubescent twigs and lower leaf surfaces, while var. *glabrata* is nearly glabrous throughout.

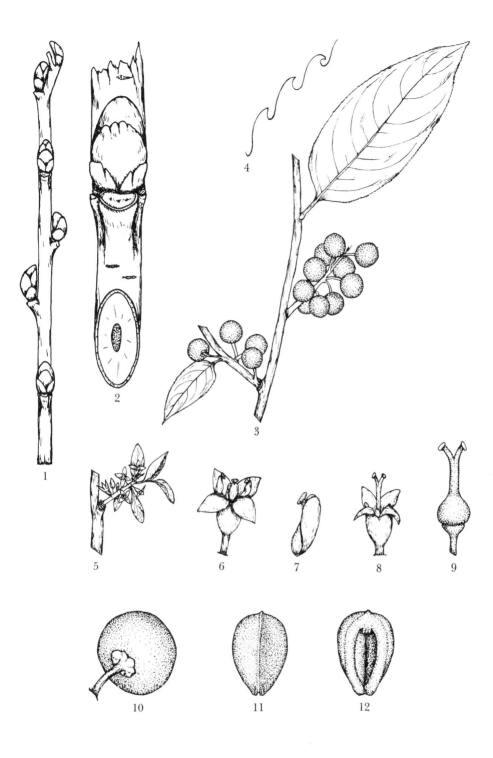

1. Winter twig
2. Detail of twig
3. Leaf and fruit
4. Leaf margin
5. Staminate flower cluster
6. Staminate flower
7. Stamen and petal
8. Pistillate flower
9. Pistil
10. Basal view of fruit
11. Dorsal view of seed
12. Ventral view of seed

RHAMNACEAE
Ceanothus americanus L. var.
pitcheri T. & G.

New Jersey tea, redroot

A low shrub to 1 m high, branched near the top, often with clustered stems.

LEAVES: Alternate, simple. Broadly ovate, 4-6 cm long, 3-4.5 cm wide, 3 main veins from the base; irregularly serrate, 7-9 teeth per cm, gland-tipped, ciliate; tip obtuse to acute, base rounded; upper surface dark green, pubescent, often rugose; lower surface pale, densely soft pubescent; petiole 4-6 mm long, pubescent; stipules lance-linear, caducous.

FLOWERS: May; perfect. Peduncles axillary on new growth, 4-10 cm long, deciduous in the autumn, the longest peduncles at the base, pubescent, leafless or with 2 bracts subtending the panicle; panicles short-cylindric, 3-5 cm long, 2.5-3 cm wide, the flowers clustered at definite nodes on the panicle branches; pedicels 8-13 mm long, white, glabrous, slender; hypanthium broadly obconic, 2-3 mm wide, white, glabrous; calyx lobes 5, sharply incurved, ovate, acute, 1.2-1.4 mm long, 1 mm wide, white, glabrous; petals 5, pipe-shaped, tapered to a filiform base, total length 1.6 mm, claw 1 mm long, cup 1 mm deep; the cup enfolds the anther in the bud; stamens 5, usually incurved, filaments 1.3 mm long, white, tapered; anthers yellow; central disk brownish, 1.4 mm diameter, 10-lobed; ovary surrounded by the disk, only a green, 3-lobed top exposed, 1 ovule per lobe; style 1.6 mm long, greenish-white; stigmas 3, linear.

FRUIT: Mid-July. Panicles, 5 cm long, 4 cm wide; pedicels 10-13 mm long; capsules 4-5 mm wide, 3-lobed, black, each lobe crested; hypanthium persistent after the seed dispersed. Seeds dark red-brown, 2.8-3 mm long, 2 mm wide, 1.8 mm thick, glossy, smooth, rounded on the dorsal side, slightly keeled with 2 flattish surfaces on the ventral side.

TWIGS: 1-1.4 mm diameter, flexible, densely pubescent, dark gray-green; leaf scars raised, crescent-shaped with ends and center enlarged; 3 bundle scars; pith white, continuous, one-third of stem. Buds narrowly ovoid, brown and with long, pale hairs; the inner scales densely hairy, the long hairs exserted from the bud tip.

TRUNK: Bark greenish-brown with wide splits showing a brown inner bark; old stems brown, sometimes flaky. Wood soft, white.

HABITAT: Upland rocky prairies, open woods, thickets, railroad banks, and roadsides.

RANGE: Georgia to Texas, north to Kansas and Iowa, and east to Indiana.

Ceanothus americanus is easily confused with *C. herbaceus,* both growing in the same habitat and often close together. The acute leaves, the cylindric flowering heads with the flowers at definite nodes, the axillary, long and leafless peduncles, and the crested fruits of *C. americanus* should be of help in making the distinction. In autumn the long fruiting peduncles are deciduous, even before the leaves, so no fruiting bodies remain on the plant during the winter. This is not so with *C. herbaceus.*

It is not as common, nor is its range as wide, as *C. herbaceus,* the most concentrated part of its range being in eastern Kansas and southeastern Nebraska. Nor does it appear to be as hardy, for dead branches are commonly found among the living ones.

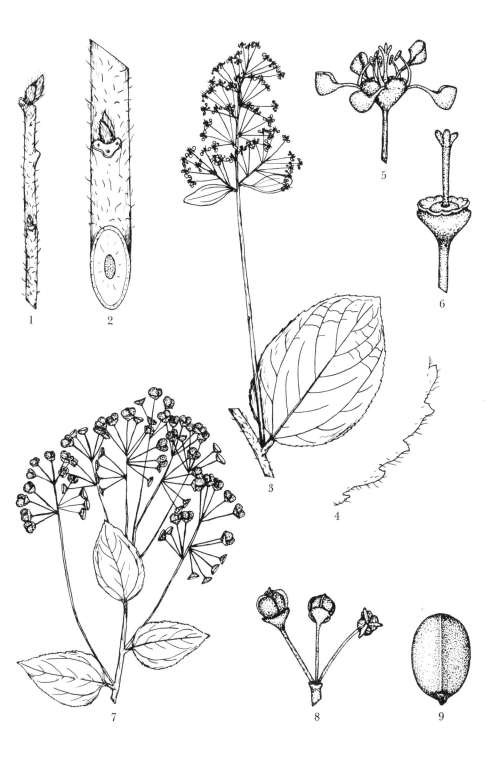

1. Winter twig
2. Detail of twig
3. Leaf with inflorescence
4. Leaf margin
5. Flower

6. Pistil, showing lobed disk
7. Fruiting panicles
8. Fruit
9. Seed

RHAMNACEAE
Ceanothus fendleri Gray

C. subsericeus Rydb.

Snow brush, deer brush, deer brier

A low, spiny, often dense bush to 50 cm high, stems often clustered and forming low thickets.

LEAVES: Alternate, tardily deciduous, simple. Three main veins from the base; thin, firm; ovate to elliptic, 1-2.5 cm long, 4-10 mm wide, entire, tip acute to rounded, mucronulate, the base cuneate; upper surface dark green with a few appressed hairs, lower surface whitened with a dense tomentum of short hairs beneath the long, straight hairs; petiole 2-3 mm long, finely pubescent, flattened above; stipules lanceolate, 2-2.5 mm long, pubescent, green, but soon brown and deciduous.

FLOWERS: July-August; perfect. Flowers in umbel-like clusters terminating the main stem or a branch. Pedicels white, glabrous 4-6 mm long; calyx lobes 5, white, ovate, acute, 1.5 mm long and as wide, glabrous, sharply incurved between the petals; petals 5, white, spreading, pipe-shaped, 1.5 mm long, claw about half of the total length, cup 1 mm high; stamens 5, opposite the petals, filaments white, tapered, incurved, the anthers yellow; central disk 2 mm across, 6-lobed, yellowish; 3 brown glands at the base of the style; style 1 mm long, white, with 3 linear stigmas about 1 mm long.

FRUIT: August-September. Pedicels 8-10 mm long, ascending, hypanthium persistent, saucer-shaped, 1-1.5 mm high, 4-5 mm across; fruit 3-lobed, 3-4 mm high, 5-6 mm across, lobes slightly keeled, the surface rough, dark brown; each lobe with 1 seed. Seeds brown, glossy obovoid, 2.5-3 mm long, 2-2.25 mm wide, 1.75 mm thick.

TWIGS: 1.3-1.6 mm diameter, rigid, canescent, pruinose, definite gray-green color, greenish-brown when the wax and pubescence are removed; tips of branches spinescent and often with blister-like glands scattered over the stems; lenticels elliptic, light-colored, noticeable only on the lower stems; leaf scars raised, nearly circular to broadly elliptic, small; 1 bundle scar; stipule scars large, noticeable; pith white, continuous, one-third of stem. Buds naked, consisting of a series of graduated embryonic, densely pubescent leaves folded together, occasionally with an outer, brown, scale-like structure, the whole unit 1.5 mm long, 1 mm wide, 0.5 mm thick, those near the tip of a spine much smaller.

TRUNK: Bark of young stems and shoots gray-green, that of old stems red-brown and minutely pimpled and wrinkled but tight. Wood hard, fine-grained, greenish-white.

HABITAT: Open valleys, hillsides in open woods, usually in rocky soil, and occasionally along rock ledges.

RANGE: Western Texas, New Mexico, Arizona, Colorado, and South Dakota.

This is a southwestern species, and the two known stations in the Black Hills are several hundred miles north of the main range. The South Dakota herbarium specimens were examined, but the drawings and descriptions are from fresh material in Las Animas County, Colorado. In some areas of the southwest, this is an important part of the forage of deer; probably the young sprouts are utilized before the spines of the branches harden. A few leaves remain on the plant until midwinter, but the plant cannot be called evergreen. It does, however, catch and hold an abundance of dead leaf material which forms excellent cover for small mammals and birds.

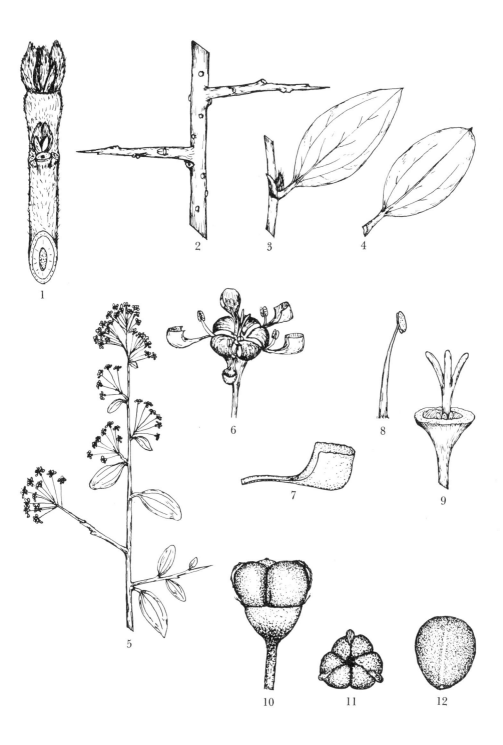

1. Winter twig
2. Twigs with spine tips
3. Leaf
4. Leaf variation
5. Inflorescence
6. Flower

7. Petal
8. Stamen
9. Cutaway showing pistil
10. Fruit, side view
11. Fruit, end view
12. Seed

RHAMNACEAE
Ceanothus herbaceus Raf.

Ceanothus ovatus Desf., incl. var.
pubescens (T. & G.) Shinners

New Jersey tea, redroot

An open shrub to a height of 70 cm, with single or clustered stems.

LEAVES: Alternate, simple. Ovate to elliptic, 4-6 cm long, 1-2.5 cm wide; finely serrate with callous-tipped teeth, 7-9 teeth per cm; leaf tip obtuse or acute, base rounded or obtuse; upper surface dark green, dull or semiglossy, glabrous or pubescent; lower surface paler, pubescent with crinkly hairs; petiole 3-5 mm long, pubescent; stipules lanceolate, 2-4 mm long, pubescent.

FLOWERS: April-May; perfect. Flowers in loose, globose panicles composed of several umbel-like clusters, terminal on new, leafy stems; peduncle 1-2 cm long, green, pubescent; pedicels 10-12 mm long, white, glabrous; hypanthium saucer-shaped, 2-2.5 mm across, white; calyx lobes 5, broadly ovate, 1.6-1.7 mm long, 1.5-1.7 mm wide, acute, sharply incurved between the petals, white, glabrous; petals 5, pipe-shaped, white, tapered to a filiform base, total length 1.5 mm, cup 1 mm deep; stamens 5, opposite the petals and attached to their base, each anther surrounded by a petal in the bud; filaments 1 mm long, white, tapered; anthers yellow; ovary 3-lobed, surrounded by the disk; style 2 mm long, white with purplish base; stigma 3-lobed, yellowish; central disk 1.5 mm across, 10-lobed, brownish.

FRUIT: Mid-June. Terminal round-topped or globose clusters of 20-30 fruits; pedicels 10-12 mm long, dark brown; hypanthium saucer-shaped, persistent after seed dispersal; capsule 3-4.5 mm wide, 3 lobes without crests, black, granular; each lobe 1-seeded; exocarp dehisces, exposing a cream-colored endocarp. Seeds red-brown, 1.9-2 mm long, 1.6-1.8 mm wide, 1.2-1.3 mm thick, smooth, glossy; ventral side slightly keeled with 2 flattish surfaces, the dorsal side rounded.

TWIGS: 1-1.5 mm diameter, brittle, densely pubescent, red-brown; leaf scars half-round, small, raised; 1 bundle scar; pith white, becoming brown, continuous, one-fourth of stem. Buds woolly, 2-2.5 mm long, narrowly ovoid, obtuse, red-brown beneath the hairs.

TRUNK: Bark gray-brown, thin, the outer bark cracked into short slits. Wood medium-hard, pale red-brown.

HABITAT: Rocky soils, open woods, prairies, roadsides, and pastures.

RANGE: Quebec to Manitoba, south to Colorado and Texas, east to Georgia, and north to Massachusetts.

Ceanothus herbaceus is a low-growing shrub 6-7 dm high, usually with many branches. Occasionally a plant is found with a single central stem and many short, lateral branches, forming a columnar shrub. At flowering time, each of these branches terminates in a cluster of flowers, producing a somewhat solid, columnar mass of white flowers. The flowers themselves are unique and well worth placing under magnification. The plant is not used for home planting but would make an attractive border plant.

This is the most common *Ceanothus* throughout the southern part of our range, being quite abundant in eastern Kansas, north central Nebraska, and in the Black Hills. It does not form colonies but often is quite thick on prairie hillsides.

The two varieties have been placed together here since the key difference is the amount of pubescence on the lower leaf surface. This is a variable character, and even on one hillside in our area the plant may grade from nearly glabrous to rather densely pubescent and no definite line can be drawn between the two conditions.

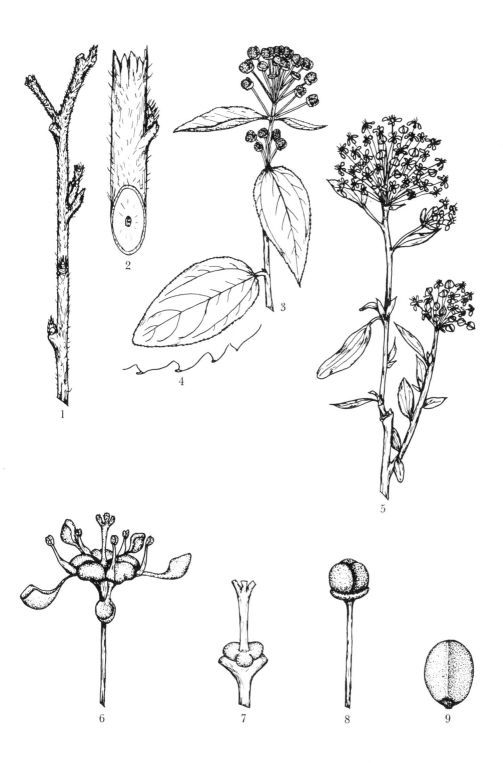

1. Winter twig
2. Detail of twig
3. Leaves with fruit
4. Leaf margin
5. Inflorescence
6. Flower
7. Cutaway, showing pistil and disk
8. Fruit
9. Seed

RHAMNACEAE
Ceanothus velutinus Dougl.

C. laevigatus Hook.; *C. velutinus* var. *laevigatus* T. & G.

Sticky laurel, greasewood

A low shrub 1-1.5 m high and as broad, the basal branches low and often procumbent for 4-5 dm.

LEAVES: Alternate, simple, evergreen, thick, coriaceous, 3 main veins. Ovate to oval, 5-7 cm long, 3.5-4.5 cm wide, tip rounded to obtuse, base rounded to subcordate, margin finely gland-tipped serrate, 10-13 teeth per cm; young leaves short pubescent on both sides, upper surface of mature leaves glabrous, waxy and glossy, the lower surface densely, minutely puberulent (use 25× magnification), appressed pubescent on the veins; petiole 10-14 mm long, stout, pubescent; stipules lanceolate, 2-3 mm long, the base triangular in section, soon dark brown but persistent.

FLOWERS: Mid-June, often continuing into summer. Thyrse from a peduncled, naked bud formed the previous season, 1-5 main branches with a lanceolate bract at the base of each; thyrse 5-7 cm long, 3-4 cm wide, the axis pubescent; pedicels slender, 5-7 mm long, glabrous, white; flowers 6-8 mm across; hypanthium saucer-shaped, 1.5-2 mm across with 5 small, triangular lobes alternating with the calyx lobes and becoming larger and green in fruit; calyx lobes 5, white, triangular, 1.5-2 mm long, 1.5 mm wide, curved sharply between petals and over the center of the flower; petals 5, white, pipe-shaped, 2-2.5 mm long, usually curved downward, cup 1 mm high, narrowed abruptly to a long claw; stamens 5, opposite the petals, inserted below the disk, filaments 2.5 mm long, curved inward; yellow anthers; ovary dark green, 1 mm across, mostly embedded in the disk, the 3 lobes with small keels; style stout, 1.5 mm long divided near the tip into 3 stigmatic lobes; disk 10-lobed, yellow.

FRUIT: September. Pedicels 9-11 mm long, hypanthium shallow, 5-lobed, persistent; fruit 5-6 mm across, 4 mm high, 3-lobed, each with a blister-like keel which dries to a rough ridge; 1 seed per lobe. Seeds obovoid, 2.5 mm long, 1.9 mm wide, 1.6 mm thick, dark brown, smooth, glossy; the ventral side with 2 flattened faces with the raphe ridge between; dorsal side rounded.

TWIGS: 2-2.4 mm diameter, rigid, slightly flattened, olive-green, puberulent, stipules brown, persistent, but leaving a triangular scar if broken off; leaf scars (on lower stem) bean-shaped, dark; 3 bundle scars; often an adjacent oval fruiting scar directly above; pith white, continuous, one-half of stem. Leaf buds axillary, naked, 2 mm long, 2 mm wide, consist of a series of small bract-like leaves; flower buds naked on pubescent peduncles 3-12 mm long, axillary near the end of the stem, an embryonic thyrse loosely and partially covered by thick, brown, pubescent bracts 3.5-4 mm long.

TRUNK: Bark red-brown, finely furrowed and ridged, tight; upper branches yellow-green, smooth. Wood hard, fine-grained, white.

HABITAT: Open woods, hillsides; dry, rocky soil.

RANGE: British Columbia, east to the Black Hills of South Dakota, south to Colorado, west to northern California, and north to Washington.

This plant would not be called common in the Black Hills, but in the few locations where it grows it may form rather dense thickets. In a few areas, a whole hillside may be covered. At flowering time these hillsides appear white, even from a distance. It is especially common in the Deadwood area, but only on certain hills. It does not grow in any other part of our range.

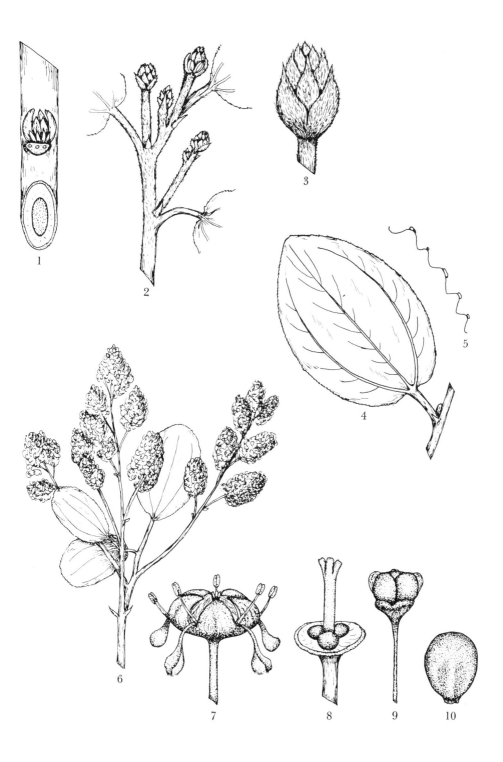

1. Winter twig
2. Twig with winter flower buds
3. Winter flower bud
4. Leaf
5. Leaf margin

6. Inflorescence
7. Flower
8. Cutaway, showing pistil and disk
9. Fruit
10. Seed

VITACEAE
Ampelopsis cordata Michx.

Cissus ampelopsis Pers.

Raccoon grape, false grape

A vine climbing by tendrils to a length of 20 m, the trunk becoming 8-15 cm in diameter.

LEAVES: Alternate, simple. Ovate to deltoid, 7-11 cm long, 6-9 cm wide; 3-5 palmate veins; coarsely serrate with sharp-tipped teeth; leaf tip acute to acuminate, the base broad, rounded or truncate; upper surface dark yellow-green, glabrate; lower surface paler, sparsely coarse pubescent on the veins; petiole 3-6 cm long, often enlarged and sharply bent at the base; stipules variable from reniform to lanceolate, about 1 mm long, toothed, caducous.

FLOWERS: Early June; dioecious. Staminate flowers in bracted, glabrous, cymose clusters opposite a leaf on new growth; peduncle 2.5-4 cm long, green; pedicels 3-4 mm long, green; calyx lobes 5, minute, distinct, obtuse; petals 5, ovate, 2.5 mm long, reflexed, persist until after anthesis; stamens 5, opposite the petals, attached at the base of the disk; filaments 2.5 mm long, white, erect; anthers yellow; disk cup-shaped, irregularly lobed, whitish; vestigial pistil usually present. Pistillate flowers in cymose clusters opposite a leaf on new growth; peduncle 4-6 cm long, green glabrous, commonly 2-branched, each cluster 2-3 cm across; pedicels 3-4 mm long, green, glabrous; calyx lobes 5, distinct, 0.2 mm long, obtuse, green with whitish margin; petals 5, green, 2.5 mm long, 1-1.2 mm wide, deciduous as the flower opens; stamens 5, vestigial, lightly fastened to the petals; ovary 0.8 mm long, ovate, half-imbedded in the disk; style 1.8 mm long, white; stigma greenish; disk cup-shaped, white, 2 mm wide, 1.3 mm high, irregularly lobed.

FRUIT: August. Peduncles 4-6 cm long; pedicels 5-6 mm long; 8-15 fruits per cluster; fruit depressed globose, 7-8 mm long, 8-11 mm wide, the surface dotted with small pustules; flesh milky, becoming dry; fruit changes color several times, causing one cluster to have several different colors at the same time—from green, to orange, to rose-purple, and finally turquoise blue; 1-3 seeded. Seeds similar to those of the grape; globose if only 1 seed or with flattened surfaces if 2 or more seeds; 4.6-5 mm long, 4.8-5.3 mm wide, red-brown to yellow-brown, granular; 2 small grooves on the ventral side with a ridged raphe between and extending over the apex, ending in a wide, flattened area on the dorsal side; a small groove from this widened raphe end to the tip of the seed.

TWIGS: 2.6 mm diameter, flexible, slightly angular, with ridges extending along the stem from the leaf bases, yellow-green, glabrous; lenticels longitudinal, warty; few tendrils opposite the leaves; twigs often die back in winter; leaf scars nearly circular, slightly flattened or concave on the upper margin; 5 bundle scars; pith white, continuous, one-half of stem. Buds concealed beneath the bark.

TRUNK: Bark on young stems pale brown, not fissured, lenticels raised; bark of old trunks, dark brown, deeply fissured, the ridges long, flat-topped, and anastomosed.

HABITAT: Fence rows, hillsides, rocky slopes, rich bottom lands, flood plains, and stream banks.

RANGE: Mexico, Texas to Florida, north to Virginia, west to Nebraska, and south to Oklahoma.

In the vegetative condition, this plant is often confused with *Vitis*. The deltoid leaves are quite different from the broad, often rotund leaves of *Vitis*.

376

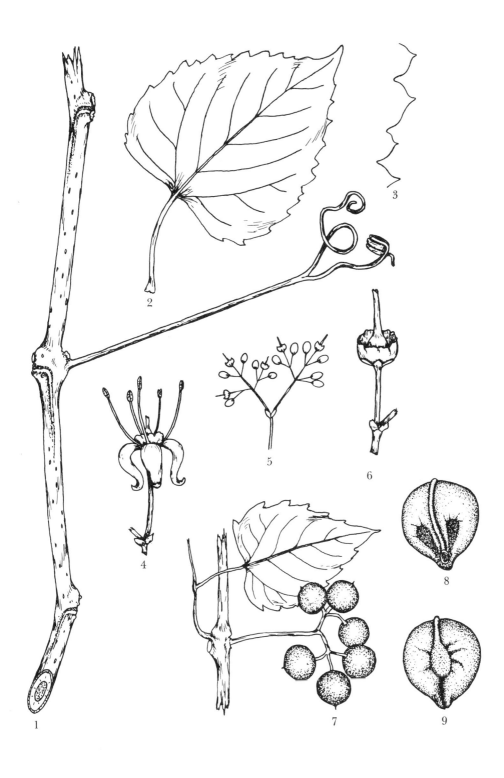

1. Winter twig
2. Leaf
3. Leaf margin
4. Staminate flower
5. Pistillate cymose flowering cluster
6. Pistillate flower
7. Fruiting cluster
8. Ventral view of seed
9. Dorsal view of seed

VITACEAE
Cissus incisa (Nutt.) Des Moulins

Possum grape, marine vine

A climbing or trailing vine to 10 m high, climbing by tendrils.

LEAVES: Alternate, trifoliolate or deeply 3-lobed; 5-8 cm long, 6-7 cm wide, thick and somewhat succulent, glabrous, smooth, but appearing granular under low magnification; end leaflet somewhat rhombic, 4-5 cm long, 3-4 cm wide, basal margins nearly straight, apical margins coarsely toothed; lateral leaflets fan-shaped, 3-3.5 cm long and as wide, apical side straight, the basal side coarsely toothed, the teeth mucronate; midrib of leaflets sharply raised near the base on the upper surface; leaves often gray-green or purplish, especially late in the season, and, when collected, have a tendency to fall off during the drying process; petiole 2-4 cm long, stout, with a narrow groove above; stipules lanceolate, 3-4 mm long, finely toothed, persistent; occasionally a slender tendril opposite a leaf.

FLOWERS: July; perfect or polygamous. Flowers in bracted, umbelliform cymes, 4-5 cm across, round or flat-topped, 3-5 rays, 50-80 flowers, peduncle 4-6 cm long, rays 1-2 cm long, glabrous. Pedicels green or reddish, 3-7 mm long, glabrous; calyx tube broadly campanulate with 4 shallow lobes, the whole calyx 1.5 mm long, 2 mm wide, green, glabrous; 4 petals, green, glabrous, cucullate, thick-walled, and cupped over a stamen, distinct, fall as the flower opens; 4 stamens opposite the petals, filaments inserted below and between the 4 lobes of the green disk; filaments white, tapered, 1-1.2 mm long, curved at the apex and attached to the center of the anther; anthers pale yellow, 1.4 mm long, stamens usually fall soon after the flower opens; ovary globular, 1 mm diameter, green, glabrous, partly enclosed by the disk; style stout, 1 mm long, tapered, the stigma not enlarged.

FRUIT: October. Cyme purplish or green, pedicels slender, 7-10 mm long, curved, expanded to a disk below the fruit; berry ovoid to globular, 9-11 mm diameter, often nearly truncate at the base, tip rounded, dark purple to black with a few light dots, smooth; pulp thin, watery, purplish; 1-seeded. Nutlet ovoid, 5-7 mm long, 4-5 mm thick, hard, brownish, mottled, muricate, the 2 deep grooves at the apical end usually covered with scurfy pulp.

TWIGS: 1-2 mm diameter, flexible, somewhat 6-ridged, gray-green to brown, with obvious, orange-colored lenticels; tip often dies back during the winter; leaf scars circular, pale brown; a ring of bundle scars; pith green, continuous, porous, one-half of stem; fruit scars circular, similar to the leaf scars and directly above, usually raised more than the leaf scar. Buds concealed under the bark, appear as a pimple-like bump above the leaf scar, spongy and indefinite when uncovered; stipule scars half-round, obvious.

TRUNK: Bark thin, soft, light tan-gray, with numerous, raised, orange-colored lenticels; inner bark green. Wood soft, fibrous, white, with large, white pith.

HABITAT: Open woods, brushy, rocky hillsides, and ravine banks.

RANGE: Texas, Oklahoma, Missouri, Arkansas, Louisiana, and east along the Gulf Coast; one record for Kansas.

Only one herbarium specimen was located from our area; it is purely vegetative, but definitely of this species. Since the locality could not be found, the description given above and the drawings were made from fresh specimens in Oklahoma. In some areas of eastern Oklahoma it is a common vine growing along sandstone ledges and sprawling over the sides of small cliffs.

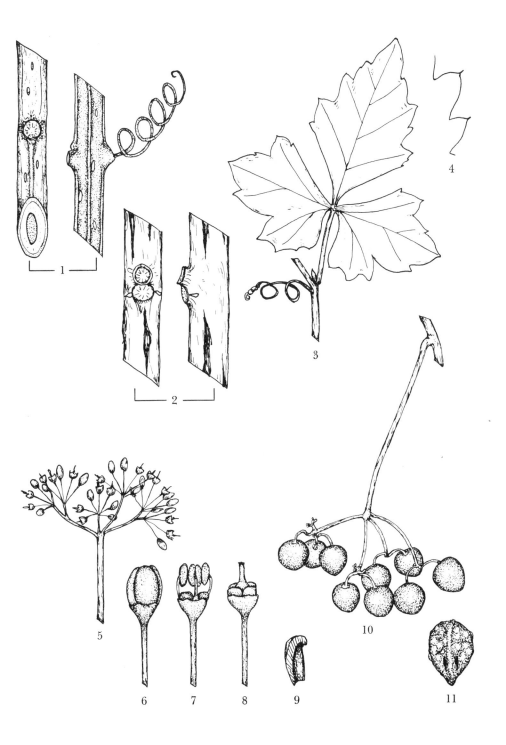

1. Winter twig
2. Winter twig with fruit and leaf scar
3. Leaf
4. Leaf margin
5. Flowering cyme
6. One flower
7. Staminate flower
8. Pistillate flower
9. One petal
10. Fruit cluster
11. Seed

VITACEAE
Parthenocissus quinquefolia (L.)
Planch.

Psedera quinquefolia (L.) Greene

Virginia creeper, five-leaved ivy

A vine, climbing by tendrils with sucker disks and by aerial roots to a height of 25 m.

LEAVES: Alternate, palmately compound, 5 leaflets. Leaflets ovate to obovate, 8-13 cm long, 4-6 cm wide, coarsely toothed, the teeth often prolonged; leaflet tips acuminate, base cuneate; sessile or with petiolules to 1 cm; upper surface dark green, dull, glabrous; lower surface paler, glabrous except for some pubescence on the veins; petiole 10-20 cm long, glabrate, base enlarged and often bent sharply; stipules lanceolate 8-12 mm long, 3-4 mm wide, caducous.

FLOWERS: June-July; perfect. Panicles opposite a leaf near the end of a short stem of the season; panicle 6-12 cm long with a well-defined axis, glabrous, 25-75 flowers; pedicels 2-3 mm long; calyx shallow, saucer-shaped, 2 mm wide, not lobed; petals 5, oblong, 3-3.5 mm long, 1.4-1.5 mm wide, tip acute, greenish to bronze, attached at the base of the ovary; stamens 5, opposite the petals; filaments white, 2.6 mm long; anthers yellow, the filament attached at the center; ovary conic, 2 mm long, 1 mm wide, green; style and stigma short, not distinctly different.

FRUIT: September. Clusters 9-13 cm long, 4-6 cm wide, open, axis and pedicels red, glabrous; pedicels 3-4 mm long; fruit subglobose, 5-7 mm long, 6-8 mm wide, dark purple with a bloom; 4 seeds per fruit. Seeds ovoid, 4-5 mm long, 3.6-3.9 mm wide, with a flat surface on each side of the raphe ridge, minutely granular, dark chocolate brown, glossy; a narrow depression on each flattened surface; the broad end indented where the raphe extends over it, the raphe then widens to a circular marking on the dorsal side.

TWIGS: 1-1.2 mm diameter, flexible, orange-brown, finely pubescent; tendrils 3-5 cm long, branched, ending in sucker disks; coarse adventitious roots without disks along the older stems; leaf scars nearly circular with the upper margin flattened; 5-10 bundle scars; fruit scars circular; pith green to white, continuous, one-tenth of stem. Buds conical, 2-2.5 mm long, red-brown, acute, glabrate, the outer scales often keeled.

TRUNK: Medium-sized stems light brown, with shallow fissures and few adventitious roots; old trunks dark brown, deeply fissured, the ridges broad and rounded; inner bark fibrous. Wood light, soft, porous, pale brown, with a wide, pale sapwood.

HABITAT: Open, rich woods, fence rows, rocky wooded hillsides, ravines, and bluffs.

RANGE: Texas to Florida, north to Maine, west to Minnesota, south to Oklahoma. Also Mexico and Guatemala.

Virginia creeper is a common vine in the eastern part of our range, climbing to the tops of trees 20 meters high. Where trees are not available, it clambers over rocks and low bushes, covering them completely. The autumn foliage is crimson and scarlet but the leaflets drop soon after the color change takes place.

One variation has been described, var. *hirsuta* (Donn) Fern. This variety has pubescence on the under leaf surface as well as on the veins. There also appears to be an integration, or hybridization, with *P. vitacea,* for plants are often found which have characteristics of each species.

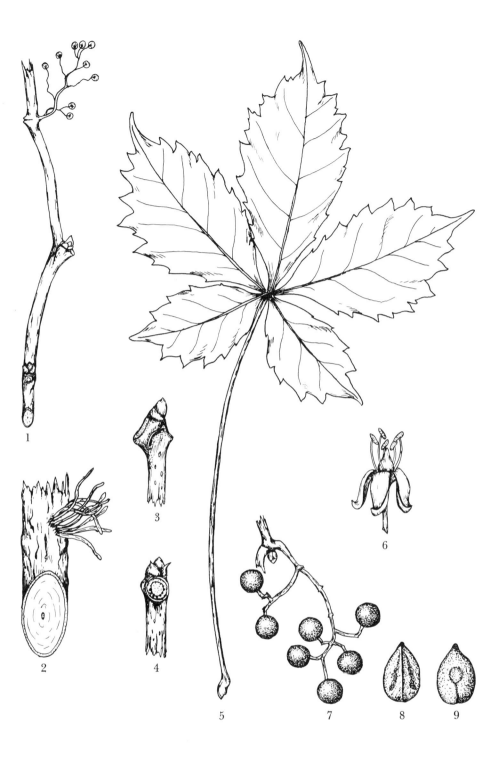

1. Winter twig
2. Portion of twig with adventitious roots
3. Detail of twig
4. Twig with fruit scar
5. Leaf
6. Flower
7. Fruit cluster
8. Ventral view of seed
9. Dorsal view of seed

VITACEAE
Parthenocissus vitacea (Knerr)
Hitch.

Parthenocissus inserta (Kerner) K.
Fritsch; *Psedera vitacea* (Knerr) Greene

Woodbine

A vine, climbing by tendrils without sucker disks to a height of 10 m, usually found sprawling over bushes and rocks.

LEAVES: Alternate, palmately compound, 5 leaflets. Leaflets subcoriaceous, ovate to obovate, 6-10 cm long, 4-6 cm wide; coarsely toothed; leaflet tip acute to short acuminate, base cuneate; upper surface dark green, lustrous, glabrous except on the veins; lower surface paler, glabrous with pubescent veins; petiole 8-12 cm long, glabrous, the lower end abruptly enlarged; petiolules 5-10 mm long; stipules lanceolate, 8-12 mm long, 3-3.5 mm wide, caducous.

FLOWERS: Mid-June; perfect. Dichotomous panicles with 10-60 flowers, on new growth; peduncle 3-7 cm long, secondary branches 1-3 cm long, pedicels 3.5-3.8 mm long, glabrous; calyx saucer-shaped, with 5 broadly rounded lobes about 0.5 mm long; petals 5, reflexed, elliptical, 3.3-5 mm long, 1.6-2 mm wide, yellow-green, tip thick and incurved, acute; stamens 5, erect, opposite the petals, filaments white, 2.9-3 mm long; anthers yellow; ovary conic, 2 mm long, 1.8 mm wide, green; the style and stigma merely an obtuse tip on the ovary.

FRUIT: August. Dichotomously branched, somewhat flat-topped cluster, 10-20 fruits, the branches red and glabrous. Fruits depressed globose, 6-9 mm long, 8-10 mm diameter, smooth, deep blue-purple with a slight bloom which does not rub off easily; 4-seeded. Seeds ovoid, 4.8-5 mm long, 4-4.7 mm wide, light brown, acute tip; ventral side sharply ridged, with a flattened surface on each side of the ridge and a small depression on each surface; the raphe extends through a groove on the rounded apex and ends in a raised, light-colored marking on the rounded, dorsal side.

TWIGS: 2-2.5 mm diameter, flexible, glabrous, gray-brown, lenticels prominent; many short, spur branches; tendrils branched, seldom with sucker disks;

leaf scars half-round; 10 bundle scars; fruit scars nearly circular, usually pitted in the center; pith white, continuous, one-fourth of stem. Lateral buds small and buried in scaly tissue; buds at the end of spurs ovoid, acute to obtuse, red-brown, glabrous, 2-3 mm long.

TRUNK: Bark of young trunks brown, irregularly broken into small, peeling plates; bark of old trunks dark brown, tight, with shallow fissures and short ridges. Wood soft, pale brown, with a wide, light sapwood.

HABITAT: Fence rows, brushy slopes, rich wood, bluffs, and stream banks.

RANGE: Nova Scotia to Ontario and Montana, south to Wyoming and Texas, northeast to Missouri, through Indiana to Pennsylvania and New England. Also in New Mexico, Arizona, and California.

P. vitacea is quite similar to *P. quinquefolia,* and the two are easily confused. The most outstanding differences are that *P. vitacea* has a dichotomously branched inflorescence and is without sucker disks on the tendrils. The other differences are comparative: *P. vitacea* does not climb as high, the leaves are somewhat glossy, and it flowers and fruits earlier.

The ranges of the two species are about the same in our area but *P. quinquefolia* is more common in the eastern section, diminishing westward. *P. vitacea* is common in the west and becomes less common eastward.

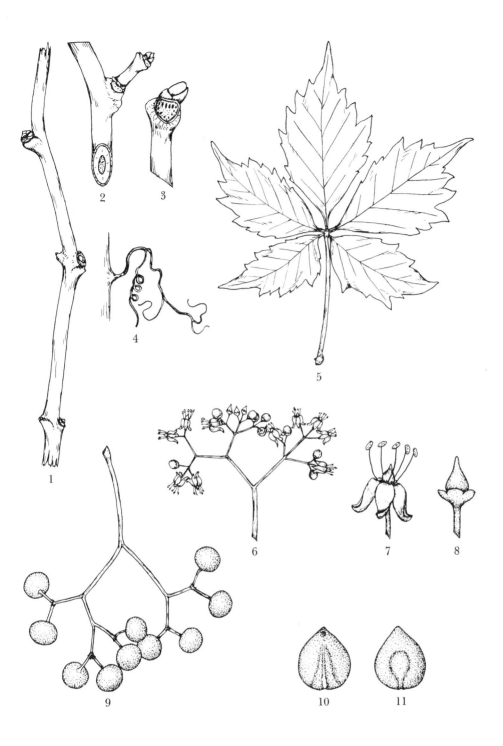

1. Winter twig
2. Twig with spur branch
3. Twig with fruit scar
4. Tendril
5. Leaf
6. Flowering panicle

7. One flower
8. Pistil and calyx
9. Fruit cluster
10. Ventral view of seed
11. Dorsal view of seed

VITACEAE
Vitis acerifolia Raf.

V. longii Prince; *V. solonis* Planch.;
V. doaniana Munson ex Viola; *V. longii*
Prince var. *microsperma* Munson

Bush grape

A bushy vine, the branches occasionally to 6 m long.

LEAVES: Alternate, simple, thick, firm. Broadly ovate to subreniform, 6-9 cm long, 8-11 cm wide; with or without lobes, sharply coarse-toothed, the tooth sinuses acute; leaf tip acuminate; basal lobes rounded with a broad sinus; upper surface yellow-green with cobwebby pubescence; lower surface paler, strongly pubescent with short hairs and white, cobwebby tomentum; petiole 3-5 cm long, cobwebby; stipules rotund, toothed, membranaceous, caducous. Leaves often tend to fold lengthwise.

FLOWERS: Mid-May; dioecious. Staminate flowers in panicles opposite a leaf on new growth; panicle 4-6 cm long, 2-2.5 cm wide, the axis tomentose; pedicels 2.5-3 mm long, green, glabrous; calyx saucer-shaped with a membranaceous rim, indistinctly 5-lobed; petals 5-6, ovate, 1.8-1.9 mm long, 0.8 mm wide, coherent at the tip, falling as a unit as soon as the flower opens; stamens 5-6, attached at the base of the disk; filaments white, 2.5 mm long; anthers yellow; disk 5-lobed, yellow-green, 0.6 mm high, larger than in other species of *Vitis*. Pistillate flowers in panicles opposite a leaf on new growth; panicles 4-5 cm long, 1.5-2 cm wide, the axis tomentose; pedicels 1.6-2 mm long, pubescent; calyx saucer-shaped, low, flat, scarcely lobed, green with a whitish rim; petals 5-6, ovate, 1.8 mm long, 0.9 mm wide, green, glabrous, coherent at the tip, falling as a unit; stamens 5-6, vestigial, attached below the disk and promptly reflexed; disk yellow-green, 5-lobed; ovary green, ovoid, 1 mm long, 1 mm wide; style 0.3 mm long; stigma 0.7 mm wide, green, 5-lobed.

FRUIT: Early August. Clusters 5-6 cm long, the axis with a few cobwebby hairs; pedicels 3-5 mm long, enlarged near the fruit; fruit globose, 5-8 mm diameter, purple with a bloom, the flavor strongly acrid; 2-5 seeds. Seeds broadly ovoid, 5.2-6 mm long, 4.5-5 mm wide, 3.4-3.8 mm thick, pale brown; raphe ridged on the ventral side, extending over an indentation at the apex of the seed and ending in an ovate marking in the center of the rounded dorsal side.

TWIGS: 2-2.5 mm diameter, semirigid, tomentose when young, later glabrous, red-brown to grayish; tendrils scarce, short, simple or with short branches; nodal diaphragm 0.5-0.8 mm thick; leaf scars oval, rough, bundle scars indistinct; pith brown, continuous, one-third of stem. Buds ovoid, 2-4 mm long, obtuse, red-brown, a white tuft of hairs at the tip.

TRUNK: Bark red-brown, thin, not shredded in the first year; old trunks with long, thin shreds. Wood soft, porous, nearly white.

HABITAT: Dry hillsides, rocky bluffs, in open or brushy areas in prairie ravines.

RANGE: New Mexico and Texas to western Kansas and southeastern Colorado.

Vitis acerifolia is a bushy type of grape with short, crooked stems, often bulky with stubby branches in the axils of the main branches. If bushes or rocks are near, it may climb over them, attaining a length of 6 meters. It often forms thickets so dense that it becomes impossible to get through them. This makes an excellent cover for birds and mammals and helps to control erosion. It occurs only locally but may be common in those locations. The fruits remain acrid until late, often drying before becoming sweet.

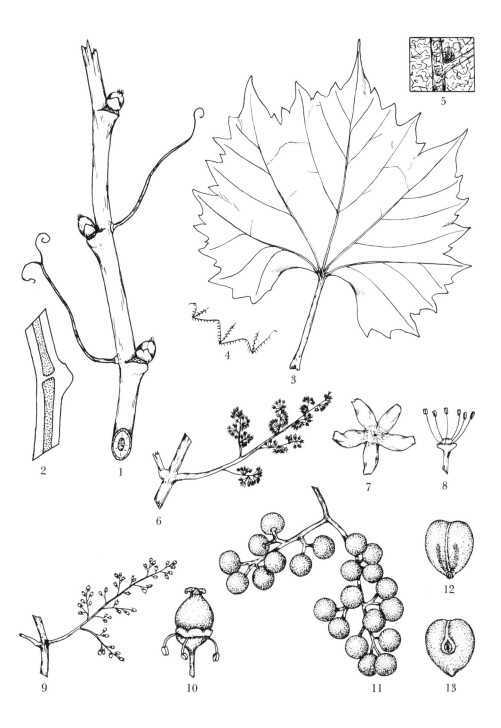

1. Winter twig
2. Longitudinal section through node
3. Leaf
4. Leaf margin
5. Portion of leaf underside
6. Staminate panicle
7. Staminate corolla
8. Stamens
9. Pistillate panicle
10. Pistillate flower, corolla removed
11. Fruit cluster
12. Ventral view of seed
13. Dorsal view of seed

VITACEAE
Vitis aestivalis Michx.

var. *aestivalis* = *V. lincecumii* Buckl.
var. *argentifolia* (Munson) Fern. = *V. lincecumii* var. *glauca* Munson; *V. lecontiana* House; *V. aestivalis* var. *bicolor* (LeConte) Britt. & Brown; *V. argentifolia* Munson

Summer grape

A vine climbing to a height of 10 m by means of tendrils, or sprawling over low bushes and trees.

LEAVES: Alternate, simple. Ovate to subrotund, 12-18 cm long, and about the same width; 3-5 lobes, deep or shallow, each coarsely toothed; lobe sinuses narrow and rounded, tips acute to short acuminate; basal lobes rounded with a narrow sinus; upper surface yellow-green with a few hairs on the veins; lower surface whitish, with light rusty, cobwebby hairs, the veins with straight hairs, some of which have a dark base; petiole 8-12 cm long, often red, glabrous or with a few cobwebby or straight hairs; stipules deltoid, thin, pubescent, caducous. Blade and stipules densely cobwebby as they unfold, petioles soon glabrate.

FLOWERS: Mid-May; dioecious. Fruits in dense pyramidal clusters 7-12 cm long, 3-4 cm wide, often a short tendril on the axis; axis finely pubescent, reddish; flower clusters opposite a leaf on two successive nodes, the third node without flowers; flowers in small umbels with a bract at the base; pedicels 3.5-4 mm long, green, glabrous. Staminate flowers with a shallow, saucer-shaped calyx, indistinctly 5-lobed, reddish; disk 5-lobed, yellow-green; petals 5, oblong to elliptic, 2.2-2.3 mm long, 1 mm wide, acute, white but often reddish at the tip, attached below the disk, coherent at the tip, and fall as a unit immediately upon opening; stamens 5, filaments 2.8-3 mm long, white, spreading, attached at the base of the disk opposite a petal; anthers yellow; the vestigial ovary depressed globose. Pistillate flowers in smaller clusters, the calyx, disk, and petals similar to those of the staminate flower; vestigial stamens attached between the lobes of the disk, soon sharply reflexed; ovary in the center of the disk, green, ovoid, 1.1-1.2 mm long, glabrous; style 0.3-0.6 mm long; stigma disk-like, unevenly lobed, green or pinkish.

FRUIT: Late August. Drooping clusters 7-12 cm long, 5-7 cm wide; pedicels 4-5 mm long, glabrous; fruits globose, 11-14 mm diameter, purple with a bloom, sweet; 3-6 seeds. Seeds pyriform, 6-6.5 mm long, 3.9-4.7 mm wide, abruptly narrowed to an obtuse base, red-brown; ventral surface flattened, 2 narrow depressions, the raphe extending over the rounded apex to a small, ovate marking on the rounded dorsal side.

TWIGS: 3-3.5 mm diameter, flexible, glabrous, definite purplish color, glaucous, finely ridged; nodal diaphragm 0.5-1 mm thick; leaf scars large, broadly crescent-shaped to shield-shaped; 3 indistinct bundle scars; pith brown, continuous, one-half of stem. Buds ovoid, obtuse or acute, 4-5 mm long, 3-4 mm wide, the outer scales purplish, glaucous, glabrate, often at right angles to the twig.

TRUNK: Bark of the branches exfoliating early; bark of old stems brown with long papery shreds. Wood tough, porous, pale brown, with a large pith.

HABITAT: Dry, rocky uplands in thickets, open woods, or along bluffs and fence rows; seldom in rich bottom-land soils.

RANGE: Massachusetts, west to Minnesota, south to Texas, east to Georgia, and north to New York.

The amount of pubescence varies and the two varieties are hardly distinguishable in our area. Var. *argentifolia* has fewer hairs and the leaf underside is glaucous.

386

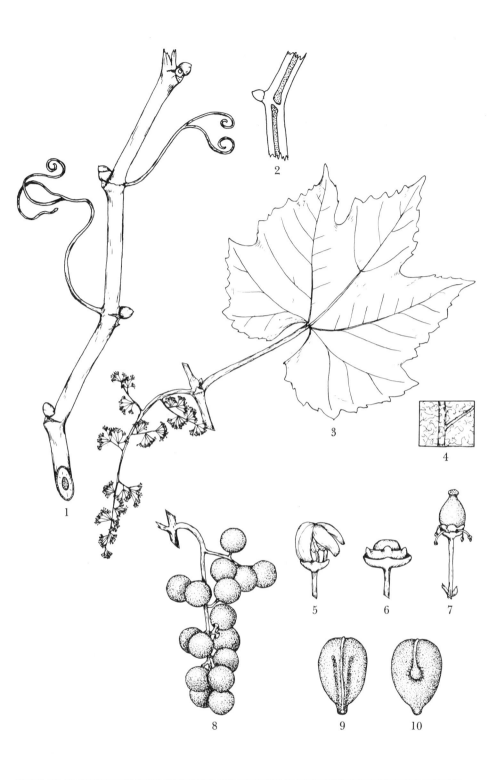

1. Winter twig
2. Longitudinal section through node
3. Leaf and staminate panicle
4. Portion of leaf underside
5. Staminate flower with petals falling
6. Staminate flower, petals and stamens removed
7. Pistillate flower, petals removed
8. Fruit cluster
9. Ventral view of seed
10. Dorsal view of seed

VITACEAE
Vitis cinerea Engelm.

V. cinerea var. *canescens* (Engelm.) Bailey

Grayback grape, winter grape, pigeon
grape

A vine climbing by tendrils to a
height of 15 m.

LEAVES: Alternate, simple. Broadly
ovate, 15-20 cm long, 12-15 cm wide;
lobeless or with 2 small, lateral lobes,
coarsely toothed, the teeth somewhat
rounded, ciliate; leaf tip acute to acuminate; basal lobes broadly rounded with a
narrow sinus; upper surface dark green
with a few cobwebby hairs, often rugose;
lower surface covered with white, cobwebby, and straight hairs; petiole 7-10
cm long, pubescent with both straight and
cobwebby hairs; stipules minute, pubescent, caducous.

FLOWERS: Mid-June; dioecious.
Inflorescence a paniculate cluster opposite a leaf on new growth, 10-15 cm
long, 3-5 cm wide, the axis with a red
tinge, tomentose, often a tendril on the
base of the peduncle. Staminate flowers
with pedicels 2.5-3 mm long, green, glabrous; calyx shallow, saucer-shaped, 0.8-1
mm wide, indistinctly lobed, white to
reddish at the top, glabrous; petals 5-6,
elliptic, 2.5-2.8 mm long, 0.8-1 mm wide,
green, the tips coherent, break at the base
and fall immediately as a unit; stamens
5-6, spreading, attached below and between the lobes of the disk; filaments
white, 2-2.3 mm long; anthers yellow;
the vestigial pistil a rounded hump in
the center of the disk; disk 5-lobed,
orange. Pistillate flowers on pedicels 2-3
mm long; calyx saucer-shaped, shallow,
indistinctly 5-lobed, reddish at the tip,
glabrous; petals elliptic, 2.5-2.8 mm long,
green, break and fall as a unit; stamens
5, vestigial, strongly reflexed; disk of 5
distinct lobes, pale orange-green, adnate
to the ovary; ovary green, ovoid, 1 mm
long, glabrous; style 0.3 mm long, green;
stigma yellowish, with 5 small, spreading
lobes.

FRUIT: Mid-September. Long, loose
clusters 8-15 cm long, 4.5-5 cm wide; axis
pubescent; pedicels 3-5 mm long, green
with reddish spots; fruit small, globose,
5-8 mm diameter; deep purple or blackish with a slight bloom that is easily removed; acrid until late autumn; 1-2 seeds.
Seeds ovoid, red-brown, 4-4.4 mm long,
3.7-3.8 mm wide, 2 narrow depressions on
the ventral side; raphe ridge extending
from the tip of the seed on the ventral
side over the broad end and widening
below the middle on the rounded dorsal
side.

TWIGS: 1.6-2 mm diameter, gray-
brown, striate, the thin, white tomentum
often in patches; nodal diaphragm 2.5-2.8
mm thick; leaf scars irregular, crescent-
shaped to oval; bundle scars indistinct;
pith brown, continuous, one-half of stem.
Buds conical, 2-2.5 mm long, red-brown
with some tomentum, especially at the
tip.

TRUNK: Bark of old trunks red-
brown, shredding into thin, long, loose
strips. Wood porous, lightweight, pale
brown.

HABITAT: Low woods or upland
in open areas, thickets, fence rows, or
stream banks; rich or rocky soil.

RANGE: Indiana to southern Wisconsin, west to Nebraska, south to Texas,
east to Florida, and north to Virginia and
Ohio.

V. cinerea is the latest flowering of
our wild grapes; the fruits are small and
do not become sweet until late autumn.
They are seldom used in the home, possibly because people have forgotten about
making jelly by the time they are ready.

Var. *cinerea* and var. *canescens* are
so intergraded that they cannot be separated; the amount of pubescence and the
shape of the leaf are quite variable.

388

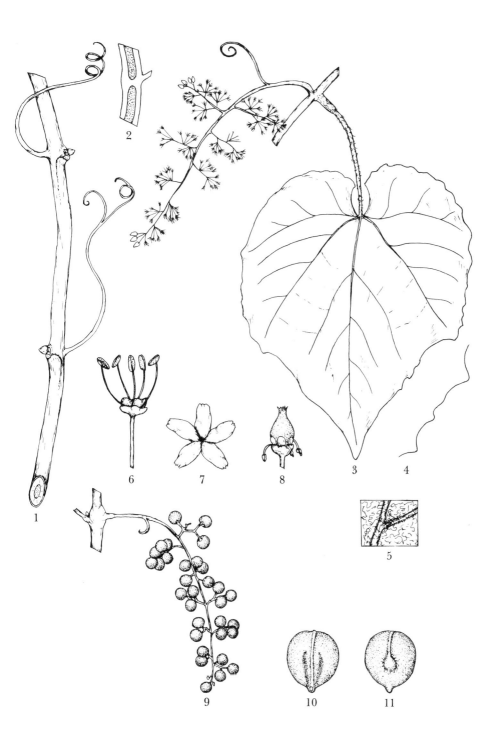

1. Winter twig
2. Longitudinal section through node
3. Leaf and staminate inflorescence
4. Leaf margin
5. Portion of leaf underside
6. Staminate flower, corolla removed
7. Corolla
8. Pistillate flower, corolla removed
9. Fruit cluster
10. Ventral view of seed
11. Dorsal view of seed

VITACEAE
Vitis riparia Michx.

Riverbank grape

A vine climbing to 25 m high by means of tendrils.

LEAVES: Alternate, simple. Ovate to rotund, 10-14 cm long, 9-13 cm wide; 2 lateral lobes, coarsely toothed, sides of the teeth mostly concave, ciliate; tips of lobes acuminate; basal lobes rounded, with a broad sinus, the leaf margin in the sinus often straight; upper surface yellow-green, glabrous; lower surface paler, pubescent on the veins and in the vein axils; petiole 4-6 cm long, glabrate, some pubescence near the blade; stipules ovate, 3-5 mm long, 2-3 mm wide, glabrate, ciliate at the tip.

FLOWERS: Mid-May; dioecious. Staminate panicles 4-12 cm long, opposite a leaf on new growth, peduncle and axis with a few cobwebby hairs at first; pedicels 2.3-2.6 mm long, green, glabrous, calyx shallow, saucer-shaped, green, 0.4 mm wide, indistinctly 5-lobed; petals 5-6, elliptical, 2.2-2.4 mm long, yellow-green, connate at the tip, break and fall as a unit; disk 5-lobed, orange-yellow; stamens 5-6, attached below the disk; filaments 1.6-1.9 mm long, white; anthers yellow. Pistillate panicles 4-9 cm long; pedicels 2-2.4 mm long, green, glabrous; calyx shallow, saucer-shaped, with 5 small acute lobes; disk 5-lobed, the lobes distinct and rounded, yellow-green, surrounding the ovary; ovary green, ovoid, 1-1.3 mm long, glabrous; style short; stigma with 3-5 large lobes.

FRUIT: July-August. Compact or loose clusters 5-12 cm long, 3-4 cm wide, the axis pubescent, pedicels 3-5 mm long, glabrous; fruits globose, 6-11 mm diameter, purple-blue with a bloom; sweet, edible. Seeds broadly ovoid, 4.5-5.5 mm long, 3.5-4 mm wide, 2.9-3.2 mm thick, base short-tipped; red-brown, the surface granular; raphe extends over the apex to a broad end on the dorsal side; 2 grooves on the ventral side.

TWIGS: 1.8-2 mm diameter, flexible, glabrous, tan-brown, minutely ridged; tendrils lacking on every third node; nodal diaphragm 0.5 mm thick; leaf scars wide crescent-shaped; bundles indistinct; pith brown, continuous, one-third of stem. Buds ovoid, 2.5-3.5 mm long, 3 mm wide, the outer scales keeled, red-brown, glabrate; a tan tomentum on the inner scales often shows at the tip.

TRUNK: Bark red-brown, shredding early, the shreds thin, long, and wide. Wood soft, porous, brown, with a light sapwood.

HABITAT: Upland, rocky soils, fence rows or ravines; low, rich woodlands, banks of streams, often in moist soil.

RANGE: Quebec, west to Montana, south to Texas and New Mexico, east to Arkansas, Tennessee and Virginia and north to New England.

Three varieties have been described and may be distinguished; however, there is some intergradation, and positive identification of varieties may be difficult: var. *syrticola* (Fern. & Wieg.) Fern., with densely hairy petioles and lower leaf surfaces, flowering in May or June and fruiting in August; var. *praecox* Engelm., with glabrate petioles and lower leaf surfaces, flowering in April and fruiting in June; var. *riparia,* with glabrate petioles and lower leaf surfaces, flowering in May or June and fruiting in late summer. All three may be found along the east edge of our area, with var. *riparia* extending much farther west.

V. riparia is commonly confused with *V. vulpina.* The larger teeth with concave sides and the narrow, nodal diaphragm of *V. riparia* are usually a good combination of characteristics for distinguishing the two species.

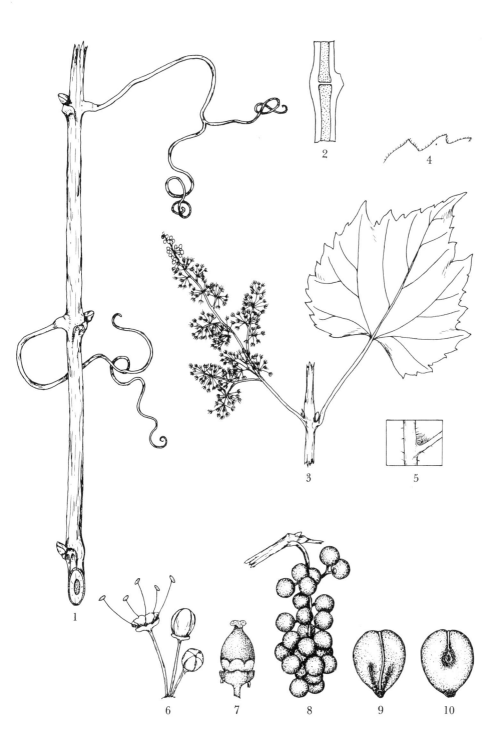

1. Winter twig
2. Longitudinal section through node
3. Leaf and staminate inflorescence
4. Leaf margin
5. Portion of leaf underside
6. Staminate flowers, petals removed from open flower
7. Pistillate flower, petals removed
8. Fruit cluster
9. Ventral view of seed
10. Dorsal view of seed

VITACEAE
Vitis vulpina L.

V. cordifolia Michx.; *V. baileyana* Munson

Winter grape, frost grape

A vine climbing to 20 m high by means of tendrils.

LEAVES: Alternate, simple. Rotund to ovate, 10-15 cm long, 8-10 cm wide, unlobed or with shoulder-like lobes; coarsely toothed, the teeth wider than long and with convex sides; leaf tip acuminate; basal lobes rounded and often overlapping, the sinus narrow; upper surface dark green, glabrous; lower surface paler, with short, straight hairs on the veins and small tufts in the vein axils; petiole 4-7 cm long, glabrous; stipules ovate, membranaceous, 4-5 mm long, caducous.

FLOWERS: Mid-May; dioecious. Loose, pendent panicles on new growth opposite a leaf at 2 successive nodes, the third node without flowers; clusters 10-14 cm long, 4-6 cm wide; peduncle and the main axis reddish, glabrous or with fine hairs; pedicels 5-6 mm long, green, glabrous; a small bract at the base of each flower umbel; calyx shallow, saucer-shaped, 5-lobed, reddish-green; petals 5, oblong to elliptic, 2 mm long, 1 mm wide, green, break at the base, often curl sharply and fall as a unit. Staminate flowers with a 5-lobed disk, bright orange in the bud, becoming orange-green as the petals fall; stamens 5, spreading, attached below the disk and between its lobes; filaments 2 mm long, white; anthers yellow; ovary vestigial, embedded in the disk. Pistillate flower disk orange, 5-lobed; stamens vestigial and sharply reflexed as soon as the petals fall; ovary green, ovoid, 1 mm long, 0.9 mm wide, glabrous, partly embedded in the disk; style 0.5 mm long; stigma capitate, greenish.

FRUIT: September-October. Loose clusters 10-15 cm long, 5-8 cm wide, the basal branches the longest, finely pubescent; pedicels 5-6 mm long, glabrous, green with red-brown spots; fruits globose, 7-10 mm diameter, often depressed, black, glossy, may have a slight bloom when immature; 1-3 seeds. Seeds pyriform, 4.5-5.4 mm long, 3.8-4.4 mm wide, long basal tip, gray-brown, 2 narrow depressions on the ventral side; raphe ridge on the ventral side and extending over the apex of the seed and widening to an ovate marking in the center of the rounded dorsal side.

TWIGS: 2.8-3.4 mm diameter, glabrous, gray-green, becoming red-brown, finely ridged; nodal diaphragm 2 mm thick; tendrils 2 dm long, branched, lacking at every third node; leaf scars half-round to shield-shaped, irregular; indistinct bundle scars; pith white, brown by midwinter, continuous, one-half of stem. Buds ovoid, 4-5 mm long, acute, the outer scales keeled, dark brown, one large scale covers most of the bud; inner scales densely brown tomentose.

TRUNK: Bark gray-brown, furrowed, very little shredding. Wood porous, light tan.

HABITAT: Usually in moist soils along streams and in low woods, occasionally on moist slopes.

RANGE: Southern New York, west to Wisconsin, southwest to Nebraska, Kansas, and Texas, east to Florida, and north to New Jersey.

V. vulpina seldom sprawls over low bushes, probably because it usually grows among trees, but the tendrils are rigid and strong and wrap firmly around small tree branches. The fruit remains acrid until late autumn and in dry seasons often dries and shrivels before attaining sweetness. The main distinguishing characteristics are: convex sides on the leaf teeth; long, loose inflorescence and fruit clusters; slight shredding of the bark; and the lack of a bloom on the mature fruit.

392

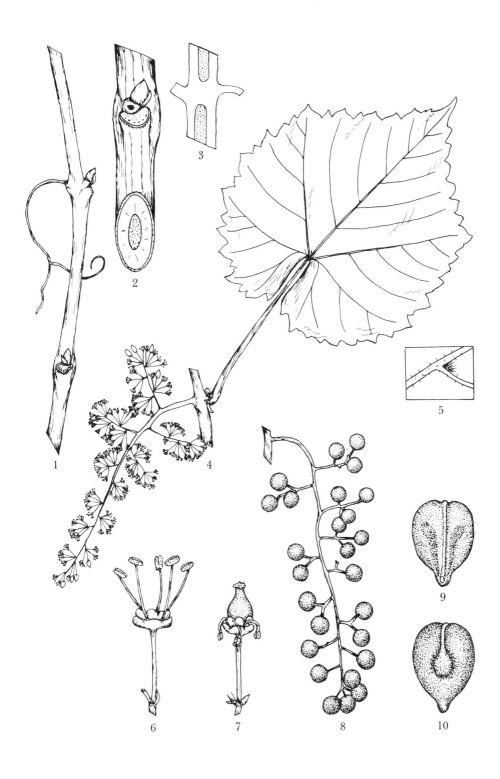

1. Winter twig
2. Detail of twig
3. Longitudinal section through node
4. Leaf and staminate inflorescence
5. Portion of leaf underside
6. Staminate flower, petals removed
7. Pistillate flower, petals removed
8. Fruit cluster
9. Ventral view of seed
10. Dorsal view of seed

TILIACEAE
Tilia americana L.

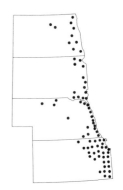

T. glabra Vent.; *T. neglecta* Spach;
T. americana L. var. *neglecta* (Spach)
Fosberg; *T. palmeri* Bush ex F. C. Gates

Basswood, linden, linn tree, whitewood

A tree to 20 m high, the crown cylindric if in the open.

LEAVES: Alternate, simple. Broadly ovate, 10-14 cm long, 9-13 cm wide; sharply serrate, the teeth often prolonged, 2-3 per cm, ciliate; leaf tip abruptly short acuminate; base oblique, variable, often one side deeply rounded, the other straight and at an acute angle with the midrib; upper surface dark green, glabrous; lower surface paler, glabrous or pubescent with simple or stellate hairs; the tufts in the vein axils often buffy; petiole 4-7 cm long, glabrous; stipules oblong or lanceolate, 10-15 mm long, 3-4 mm wide, ciliate, pubescent on the outer side, caducous.

FLOWERS: Mid-June; perfect. Sweet-scented. Inflorescence a cyme on a long, pendent peduncle attached at midsection of a foliaceous, axillary bract; bract 9-10 cm long, 2-2.5 cm wide, oblanceolate, sessile or with a petiole up to 1 cm long, glabrous or stellate pubescent, pale yellow-green; peduncle 3.5-5.5 cm long, green, glabrous; pedicels dark green, 7.5-10.9 mm long, glabrous or pubescent; calyx lobes 5, ovate, distinct, 6.9-8.8 mm long, 3.4-3.6 mm wide, thick, pale yellow, puberulent outside and with longer hairs inside, the lobes curved inward between the petals; petals 5, oblong, 7-9 mm long, 3.2 mm wide, cream-white, often with a pinkish base, glabrous, erect or slightly divergent, margins usually curled inward, tip truncate, toothed or erose; staminodes 5, oblanceolate, tapered, 6-7 mm long, 1.7-2 mm wide, petal-like, white, glabrous, opposite the petals; stamens many, often in 5 clusters opposite the petals; filaments white, 4-4.8 mm long; anthers yellow; ovary globular, 1.8-2 mm diameter, white, pubescent; style 4.9-5.2 mm long, white, pubescent at the base; stigma capitate, greenish, 5-lobed.

FRUIT: Late July. Pendent from the center of axillary bracts 10-15 cm long; peduncle glabrous, 4-6 cm long; pedicels 7-11 mm long, usually pubescent; fruit globose, 8-9 mm long, 7-8 mm thick, slight tip at the apex, tan color, finely puberulent, hard, the wall 1 mm thick; 2-7 fruits per bract, each fruit usually 1-seeded, the abortive seeds flattened against the inner wall of the fruit. Seeds ovoid, dark brown, 4.5-4.7 mm long, 3.7-3.8 mm thick.

TWIGS: 1.9-2.5 mm diameter, flexible, glabrous, red-gray, the lenticels not obvious; leaf scars half-round; bundle scars scattered or in 3 groups, indistinct; pith white, continuous, one-fourth of stem. Buds red, ovoid, acute, glossy, glabrous; an occasional tree with gray-brown buds tipped with hairs.

TRUNK: Bark light gray and smooth on young trees; dark gray with shallow furrows and flat-topped ridges on old trees. Wood, light, soft, nearly white, the sapwood hardly distinguishable.

HABITAT: Rich woods, hillsides, stream banks, and occasionally on rocky hillsides.

RANGE: New Brunswick, west to Manitoba, south to North Dakota, Kansas, and Texas, east to Alabama, Tennessee, and North Carolina, and north to New England.

Tilia americana has been divided into several varieties mainly on the pubescence of the leaf, but the material examined from our area could not be sharply separated. For this reason all variations have here been placed under one taxon without recognition of the varieties.

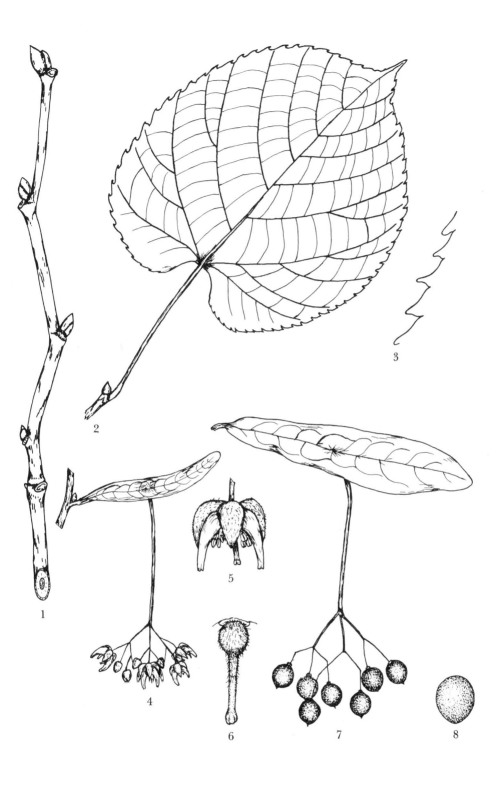

1. Winter twig
2. Leaf
3. Leaf margin
4. Flowers with bract
5. Flower
6. Pistil
7. Fruit
8. Seed

GUTTIFERAE

Ascyrum hypericoides L., var.
multicaule (Michx.) Fern.

A. multicaule Michx.

St. Andrew's Cross

A suffrutescent shrub to 30 cm high, forming clusters 40-50 cm across.

LEAVES: Opposite, simple, ever-green, often punctate, usually two small leaves in each leaf axil; leaves elliptic to obovate, 1.8-2.3 cm long, 8-9 mm wide; margin entire, often slightly revolute; tip rounded, base cuneate; both surfaces green, glabrous; sessile; stipules merely brown, knob-like structures.

FLOWERS: Early July; perfect. Terminal, on main stem or on a short, axillary new stem, usually in 3's; pedicels 2-4 mm long with 2 linear bracts below the flower; calyx green, glabrous, di-morphic, the 2 outer lobes broadly ovate, 9-11 mm long, 6-7 mm wide with acute tip, arranged at right angles to the linear bracts just below; the 2 inner lobes ovate, 1 mm long, at right angles to the large, outer lobes; petals 4, elliptic or obovate, 10-11 mm long, 3-5 mm wide, pale yellow, rounded tip, thick at the base, easily shaken off; stamens many, attached at the base of the ovary, filaments yellow, 3-3.5 mm long with yellow anthers; ovary ellip-tical, flat, 3-3.5 mm long; styles 2, diver-gent, 0.5 mm long; stigmas linear.

FRUITS: Mid-September. Pedicels 3.5-3.7 mm, glabrous, minutely winged; broad, outer calyx lobes with 3 palmate veins, punctate, folded together vertically around the fruit; fruit ovoid, 2-valved, dehiscent, 5-6 mm long, 2.4-2.8 mm wide, mucronate tip, and small, brown projec-tions scattered over the surface, the pu-bescence consists of few light, cobwebby hairs and dark brown, long, often clus-tered hairs; 70-80 seeds per fruit. Seeds oblong, chocolate-brown, 1 mm long, 0.6 mm wide, appear pitted, but this is a fibrous coating which can be rubbed off, exposing the smooth, brown seed with loose, fine, curly fibers extending from the surface.

TWIGS: 0.5 mm diameter, flexible, glabrous, red-brown, glossy; a sharp, thin wing on two sides and alternating at right angles to those of the adjacent internode (decussate); pith green, con-tinuous, one-third of stem. Buds ovoid, obtuse, 0.3-9.4 mm long, red-brown, gla-brous.

STEMS: Bark thin, red-brown, split-ting and peeling into long, thin strips fastened at one edge. Too small to be called true wood.

HABITAT: Dry, rocky, open woods, slopes and ridges, wooded hillsides, mostly acid soils, especially common in chert-rock soils.

RANGE: Massachusetts, New Jersey, west to Missouri and southeast Kansas, south to Texas, east to Georgia, and north to West Virginia.

This little shrub is found in dense clusters in open woods. It is evergreen but is commonly covered by the fallen leaves of the trees above. As the varietal name implies, as many as 50 stems may arise from one base. It is often used as a rock-garden plant, flowering over a long period but never densely flowered.

Var. *hypericaule* is a taller plant, usually with a single stem and narrow, linear leaves. It has not been reported for our area.

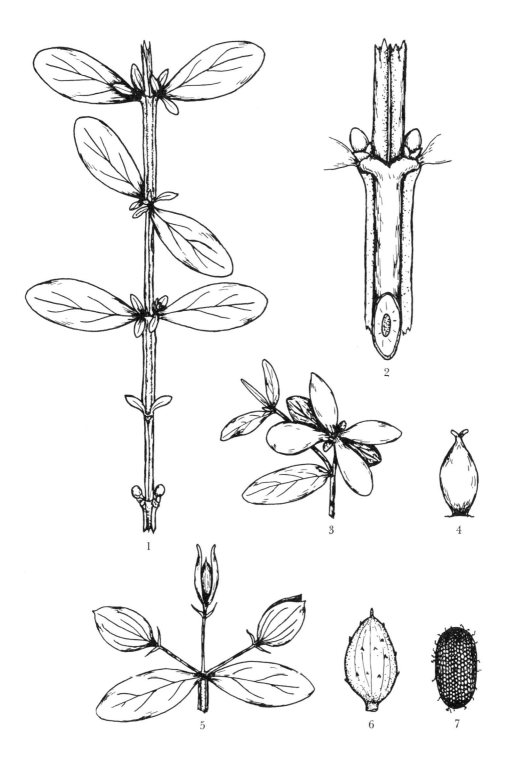

1. Winter twig with leaves
2. Detail of twig
3. Flower
4. Pistil
5. Group of fruit enclosed in bracts
6. Fruit, bracts removed
7. Seed

TAMARICACEAE
Tamarix ramosissima Ledeb.

T. *pentandra* Pall.; erroneously called
T. *gallica* L.

Tamarisk, salt cedar

A dense shrub to 5 m, with many slender branches and often several trunks.
LEAVES: Alternate, simple. Scale-like, trough-shaped, appressed, 0.8-1 mm long, somewhat triangular, sharp-pointed, the base clasping; margin entire; green or gray-green, glabrous, thick; no stipules; the small twigs and fruiting branches often deciduous with the leaves.
FLOWERS: Late May to October; perfect. Flowers at the end of a new branch in many spike-like racemes of 20-60 flowers, the whole flowering portion of the branch 10-25 cm long; racemes 1-5 cm long, 1 cm diameter; each flower with a subulate bract, often deciduous; pedicels 0.9 mm long; calyx tube short, obconic, 0.5 mm long; lobes 5, ovate, 0.8 mm long, acute to rounded, green with light, denticulate margins; disk purple, 5-lobed, the lobes often emarginate; petals 5, obovate, 1.8-2 mm long, 1-1.3 mm wide, pink, often cupped, attached below the disk; stamens 5, inserted between and below the lobes of the disk; filaments white, 1.8-2.2 mm long, slender, weak; anthers pink; ovary ovate-acuminate, 1.3-1.5 mm long, 0.8 mm wide at the base, white, glabrous; styles obsolete, the 3 stigmas clavate, spreading.
FRUIT: July-October. Capsules ovate, abruptly or gradually acuminate, 3.5-4 mm long, 1 mm thick at the base, 3-angled, yellowish or pinkish, glabrous, petals and calyx lobes usually persistent; many seeds in each capsule. Seeds irregularly cylindric, 0.5 mm long, smooth, yellow, tipped by a tuft of white hairs with a central axis 1 mm long, the hairs 1-1.3 mm long.
TWIGS: 0.5 mm diameter, slender, flexible, smooth, red, becoming grayish; roughened only by the light-colored, raised leaf scars; upper twigs break off easily and are eventually deciduous; leaf scars only a narrow line; pith white, continuous, one-third of stem. Buds reddish, 1 mm long, nearly globular, set into a shallow cavity in the stem, brownish, often 3 at each node.
TRUNK: Bark of young limbs red with light leaf scars; bark of old trunks dark brown, exfoliating into thin scales or strips. Wood soft, white.
HABITAT: Sandy, moist soil, especially in alkaline soils, along streams or in low, undrained areas, and around lakeshores; especially common in the Arkansas and Cimarron river valleys; occasionally on dry hillsides.
RANGE: Native of Eurasia. Spreading from the southern states north to Massachusetts, Indiana, Missouri, Kansas, Colorado, and Oklahoma, and west to California. One collection from North Dakota.

Tamarisk is an attractive, ornamental plant, forming a shrub to 5 meters high and nearly as broad. The branches are numerous, erect, slender, and often bright red. It is an aggressive plant and, if not controlled, will soon take over a low pasture or field. In some areas on the flood plain of the Arkansas River it forms dense groves, so dense that even the larger mammals use it as protective cover. It is often used in windbreaks, but should not be used unless some method of control is employed.
T. *parviflora* DC. is also found in Kansas and can be distinguished by the 4-merous flowers. It is not common and appears only as widely scattered plants, often with T. *ramosissima*.

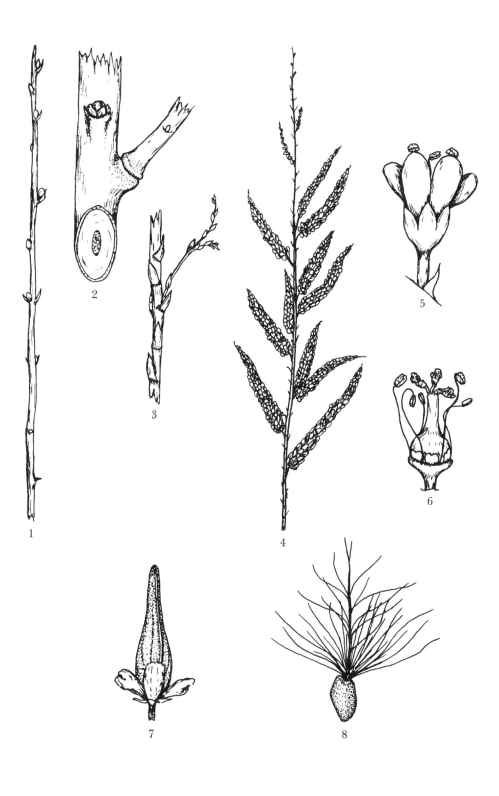

1. Winter twig
2. Detail of twig
3. Leaves
4. Inflorescence
5. Flower
6. Flower with petals and sepals removed
7. Fruit
8. Seed

CISTACEAE
Hudsonia tomentosa Nutt.

False heather, beach heath

A suffrutescent shrub, 10-20 cm high, densely branched; often half-buried in the sand.

LEAVES: Alternate, simple, persistent, but turning brown. Leaves lanceolate, 3-4 mm long, green, thick, densely pubescent on both sides, leaves appressed and imbricated; tip acute; sessile; no stipules.

FLOWERS: June-July; perfect. Flowers single, mostly terminal on a short, leafy stem 3-9 mm long, occasionally axillary on the same shoot; pedicels 2-2.5 mm long, pubescent; calyx lobes`5; however, the two outer lobes usually fused with two of the inner lobes, giving the casual appearance of 3 lobes; occasionally only one of the outer lobes fuses with an inner lobe, giving the appearance of 3 large lobes and one small one on the outside; lobes pubescent on the dorsal side and glabrous on the ventral side, green, tip rounded or somewhat cucullate and often reddish, those with fused lobes appear to have a tooth near the apex, but careful examination shows a vascular trace going from the base to the tip of the tooth and another trace to the lobe tip; petals 5, yellow, oblong or narrowly obovate, 3.5-4.5 mm long, 1-1.5 mm wide, spreading, soon deciduous; stamens about 15, attached at the base of the ovary; filaments slender, 2.5-3 mm long, glabrous; anthers yellow; ovary green, about 1 mm long, ovoid; style 1.5-2 mm long, glabrous, the stigma capitate.

FRUIT: The capsule enclosed in the calyx; capsule ovoid, 3-angled, 2.5-3 mm long, glabrous, yellow-brown; 5-6 ovules, but usually no more than 2 develop, these about 1 mm long and 0.8 mm wide, ovoid or ovoid with one oblique end, brown, granular, glabrous; the undeveloped ovules ovate and small, on either a short or long funiculus.

TWIGS: Twigs 1 mm diameter, tomentose, erect, bark red-brown when the tomentum is removed; no leaf scars because of persistent leaves; pith white, continuous, one-sixth of stem. Buds naked, about 2 mm long and 0.75 mm wide, consist of a series of densely tomentose, tightly packed, small leaves.

TRUNK: Bark thin, brown, smooth, but with some patches of tomentum remaining. Wood white, medium-hard.

HABITAT: Sand dunes, beaches, and sandy grasslands.

RANGE: Quebec across Canada to Alberta; Minnesota, Illinois, and northwest Indiana; one station in North Dakota; along the coast from Gaspé Peninsula to North Carolina.

Hudsonia tomentosa is a small, diffusely branched, woody plant, forming rounded clumps about 20 centimeters high and twice that wide. The leaves are evergreen but become quite brownish in the fall and evidently persist for several seasons.

The family Cistaceae is normally tropical or subtropical, extending into the warmer areas of the temperate zone. Obviously, *Hudsonia tomentosa* is tolerant of winter cold, for it grows in such places as the low sandhills of Ransom County, North Dakota, and along the shore of Lake Superior at Duluth, Minnesota. Both of these areas may have prolonged periods of subzero weather. Growing as it does in the semibarren sand dunes, there is little to protect it from the winter winds.

Two collections of several years ago were made in North Dakota, but the sites given on the herbarium labels were general and the plant could not be located in the area. The descriptions and drawings were made from fresh material along the shore of Lake Superior at Duluth.

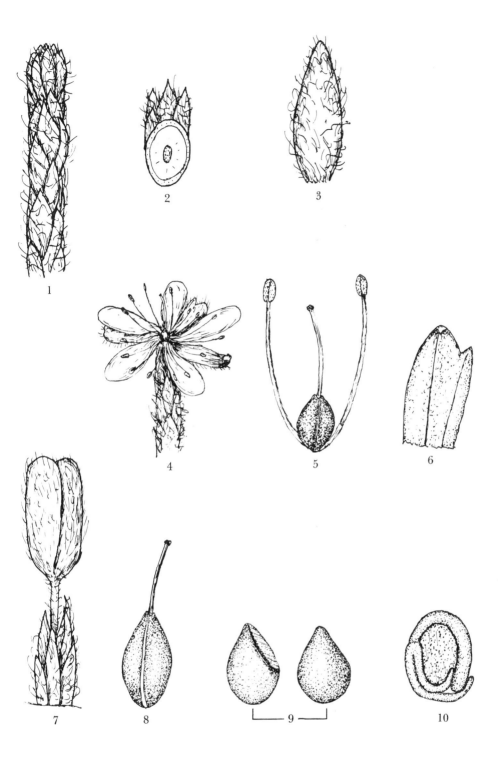

1. Twig in late fall
2. Section of twig
3. Leaf
4. Flower
5. Pistil and two stamens
6. Fused calyx lobes showing vascular trace
7. Mature fruit surrounded by calyx
8. Mature fruit
9. Seeds
10. Embryo

CACTACEAE
Opuntia imbricata (Haw.) DC.

O. arborescens Engelm.

Tree cactus, cane cactus

A dense shrub to 2 m high and about as wide, the main branches close to the ground.

LEAVES: Alternate, simple. Nearly terete, 1-2 cm long, 2-3 mm diameter, succulent, curved, tapering to a pointed or rounded tip, the base merging into a tubercle; green, early deciduous, glabrous. Not very functional as leaves.

FLOWERS: June-July; perfect. Clustered at the end of a stem; involucre obconic to cylindric, 1.5-3 cm long, 1.5-2 cm diameter, tuberculate, a few spines and leaves present at flowering time; sepals and petals several, obovate, 2-3 cm long, 1-3 cm wide; apex rounded or truncate, mucronate or emarginate; cupped; rose-pink to purple, outer tepals greenish; stamens many, filaments 1.4-1.6 cm long, purple; anthers yellow; ovary embedded in the involucre, 1-locular; style stout, cylindric, 2.8-3.2 cm long, 3-3.5 mm diameter; stigmas 6-9, white, cylindric, 5-6 mm long, erect.

FRUIT: October. Short cylindric, 2.5-3.5 cm long and about as thick, depressed at the summit, yellow, tuberculate, with glochids and often a few spines at the tip of the tubercles; 50-300 seeds. Seeds circular, flat, 3.3-3.6 mm wide, 1.7-1.9 mm thick, yellowish-gray, smooth.

TWIGS: End branches 1.5-3 cm diameter, rigid, green, terete, erect in summer, somewhat drooping in winter; tubercles 1.8-2.3 cm long, 5-6 mm wide, 6-7 mm deep, spines radiating from the upper end; branches verticillate, 3-5 per whorl; spines 10-20 from each areole, the spines 1-2.5 cm long, slender, straight, barbed, white to purple, sheathed in a white, papery covering with a yellowish tip. Areoles depressed, the spines from the basal portion, their base surrounded by white hairs, the glochids in the upper portion. Buds not visible. Leaf scars circular, soon disappearing. End of stems whitish in the center, spongy and succulent.

TRUNK: Cortex of old stems dry, brownish, covered with thin, papery scales of old cortex and spines; branches tuberculate, green, with a reddish cast in winter. Wood a hollow framework with large oval holes arranged spirally, medium-hard, fibrous, splits easily; the soft tissue decomposing shortly after the branch dies, leaving only the hollow skeleton.

HABITAT: Dry, sandy areas, pastures, weedy sand hills and flats.

RANGE: Colorado, Kansas, Oklahoma, Texas, and New Mexico.

The tree cactus grows to a height of about 2 meters and spreads over an area 2.5 meters wide, but most of our plants are smaller than this. It is definitely native in the southwestern corner of Kansas in at least two locations and possibly more. Most of the plants that are in yards around homes have been brought in from other states, but it is not uncommon to see them in towns in western Kansas.

Animals do not eat the branches of the plant, but the old spines on the ground are picked up and carried to the dens of wood rats. There, they are placed over the top of the den and around the openings. The seeds are eaten by wood rats and mice after the fruits have fallen to the ground and the fleshy part has decayed. This may be some time after maturation, since the fruits usually remain on the plant until midwinter.

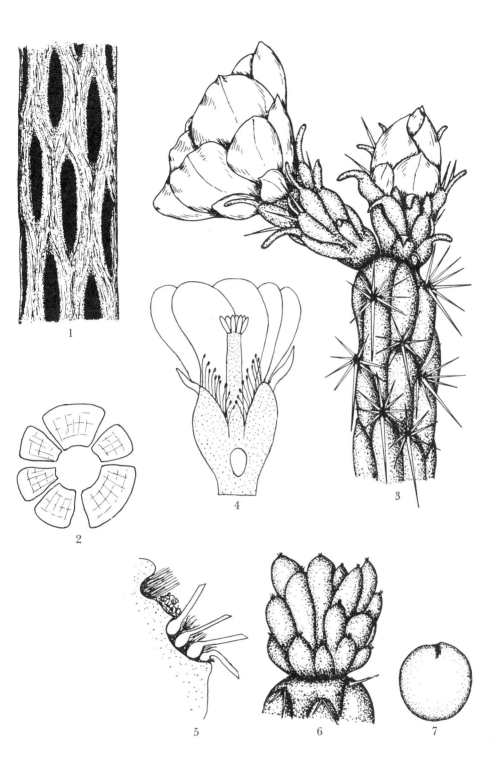

1. Old dead branch
2. Transverse section of dead branch
3. Stem with flowers and leaves
4. Diagrammatic section of flower

5. Diagrammatic section through aerole
6. Fruit
7. Seed

ELAEAGNACEAE
Elaeagnus angustifolia L.

Russian olive, oleaster

A dense tree to 6 m high, with low branches, the trunk often reclining.

LEAVES: Alternate, simple. Lance-olate to narrowly ovate, 4.5-9 cm long, 1.2-2.5 cm wide, entire, often undulate; tip acuminate to obtuse, base cuneate; upper surface dark green, covered with scale-like stellate pubescence, the radiat-ing branches of the hairs on the midrib being longer than those of the main surface; lower surface silvery white, densely covered with the scale-like stel-late hairs; petiole 8-12 mm long, densely scaly and with scattered, coarse hairs; no stipules.

FLOWERS: May-June; perfect. Flowers in clusters of 1-3 in the leaf axils of the current year's growth; pedicels 2-2.5 mm long, silvery scaly; hypanthium tubular, 5-6.4 mm long, abruptly broad-ened above the ovary, the upper portion 4-5 mm long, 2.5-3 mm wide, silvery-yellow; calyx lobes 4, deltoid, 3.2-4 mm long, spreading or reflexed, silvery scaly on the outside, the inside yellow and stellate pubescent near the tip; petals none; stamens 4, the sessile anthers 1.5-2 mm long, attached just below the rim of the hypanthium and scarcely exserted; style 4 mm long; disk yellowish, abruptly narrowed and surrounding the base of the style for about 1 mm; stigma linear, 1.7-1.8 mm long, recurled, usually ex-serted.

FRUIT: September. Oval, 10-14 mm long, 8-11 mm thick, the tip mucro-nate; yellow to brown, but densely cov-ered with silvery scales; flesh pithy and dry. Stone oblong, acute at both ends, 9-11 mm long, 4.1-4.8 mm thick, brown, with darker longitudinal stripes, the sur-face fairly smooth.

TWIGS: 1-1.4 mm diameter, flexible, reddish, coated with a gray, scaly pubes-cence, becoming glabrous; twigs often ending in a short spine; leaf scars small, half-round; 1 bundle scar; pith salmon-colored, continuous, one-third of stem.

Buds ovoid, 2.5 mm long, obtuse, gray, scaly.

TRUNK: Bark thin, dark gray-brown, with shallow fissures and exfoliat-ing into long strips. Wood lightweight, dark brown, with a light sapwood.

HABITAT: Planted in yards and windbreaks; escaping in many places in our area, especially in low, moist, sandy pastures.

RANGE: Native of Eurasia, and es-caping in most of the states.

Russian olive is of most importance in the prairie states as a windbreak, being planted as an understory with the taller trees. It reaches a height of 6 meters and forms a dense, rounded crown. It is also planted in yards and along highways as part of the landscape plan, its silvery leaves contrasting sharply with the green leaves of other plants.

The plant is hardy and can with-stand drouth or extreme cold. It requires very little water in the dry areas, but does well in low, moist pastures, where the growth rate is rapid.

Several birds eat the fruits and use the branches as nesting sites, especially when the trees are planted close together in a windbreak. Since it is a food plant for birds, a cover plant for birds and mammals, a dense windbreak, and has a strong root system, it is highly desirable in land-management planting. However, it sends up suckers from the roots and may have a tendency to spread to other areas where it is not desired.

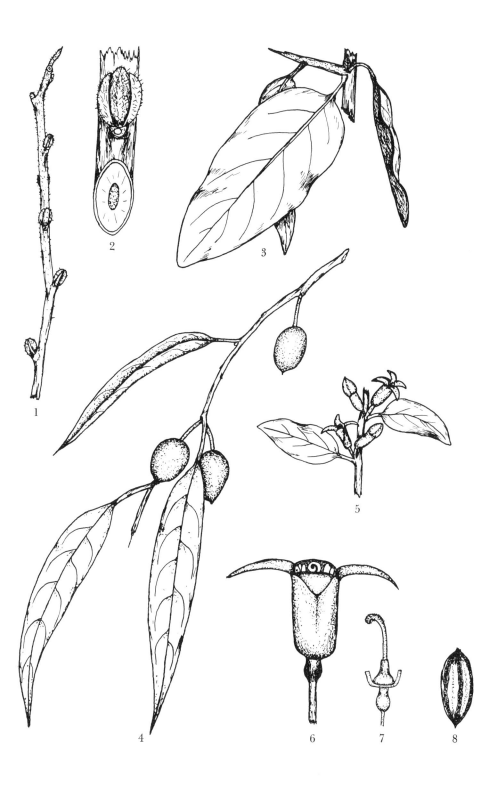

1. Winter twig
2. Detail of twig
3. Leaf
4. Leaf variation and fruit

5. Flower clusters
6. One flower
7. Cutaway to show disk and style
8. Seed

ELAEAGNACEAE
Elaeagnus commutata Bernh.

E. argentea Pursh, not Moench.

Silverberry

A stoloniferous shrub 1-4 m high, usually in large, loose colonies, the stems single.

LEAVES: Alternate, simple, soft, leathery, often folded lengthwise, the lateral veins obscure. Broadly elliptic, 3-6 cm long, 1-2.5 cm wide, tip acute or obtuse, base cuneate to somewhat rounded; margin entire, undulate; both surfaces completely covered with silvery, peltate scales with a few reddish scales below; petiole 4-6 mm long, usually nearly parallel to the stem and covered with silvery scales and a few reddish scales; no stipules.

FLOWERS: Mid-June; perfect. Flowers fragrant, 1-3 in the leaf axils of the current season's growth, erect or drooping; pedicels 2.5 mm long, silvery scaly; hypanthium tubular, 7-7.5 mm long, 2 mm wide, constricted above the ovary, 4 rounded angles extending from the base to the lobe sinuses; calyx lobes 4, triangular, 2.5-3 mm long, tube and lobes silvery scaly outside, yellow inside; no petals; stamens 4, filaments short, attached to the throat of the calyx just below the lobe sinuses; anthers tan, 1 mm long, not exserted; style 5-6 mm long, scaly at least on the lower half, stigma capitate, irregular, yellow-brown. Hypanthium tube persistent until fruit partly grown.

FRUIT: Pendent on curved pedicels 3-5 mm long, oval, 10-15 mm long, 8-11 mm thick, covered with silvery peltate scales and a few red scales; flesh dry, 1-seeded. Stone ellipsoid, dark brown, with 8 longitudinal, low ridges and yellow stripes, 8-12 mm long, 4.5-5.5 mm thick, surface scurfy, the wall hard and densely white woolly inside.

TWIGS: 1-2 mm diameter, slightly compressed, densely covered with peltate scales which are fulvous in the center and with a white, erose margin; leaf scars half-round, small; 1 bundle scar; pith brown, continuous, one-third of stem. Bud scales valvate, red-brown, scaly inside and out; end bud deltoid, compressed, 4-6 mm long, 3-3.5 mm wide, acute; lateral buds 3-4 mm long and as wide, flattened.

TRUNK: Bark gray-brown, somewhat scaly on old trunks, inner bark fibrous; wood lightweight, medium-hard, dark brown, with a wide, yellowish sapwood.

HABITAT: Dry prairie hillsides, usually in poor soil, or along brushy ravine banks in dry areas.

RANGE: Central Alaska, south through British Columbia to the mountains of Idaho, Montana, and Utah, northeast across Wyoming to North Dakota and northern Minnesota, east to Quebec, and west across northern Manitoba to Yukon.

E. commutata is our only native *Elaeagnus,* and in the central plains states it is found only in North Dakota. It has been reported for the western slopes of the Black Hills and one Rydberg specimen was located, but even though the area given in the data was searched intensively no plants could be found.

This is a hardy plant and grows on the most exposed hillsides, and often in dry sandy or gravelly soil. In such locations the plants grow only to 1.5 meters high and form loose, open colonies. Occasionally a small colony is found in the more moist soil of a ravine. Here, the plants reach 4 meters and the colonies become impenetrable. The odor of the flowers can be detected at quite some distance from the plants.

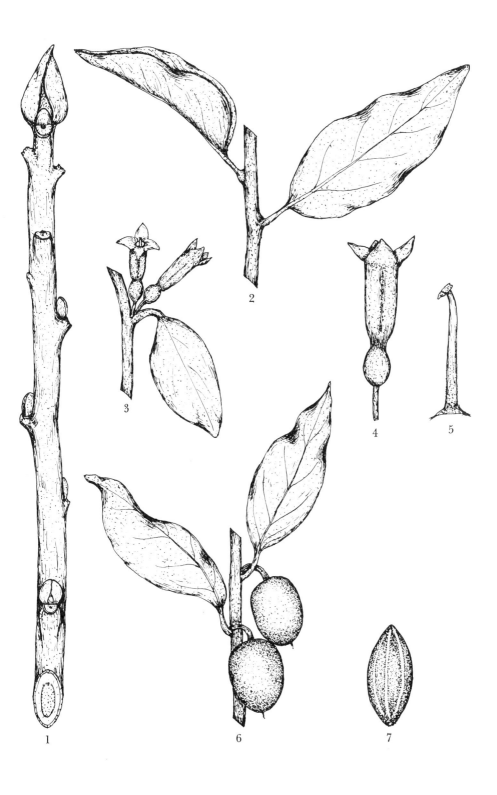

1. Winter twig
2. Leaves
3. Flower cluster
4. One flower
5. Pistil
6. Fruit
7. Seed

407

ELAEAGNACEAE
Shepherdia argentea (Pursh) Nutt.

Hippophae argentea Pursh; *Lepargyraea argentea* Greene; *Elaeagnus utilis* A. Nels.

Buffalo berry, rabbit berry

A densely branched shrub to 4 m high, occasionally in dense stands.

LEAVES: Opposite, simple, the lateral veins indistinct. Oblong, 3-5 cm long, 7-10 mm wide; tip rounded, base cuneate; margin entire; both surfaces completely covered with erose-margined, silvery, peltate scales about 0.2 mm across, the scales often with a reddish center; petiole 4-6 mm long, covered with scales; no stipules.

FLOWERS: Late April, the leaves just started; dioecious. Staminate flowers clustered on short spur branches, sessile or on scaly pedicels about 1 mm long. Hypanthium broadly campanulate, 1-1.2 mm long, 2 mm across, covered with silver and red peltate scales with fimbriate margins; calyx lobes 4, ovate, 2-2.5 mm long, 1.5 mm wide, spreading or sharply reflexed, yellow on the inside; stamens 8, filaments tapered, 2 mm long, white, glabrous; anthers 0.7 mm long, pale yellow; disk consists of golden-yellow glands alternating with the stamens, these may be in contact with each other but are distinct, each 0.5 mm long; a white pubescence around the glands and inside of the hypanthium. Pistillate flowers on stout pedicels 1-1.2 mm long. Hypanthium campanulate, 1.5 mm long, 1 mm wide, thick-walled, externally covered with silver and red peltate scales, internally green; calyx lobes erect, ovate, 0.75-1 mm long, scaly; disk of 8 golden-yellow glands on the hypanthium rim; ovary green, ovoid, 1 mm long, 0.75 mm wide, pubescent at the summit; style stout, 1 mm long, green, glabrous; stigma capitate, yellow-brown.

FRUIT: August. Single or clustered on short branches, the leaves often having fallen from the branch. Fruit drupelike, subglobose, 6-8 mm long, 7-9 mm wide, bright red, glossy, smooth, and with a few scattered silvery or reddish scales; calyx and style often persistent; 1-seeded.

Seeds light or dark brown, flattened ovoid, 3.2-4 mm long, 2.4-2.8 mm wide, 1.8-2 mm thick, smooth or minutely pitted, glossy, a shallow groove on each flattened surface. One collection from Harding County, South Dakota, has seeds up to 5 mm long.

TWIGS: 1.8-2 mm diameter, rigid, red-brown, and densely covered with silvery-white peltate scales toward the tip; some twigs end in spines; leaf scars small, broadly crescent-shaped; 1 bundle scar; pith brown, continuous, about one-third of stem. End bud oblong, obtuse, the scales valvate, thick, covered with silvery and reddish peltate scales; staminate flower buds usually in clusters of 4, about 1.5 mm diameter, on stalks 0.75 mm long, the stamens well developed within them; lateral buds oval, 1-1.25 mm long, flattened.

TRUNK: Bark gray-brown, exfoliating into long, thin strips; bark of main branches with shallow furrows and flat-topped ridges, the smaller branches gray-brown and smooth. Wood lightweight, fairly hard, open-grained, light brown, with a narrow, yellow sapwood.

HABITAT: Prairie valleys, banks of streams or on steep, eroded, dry hillsides; often in poor soil.

RANGE: British Columbia, Alberta, southeastern Oregon, northeastern California, and an area north of Death Valley in California; Arizona and New Mexico, north through western Colorado to Montana, east to North Dakota, Manitoba, and Minnesota, south to northwestern Iowa, and west across Nebraska to Wyoming. Although listed for Kansas, no specimens could be found. Within our area the distribution is spotty.

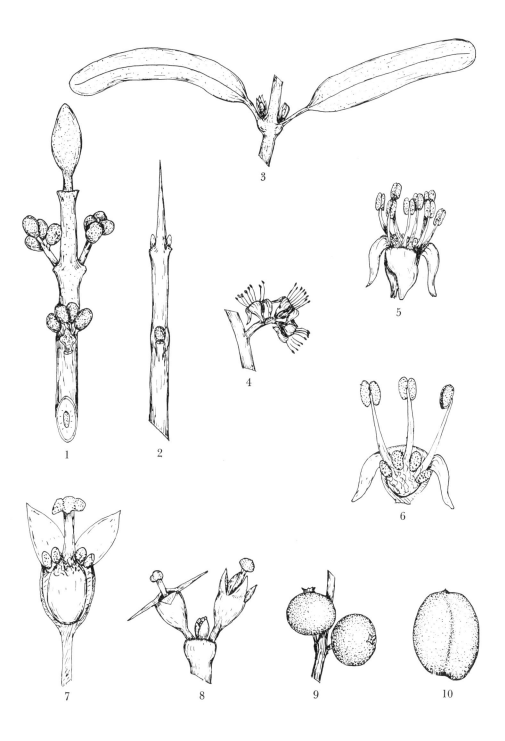

1. Winter twig with flower buds
2. Twig with leaf buds
3. Leaves
4. Staminate flower cluster
5. Staminate flower
6. Diagrammatic cutaway of staminate flower
7. Diagrammatic cutaway of pistillate flower
8. Pistillate flowers
9. Fruit
10. Seed

ELAEAGNACEAE
Shepherdia canadensis (L.) Nutt.

Hippophae canadensis L.; *Lepargyraea canadensis* Greene; *Elaeagnus canadensis* A. Nels.

Buffaloberry, soapberry

A low, open shrub to 1 m high, occasionally freely branched.

LEAVES: Opposite, simple, the lateral veins indistinct. Ovate to narrowly ovate, 4-6 cm long, 1.5-3 cm wide; tip rounded to obtuse, base rounded; margin entire; upper surface dark yellow-green, with scattered white stellate hairs; lower surface densely covered with silvery, long-ciliate, peltate scales and scattered red-brown, lobed scales; petiole 5-8 mm long, densely covered with both silvery and red peltate scales; no stipules.

FLOWERS: May-June; dioecious or monoecious. Staminate flowers clustered on a short branch of the season, pedicels 1-1.5 mm long, with silvery and red scales; hypanthium broadly turbinate, 1 mm long, 1 mm wide, red scaly; calyx lobes 4, ovate, 3 mm long, 1.5 mm wide, red scaly outside, glabrous and yellow-green inside with 3 darker green veins and sharply reflexed or spreading; no petals; center of flower pubescent; no pistil; disk of 8 greenish, fleshy lobes along the rim of the hypanthium; stamens 8, inserted on the calyx between the fleshy lobes of the disk, filaments 1-1.5 mm long, white, usually curved inward; anthers pale yellow. Pistillate flowers similar, ovary ovoid, 1.5-2 mm long, scaly with both silvery and red scales; pedicel 0.25 mm long; calyx lobes 1.5 mm long, scaly outside, glabrous inside with 3 reddish or green veins; an 8-lobed disk attached to the hypanthium around the pistil; style 0.5 mm long, green; stigma elliptic, flattened, curved.

FRUIT: July-August. Single or clustered, sessile or on a short, curved pedicel; fruit ovoid to depressed globose, 8-10 mm diameter, bright red, glossy, smooth and with a few short-stalked, red-brown, peltate scales; fruit juicy and 1-seeded. Seeds ovate, flattened, 4.8 mm long, 2.9 mm wide, 1.8 mm thick, dark brown, glossy, smooth or minutely pitted, a longitudinal groove on each side.

TWIGS: 1.5-1.7 mm diameter, flattened, densely covered with overlapping peltate scales that are red in the center and have a transparent or red, erose margin; twig green beneath the scales; leaf scars small, half-round; 1 bundle scar; pith brown, continuous, one-third of stem. Terminal buds ovoid, obtuse, 5-7 mm long, and on a stalk about 2 mm long, densely covered with silvery and red scales; the inside of the bud scales densely white villous; lateral buds similar but smaller; flower buds clustered on short spur branches, globose, 1-1.5 mm diameter.

TRUNK: Bark gray-brown, smooth, but with minute horizontal cracks and lenticels. Wood hard, fine-grained, light brown, with a wide, light sapwood and dark brown pith; annual rings obvious.

HABITAT: Open hillsides or ravine banks, open woods, either dry or moist, and in either rocky or sandy soil.

RANGE: Nova Scotia, southwest across Maine to western New York and to Ohio, west to the Black Hills and Wyoming, south to New Mexico and Arizona, north to Oregon and Alaska, and east across northern Canada to Newfoundland.

This small shrub enters our area from the Northern Conifer Province and it is quite common through the woods and at the margins of wet tundra areas. It is not easily mistaken for any other shrub, since no other in our area is so completely covered with the silvery and red scales.

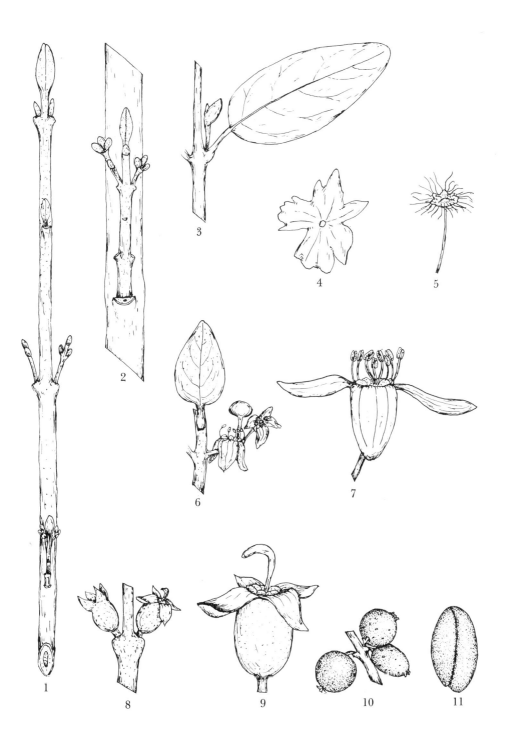

1. Winter twig
2. Detail of twig
3. Leaf
4. Red peltate scale
5. White peltate scale
6. Staminate flower cluster
7. Staminate flower
8. Pistillate flowers
9. Pistillate flower
10. Fruit
11. Seed

ONAGRACEAE
Oenothera serrulata Nutt.

Calylophus serrulatus (Nutt.) Raven;
Meriolix serrulata (Nutt.) Walp.;
M. intermedia Rydb.

Evening primrose

A suffrutescent shrub to 40 cm high, woody on the basal 20 cm or herbaceous to near the ground.

LEAVES: Alternate, simple; young leaves folded lengthwise, the midrib outstanding below, laterals indistinct. Lanceolate to linear, 2-5 cm long, 3-5 mm wide; tip acute and often calloused; base tapered, sessile; margin entire or serrate, 1-5 teeth per cm, mostly callous-tipped; young leaves variable from densely white strigose on both sides to glabrous above and slightly pubescent below, older leaves usually glabrate above and sparingly strigose below, the midrib often densely pubescent; no stipules.

FLOWERS: Late May-June; perfect. Axillary near the end of a stem. Hypanthium 2-3 cm long, 4-angled, the ovary part cylindric, 1-2 cm long, 1-2 mm diameter, densely strigose, greenish-brown, hypanthium above the ovary funnelform, narrow at the base, flared to 5-6 mm across at the top, yellow, heavily 4-ribbed, strigose, especially on the ribs; calyx lobes 4, triangular, 5-7 mm long and 2-2.5 mm across at the base, yellow, glabrous or strigose, strongly ribbed, reflexed; petals 4, rounded to obcordate, 10-20 mm long and about as wide, margin often undulate, bright yellow; stamens 8, the 4 opposite the calyx lobes with yellow filaments 3-4 mm long, the 4 opposite the petals with filaments 0.5-1 mm long, all fastened at the rim of the flared hypanthium; anthers 5-6 mm long, with the filament attached at the center; style slender, 17-19 mm long, greenish, glabrous; stigma capitate, lobed, about 4 mm across, usually oblique.

FRUIT: July-August. 4-angled, cylindric, 1.5-2.5 cm long, 2.5-3 mm diameter, brown, appressed pubescent, the 4 sections splitting apart at the apex when mature, releasing the many seeds from around the central placenta. Seeds brown, rhombic to ovoid, 1.25 mm long, 0.5 mm wide, smooth, angular.

TWIGS: 0.8-1.5 mm at the point of dying back; gray-brown, flexible, the outer bark exfoliating into long strips, inner bark yellow-brown; twigs die back to below the lowest flower, which is often as much as 20 cm, but still some distance above the ground; leaf scars half-round, raised; 1 distinct bundle scar; pith brown, continuous, one-third of stem. Buds naked, consist of a series of pubescent leaves ranging in size from mere knobs to 5 mm long.

TRUNK: Bark red-brown, thin, exfoliating into thin, papery strips. Wood fairly hard, fine-grained; 2-year-old wood brown, 1-year-old white, with a dark brown pith.

HABITAT: Open prairies, pastures, and brushy hillsides, on either rich or poor soil, but usually on poor soil in dry areas.

RANGE: Manitoba to Montana, south to New Mexico, Arizona, and Texas, and north through the Great Plains to Minnesota.

O. serrulata is usually listed as herbaceous, and a great many of the plants definitely go into that category. Since a large proportion of them are suffrutescent with the overwintering buds 10-20 cm above ground, it is included here. These winter buds are naked, but it is from them that the new vegetative and flowering branches appear the following spring.

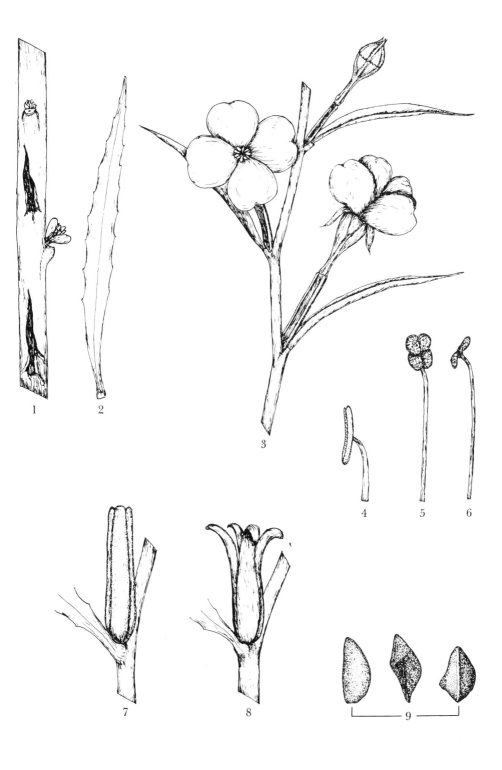

1. Portion of winter twig
2. Leaf
3. Flowering branch
4. Stamen
5. Face view of style and stigma
6. Side view of style and stigma
7. Fruit capsule
8. Capsule opening
9. Seeds

CORNACEAE

Cornus amomum Mill. ssp. obliqua
(Raf.) J. S. Wilson

C. obliqua Raf.; *C. purpusii* Koehne

Swamp dogwood, silky dogwood,
pale dogwood, red dogwood

A shrub to 2.5 m high, with clustered, upright trunks.

LEAVES: Opposite, simple. Elliptic to narrowly ovate, 7-10 cm long, 2-3 cm wide; margin entire; tip long or abruptly acuminate, the base cuneate to obtuse; upper surface dark yellow-green, glabrate; lower surface paler, glaucous, pubescent with white, appressed malpighiaceous hairs; petiole 1.5-2 cm long, pubescent, flattened above; stipules absent.

FLOWERS: May-June; perfect. Terminal round-topped, naked cymes on new growth, all parts with appressed malpighiaceous hairs, the rays and pedicels usually with long red hairs; peduncle 4-6 cm long, green; rays 1-2.5 cm long, green; pedicels 5-7 mm long, pale green; hypanthium ovoid, 2 mm long, 1.4 mm wide, yellow-green, pubescent; calyx lobes 4, lanceolate to triangular, 1.5 mm long, yellow-green, pubescent; petals 4, narrowly ovate to oblong, 6-7 mm long, 1.9-2 mm wide, acute or acuminate, white, rather thick; stamens 4, opposite the calyx lobes, filaments white, 4.6-4.8 mm long; anthers yellow; disk 4-lobed, dark red, the lobes opposite the petals; ovary inferior; style from the center of the disk, 3.5 mm long, green, widened just below the stigma; stigma yellowish, indistinctly 4-lobed, capitate.

FRUIT: August. Cymes drooping; peduncle 4-6 cm long, green; pedicels pinkish, 5-7 mm long; fruit globose, 7-11 mm diameter, blue, often slightly indented at the apex, sparingly pubescent with tightly appressed malpighiaceous hairs; 1-seeded; calyx lobes persistent. Seeds subglobose to obovoid, irregular, 6-7 mm long, 5-5.5 mm wide, greenish-white becoming pale brown, rough with 5-7 irregular ridges.

TWIGS: 1.2-1.5 mm diameter, flexible, red, the lower branches gray-red with shallow fissures; pubescent with tan-colored malpighiaceous hairs, some of which are short and appressed, others with long stalks and ascending, crinkly branches; leaf scars narrowly crescent-shaped; 3 bundle scars; pith brown, continuous, one-fourth of stem. Terminal bud narrowly ovoid, gray-brown, 4-6 mm long, pubescent, obtuse tip; lateral buds 2.5 mm long, flattened, acute or blunt, pubescent.

TRUNK: Bark tight on most stems, red, with tan horizontal lenticels; some stems with longitudinal splits in the outer bark. Wood hard, fine-grained, white.

HABITAT: Low woodlands, wet thickets, rocky streams, swampy sloughs in prairies, and occasionally on rocky uplands.

RANGE: Maine, west to Michigan and Minnesota, south to Nebraska and Oklahoma, east to Arkansas, northeast through Illinois and Ohio to New York.

The plants of our area are ssp. *obliqua,* having the leaf bases cuneate, the lower surface glaucous and with white hairs. Ssp. *amomum* has rounded leaf bases and a green under leaf surface with brown or reddish hairs; its range is from Pennsylvania to Mississippi and to the east coast.

Our plants are most quickly recognized in the winter when the clustered growth habit and the red limbs can be seen. During the fruiting season, the clusters of blue drupes are a good character, being the only blue-fruited *Cornus* in our region. The leaves are variable, from quite narrow to fairly wide, and may resemble those of *C. drummondii,* but lack the scabrous upper surface.

414

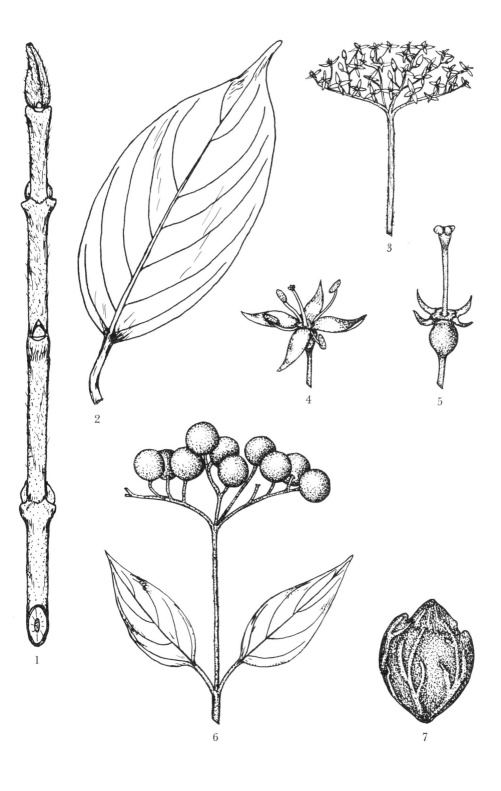

1. Winter twig
2. Leaf
3. Inflorescence
4. One flower

5. Pistil
6. Fruiting cyme
7. Seed

CORNACEAE
Cornus drummondii Meyer

C. asperifolia of auth., not Michx.; *Svida asperifolia* (Michx.) Small

Rough-leaved dogwood

A shrub to 4 m high, thicket-forming, most of the trunks 1-5 cm diameter at the base.

LEAVES: Opposite, simple, 3-5 lateral veins on each side. Ovate to elliptic, 4-9 cm long, 3-4.5 cm wide, entire; tip acuminate, base obtuse to rounded; upper surface dull green, scabrous with short, stiff malpighiaceous hairs; lower surface paler, densely woolly, the hairs longer than on the upper surface; petiole 0.5-1 cm long, pubescent; stipules absent.

FLOWERS: May-June; perfect. Terminal cymes 4-6 cm across on new growth; all parts, including the perianth, with malpighiaceous hairs; peduncle 2-3 cm long, green; rays 3-5, green, 1.5-1.8 cm long; pedicels 1-2.6 mm long, white; hypanthium ovoid to obovoid, 1.8-2 mm long, white; calyx lobes 4, deltoid, 0.5 mm long, white; petals 4, ovate, 3.8-5 mm long, acute, creamy-white, spreading, finely appressed pubescent outside, glabrous inside; stamens 4, opposite the calyx lobes, filaments attached below the disk, white, tapered, 3.8 mm long; anthers yellow; disk 4-lobed, yellow at anthesis, becoming pinkish, lobes opposite the petals; ovary inferior; style 2-2.5 mm long, white, often gradually enlarged upward, sparsely appressed pubescent with longer hairs at the base; stigma yellowish, irregularly 4-lobed, capitate.

FRUIT: September-October. Flat to round-topped cymes terminating branches; pedicels 1-5 mm long, often red; fruits globose, 6-7 mm diameter, white, appressed pubescent, the thin pulp with a milky juice; 1-seeded. Seeds greenish-white, becoming tan-brown, globose, 3.9-4.4 mm diameter, smooth but slightly indented along the longitudinal vein lines.

TWIGS: 2 mm diameter, flexible, tough, pubescent, brown to reddish-gray, often red when young; leaf scars narrowly crescent-shaped with enlarged ends and center; 3 bundle scars; pith white to salmon, continuous, one-third of stem. Terminal bud 4-5 mm long, 2.7-2.9 mm wide, swollen below the middle, 2 pubescent valvate scales; lateral buds ovoid, 2-3 mm long, obtuse, reddish, pubescent.

TRUNK: Bark gray-brown, with shallow fissures and short, thin plates. Wood hard, white.

HABITAT: Dry uplands, roadsides, fence rows, margins of woods, pastures; in rocky or clay soils, and occasionally in moist soil.

RANGE: Michigan, west to South Dakota, south to Texas, east to Mississippi, and north to Ohio.

C. drummondii is a thicket-forming shrub usually about 2 meters high, but an occasional, isolated plant may reach 5 meters. It spreads by underground stems, sending up sprouts at the margin of the thicket. It is one of the most hardy of our shrubs and will withstand drought or extreme cold, even though its range does not extend to the far north.

It is seldom planted in windbreaks or around farmsteads, but its dense thickets are excellent cover for birds and small mammals and give good protection to the soil. Many birds nest in the thickets, and Bell's vireo seems to have a special preference for the low, outside branches, often so low that a nest full of young birds will weigh the branch to the ground.

The ranges of *C. drummondii*, *C. stolonifera*, and *C. amomum* ssp. *obliqua* overlap in eastern Nebraska, but the species are distinct enough and are quite easily identified.

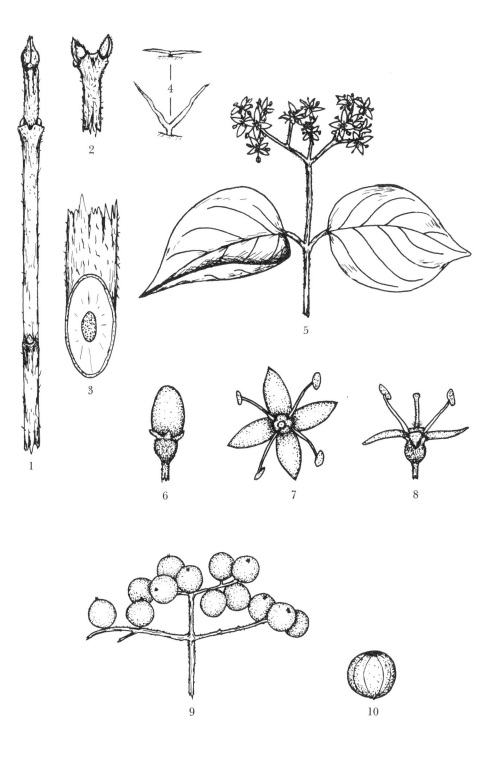

1. Winter twig
2. Twig with two end buds
3. Cutaway of twig
4. Malpighiaceous hairs of stem
5. Leaves and inflorescence
6. Flower bud
7. Face view of flower
8. Side view of flower
9. Fruiting cyme
10. Seed

CORNACEAE
Cornus florida L.

Cynoxylon floridum (L.) Raf.

Flowering dogwood

A tree to 12 m high, usually only to 7 m in our area, branching above the middle.

LEAVES: Opposite, simple. Ovate to elliptic, 9-15 cm long, 5-9 cm wide; entire, abruptly acuminate or occasionally acute, the base broadly cuneate to attenuate; upper surface dark green, with short appressed malpighiaceous hairs; lower surface paler, with short or long malpighiaceous and stellate hairs, the veins more densely pubescent; petiole 3-8 mm long, appressed pubescent; stipules absent.

FLOWERS: April-May, with the leaves; perfect. Flowers terminal from buds formed the previous year; the flowers in clusters of 25-30 subtended by 4 white bracts deeply notched at the apex, with or without tightly appressed hairs; the bracts deciduous; outer bracts nearly circular, 3-3.5 cm long, 3.5-4 cm wide; inner bracts obovate, 3.5-4 cm long, 3-3.5 cm wide; calyx cylindric, 3.5-4 mm long, green, white pubescent outside; lobes 4, acute, 0.8 mm long; petals 4, attached inside the calyx tube, spreading or tightly recurled, 4.5-5 mm long, greenish-yellow, oblong, acute, pubescent outside toward the tip; stamens 4, filaments white, 4-4.5 mm long, alternate with the petals; anthers yellow; ovary 2-2.5 mm long, 2-celled, 1 ovule per cell; style green, pubescent, 2-2.5 mm long; stigma capitate, yellow-green; the central disk small and yellow.

FRUIT: Late September. The drupes sessile in clusters of 2-6; peduncle 2.5-3 cm long, appressed pubescent, usually terminal between two branches, each with 2 leaves; drupe ellipsoid, 8-10 mm long, 7-8 mm thick, brilliant orange-red, smooth, glossy, minutely puberulent; flesh 1 mm thick, dry, firm, yellow-orange; calyx lobes persistent; 1 seed per fruit. Seeds ellipsoid, 7-9 mm long, 4.4-5.4 mm wide, cream-colored, 5-7 shallow longitudinal grooves.

TWIGS: 2-2.5 mm diameter, flexible, reddish-gray to purplish, or greenish with red dots, pubescent with white appressed malpighiaceous hairs; leaf scars crescent-shaped; 3 groups of bundle scars; pith white, continuous but not solid, later with many pale brown, hard disks, one-third of stem. Flower buds terminal, peduncle 8-10 mm long, 2 bracts at the base, buds depressed globose, 5 mm high, 6-8 mm diameter, bracts gray-pink, densely appressed pubescent, slightly keeled at the summit; these bracts enlarge and become the white bracts subtending the flowers. Leaf buds compressed ovoid, 3-4 mm long.

TRUNK: Bark dark gray-brown, with shallow fissures and thin, squarish plates. Wood hard, not easily dented, heavy, brownish, with a light sapwood.

HABITAT: Acid soils, gravel, sandstone, or limestone; wooded slopes, ravines, and bluffs.

RANGE: Maine, west to Michigan, and south to southeastern Kansas and Texas, east to Florida, and north to New England.

The large, white "flowers" make *Cornus florida* one of the most conspicuous and attractive flowering trees in our area. It often occurs scattered through the woods or covering a small hillside in the southeastern corner of our range. The wood is not used commercially from our area, but elsewhere it is made into shuttles, roller-skate wheels, heads of golf clubs, mallets, chisel handles, pulleys, and knitting needles. In late September when the fruits are a brilliant orange-red, the leaves become deep red. It is often used as an ornamental tree in yards and for street planting, but most of those trees are horticultural varieties produced in a nursery, many of them are the pink form.

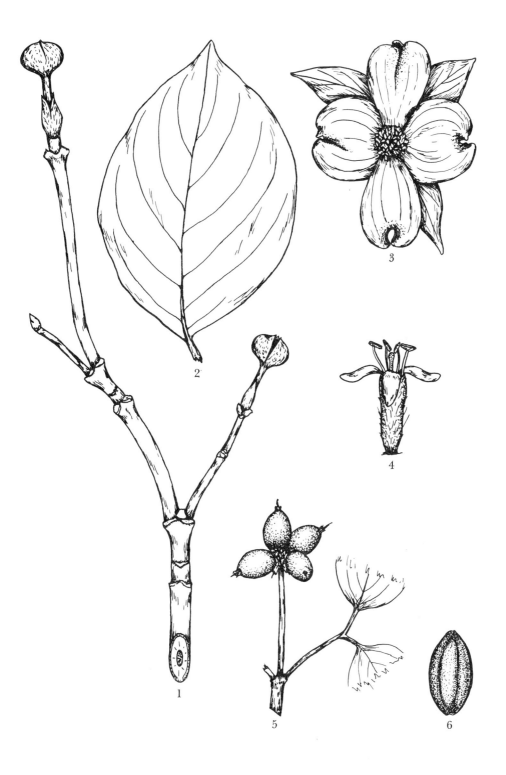

1. Winter twig with terminal flower buds
2. Leaf
3. Cluster of flowers subtended by the four bracts
4. One flower
5. Cluster of fruit
6. Seed

CORNACEAE
Cornus foemina Mill. ssp. *racemosa*
(Lam.) J. S. Wilson

Cornus racemosa Lam.; *C. paniculata*
L'Her.; *C. candidissima* Marsh., not Mill.;
Svida foemina Small

Gray dogwood

A shrub to 3 m high, forming small thickets.

LEAVES: Opposite, simple. Elliptic, 4-9 cm long, 1.5-4 cm wide, with 3-4 lateral veins on each side; margin entire, tip acuminate, base cuneate; both surfaces with appressed malpighiaceous hairs, lower surface microscopically papillate; petiole 5-10 mm long, flat on top, appressed pubescent; no stipules.

FLOWERS: Early July; perfect. Inflorescence paniculate with a definite axis, rounded to somewhat pyramidal, 3-4.5 cm long, 3-4 cm wide, branches and pedicels greenish-white to pinkish and with appressed malpighiaceous hairs. Hypanthium ovoid, 1.5 mm long, 1.2 mm wide, white, appressed pubescent; calyx lobes 4, triangular, 0.5 mm long, inconspicuous; petals 4, oblong, 4-4.5 mm long, 1.5 mm wide, white, spreading or recurved, pubescent outside and minutely papillate inside; stamens 4, filaments white, 4 mm long, narrowed to a slender attachment at the center of the anther; anthers pale yellow, 1.25 mm long; filaments inserted in the sinuses of the 4-lobed, pale yellow disk; style about 2 mm long, stout, abruptly enlarged to the capitate stigma.

FRUIT: September. The paniculate cyme 4-5 cm long, 5-7 cm wide, branches pink or red, glabrous or sparingly pubescent, peduncle 3-4 cm long, pedicels 2-4 mm long, abruptly expanded at the summit; fruit subglobose, 6-7.5 mm long, 6-9 mm wide, white, semiglossy, minutely appressed pubescent, style often persistent; 1-seeded. Stone white to pale brown, ovoid or oblique-ovoid, 5-6 mm long, 4.5-5 mm diameter, with 8 longitudinal veins adhering tightly.

TWIGS: 1.4-2 mm diameter, flattened, enlarged at the nodes, flexible, tan to gray-red, smooth, glabrous except for a few appressed malpighiaceous hairs near the end; lenticels small, elliptic, and inconspicuous; leaf scars U-shaped, the ends often incurved, dark color; 3 bundle scars; pith white the first year and brown by the second year, continuous, one-half of stem. Buds ovoid, 3-4 mm long, 1.5-2 mm wide, acute, the scales valvate, red-brown and appressed pubescent; usually a small bud on each side of the terminal bud.

TRUNK: Bark gray or gray-brown, tight, roughened by the lenticels; old bark with small, squarish, thin flakes. Wood hard, fine-grained, white, the sapwood not distinguishable.

HABITAT: Moist creek banks, rock ledges, rocky-clay ravine banks; usually in wooded or brushy areas.

RANGE: Maine through southern Canada to Manitoba and North Dakota, south to Nebraska and Missouri, east to Kentucky and Virginia, and north to New England. No stations are known in Kansas.

The ssp. *foemina* does not enter our area but is generally limited to an area along and south of the Ohio River and west of the Mississippi River to eastern Oklahoma and Texas. It has dark red branches, a rounded inflorescence and blue fruits. Ssp. *racemosa* is found north of that line and in our area is restricted almost entirely to South and North Dakota. This would mean, generally, that it would be confused only with *C. stolonifera*, since our other two similar species are southern in range. *C. stolonifera* has a flat or rounded inflorescence, the young twigs are red with a white pith. *C. foemina* ssp. *racemosa* has a pyramidal inflorescence and tan or grayish twigs with a white pith the first year but brown by the second year.

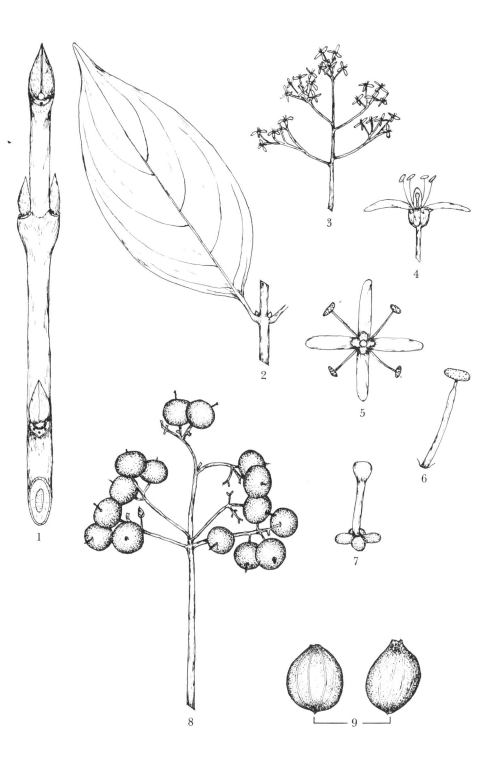

1. Winter twig
2. Leaf
3. Inflorescence
4. Side view of flower
5. Face view of flower
6. Stamen
7. Pistil and disk
8. Fruit cluster
9. Seeds

CORNACEAE
Cornus stolonifera Michx.

C. alba ssp. stolonifera Wang.; C. insto-
lonea A. Nels.; C. sericea L. ssp. stoloni-
fera Fosberg; C. interior (Rydb.) Peters.;
C. nelsoni Rose; C. stolonifera var. baileyi
(Coult. & Rose) Drescher; Svida stoloni-
fera Rydb.; S. interior Rydb.; S. baileyi
(Coult. & Evans) Rydb.; Ossea interior
Lunell

Red dogwood, red osier

A shrub to 3 m high, with clustered stems.

LEAVES: Opposite, simple, 5-6 lateral veins on each side and becoming obscure near the margin. Ovate to ovate-lanceolate, 6-9 cm long, 2-4 cm wide, entire, the tip acute or acuminate, base cuneate; upper surface dark green, not scabrous but with tightly appressed malpighiaceous hairs; the lower surface whitened with microscopic papillae, the pubescence variable from appressed malpighiaceous hairs only, to these plus an abundance of long, spreading clustered or stellate hairs and an occasional red hair; petiole 1-1.5 cm long, with appressed or spreading hairs, flattened above, often red; no stipules.

FLOWERS: June, but continuing through the summer; perfect. Cymes flat-topped or slightly rounded, 3-6 cm across, compact; peduncle 2-4 cm long, pubescent with either appressed or spreading hairs; pedicels 4-6 cm long; hypanthium ovoid, 2 mm long, green, densely appressed pubescent; calyx lobes 4, triangular, 0.5 mm long, spreading or erect between the petals, green, pubescent with white and red hairs; petals 4, ovate, 3-4 mm long, white, appressed pubescent outside and minutely papillose inside; stamens 4, opposite the sepals, filaments 3-3.5 mm long, white, attached at the center of the pale yellow anthers; disk lobed, yellow-green; style white, 2 mm long, stout, appressed pubescent, enlarged below the capitate stigma.

FRUIT: August-September. Round-topped cymes of 10-30 fruits, branches reddish-gray, strigose; fruit globular, 7-8 mm diameter, depressed at the pedicel, dull white with a few appressed hairs; style persistent; 1 seed per fruit. Stones pale or dark brown with 8 white, longitudinal veins, obliquely ovoid, about 5 mm long, 4 mm wide, and 3 mm thick, smooth, somewhat pointed at both ends and with a shallow groove along each margin.

TWIGS: 1.5-1.7 mm diameter, purple-red, appressed or spreading pubescent, the lenticels elliptic, raised, light-colored; leaf scars narrowly crescent-shaped, dark-colored; 3 bundle scars; pith white, continuous, one-third of stem. Terminal buds on stalks 1-1.5 mm long; buds ovoid, compressed, 3-6 mm long, 2-2.5 mm thick, swollen near the base, acute; the scales valvate, red-purple, pubescent with white and reddish appressed hairs, the inside densely pubescent at the tip and glabrous at the base; flower buds larger than the leaf buds; lateral buds 1.5-5 mm long, compressed, rounded or acute.

TRUNK: Bark usually red, occasionally greenish, smooth but with obvious horizontal lenticels. Wood hard, fine-grained, white, with a white pith.

HABITAT: Moist soil along stream banks or in meadows; in wooded or open areas, rocky or good soil.

RANGE: Alaska across central Canada to Labrador and Newfoundland, south to the District of Columbia, west to Ohio, Missouri, Nebraska, and Colorado, south and west to New Mexico, Arizona, and California, and north to British Columbia.

A few clusters of plants in Roberts County, South Dakota, were found to start flowering about ten days later than the other plants growing nearby. This was observed in three separate seasons. The fruits of those plants were smaller, matured later, and were much firmer.

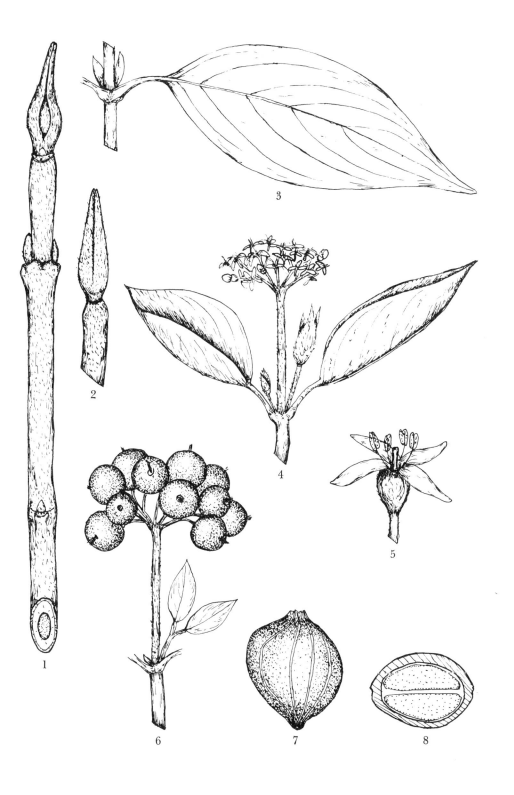

1. Winter twig with terminal flower bud
2. Twig with terminal leaf bud
3. Leaf
4. Leaves and inflorescence
5. One flower
6. Fruit cluster
7. Seed
8. Transverse section of seed

ERICACEAE
Chimaphila umbellata (L.) Bart.
var. *occidentalis* (Rydb.) Blake

Chimaphila occidentalis Rydb.

Prince's pine, wintergreen, wax flower,
pipsissewa

A low plant with creeping rootstalk, above ground stem only slightly woody, 10-20 cm high, leaves on the upper portion.

LEAVES: Subverticillate, simple, evergreen, thick, coriaceous, the lateral veins impressed above and inconspicuous below. Oblanceolate, 4-7 cm long, 1-1.7 cm wide, tip acute, the base tapered, margin sharply serrate, 8-18 callous-tipped teeth per side; upper surface dark, glossy green, glabrous, the lower surface pale green, glabrous; petiole 5-7 mm long, grooved above, usually forming a narrow, acute angle with the stem; no stipules.

FLOWERS: July-August; perfect. Inflorescence racemose, terminal; peduncle 4-5 cm long, 2-7 flowers, pedicels 10-13 mm long, glabrous or puberulent, purplish, an acicular bract 4 mm long below the flower; flowers nodding, 11-13 mm across; hypanthium shallow saucer-shaped, green, 2.5 mm across, 1-1.5 mm deep; calyx lobes 5, broadly ovate, purplish-green, usually with minute, transverse wrinkles, spreading or reflexed, the margin whitish, minutely erose and ciliate; petals 5, pale rose, obovate, 7 mm long, 5 mm wide, concave with the rounded tip recurved, the margin whitish, finely erose or ciliate; stamens 10, filaments pink, 3-3.5 mm long, the basal portion enlarged, flat and ciliate, attached at the base of the anther body above the two tubes bearing apical pores, the anther is inverted, which gives a terminal appearance to the pore, anthers brownish-purple, about 2 mm long including the tubes. Ovary 5-lobed, 3 mm long, 5 mm wide, green; style short, stout, mainly embedded in the top of the ovary; stigma a flattened, spongy disk, 2 mm across, dark green becoming red-brown.

FRUIT: Peduncle 5-7 cm long, pedicels 15-20 mm long, ascending, slender; fruits subglobose, 5-loculed, splitting loculicidally, dark brown; innumerable seeds. Seeds golden brown, total length less than 1 mm, the embryo portion 0.1 mm long in the center of a thin, linear membrane, the end of which is attached to the placenta. Often the fruits of the preceding year are still present at the time of flowering.

TWIGS: 1.7-2 mm diameter, green; young twigs with deciduous triangular to lanceolate bracts; leaf scars on the lower stem half-round, appearing calloused; 1 central bundle scar; bract scars narrowly crescent-shaped; pith white, coarsely porous, three-fourths of stem. Buds axillary at the apex of the stem, ovoid, 6-7 mm long, 1 mm wide, the scales green, thin, coarsely ciliate, often apiculate.

TRUNK: "Bark" mostly green, brownish toward the base, glabrous. Wood soft, white, with a large, white pith.

HABITAT: Moist areas on wooded hillsides in the Black Hills.

RANGE: Quebec across Canada to Alaska, south to southern California, east to New Mexico, north through Colorado to central Manitoba, and around the Great Lakes in the United States. Isolated in the Black Hills.

Chimaphila is a small suffrutescent plant, the glossy, evergreen leaves standing out sharply after a light snow. In the Black Hills it grows in moist valleys beneath a mixed stand of *Pinus ponderosa* and broad-leaved trees. In all the locations found, it was in granite soil and often around granite boulders near the bottom of a shallow ravine on a hillside.

424

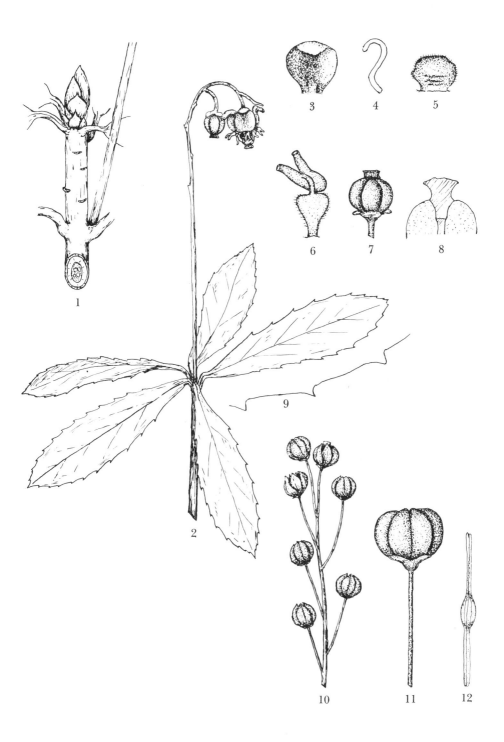

1. Winter twig, indicating leaves and old fruit stalk
2. Leaves and flowers
3. Petal
4. Diagram of longitudinal section of petal
5. Sepal
6. Stamen
7. Pistil
8. Diagram of longitudinal section through style
9. Leaf margin
10. Fruit cluster
11. One fruit
12. Seed

ERICACEAE
Arctostaphylos uva-ursi (L.) Spreng.

Uva-ursi uva-ursi (L.) Britt.; *Uva-ursi procumbens* Moench; *Arbutus procumbens* Patze.; *Arbutus uva-ursi* L.

Bearberry, kinnikinick, mealberry

A prostrate, vine-like plant, often densely covering an area 3 m across, rooting at the nodes.

LEAVES: Alternate, simple, evergreen, becoming reddish in winter; coriaceous; pinnately veined with obvious veinlets. Spatulate to obovate, 1.5-2.5 cm long, 7-9 mm wide, tip rounded, the base long-tapered; margin entire, finely and often densely ciliate; glabrous on both sides, dark green and glossy above, paler beneath; petiole 2-4 mm long, tomentulose; no stipules.

FLOWERS: Early June; perfect. Racemes terminal, 1-2 cm long, with 3-6 flowers; peduncle 6-10 mm long, curved, the bracts at the base of the pedicel lanceolate to ovate, 2-4 mm long, green; pedicels 2-3 mm long, white, glabrous; calyx lobes 5, rounded, 1.5-2 mm long, 1.5 mm wide, white or pink, distinct; corolla urn-shaped, 6-7 mm long, 5 mm wide, white, glabrous outside, pubescent inside, lobes 5, broadly rounded, 0.7 mm long, 2 mm wide; disk purplish and lobed; stamens 10, filaments white, 2 mm long, inserted below the disk, a pubescent swelling on the filament near the base and 2 reflexed, muriculate awns at the junction of the filament and anther; anthers broad, oval, red-purple, inverted so the filament appears to be attached at the outer end; ovary green, broadly ovoid, 1 mm long, 1.25 mm wide; style 3 mm long, stout, wider at the apex, green; the stigma capitate.

FRUIT: September. Pedicels 3-4 mm long, green, glabrous; fruit globose or slightly depressed, red, smooth, semiglossy, 7-10 mm diameter, mealy and not palatable; 5-8 seeds per fruit. Seeds brown, the shape of an orange section, 4.3-4.8 mm long, 2.9-3.1 mm wide, 2-3.6 mm thick, longitudinally ridged, the surface rough; often the seeds do not break apart but remain as a depressed sphere 3.5-4 mm long, 5-6.5 mm broad, or tear-drop-shaped 5.8-6.3 mm long and 4.8-5.5 mm thick.

TWIGS: 1 mm diameter, reddish-brown, tomentulose or glabrous, flexible; bark smooth for 3-4 seasons and then exfoliates into thin flakes; leaf scars crescent-shaped with the ends rounded; pith greenish, continuous, one-third of stem. Buds 1 mm long, ovoid, slightly pubescent, light red; flower buds in the axils of the bracts at the tip of the stem, globular, 0.5 mm diameter.

TRUNK: Bark exfoliating as thin, brown sheets. Wood hard, white, fine-grained.

HABITAT: Open hillsides, open woods, usually on rocky or sandy soil in semidry areas.

RANGE: Subarctic regions from Yukon to Newfoundland, south to Virginia, west to Illinois, South Dakota, Colorado, and New Mexico, northwest to northern California and British Columbia. Although the range is broad, there are large areas within it where no plants exist.

This is the only species of *Arctostaphylos* in our area and also the only vine-like, evergreen plant with broad leaves and forming large, dense mats. This, combined with the spatulate leaves and urn-shaped flowers or the red fruits, should be sufficient for identification. It enters the Prairie Province from the Northern Conifer Forest but is also quite common in the Rocky Mountains.

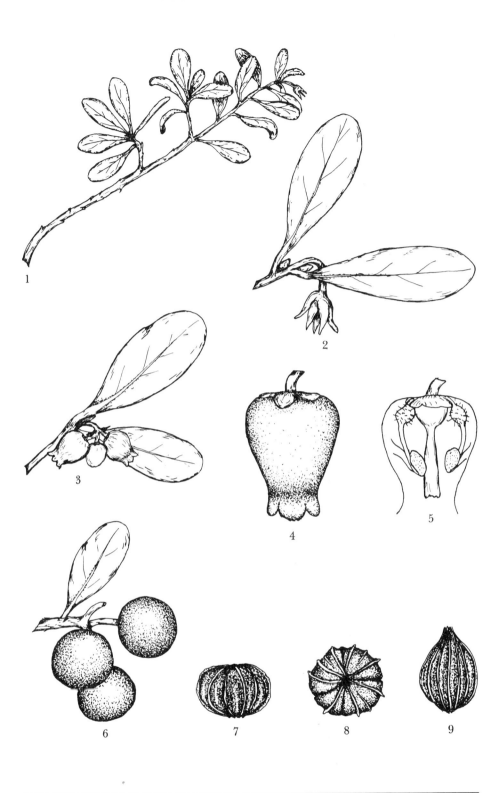

1. Winter twig
2. Tip of twig
3. Leaves and flowers
4. One flower
5. Diagram of flower
6. Fruits
7. Side view of seed group
8. End view of seed group
9. Side view of seed group variation

ERICACEAE
Vaccinium arboreum Marsh.

V. arboreum Marsh. var. *glaucescens* (Greene) Sarg.; *Batodendron glaucescens* Greene; *B. andrachneforme* Small; *B. arboreum* (Marsh.) Nutt.

Farkleberry

A shrub to 3 m high, often with crooked, zigzag branches.

LEAVES: Alternate, simple; thick, coriaceous. Obovate to elliptic, 3-6 cm long, 1.5-3 cm wide; margin entire; tip rounded or acute, base obtuse to cuneate; upper surface dark green, often with some red, lustrous, the lower surface paler green, both sides with a few hairs on the midrib; petiole 3-3.5 mm long, pubescent. Leaves quite variable in size and shape, even on the same branch, whether it is a flowering or a foliage branch; no stipules.

FLOWERS: Late May; perfect. Flowers in loose, axillary, leafy racemes 5-8 cm long, the leaves or leaf-like bracts of various shapes; flowers pendent on slender, glabrate pedicels 7-8 mm long, jointed at the summit and with a lanceolate, deciduous bract about the middle. Hypanthium obovoid, 1-1.5 mm long, pale green, glabrous; calyx lobes 5, deltoid with the sides slightly concave, 1 mm long, nearly white, ciliate near the tip; corolla urceolate, white, 4.5-4.7 mm long, 4.3-4.4 mm wide; lobes 5, acuminate but only 2.5-2.7 mm long, sharply recurled, the tip occasionally rounded; stamens 10, not exserted, attached to the corolla tube; filaments 1.2-1.5 mm long, flattened, gibbous on the ventral side, pubescent along the sides, 2 awns 0.5-1 mm long on the dorsal side at the point of attachment with the anther; anthers yellow, 2-loculed, 1.4 mm long, opening on the ventral side at the outer end of the tubular tips; disk dark green, 10-lobed; style 4 mm long, slightly exserted, white, greenish on the outer end, tapered; stigma green, hardly larger than the style summit.

FRUIT: September-October. Racemes 5-8 cm long, the bracts usually deciduous by the time the fruit is ripe; pedicels 7-10 mm long, glabrate or glabrous; fruit depressed globose, 8-9 mm long, 9-10.5 mm wide, dark blue-purple with a slight bloom, the old calyx lobes appressed to the fruit; flesh mealy, edible but not palatable; many seeded. Seeds of various shapes, generally oval but with odd angles and flattened sides, 1.5-2.6 mm long, 1.1-1.8 mm wide, 0.7-0.9 mm thick, golden-brown, glossy, deeply pitted with small, oval pits.

TWIGS: 1.2-1.6 mm diameter, rigid, red-brown, pubescent with slightly curled, white hairs; leaf scars crescent-shaped; 1 bundle scar; pith pale brown, continuous, one-third of stem. Buds short ovoid, 1.4 mm long, red-brown, acute, glabrous, often glaucous.

TRUNK: Bark dark brown, with fine fissures exposing the reddish inner bark; younger branches reddish, with brown outer bark peeling off in flat, thin sheets. Wood hard, fine-grained, light red-brown.

HABITAT: Acid soils, rocky open woods, upland slopes or along creek banks.

RANGE: Illinois, west to southeastern Kansas, south to Texas, east to Alabama, and north to Kentucky.

Vaccinium arboreum differs from the other *Vaccinium* plants of our area by being a tree-like shrub growing to a height of 3 meters. It is usually found in loose thickets. Some of the plants are tall, with a rounded crown, and others are somewhat flat-topped and have crooked, zigzag branches. It flowers abundantly, but is sparsely fruited, the fruits ripening over a long period. It is not a dense shrub, and birds seldom use it for nesting; however, birds do eat the fruits as soon as they ripen.

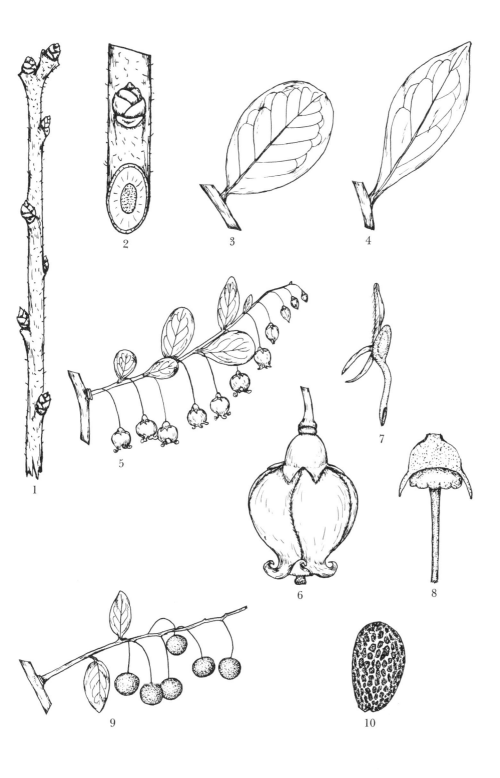

1. Winter twig
2. Section of twig
3. Leaf
4. Leaf variation
5. Flowering raceme
6. Flower
7. Stamen
8. Pistil and disk
9. Fruit
10. Seed

ERICACEAE
Vaccinium membranaceum Dougl.
ex Hook.

V. macrophyllum Piper; *V. myrtilloides*
Michx. var. *macrophyllum* Hook.

Bilberry, mountain huckleberry

A low shrub to 40 cm high, often densely branched and in small colonies.

LEAVES: Alternate, simple. Broadly elliptic to occasionally oval, 2.5-4.5 cm long, 1.5-2.5 cm wide, veins slightly impressed above; tip acute or occasionally rounded, base cuneate; margin finely toothed with low teeth 6-8 per cm, a linear gland pointing forward on the tip of each tooth; both surfaces glabrate with an occasional red gland; petiole 2-3 mm long, broadly grooved above, sparsely puberulent; no stipules.

FLOWERS: Early June; perfect. Flowers single, axillary on new growth; pedicels usually curved, 6-8 mm long, green, glabrous, broadened toward the summit, not jointed; hypanthium 1-1.5 mm long, 4 mm wide, expanded to an undulate, thin, green, disk-like calyx; corolla urn-shaped, 5-5.5 mm long, 5.5-6 mm wide, or depressed urn-shaped, 3-3.75 mm long, 6-7 mm wide, the opening 2-3 mm across, the flowers pink, glabrous, with 5 broad, sharply recurved lobes 0.5 mm long and 1.5 mm broad; disk thick, greenish, 10-lobed; 10 stamens attached below and between the disk lobes; filaments tapered, 1.5 mm long, flat, white, glabrous; anthers orange, the main body 0.75 mm long, the 2 tubes 0.75 mm long, and the 2 awns 0.75-1 mm long at the base of the tubes and curved upward; style stout, 5 mm long, green, glabrous; stigma capitate.

FRUIT: July. Pedicels drooping, 7-10 mm long, glabrous; berry depressed globose, 6-8 mm long, 8-10 mm wide, the outer end flattened and with a trace of the old calyx, fruit dark red-purple, smooth, semilustrous or with a very thin bloom, flesh mealy but delicious; 15-30 seeds. Seeds the shape of an orange section, 1.3-1.5 mm long, 0.6-0.7 mm across, yellow, minutely striate.

TWIGS: 1.3-1.8 mm diameter, angular, orange-brown, finely puberulent, the tip of the twig often extended as a blunt point about as long as the end bud; lenticels small and dark; leaf scars broadly crescent-shaped, raised, darker than the twig; 1 bundle scar; a ridge extending downward from each side of the leaf scar; pith green to brownish, continuous, one-third of stem. Buds ovoid, 2-3 mm long, 1-1.5 mm wide, the two outer scales keeled, orange-brown, glabrous, the end bud largest.

TRUNK: Bark of most of the plant green or brownish, but red-brown toward the base, the whole thing usually drying to a red-brown; bark tight but slightly fissured. Wood greenish-white, porous, no differentiation of sapwood, but with a small, firm, greenish, central pith.

HABITAT: Rocky hillsides and mountain slopes, usually in open woods, either moist or dry soil.

RANGE: Ontario, west across northern Michigan to British Columbia, south to California, east to Montana, and in the Black Hills of South Dakota.

Vaccinium membranaceum is not common in the Black Hills, being found mostly in the Lead-Deadwood area. It may be in small, open colonies or as scattered plants toward the top of a hill and above 5000 feet elevation. On Terry Peak it grows in the open woods near the top where the soil is quite rocky and becomes dry in the summer. However, the only spots in which it was located were on the north slopes or in shallow valleys where the snow becomes deep and melts late. Near Deadwood, it was located in a rather dense woods and in moist soil.

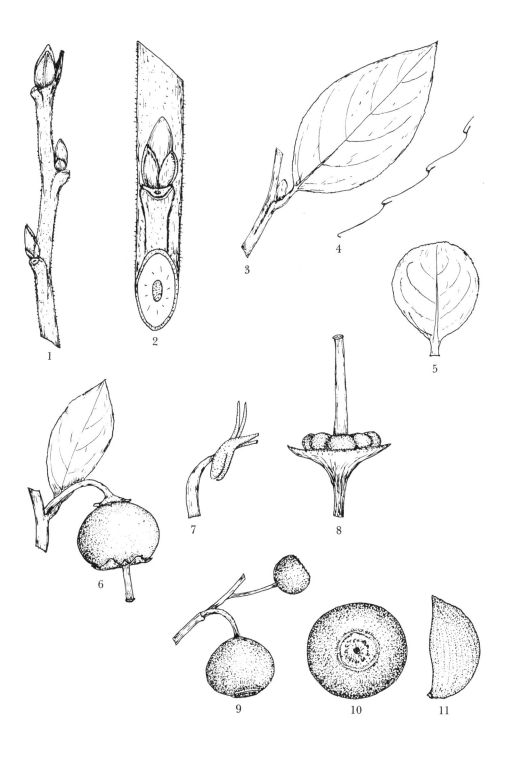

1. Winter twig
2. Detail of twig
3. Leaf
4. Leaf variation
5. Leaf margin
6. Flower

7. Stamen
8. Calyx, disk, and pistil
9. Fruit
10. Apex of fruit
11. Seed

ERICACEAE
Vaccinium scoparium Leiberg

V. myrtillus L. var. *microphyllum* Hook.;
V. microphyllum Rydb.; *V. erythrococcum* Rydb.

Grouseberry, whortleberry

A low undershrub to 35 cm high, with many small branches, and forming large colonies.

LEAVES: Alternate, simple, the lateral veins diffusely branched with many transverse, connecting veinlets. Ovate to elliptic, 10-15 mm long, 6-9 mm wide; tip acute, base rounded or cuneate; finely serrulate, 12-14 teeth per cm, teeth low with a long basal side, the tip incurved and with a transparent, linear gland at the tip; glabrous or nearly so on both sides, hardly paler beneath, occasionally a few short, curled hairs on the midrib; petiole 1-1.5 mm long, flattened above, blade attenuate most of its length; no stipules.

FLOWERS: Early June; perfect. Flowers single, axillary; pedicels 1-2 mm long, glabrous, not jointed; hypanthium shallow turbinate, 1-1.5 mm long, flared at the top to 3 mm wide, green and glabrous, lobes, if present, shallow and broad; disk green, lobed, surrounding the style; corolla urn-shaped, 4-5 mm long, 4-5 mm wide, pink and white, lobes ovate, 0.5-1 mm long, sharply reflexed or recurled; stamens 10, inserted below the disk; filaments 1 mm long, flat, greenish, curved, glabrous, attached to the anther just below the center; anthers orange-yellow, 1.5 mm long, with 2 terminal tubes 1.5 mm long, the pores at their ends, 2 spreading prong-like awns 1 mm long at the base of the tubes; style 4 mm long, stout, glabrous, slightly tapered, green; stigma capitate.

FRUIT: July-August. Pedicels drooping; berry depressed globose, 6-7 mm long, 7-9 mm diameter, deep red, glabrous, glossy, 10-20 seeds. Seeds the shape of an orange section, yellow-brown, longitudinally striate, 1.3-1.4 mm long, 0.7 mm wide.

TWIGS: 0.8 mm diameter, greenish or reddish, flexible, ridged, and broadly grooved, appearing twisted, glabrous or with a few short, curled hairs; the tip of the twig extending as a point along the end bud; lenticels small, brown, inconspicuous; leaf scars a half circle, small, raised; 1 bundle scar; pith green, continuous, one-half of stem. Buds ovoid, acute, the scales thin, glabrous, pale green; end buds 2.5-3 mm long, 1.5 mm wide, buds contain the embryonic stem and leaves and at least one well-formed flower; lateral buds 1.2 mm long, 0.6 mm wide, flattened, the tip obtuse.

TRUNK: Bark of branches remain green, that of the main trunk red-brown, tight but with small slits. Wood greenish-white, soft, porous, with a small, firm, green pith.

HABITAT: Moist, wooded hillsides or mountain sides and ravine banks; also in open woods in dry, rocky soil at the top of some of the peaks in the Black Hills.

RANGE: British Columbia, east to Alberta, Idaho and Montana, south through the Rocky Mountains of Colorado; and in the Black Hills of South Dakota.

Vaccinium scoparium forms large colonies and often covers the floor of a woods over an extensive area. It is most abundant on moist slopes in an open woods, often where the snow accumulates and melts late in the spring. The berries are eaten by such mammals as the chipmunk, and small birds are seen frequenting the shrubs during fruiting time.

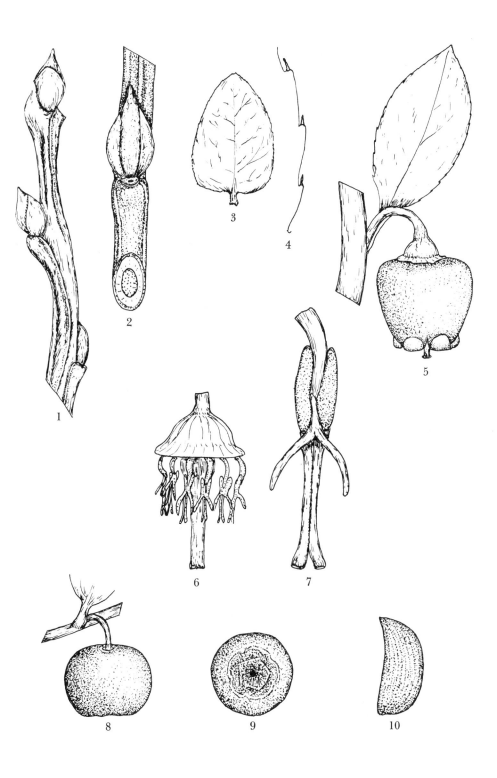

1. Winter twig
2. Detail of twig
3. Leaf
4. Leaf margin
5. Flower
6. Flower with corolla removed
7. Stamen
8. Fruit
9. Apex of fruit
10. Seed

ERICACEAE
Vaccinium stamineum L.

Polycodium stamineum (L.) Greene

Buckberry, squaw huckleberry, deerberry

A diffusely branched shrub to 1 m high.

LEAVES: Alternate, simple, veins slightly impressed above. Elliptic to slightly broader above the middle, 4-8 cm long, 1.5-3 cm wide; tip acuminate to acute, the lower portion tapered to a narrow, rounded base; margin entire, ciliate; upper surface yellow-green, glabrous with a few hairs on the midrib, lower surface paler, glabrous or nearly so, with the veins pubescent; petiole 2-3 mm long, flattened to trough-shaped above, glabrous or pubescent; no stipules.

FLOWERS: Late April; perfect. Floral branches often without regular leaves, only bract-like leaves. Racemes 4-6 cm long, finely pubescent, 3-10 flowers, bracts narrowly ovate, 4-17 mm long, 2-6 mm wide, ciliate; pedicels pendent, slender, 8-14 mm long, not jointed, a few scattered hairs; hypanthium turbinate, green glabrous, 2 mm long, 3 mm wide; calyx lobes 5, triangular, rounded or acute, ciliate; corolla open campanulate, 5-6 mm long, white to pinkish, lobes 5, ovate, 2 mm long, spreading; stamens 10, inserted on the base of the corolla; filaments broad, 2 mm long, flat, green, pubescent; anthers orange-brown, 1.5 mm long, with 2 yellow-brown tubes 4 mm long, the pores at their ends, 2 spreading awns at the base of the tube 1-1.5 mm long, orange; the tubes extruded beyond the corolla; exposed top of the ovary 1 mm high, 2 mm wide, green; style tapered, 7-9 mm long, longer than the anther tubes; stigma no wider than the style.

FRUIT: July-August. Racemes pendent. Fruits globular, 9-12 mm diameter, the base often somewhat conical, red-purple to dark purple, with or without a bloom, glabrous, sweet; calyx persistent; 4-8 mature seeds per fruit. Seeds irregularly ovoid, 1.5 mm long, 1 mm wide, 0.8 mm thick, golden-brown, finely pitted.

TWIGS: 1-2 mm diameter, dark red, glabrous or finely curled pubescent, flexible; lenticels small and circular or large and elliptic; leaf scars half-round, light-colored; 1 bundle scar; pith greenish, firm, continuous, one-fourth of stem. Buds short ovoid, 1.6-2 mm long, 1.6-1.9 mm wide, obtuse, usually at a 45° angle, scales red to red-brown, often with a grayish tip, tight except for the spreading apiculate tip of the outer scales.

TRUNK: Outer layer of bark gray-brown, split into long, narrow, papery strips with loose margins, inner bark red-brown. Wood soft, fine-grained, white to pale brown, the rays obvious.

HABITAT: Dry, wooded hillsides and open areas in the woods, often on chert rock soils.

RANGE: Maine to Indiana, Missouri, and southeast Kansas, south through eastern Oklahoma to Louisiana, east to Florida, and north to New England; also southern Ontario.

Since no fresh plants could be located in our area, the above description and accompanying drawings were made from Oklahoma specimens a few miles south of the Kansas border. Only one herbarium specimen was located from our area and the location given for it is at the Oklahoma line. That specimen is var. *interius* (Ashe) Palmer & Steyerm., with glabrous calyx and under leaf surface and puberulent twigs. Two other varieties often listed are var. *melanocarpum* Mohr., with pubescent leaves and white tomentose calyx; and var. *neglectum* (Small) Deam, with glabrous branches and leaves, the leaves glaucous beneath. The latter two have not been reported for our area.

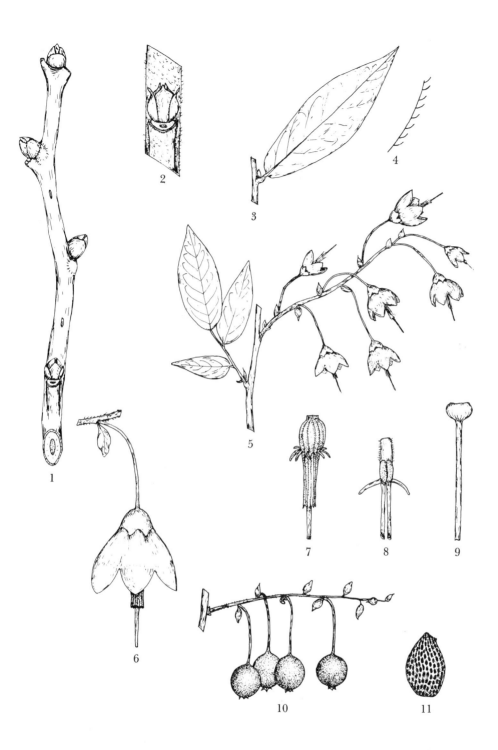

1. Winter twig
2. Detail of twig
3. Leaf
4. Leaf margin
5. Flowering raceme
6. Flower
7. Stamens surrounding the pistil
8. Stamen
9. Pistil
10. Fruit
11. Seed

ERICACEAE
Vaccinium vacillans Torr.

V. pallidum Ait.; *Cyanococcus vacillans* (Kalm) Rydb.; *C. pallidus* (Ait.) Small

Lowbush blueberry, lowbush huckleberry

A colonial shrub to 50 cm high.

LEAVES: Alternate, simple. Elliptic, 25-43 mm long, 12-20 mm wide; margin ciliate, entire or minutely serrulate, with a linear gland at the tip of each tooth; tip acute or short acuminate, mucronate; base cuneate; upper surface yellow-green, often with a reddish margin, glabrous, the veins often more obvious above than below; lower surface pale green, pubescent or glabrous; petiole 1-2 mm long, pubescent; no stipules.

FLOWERS: Mid-April; perfect. Flowers clustered on short racemes from buds on wood of the previous year, usually near the end of the branch; pedicels 2-2.5 mm long, green, glabrous, an obovate bract 3 mm long at the base and 2 bracts on the pedicel, these 1.5 mm long, ovate and acuminate; hypanthium campanulate, about 0.5 mm long, green, glabrous; calyx lobes 5, acute, 1.2-1.5 mm long, green, glabrous, ciliate, often a minute tooth on each side near the base; corolla tubular or barrel-shaped, 5-7 mm long, 3-4 mm diameter, white, pink, or red, lobes 5, obtuse, 1 mm long, sharply recurled; stamens 10, not exserted; filaments flattened, 2.2 mm long, pale green, pubescent, incurved so the anthers are close around the style; anthers 2-loculed, 2.2 mm long, each locule tapering to a slender tube with a lateral opening at the end; stamens inserted on the corolla and deciduous with it; disk flattened, 2 mm wide, 5-lobed, green; ovary embedded in the disk; style from the center of the disk, 5.5 mm long, tapered; stigma lobed, exserted.

FRUIT: Early July. Pedicels 2-3 mm long; berry depressed globose, 6-9 mm diameter, clustered on a short raceme, dark blue or nearly black with a bloom; sweet, palatable, ripening over a long period; calyx lobes persistent; many seeds per fruit. Seeds irregular and variable, 1-1.6 mm long, 0.7-1.3 mm wide, glossy, red-brown, pitted.

TWIGS: 1.4-1.7 mm diameter, flexible, green or red, rough with minute dots, the curly pubescence heaviest in a line extending down from each side of the leaf scar; leaf scars narrowly crescent-shaped, fringed with white hairs on the lower margin; 1 bundle scar; pith green, continuous, one-half of stem. Buds red, the outer scales keeled and apiculate, the tip often spreading; flower buds 3 mm long, plump; leaf buds 1.9 mm long, flattened.

TRUNK: Bark greenish-brown or red, smooth, often slightly ridged. Wood soft, white.

HABITAT: Upland ridges and slopes; dry, rocky, open woods; acid soils.

RANGE: Nova Scotia, southwest to Michigan and Iowa, south to southeast Kansas and Oklahoma, east to Georgia, and north to New England.

Two varieties have been described for our area, but the characteristics given do not always hold true. Var. *vacillans* has glabrous twigs and leaves, while var. *crinitum* Fern. has pubescent twigs and leaves. Most of our plants come closer to var. *crinitum*.

This plant forms dense colonies on rocky hillsides. It branches freely, the branchlets often at right angles to the main stem. It produces an abundance of fruit, but the long ripening period makes it inconvenient for people who wish a quantity at a time. This is, however, an advantage to birds and small animals. The fruits are eaten raw or made into pies and jams.

During the winter months the bushes may be partly hidden by the mass of tree leaves which lodge in the thickets. This mulch affords good cover for small mammals and also saves the soil moisture and retards erosion.

436

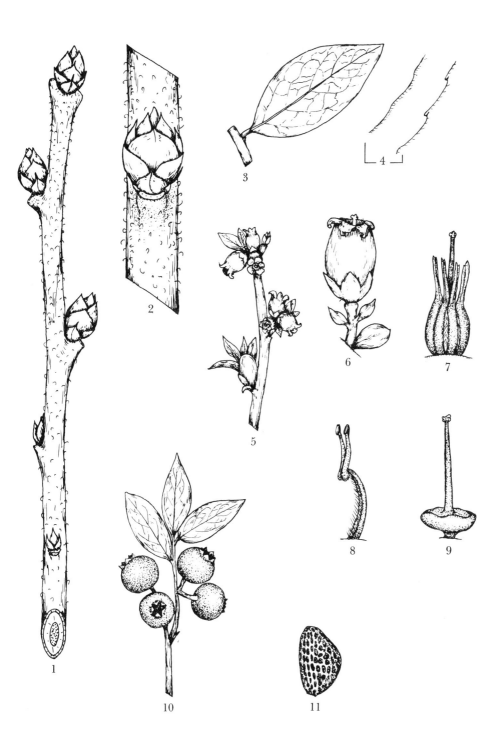

1. Winter twig
2. Detail of twig
3. Leaf
4. Leaf margin variation
5. Flower clusters
6. Flower
7. Stamens and pistil
8. Stamen
9. Pistil and disk
10. Fruit
11. Seed

SAPOTACEAE
Bumelia lanuginosa (Michx.) Pers.
var. *oblongifolia* (Nutt.) R. B. Clark

B. lanuginosa var. *albicans* Sarg.

Woolly buckthorn, false buckthorn,
southern buckthorn, chittim-wood

A rugged tree to 12 m high, with rounded crown.

LEAVES: Alternate or clustered at the end of a spur, simple. Obovate to spatulate, 4-8 cm long, 2-3.5 cm wide, entire; tip rounded, occasionally with a minute mucro; base cuneate or long-tapered; upper surface dark green, glossy, with some pubescence on the midrib; lower surface densely woolly with rusty or nearly white hairs; thick, leathery; petiole 5-10 mm long, tomentose with buffy hairs; no stipules.

FLOWERS: Early July; perfect. In dense clusters of 20-30 flowers, axillary along the branches or on the end of spurs; a sticky, milky juice in the pedicels and calyx, pedicels green, 6-8 mm long, rusty pubescent; calyx lobes 5, distinct, oval, 2.8 mm long, 1.9 mm wide, stiff and thick but with thin edges, incurved, a rusty pubescence outside; petals 5, white, 4 mm long, connate for 1.5 mm; each petal 3-lobed, the center lobe ovate, the laterals narrowly ovate with irregular tips; staminodes 5, white, 2 mm long, 1.5 mm wide, petal-like, alternate with the petals, attached at the upper end of the corolla tube, somewhat keeled, the margins erose; stamens 5, the anthers yellow; ovary globose, 1 mm long, densely pubescent with long, straight, white, somewhat appressed hairs; style and stigma 1.7 mm long, greenish, glabrous.

FRUIT: October. Single or clustered; pedicels 6-8 mm long, enlarged at the summit, calyx lobes persistent; fruit ovoid to obovoid, 10-13 mm long, 8-8.5 mm thick, glabrate when green, becoming glabrous, smooth and glossy black; 1-seeded. Seeds obovoid, 7.8-9.1 mm long, 4.6-5.3 mm thick, glossy, rich brown, with a light scar.

TWIGS: 2.5-3 mm diameter, rigid, tough, dark gray; usually a stout spine at the leaf base, or the short branches ending in a spine; young twigs rusty tomentose; leaf scars short, crescent-shaped, often hidden by the hairs; 5 bundle scars; pith hard, indistinct, one-fifth of stem. Buds 2-3 mm long, lateral on the main twig or on spurs, broadly ovoid, red-brown, densely pubescent.

TRUNK: Bark nearly black, with shallow furrows, the ridges blocky and anastomosed. Wood hard, heavy, yellowish, with a narrow, light sapwood.

HABITAT: Bluffs; ravines; dry, open, rocky woodlands; and stream flood plains.

RANGE: Florida to Texas, north to Kansas and Missouri.

Bumelia lanuginosa is often a low, rounded tree, but occasionally it reaches a height of 12 meters. The branches are rugged and quite crooked, forming a rounded, irregular crown. It is the latest of our southern native trees to produce flowers; although the flower buds appear in early June, they do not open until July. The cave-dwelling wood rats often take quantities of the leaves into their dens. These leaves have been taken from the dens in January, thoroughly dry but green and fresh looking.

Two completely integrated varieties occur in Kansas: var. *oblongifolia*, described above; and, rarely, var. *albicans* Sarg., which has long, silvery-white hairs on the lower leaf surface. The range of var. *lanuginosa* is considered to be entirely east of the Mississippi River.

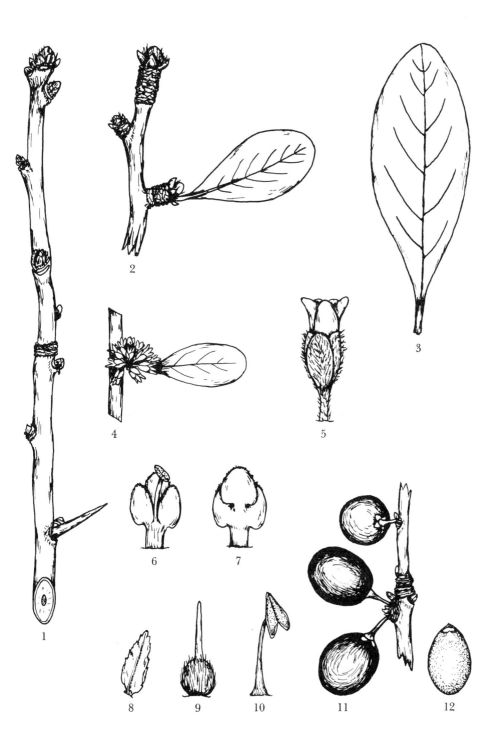

1. Winter twig
2. Twig with spur branches
3. Leaf
4. Cluster of flowers
5. One flower
6. Petal, ventral view
7. Petal, dorsal view
8. Staminode
9. Pistil
10. Stamen
11. Fruit
12. Seed

EBANACEAE
Diospyros virginiana L.

Persimmon, date plum

Tree to 20 m high, but more often found as an open tree up to 10 m high, either single or in colonies.

LEAVES: Alternate, simple. Ovate to oblong, 7-13 cm long, 3.5-8 cm wide, entire, often ciliate; tip short acuminate, base cuneate or obtuse; upper surface dark green, glabrous; lower surface pale, glabrate or pubescent; petiole 7-10 mm long, glabrous or pubescent; stipules absent.

FLOWERS: Late May; dioecious. Staminate flowers axillary on new growth, solitary or in clusters of 2-3; peduncle 2 mm long, green, pubescent; pedicels 3 mm long, green, pubescent; calyx lobes 4, connate at the base, 2.8-3.1 mm long, acute, green, pubescent; corolla urceolate, pale yellow, with 4 darker yellow lobes, the tube 3-3.5 mm long, lobes 2.5-3 mm long, acute, spreading or recurved; stamens 16, attached in 2 rows at the base of the corolla tube; filaments 0.5-1.2 mm long, white, sharply curved, pubescent at the top, those of the outer row longest; anthers 4.5-5 mm long, yellow, pointed, glabrous or pubescent, opening on both sides near the tip; disk 8-lobed, brownish; vestigial pistil in the center. Pistillate flowers axillary on new growth, solitary; pedicels, 1.5-2 mm long, stocky, green, pubescent; 2 pubescent, lanceolate bracts 6-7 mm long on the pedicel; calyx lobes 4, connate at the base, 7.4-7.5 mm long, 6.9-7 mm wide, tip blunt, silvery pubescent inside, ciliate; corolla urceolate, 10-13 mm long, pale yellow; lobes 4, ovate, 4.5-6 mm long, dark yellow, overlapping at the base; vestigial stamens 6-10; ovary broadly ovoid, 3-3.5 mm long, 4 mm wide, green, glabrous; styles 4, connate half their length, 5.8-6 mm long, greenish, glabrous or pubescent at the base; stigma brownish, capitate; the 8-lobed disk thin, brown, and pubescent at the tips of the lobes.

FRUIT: September-October. Pedicels 4-6 mm long, stout, pubescent; calyx accrescent, persistent; base of styles persistent; fruit globose, 3-6 cm diameter, often depressed, salmon-colored, slight bloom, smooth, but wrinkled when fully ripe; sweet, edible, do not need frost to ripen; 3-6 seeds. Seeds oval, flat, 1.8-2 cm long, 1.1-1.3 cm wide, 3.5-3.7 mm thick, red-brown, the surface finely granular.

TWIGS: 1.5-1.8 mm diameter, flexible, dark gray or brown, pubescent or glabrate, lenticels prominent; leaf scars half-round; 1 indistinct bundle scar; pith green, continuous, one-eighth of stem. Buds 2-3 mm long, 2 large scales cover most of the bud; scales dark brown, slightly pubescent, tip obtuse.

TRUNK: Bark dark gray, furrowed, the ridges short and blocky. Wood hard, dark brown, with a wide, white sapwood.

HABITAT: Rocky, dry fields, pastures, waste ground; rich bottom lands or open hillside woods.

RANGE: Connecticut, south to Florida, west to Texas, north to Kansas and Iowa, and east to New York.

Most of our trees are var. *virginiana*, with the under leaf surface glabrous or glabrate, but a few of them have pubescent under leaf surfaces. If the fruits on the pubescent forms are depressed globose and 5-6 cm in diameter, they are placed in var. *platycarpa* Sarg. If the fruits are oblong and 2-4 cm in diameter, and the leaves are cuneate at the base, they are placed in var. *pubescens* (Pursh) Dippel. These varieties are often hard to identify positively.

1. Winter twig
2. Cutaway of twig
3. Leaf
4. Staminate flower
5. Section through corolla tube, showing two stamens
6. Pistillate flower
7. Pistil
8. Fruit
9. Seed

OLEACEAE
Fraxinus americana L.

White ash

A tree to 20 m high, usually with a long, straight trunk.

LEAVES: Opposite, pinnately compound, 20-25 cm long, 5-7 leaflets; thin, leathery. Leaflets ovate to oval, 6-12 cm long, 4-6 cm wide; margin entire or serrate; tip acuminate, base cuneate to obtuse; upper surface dark green, semilustrous, glabrous; lower surface whitened, glabrous with a few hairs on the main veins; petiole 4-5 cm long, abruptly enlarged at the base; rachis enlarged above the petiolules, glabrous; terminal petiolule 2-3 cm long, laterals 5-12 mm long, mostly wingless and grooved above.

FLOWERS: Mid-April; dioecious. Staminate flowers in crowded, clustered panicles from buds of the previous year, 200-300 flowers per cluster; flowers opposite on the panicle branch, usually on a peduncle with 3 flowers; several small bracts in the inflorescence, green, often with a red fringe; immature flower clusters red-brown, becoming yellow to orange at anthesis; pedicels 3-3.5 mm long, green, glabrous; calyx minute with an irregular margin; petals none; stamens 2, the filament 0.3 mm long, anthers 2-2.5 mm long, red-brown to yellow-orange, apiculate tip. Pistillate flower clusters similar to the staminate; pedicels 2-3.5 mm long, green, glabrous; calyx 4-lobed, 1-1.4 mm long, green, glabrous; no petals; ovary dark green, ovate, 0.8 mm long, winged; style 0.7-0.9 mm long; stigma 2-cleft at the tip, reddish, 3.8-5 mm long.

FRUIT: July-August. 1-seeded samara, wing mostly terminal, extending no more than one-third of the distance along the fruit body; total length 3-4 cm, width 5-6 mm; fruit body 6-10 mm long, plump, slightly ridged, stramineous.

TWIGS: 3.5-4 mm diameter, rigid, brittle, glabrous, gray to yellow-brown; leaf scars narrow, U-shaped, slightly wider at the bottom, extending nearly to the top of the bud; row of 8-10 bundle scars;

pith white, continuous, one-third of stem. Terminal bud 4.5 mm long, 6.5 mm wide, laterals 2-3 mm long, dark brown, rough, scurfy.

TRUNK: Bark dark gray, furrowed, the ridges narrow and flat-topped. Wood tough, strong, heavy, nearly white, with a wide sapwood hardly distinguishable from the heartwood.

HABITAT: Upland in rocky woods, pastures, and along roadsides; in rich, low woods along streams and on flood plains.

RANGE: Nova Scotia to Minnesota, south to Texas, east to Florida, and north to New England.

The range of white ash in the central plains states is restricted to southeastern Nebraska and eastern Kansas. It is not as common as the green ash, although in local areas it may be abundant. The fall foliage of *F. americana* is usually a deep wine-red, but varies to orange-yellow. This is in contrast to the brilliant light yellow of *F. pennsylvanica* leaves. The two species are commonly confused; but *F. americana* has deep, U-shaped leaf scars, the plump seed portion on the samara has the wing hardly extended along its side, and the leaflets are broader.

Var. *biltmoreana* (Beadle) J. Wright is a more eastern variety, and has pubescent twigs, petioles, and leaf rachises. The specimens collected in Kansas and so labeled are questionable.

The wood is tough and is used for ball bats, tennis rackets, hockey sticks, polo mallets, tool handles, and garden furniture.

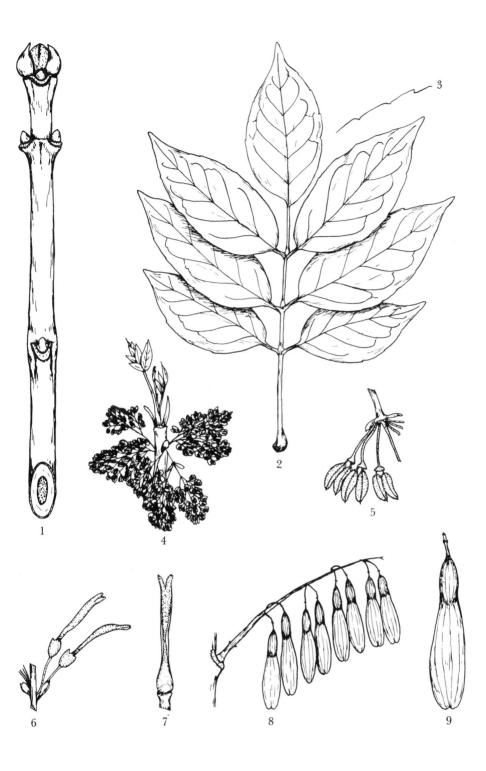

1. Winter twig
2. Leaf
3. Leaf margin
4. Cluster of staminate flowers
5. Three staminate flowers

6. Two pistillate flowers
7. Pistil
8. Portion of fruit cluster
9. One samara

OLEACEAE
Fraxinus nigra Marsh.

F. sambucifolia Lam.

Black ash, basket ash, swamp ash

A tree to 25 m high, the lower trunk usually straight and the branches high.

LEAVES: Opposite, pinnately compound. Leaves 20-30 cm long, 15-25 cm wide, 7-11 leaflets, the terminal leaflet with a petiolule to 2 cm, the lateral leaflets sessile; leaflets broadly elliptic or ovate, 8-14 cm long, 3-4 cm wide; second and third pairs the longest; margin serrate, the teeth low and blunt, 3-4 per cm; tip acuminate, the base rounded or broadly cuneate; dark yellow-green and glabrous above, the lower surface paler and glabrate, except for a buffy tomentum along the midrib; rachis narrowly grooved above, glabrous, or with occasional tufts of tomentum at the base of the leaflets; petiole 5-7 cm long, flattened or narrowly grooved on top; no stipules.

FLOWERS: May, just before the leaves; polygamous or dioecious. Axillary on growth of the previous season. Panicles 5-10 cm long, glabrous, the lanceolate bracts at the base of the branches, caducous. No calyx or corolla. Staminate flowers consist of 2 large, dark purple, acute anthers on short filaments. Pistillate flowers consist of a single ovoid pistil tapered to a long style with a 2-cleft, purplish stigma. (Fresh material not seen by the author.)

FRUIT: August. Panicles 10-20 cm long, drooping. Pedicels 5-8 mm long. Samaras oblong, 3.5-4 cm long, 8-11 mm wide, the seed portion flat, 1.5-2 cm long; wing yellow-brown, extending along the seed to the obtuse base, the apex usually notched.

TWIGS: 5-7 mm diameter, yellowish or gray-brown, glabrous; lenticels prominent, dark-colored on wood of the season, and light on older wood; leaf scars large, dark, semicircular, straight or slightly concave on the upper margin; many bundle scars forming a nearly closed ring; pith white, continuous, one-third of stem. Terminal bud conical, compressed, 6-7 mm long, 6-7 mm wide, 5-6 mm thick, the scales valvate, thick, scurfy, dark brown to nearly black, brown tomentose inside; lateral buds smaller and rounded at the tip.

TRUNK: Bark light gray, thin, scaly or flaky, occasionally with shallow furrows. Wood coarse-grained, soft, durable, easily split, brown, with a narrow, yellowish sapwood.

HABITAT: Swamps or moist, sandy soil, often at the edge of bogs; usually a tree of rather densely wooded areas.

RANGE: Newfoundland, west to Manitoba, south across the northeast corner of North Dakota to Iowa, east across Indiana to Virginia, and north to New England.

Fraxinus nigra is definitely a tree of low, wet ground and in our area appears only along the eastern edge of North Dakota. There is a possibility that it might be confused with *F. pennsylvanica,* and the best characteristic for separating the two is the broad-winged, flat fruits of *F. nigra.* The other differences are variable, such as more leaflets on *F. nigra* leaves, the leaflets are larger, and the tree grows in wetter areas.

The fruits are similar to those of *F. quadrangulata* Michx., but the two ranges do not meet in our area. Too, the twigs of *F. nigra* are terete or nearly so.

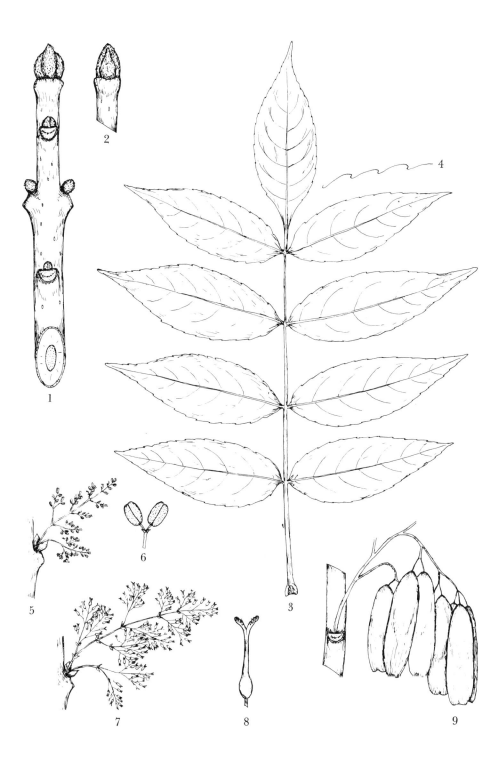

1. Winter twig
2. Variation in terminal bud
3. Leaf
4. Leaf margin
5. Staminate flower cluster
6. Staminate flower
7. Pistillate flower cluster
8. Pistillate flower
9. Fruit cluster

OLEACEAE
Fraxinus pennsylvanica Marsh. var.
subintegerrima (Vahl) Fern.

F. pennsylvanica var. *lanceolata* (Borkh.)
Sarg.; *F. lanceolata* Borkh.; *F. viridis*
Michx. f.; *F. campestris* Britt.

Green ash

A tree to 20 m high, with a large straight trunk and high branches.

LEAVES: Opposite, pinnately compound. 15-28 cm long, 5-9 leaflets. Leaflets ovate to elliptic, 6-10 cm long, 2-5 cm wide; serrate with low teeth, 2-3 per cm; tip acuminate, base cuneate, often unequal; upper surface dark green, lustrous, glabrous; lower surface paler, pubescent on the midrib and in the vein axils; thin, firm, leathery; petiole 4-5 cm long, glabrous, grooved above, rachis glabrous and grooved, slightly enlarged beyond the petiolules; terminal petiolule 1.5-2.5 cm long, lateral leaflets subsessile, the blade decurrent; no stipules.

FLOWERS: Mid-April; dioecious. Staminate flowers in dense panicles from buds of the previous year; pedicels 2.5-3.5 mm long, glabrous; calyx 4-lobed, distinct, minute, acute, green and glabrous; corolla none; stamens 3-5, filaments 0.5 mm long; anthers 3.5 mm long, reddish, usually purplish before opening, apiculate. Pistillate flowers in short panicles from buds of the previous year; panicle 4.5-5 cm long, glabrous, flowers opposite on the branches; peduncles bracted, 3-flowered; pedicels 2.5-3 mm long; calyx 2-4 lobed, 1.5-1.7 mm long, lobes distinct or united at the base, green with a yellowish fringe; ovary compressed ovoid, 0.8-0.9 mm long, green, glabrous, often granular; style 1 mm long; stigma 2-cleft, reddish, 2 mm long.

FRUIT: August-September. Pendent panicles 12-15 cm long; pedicels 5-6 mm long; fruit a terminal winged samara, total length 3-4 cm, width of wing 3.9-4.4 mm; fruit body 10-15 mm long, narrow, ridged; wing stramineous, extending at least halfway along the seed body, or as a ridge to the base, often abruptly widened beyond the seed.

TWIGS: 3.5-4 mm diameter, rigid, gray or with a yellowish cast, glabrous or pubescent, a few longitudinal lenticels; leaf scars large, half-round, upper margin straight or slightly concave; bundle scars in a half circle; pith white, continuous, one-fourth of stem. Buds 2 mm long, 3 mm wide, rounded or acute, brown, glabrous or pubescent, often with a small bud at the base.

TRUNK: Bark dark gray with shallow fissures, the ridges narrow, flat-topped, and anastomosed. Wood heavy, hard, strong, yellowish, with a wide, white sapwood.

HABITAT: Rich soils along streams and lakes or in a flood-plain woods, upland along hillsides or ravines. Often planted around farmsteads and in windbreaks.

RANGE: Maine, west to Quebec and Saskatchewan, south to Montana and Texas, east to Florida, and north to New England.

F. pennsylvanica is a variable species, and the two commonly accepted varieties intergrade to some extent: var. *subintegerrima,* as described above, and var. *pennsylvanica* (*F. pubescens* Lam., *F. darlingtoniana* Britt.), which has pubescent twigs, petioles, and leaf rachises, and the leaflets usually have a definite petiolule. Var. *pennsylvanica* is usually called red ash, a name given because of the reddish color of the inner bark.

In one small area of Sowbelly Canyon, Sioux County, Nebraska, is a group of trees whose fruits are abruptly widened just above the seed, giving a ping-pong paddle shape to the fruit.

● *Fraxinus pennsylvanica pennsylvanica*
■ *Fraxinus pennsylvanica subintergerrima*

446

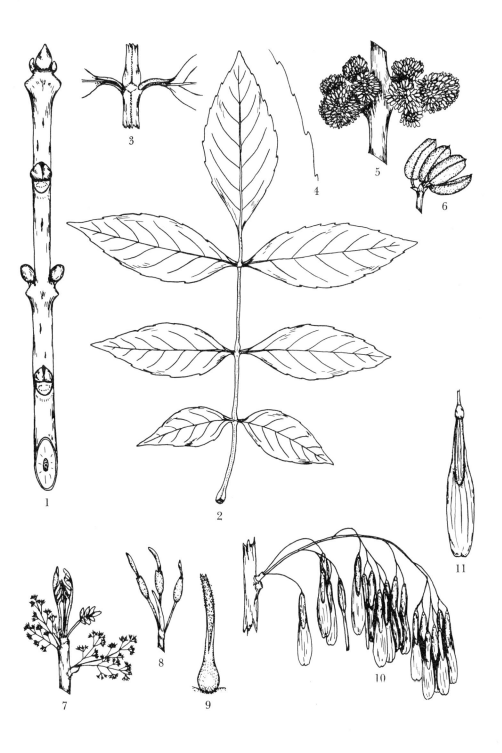

1. Winter twig
2. Leaf
3. Portion of rachis and two petiolules
4. Leaf margin
5. Cluster of staminate flowers
6. One staminate flower
7. Cluster of pistillate flowers
8. Three pistillate flowers
9. Pistil
10. Portion of fruit cluster
11. One samara

OLEACEAE
Fraxinus quadrangulata Michx.

Blue ash, square-stemmed ash

A tree to 20 m high, with long trunk and high branches, or, on rocky hilltops, low with a spreading crown.

LEAVES: Opposite, pinnately compound, 15-20 cm long, 7-9 leaflets. Leaflets ovate to lance-ovate, 5-12 cm long, 2-5 cm wide; serrate with low sharp teeth about 2 per cm; leaflet tip acuminate, often falcate; base rounded to cuneate, usually unequal, the wider and more rounded side toward the apex; upper surface dark green, lustrous, glabrous; lower surface paler, pubescent especially on the veins, the rachis often hairy at the base of the petiolules; petiole 4-6 cm long, slightly 2-ridged, pubescent or glabrous; terminal petiolule 1-2 cm long, lateral leaflets sessile or with a petiolule of 5 mm, blade often decurrent; no stipules.

FLOWERS: Early April, before the leaves; perfect. Flowers in short, crowded panicles at the base of the end bud, or in leaf axils on old wood; pedicels 2-3 mm long, calyx minute, reddish, the lobes acuminate and promptly deciduous; corolla none; stamens 2, attached below the ovary; filaments reddish, 0.5 mm long, the broad base wrapping halfway around the ovary; anthers 0.5 mm long, yellow-brown; ovary compressed, 1 mm long, green, granular; style short, the stigma 2-cleft, reddish, 1.5 mm long.

FRUIT: Late July. Pedicels 5-7 mm long, glabrate; fruit a terminal winged samara, 2.4-3.5 cm long, 6-8 mm wide; wing obovate, apex notched, or less often, mucronate, extending the full length of seed body, wing similar to that of black ash; seed body flat and not well-defined, 1.5-2 cm long.

TWIGS: 3.5-4 mm diameter, rigid, gray-brown, glabrous or finely pubescent, strongly 4-angled, the angles often corky-ridged; leaf scars crescent-shaped; bundle scars in a row; pith white, continuous, one-fifth of stem. Buds compressed, the terminal bud 5 mm long, 5 mm wide, ob-tuse, gray-yellow, densely pubescent, lateral buds 3.5 mm long and about as wide.

TRUNK: Bark light gray, irregular with shallow furrows, the ridges narrow and flat-topped. Wood hard, light yellowish, with a wide, white sapwood.

HABITAT: Dry, rocky hillsides, ravines, open woods on hills, or along creek banks.

RANGE: New York, west to Ontario and Michigan, southwest to Kansas, south to Oklahoma, east to Alabama, and north to Ohio.

The most outstanding characteristic of *F. quadrangulata* is the square twigs, often with a corky ridge on the angle. No other ash of our area has this characteristic. The two other distinguishing characteristics are the perfect flower and the broad wing and flat body of the samara, but these are seasonal characteristics. In general, *Fraxinus* species grow along moist areas, but blue ash is most commonly found growing on upland, rocky soil. Its range is southern and eastern and in the prairie states does not overlap with *F. nigra,* the only other ash of our area with broad wings and flat fruit bodies on the samaras.

Blue ash gets its name from the fact that a blue dye can be made by soaking the inner bark of the limbs in water. This dye was formerly made and used a great deal, but this is no longer a common practice.

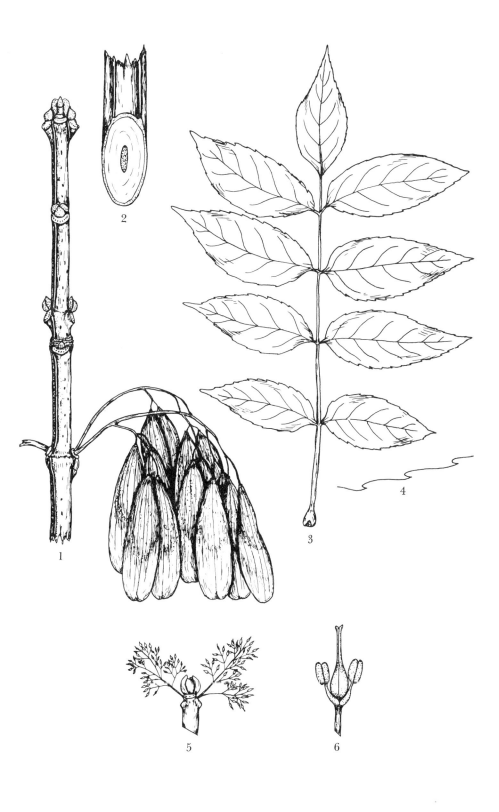

1. Winter twig and fruit cluster
2. Section of twig
3. Leaf

4. Leaf margin
5. Flower cluster
6. One flower

OLEACEAE
Forestiera acuminata (Michx.)
Poir.

Adelia acuminata Michx.

Swamp privet, swamp ash

A shrub to 6 m high, with 2-10 trunks from the base and with many fine branches.

LEAVES: Opposite, simple. Leaves thin and soft, elliptic, widest at or just below the middle, 5-8 cm long, 2.4-3.5 cm wide; margin entire or remotely fine-toothed; tip acuminate, base cuneate; upper surface yellow-green, semilustrous, glabrous, the lower surface paler and glabrous; petiole 1-1.4 cm long, with a deep, narrow groove above.

FLOWERS: Early April; dioecious or perfect. Flowers in fascicles in the axils of leaf scars of the previous year, the bud scales often persistent at the base of the cluster. Peduncles 3-4.5 mm long, 1-3 flowered; pedicels 0.5-1 mm long, green, glabrous; calyx none, corolla none; stamens 2-6, attached at the base of the ovary, filaments 1-1.5 mm long, pale yellow; anthers pale yellow; ovary 1 mm long, ovoid, green; style 1.5-2 mm long, yellow-green, gradually fading to pale yellow near the tip; stigma oval, yellow.

FRUIT: Early June. Drupe cylindric, 2-3 cm long, 3-6 mm thick, often narrowed at the apex, occasionally irregular and unsymmetrical, curved, dark blue to nearly black, but may fall while still green, flesh dry; 1 seed per fruit. Stone 14-15 mm long, 3-4 mm diameter, pale brown, hard, with several longitudinal lines of erect, stiff fibers.

TWIGS: 1 mm diameter, flexible, glabrous, light gray-brown; the shoots gray-green; lenticels light-colored, prominent; branches often spinescent; leaf scars small, half-round; 1 bundle scar; pith greenish, continuous, but becoming broken in older stems, one-third of stem. Buds small, ovoid, 1-1.5 mm long, scales reddish with yellow edges, glabrous.

TRUNK: Bark gray-brown, not fissured, but roughened with prominent, raised lenticels. Wood hard, pale red-brown.

HABITAT: Swampy ground, stream banks, wet woods, rich or rocky soils in shaded, moist areas.

RANGE: Indiana, west to southeastern Kansas, south to Texas, east to Florida, and north to South Carolina.

Swamp privet becomes a large shrub up to 6 meters high, with many trunks from one base, and may become as wide as it is high. The trunks are clustered in groups of 2-10, the outermost usually somewhat decumbent. The upper branches are numerous and have many fine, often drooping, branchlets. When these branchlets touch the ground they may take root.

Although many plants were examined in the field, only plants with perfect flowers were found—no purely dioecious plants were located. In one location along the Neosho River in Labette County, Kansas, one thicket was located where the stems were not clustered; each trunk was distinct and entirely separated from all others. These plants were tall and straight.

The flowers appear before the leaves and, although they have no perianth, the great quantity gives a yellowish cast to the shrub. Because of this, the plant can often be found easily at this season. The plant is occasionally confused with *Ilex decidua* Walt., but the opposite leaves of *Forestiera* will separate the two species.

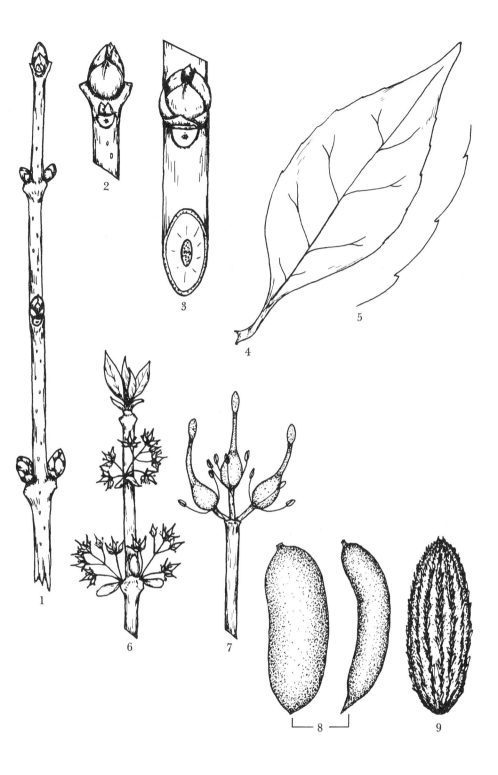

1. Winter twig
2. Detail of end bud
3. Detail of lateral bud
4. Leaf
5. Leaf margin
6. Flower clusters
7. Group of three flowers
8. Variation in the fruit
9. Seed

SOLONACEAE
Lycium halimifolium Mill.

Matrimony vine

A spreading, vine-like shrub, the branches often reaching 4 m.

LEAVES: Alternate, simple or clustered on short spurs. Elliptic to lance-ovate or oblanceolate, 3-7 cm long, 1-2.5 cm wide, those of the flowering branches smaller; margin entire, often undulate; tip rounded to acute, base acuminate, decurrent; dull, gray-green or yellow-green, and glabrous on both sides; petiole 2-8 mm long, thick, flattened on top; no stipules.

FLOWERS: June-September; perfect. Flowers axillary, 1-3 in each axil; pedicels 8-12 mm long, sparsely puberulent. Calyx tubular or funnelform, 5-7 mm long, lobes 3-5, acute, 2.6-2.8 mm long, green with pale tips; corolla tubular, attached below the ovary, tube 7-8 mm long, 2.3 mm wide; lavender, fading to pink, with dark purple lines in the throat of the tube, a line of hairs around the throat; lobes 5, ovate, 6-7 mm long, obtuse or acute tip; stamens 5, the filaments white, adnate to the corolla tube for 2 mm, the free portion 7-8 mm long, pubescent at the base; anthers yellow; ovary ovate, 2 mm long, 1.8 mm thick, cream-colored; style 1 cm long, often bent at the outer end, white, glabrous, slender, enlarged just below the bright green, discoid stigma.

FRUIT: August-October. Pedicels 10-15 mm long, glabrate, calyx persistent. Berry bright red, ellipsoid to obovoid, 8-11 mm long, 4.5-6 mm diameter. Seeds flat, circular to oval, with an indentation at the hilum, 2-2.7 mm diameter, 0.7-0.8 mm thick, yellow-brown, finely pitted.

TWIGS: 1-2 mm diameter, flexible, yellowish, glabrous, often winterkill at the tips; leaf scars narrow, upper edge nearly straight; 1 bundle scar; fruit scars circular, large, irregular; pith white, continuous, one-third of stem. Buds concealed in a cluster of old leaf bases and scars. Branches occasionally spinescent.

TRUNK: Bark gray-brown, pale, smooth, occasionally with small flakes; brown and flaky on old stems. Wood medium-hard, yellow to brown.

HABITAT: Escaped or persistent for many years around old buildings and farmsteads. Reproducing only locally.

RANGE: Introduced from Europe. Escaped throughout eastern and central United States and southern Canada.

Matrimony vine was formerly used a great deal for yard planting, but more desirable shrubs and vines have been developed and it is now seldom used. It persists around old farmsteads long after the buildings are gone, occasionally spreading into nearby unused plots. The stems reach a length of 4 meters and clamber over fences or low shrubs, occasionally arching high if there is nothing for support.

When growing without the support of a fence or shrub it forms a dense entanglement of long, vine-like branches, some of which are spinescent. This is excellent cover for game birds, songbirds, and small mammals. Cases of poisoning in sheep and cattle from eating the leaves have been reported; but from the appearance of the shrubs growing in a grazed area, the domestic animals browse heavily on it.

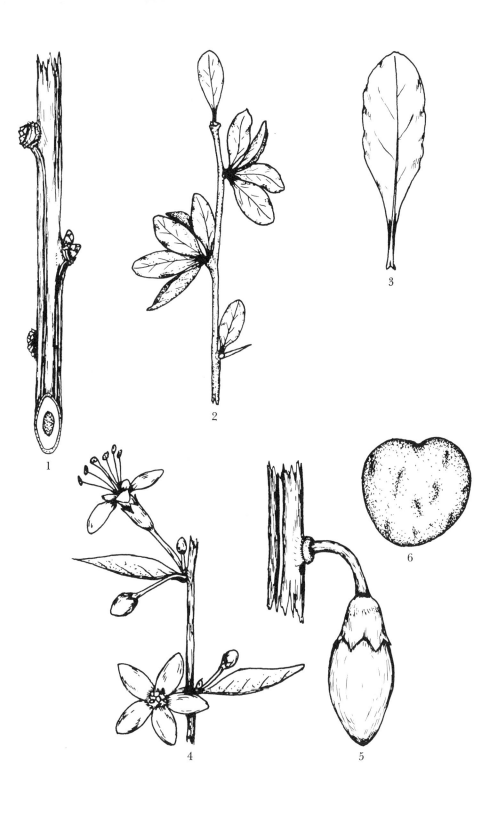

1. Winter twig
2. Vegetative branch
3. Leaf
4. Floral branch
5. Fruit
6. Seed

BIGNONIACEAE
Campsis radicans (L.) Seem.

Tecoma radicans (L.) Juss.; *Bignonia radicans* L.

Trumpet creeper, trumpet vine

A vine climbing by aerial roots to a height of 20 m, but usually seen sprawling over a fence or low bushes.

LEAVES: Opposite, pinnately compound, 20-30 cm long, 7-11 leaflets. Leaflets 5-8 cm long, 3-4 cm wide, ovate; coarsely toothed, the teeth often prolonged; leaflet tip long acuminate, base rounded or cuneate, the blade decurrent on the petiolule; upper surface dull yellow-green, glabrous, the lower surface paler with some pubescence on the veins; petiole 4-6 cm long, glabrous, grooved above, often slightly winged, petiolules 1-6 mm long; stipules are mere tufts of hair.

FLOWERS: June-July; perfect. Terminal clusters of 8-15 flowers, all parts glabrous; peduncles 8-15 mm long, usually 3-flowered; pedicels 4-6 mm long, green; calyx tubular, 20-22 mm long, reddish-green, with small pustules near the summit; calyx lobes 5, acute, 4-6 mm long, thick and coriaceous, tips often slightly recurved; corolla tubular to funnel-shaped, 5.5-6.5 cm long, 1.8 cm wide, abruptly enlarged above the calyx, orange externally, the inside yellowish with red lines; lobes 5, oval, 12-16 mm long and as wide, overlapping; stamens 5, one of which is vestigial; filaments adnate to the corolla for 2-2.2 cm, then distinct and arched for 3-4 cm, white; anthers yellow, 3-5 mm long, recurved against the filaments in bud, becoming straight as the flower opens; ovary green, ovoid, 5.5-6.4 mm long, 2.6-3.3 mm wide, ridged; style 3.5-4 mm long, white; stigma flattened, ovate, 3.7-4 mm long, green; the fleshy disk at the base of the ovary 5-lobed.

FRUIT: August-September. Pods 14-20 cm long, 2.1-2.4 cm wide, 1.8-2.2 cm thick, gibbous with a flat flange on both sutures, slightly curved or straight, tapered at both ends, tan to brown, sparingly pustuled; dehiscent on both sutures; inside divided by a flat partition fastened only at the base of the pod; numerous seeds in each pod. Seeds flat, 16-20 mm long, 5.5-8 mm wide, dark brown, with scarious wings on 2 sides.

TWIGS: 2.5 mm diameter, pale yellowish, flexible, glabrous, long internodes; adventitious roots appear first at the nodes, later along the internodes; leaf scars half-round; a ring of bundle scars; pith white, continuous, one-third of stem. Buds yellowish, 1.5-1.8 mm long, 2-2.2 mm wide, the 2 broad, outer scales glabrous outside and hairy inside.

TRUNK: Bark yellow and tight on young plants; brown and exfoliating into long shreds on old trunks. Wood soft, porous, yellow.

HABITAT: Thickets, fence rows, roadsides, stream banks, flood plains, rocky hillsides, or open woods.

RANGE: New Jersey, west to Iowa, south to Kansas and Texas, east to Florida, north to Pennsylvania; escaping in many areas.

Trumpet creeper is an aggressive plant, spreading rapidly and covering anything within reach, the trunk of large vines becoming 8 cm in diameter. Shoots often come up from the roots and form a nearly impassable tangle of vines around the parent plant. For this reason it should be planted only where it can be controlled.

Hummingbirds are regular visitors to the flowers, and songbirds often nest in the tangle of vines. It is a valuable plant as a cover for small mammals and birds, and as a deterrent to soil erosion.

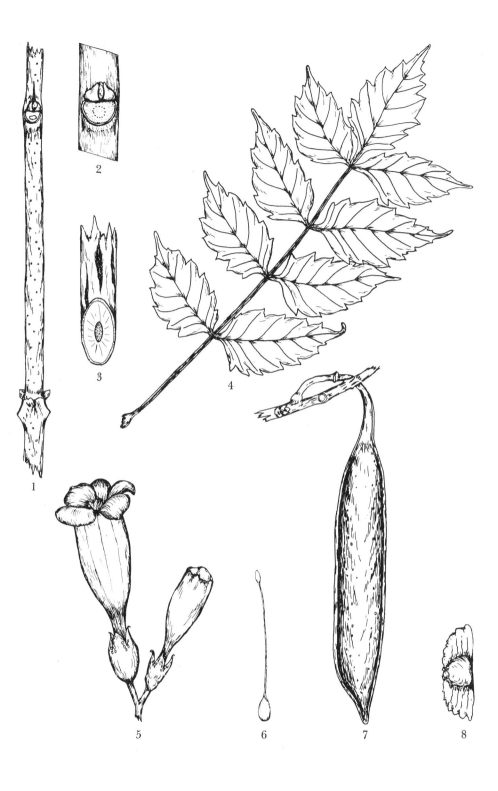

1. Winter twig
2. Detail of twig
3. Section of twig
4. Leaf

5. Flower
6. Pistil
7. Fruit
8. Seed

BIGNONIACEAE
Catalpa speciosa Warder

Catalpa, cigar tree, Catawba tree,
Indian bean

A tree to 20 m high, with large trunk, often straight for 7-8 m, occasionally with low branches.

LEAVES: Opposite, simple. Ovate 15-25 cm long, 10-20 cm wide, entire, long acuminate, truncate to broadly cordate at the base; upper surface yellow-green, glabrous, the lower surface paler, pubescent with curled or bent hairs; soft, pliable; petiole 8-11 cm long, sparingly hairy; stipules absent.

FLOWERS: May-June; perfect. Flowers in glabrous terminal panicles 15-20 cm long, 10-15 cm wide, the branches in whorls of three; irregularly spaced, linear, pubescent bracts along the floral branch; pedicels 10-12 mm long, purplish; calyx lobes 2, distinct, ovate to obovate, apiculate, purplish, 9-11 mm long, lower lobe somewhat membranaceous; corolla tubular, 4-5 cm long, the tube gibbous on the lower side, the lobes undulate, 2 above, 2 laterals, and 1 long lobe at the bottom; tube and lobes white, spotted and streaked with purple, 2 yellow ridges on the lower lobe; stamens 2, attached to the lower lobe of the corolla; filaments 2 cm long, white, purple spotted near the base, sharply bent upward at the outer end and paralleling the style; anthers 4 mm long, yellow; 3 rudimentary stamens attached to the upper lobes; ovary cylindric, 2.5-3 mm long, green, glabrous; style 1.8-2 cm long, white, arched, the stigma with 2 flat, ovate lobes, 2 mm long.

FRUIT: September. Capsule cylindric, 30-45 cm long, 1-1.3 cm thick, brown, irregularly ridged, usually curved, loculicidal; many seeds. Seeds flat, 2.8-4 cm long, 4.7-8.2 mm wide, tan-brown, winged on 2 sides, the wings terminating in a rounded tuft of hairs; the embryo area 13-16 mm long, 2-lobed; the cotyledons are so deeply divided that at first glance they appear as 4 cotyledons, the lobes of each cotyledon at 180° to each other.

TWIGS: 4-6 mm diameter, rigid, coarse, glabrous, brown, with prominent, yellow lenticels; leaf scars nearly oval, but somewhat straight across the top, 3-5 mm long, raised; several bundle scars in a circle; fruiting scars circular; pith white, continuous, two-thirds of stem. Buds 2 mm long, 4 mm wide, the scales red-brown with a lighter margin, ciliate, the outer scales keeled.

TRUNK: Bark red-brown, often pale on younger trees; the furrows shallow and the ridges short and plate-like. Wood soft, durable, light brown, with a narrow, light sapwood.

HABITAT: Stream banks, low rich woods, or uplands. Planted extensively.

RANGE: Indiana, west to Iowa, south to Texas, east to Tennessee and Kentucky. Introduced and naturalized in our area and in many other parts of the United States.

The other species of *Catalpa* common in our area is *C. bignonioides* Walt., which is quite similar to *C. speciosa*. The principal characteristics of *C. bignonioides* are: short acuminate leaf tip, inflorescence to 30 cm long, corolla conspicuously spotted, small flowers, and the tuft of hairs at the tip of the seed wing narrowed to a point.

Both species are commonly planted in groves and harvested for fence posts. The trees are cut near the ground, and sprouts arise from below the cut. These sprouts are thinned and eventually a second crop of posts may be cut from the same root stock. The wood is durable and the posts last for several years.

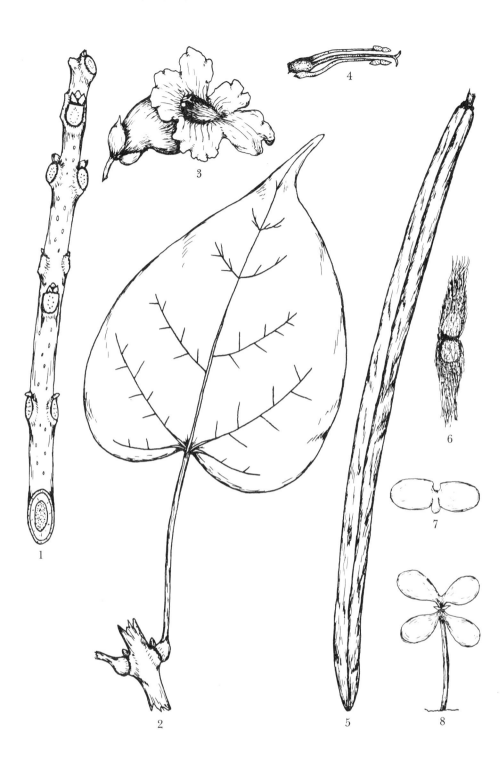

1. Winter twig
2. Leaf
3. Flower
4. Pistil and two stamens
5. Fruit
6. Seed
7. Embryo
8. Seedling, showing the deeply lobed cotyledons

457

RUBIACEAE
Cephalanthus occidentalis L.

Buttonbush, globe flower, honeyball,
swamp sycamore, pond dogwood

A broad, open shrub to 3 m high, usually several trunks from the base.

LEAVES: Opposite or in whorls of 3, simple. Narrowly ovate to ovate-oblong, 9-11 cm long, 3-4.5 cm wide; entire, often short-ciliate; tip acuminate or abruptly short acuminate; base rounded, obliquely truncate or cuneate; upper surface dark green, lustrous, glabrous; lower surface paler, a few hairs in the axils of the veins and on the veins; thick, coriaceous; petiole 0.5-1 cm long, stout, glabrous; stipules deltoid, 3-4 mm long, glandular dentate, becoming brown.

FLOWERS: Early July; perfect. Fragrant, globular clusters, 3 cm diameter, including the exserted styles; terminal or axillary on new growth; peduncle 2-4 cm long, green, glabrous; calyx obconic, 3.5-4 mm long, glabrous, the 4 lobes green, 0.3 mm long, usually incurved, thick; corolla funnel-form, 10-12 mm long, constricted at the base, white; lobes 4, rounded to obtuse, 2-2.2 mm long, pubescent inside, spreading, a minute, dark tooth between the lobes; stamens 4, alternate with the corolla lobes, not exserted, filaments adnate to the corolla tube; anthers free; ovary 1 mm long, green; style long exserted, filiform, 15-16 mm long, white; stigma cylindric, 1 mm long, yellow-brown; numerous, narrow, clavate, whitish bracts on the receptacle between the flowers.

FRUIT: October. Globose heads 2.3-2.7 cm diameter; fruits narrowly obconic, 7.1-8.8 mm long, 1-1.8 mm wide, yellow-brown, ridged, tipped with the persistent calyx, 2-locular, 1 seed per locule. Seeds obconic, 4-5 mm long, yellowish, capped with a large, white, opaque aril.

TWIGS: 1.5-2 mm diameter, semi-rigid, glabrous, gray-brown, with a few, large, light lenticels; leaf scars half-round, deltoid or circular; bundle scars often indistinct, either 3 or a crescent-shaped line of several; pith yellow, continuous, one-half of stem. Buds covered by the bark, or as a mere pimple showing through a short slit, with bark removed they can be seen as 2-ranked, a small bud directly above the leaf scar and a larger bud 1 mm above.

TRUNK: Bark dark brown, thick, heavily ridged with long, flat-topped ridges; inner bark fibrous. Wood soft, lightweight, white.

HABITAT: Low, wet ground, stream banks, lakeshores, prairie sloughs, and pond borders.

RANGE: Nova Scotia, west to Ontario, south to Texas, Mexico, and California, east to Florida, and north to New England; also the West Indies.

Cephalanthus is a southern and eastern plant, entering our area in southeastern Nebraska and eastern Kansas, where it is common around ponds, lakes, and sloughs. It is easily identified in the summer by the white balls of flowers and in the autumn and early winter by the globular seed heads. The trunks are clustered, the outer ones often leaning at an angle of 45° before becoming erect. These trunks are usually about 5 cm or less in diameter, but an occasional plant reaches 20 cm diameter and has fairly large branches near the base.

The plant is of no commercial value, and its use in conservation planting would be limited to protection of a lakeshore from wave action. Also, birds use the shrub commonly for nesting sites.

The variety *pubescens* Raf., with pubescent lower leaf surfaces and twigs, has not been reported for our area.

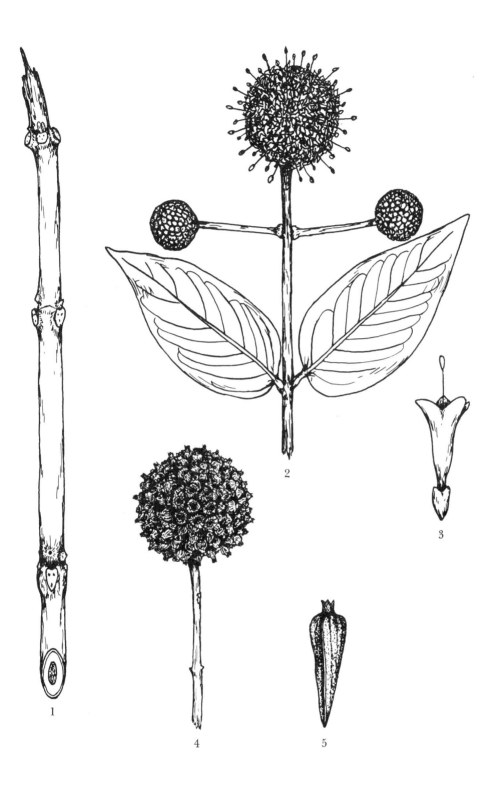

1. Winter twig
2. Leaves with flower clusters
3. One flower
4. Fruiting cluster
5. One fruit

CAPRIFOLIACEAE
Lonicera dioica L.

L. glaucescens Rydb.; incl. var.
glaucescens (Rydb.) Butters

Wild honeysuckle, limber honeysuckle

A sprawling, vine-like shrub, the branches often reaching 3 m long.

LEAVES: Opposite, simple. Oval to elliptic, 4-8 cm long, 2.5-5 cm wide; entire, often revolute; tip obtuse, mucronate, the base cuneate or rounded; upper surface yellow-green, glabrous; lower surface glaucous, glabrate or pubescent; leaves directly below the flowers connate, forming an oblong or elliptical disk with obtuse tips; lower leaves sessile or with a semiclasping petiole 2-3 mm long; no stipules.

FLOWERS: May-June; perfect. Flowers terminal, 1-3 whorls each composed of 2 clusters of 3 flowers each, the whorls separated by 2-3 mm; a broad, deltoid bract 1.5 mm long and 2 mm wide at the base of the cluster, and a minute bract at the base of each flower; peduncle above the connate leaves 6-20 mm long. Hypanthium ovoid, glabrous, 2.4-2.5 mm long; calyx lobes 5, deltoid, acute, 0.7-0.8 mm, glabrous, green; corolla tubular, bilabiate, the tube 1-1.2 cm long, gibbous on the lower side near the base, orange to purple, pubescent inside and out; lower lip 1 cm long, 3.8-4 mm wide, oblong with rounded tip; upper lip 1 cm long, 4-lobed; center 2 lobes sub-acute; the outer 2, oblong and rounded; lips slightly pubescent, lighter color than the tube; stamens 5, attached to the corolla at the throat of the tube; filaments 8-9 mm long, pubescent toward the base; anthers 4 mm long, yellow; ovary inferior; style 1.8-2 cm long, enlarged at the base, pubescent; stigma capitate, 3-lobed.

FRUIT: July-August. Fruits clustered, globose to obovoid, 6-9 mm long, red, smooth, glossy. Seeds oval, often irregular and flattened, 4-4.5 mm long, 3.4-3.8 mm wide, 1.8-1.9 mm thick, yellow-brown, smooth.

TWIGS: 1.5-1.7 mm diameter, flexible, pale to dark brown, glabrous; leaf scars triangular, extending halfway around the twig; 3 bundle scars; center usually hollow, but occasionally with a white, broken pith, one-half of stem. Buds 2.5-4 mm long, scales brown, the tip acuminate and often spreading.

TRUNK: Bark yellow-brown, shredded into thin, long strips. Wood yellowish.

HABITAT: Wooded bluffs and rock ledges; rich woodlands.

RANGE: Quebec to Manitoba and British Columbia, south to Kansas, east to Kentucky and North Carolina, and north to New York.

Lonicera dioica is divided into two varieties: var. *dioica*, with the corolla tube and the lower leaf surface glabrous, and var. *glaucescens* (Rydb.) Butters, with pubescent corolla tube and lower leaf surface. The material from Nebraska and the Dakotas falls clearly into var. *glaucescens*, but a great deal of the Kansas material is questionable. Part of the specimens from Kansas are var. *glaucescens*; but other specimens are totally glabrous and with a green (not glaucous) upper surface on the connate leaves, these leaves being definitely acute. These characteristics place the plants in var. *dioica* but the given range of that variety is much to the northeast.

A closely related species, *L. prolifera* (Kirchn.) Rehd., is similar to *L. dioica* var. *dioica* but with the connate leaves nearly circular or with definitely rounded ends and glaucous above. Apparently some of the Kansas material may be *L. prolifera*, but the glaucous coating is not pronounced.

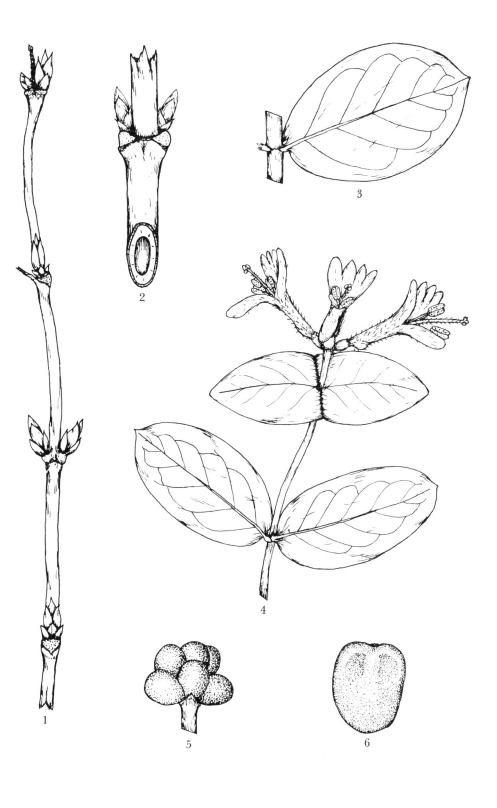

1. Winter twig
2. Detail of twig
3. Leaf

4. Flowers with leaves
5. Fruit cluster
6. Seed

CAPRIFOLIACEAE
Lonicera japonica Thunb.

Japanese honeysuckle

A climbing or sprawling vine to 10 m long; climbs by twining and becomes dense.

LEAVES: Opposite, simple, tardily deciduous, none of them connate. Ovate, 4-8 cm long, 2-3 cm wide; margin entire and ciliolate; tip acute, the base truncate or cuneate; upper surface dark green, semilustrous, pubescent, especially on the midrib; lower surface pale, glabrate, except for the pubescent midrib; soft, coriaceous; petiole 1 cm long, pubescent. Many leaves remain green all winter, some become purplish; no stipules.

FLOWERS: Mid-May, continuing for most of the summer; perfect. Peduncles axillary, 5-15 mm long, pubescent, 2-flowered, and with 2 leaf-like bracts near the peduncle summit and 2 bracts at the base of each flower, these 1 mm long, green, rounded, and long ciliate; the flowers sessile on the peduncle. Hypanthium 0.5 mm long, green, glabrous, with 5 lanceolate, ciliate lobes 1.2-2 mm long; corolla bilabiate; tube 2.5-2.6 cm long, pubescent inside and out and glandular stipitate; lips 2.4-2.6 cm long, one lip narrow, oblong, and with rounded end; the other lip 4-lobed, the oblong lobes 6-8 mm long, overlapping at the base; flowers white, cream or pinkish, becoming yellow before falling; stamens 5, adnate to the corolla tube at the base, free for 2.2-2.5 cm; filaments yellow; anthers brownish, slender, 3 mm long; ovary inferior, cylindric, 2 mm long, 1.7 mm wide, green, glabrous; style 4.1-4.3 cm long, whitish; stigma subglobose, indistinctly 5-lobed.

FRUIT: October. Berries single or paired on a peduncle 5-15 mm long, subtended by a ciliate bract and 2 small bractlets; fruit globose, 5-8 mm diameter, black, glossy, smooth, 4-10 seeds. Seeds oval, 2.1-3.3 mm long, 1.8-2 mm wide, dark brown to black, rugose, granular, ridged lengthwise.

TWIGS: 1.5-2 mm diameter, flexible, pubescent, pale red-brown, soon exfoliating and showing the straw-colored bark beneath, old leaf bases often persistent; leaf scars narrow to broadly crescent-shaped, reaching halfway around the stem; 3 bundle scars, pith white, continuous or hollow, one-sixth of stem. Buds 1 mm long, red-brown, scale tips spreading; often overwinters without buds, but with clusters of minute leaves.

TRUNK: Bark yellow-brown, exfoliating into long, papery shreds. Wood white, soft, fine-grained.

HABITAT: Escaped to fence rows, roadsides, open woods, rocky slopes, and thickets.

RANGE: Native of Asia. Well established in parts of eastern and central United States.

Lonicera japonica is one of the most aggressive woody plants in our area. It clambers over bushes and small trees, smothering them as it goes, and climbs to the top of medium-sized trees and covers the branches. It is an excellent cover for small mammals and birds and is a good plant for the control of soil erosion. However, it should not be planted unless there is some means of control. The flowers are attractive and sweet-scented, and are visited commonly by hummingbirds and a host of insects.

Several horticultural varieties have been developed and are available in nurseries. For home planting, these would be more desirable than the plant described above.

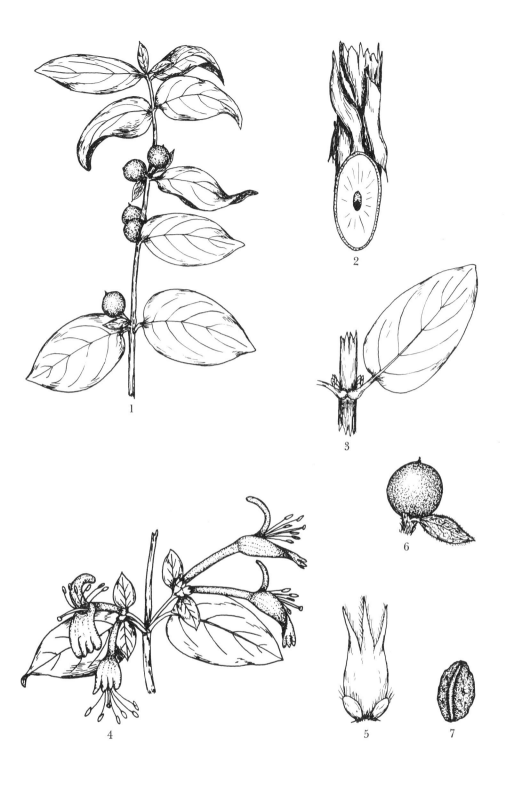

1. Winter twig
2. Detail of twig
3. Leaf
4. Flower cluster
5. Calyx with two small bracts
6. Fruit
7. Seed

CAPRIFOLIACEAE
Lonicera sempervirens L.

Trumpet honeysuckle, coral honeysuckle

A twining vine to 5 m in length, occasionally dense.

LEAVES: Opposite, simple, upper pair of leaves just below the flowers connate. Oval to broadly elliptic, 4-7 cm long, 3-4.5 cm wide, acute, base broadly cuneate; total length of connate leaves 5-6 cm, width 4-4.5 cm, each leaf deltoid, acute; upper surface of leaves dark yellow-green, semiglossy, glabrous; lower surface glaucous, glabrous; thick, coriaceous; petioles 4-8 mm long, flattened above, slightly clasping; no stipules.

FLOWERS: Late May; perfect. Terminal clusters on new growth, in 2-4 whorls of 3-6 flowers each; peduncle 2-4 mm long, flowers sessile. Hypanthium 0.5 mm long, green, glaucous, the 5 lobes short and rounded; corolla tubular, 4.5-5 cm long, often abruptly widened above the middle, red, orange, or tinged with purple, the inside yellow; corolla lobes 5, subequal, 0.5-0.6 mm long, ovate, rounded tips, the yellowish margins overlapping; lower part of the tube pubescent inside; stamens 5, the yellow filaments inserted at the base of the corolla but adnate to the tube about half its length, then free for 12-14 mm; anthers 3.5 mm long, slender, orange or red; ovary green, 2-2.3 mm long, 1 mm thick, ellipsoid, glabrous, glaucous; style 5-5.5 cm long, exserted, yellow with greenish outer end; stigma green, discoid.

FRUIT: Late July. The fruits in whorls of 3-6, ovoid, 8-10 mm long, 6-7 mm thick, bright red with a minute black tip, glabrous, 1-seeded. Seeds oval, flattened, 4-5 mm long, 3-3.6 mm wide, 1.7-1.9 mm thick, yellow to brown, 2 slight grooves with a raised area between on each flattened surface, finely pitted.

TWIGS: 1-1.3 mm diameter, yellow-brown, shredding in the second year, flexible, glabrous; leaf scars extend halfway around the stem, narrow, widened at the center, 3 bundle scars; pith only a thin layer around the hollow center and a wide diaphragm at the nodes. Buds narrowly ovoid, 3-6 mm long, 1.3-1.5 mm wide; scales brown, acute, keeled, and spreading at the tips.

TRUNK: Bark yellow-brown, shredding into long fibrous strips. Wood soft, yellow.

HABITAT: Roadside thickets, old farmsteads, fence rows, open woods.

RANGE: Maine, south to Florida, west to Texas, north to Nebraska, east to Iowa, Ohio, and New York. Escaped in many areas.

The *Lonicera sempervirens* in our area has escaped or persisted and may be found sporadically in several places. It is easily distinguished from the other honeysuckles by the long, 4-5 cm, red, trumpet-shaped corolla. The vine when in flower is quite attractive and produces an abundance of flowers if given proper care. It does not grow to as great a length as *L. japonica* and is neither as dense nor as aggressive, but does provide good cover for birds. It is not at all uncommon to see a covey of bobwhites resting or feeding beneath a sprawling vine.

In the southern states, where *L. sempervirens* is used greatly, the leaves are evergreen. This, accompanied by several other features such as the longer flowering period, makes the plant desirable for landscaping purposes. In the northern states, the plant is not as hardy as the native species, and, when planted, often does not produce flowers.

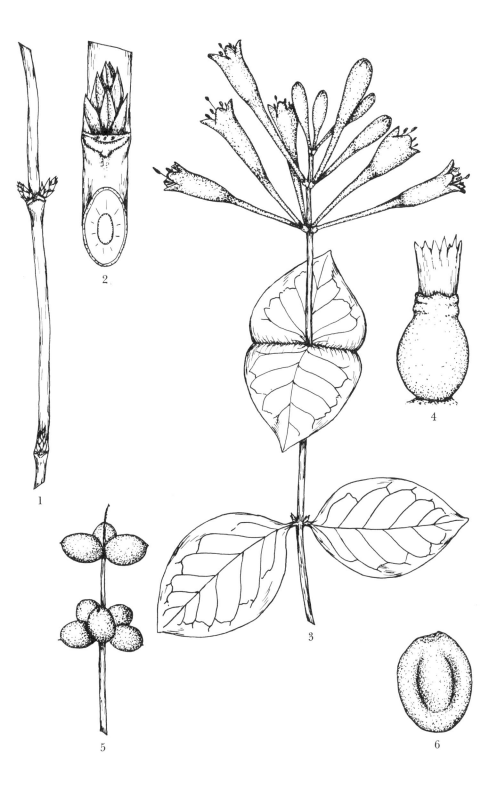

1. Winter twig
2. Detail of twig
3. Leaves and flower cluster

4. Ovary
5. Fruit clusters
6. Seed

CAPRIFOLIACEAE
Lonicera tatarica L.

Tartarian honeysuckle

A shrub to 3 m high, often dense with fine branches.

LEAVES: Opposite, simple; ovate to oblong, 2.5-4.5 cm long, 1.5-2.5 cm wide; tip obtuse to acute, base rounded or cuneate; margin entire, sparsely ciliate; upper surface yellow-green, glabrous, lower surface slightly paler and with a few hairs; petiole 3-5 mm long, broadly grooved above, glabrous or sparingly pubescent; no stipules.

FLOWERS: Late April through May; perfect. In pairs in leaf axils; peduncle 12-15 mm long, slender, green, and glabrous or slightly pubescent, with 2 green bracts 4-5 mm long at the summit, these spreading, linear-lanceolate, and glabrous; 2 small, glabrous, ovate, obtuse bracts 2 mm long attached to the base of the hypanthium. Hypanthium green, ovoid, 2 mm long, 1.5 mm wide, glabrous; lobes 5, triangular, obtuse or acute, 0.5-0.75 mm long; corolla funnel-form, bilabiate, pink or whitish, tube 7-8 mm long, gibbous at the base, gradually widened toward the tip, upper lobe deeply divided into 3-4 oblong segments, lower lobes undivided; lobes 9-12 mm long, glabrous outside, the tube pubescent inside along the adherent filaments; stamens 5, filaments slender, the free portion 5-6 mm long, pubescent toward the base; anthers 2-2.5 mm long, yellow, exserted; style slender, 6-8 mm long, pubescent; stigma capitate, lobed.

FRUIT: July-August. Fruits globose, 5-7 mm diameter, single or in pairs with the bases fused, orange-red, translucent, 3-6 seeds per fruit. Seeds oval, flattened, 2.7-2.9 mm long, 2-2.5 mm wide, 1-1.3 mm thick, yellow, granular surface, a longitudinal ridge with a depression on each side of it on both flat surfaces.

TWIGS: 0.8-1 mm diameter, flexible, brown to green-brown, glabrous, smooth, the few lenticels light-colored; leaf scars broadly triangular; 3 bundle scars near the center; pith brown around the hollow, about two-thirds of stem. Buds broadly ovoid to globose, brown, 2.5-3 mm long, 1.7-2 mm wide, often superposed, larger on sprouts, the scales minutely erose-ciliate, often slightly keeled.

TRUNK: Bark gray-brown with long, flat, thin scales and not much shredding. Wood hard, fine-grained, yellow-brown, with a wide, pale sapwood, and usually hollow in the center.

HABITAT: Often found in open woods, along stream banks or brushy pastures. Escaped.

RANGE: Introduced from Eurasia and commonly escaping in north central, northeastern, and eastern United States, as well as in southern Canada.

Lonicera tatarica is one of the plants commonly used in windbreaks or as a screen in landscaping. The many fine branches make it well adapted to such purposes, as well as furnishing an abundance of flowers. The flowering period is quite long.

The shrub often escapes and appears in unexpected places, quite commonly on wooded hillsides or the banks of small streams. It may persist for several years after being abandoned and often continues to reproduce itself if not disturbed.

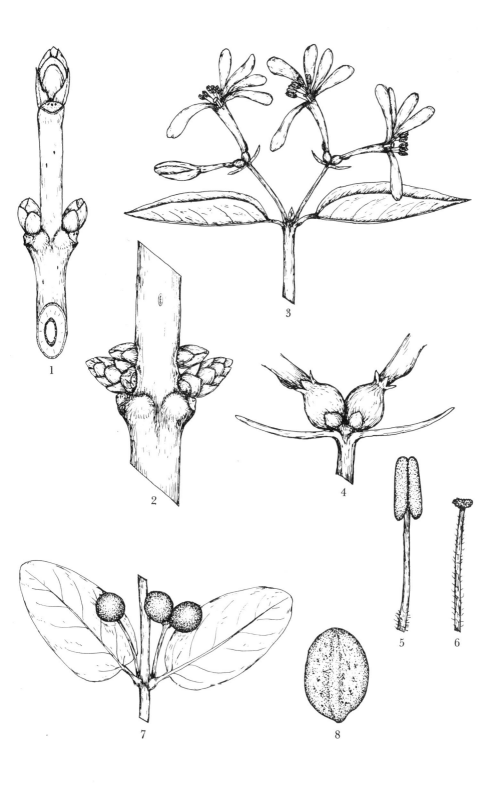

1. Winter twig
2. Detail of twig
3. Flower cluster
4. Detail of ovary
5. Stamen
6. Style and stigma
7. Fruit and leaves
8. Seed

CAPRIFOLIACEAE
Symphoricarpos albus (L.) Blake

S. racemosus Michx.; S. pauciflorus
(Robbins) Britt.; Symphora racemosa
Pursh; Vaccinium album L.; Xylosteum
album Moldenke; Lonicera racemosa
Pers.

Snow berry, wolf berry

A shrub to 50 cm high, sparingly branched with fine branches.

LEAVES: Opposite, simple. Ovate to elliptic, 2-3 cm long, 1.5-2 cm wide, acute, base broadly cuneate; margin entire and ciliate, occasionally irregularly toothed; dark green and glabrate above, paler and pubescent below; petiole 3-4 mm long, pubescent, flattened above, broad at the base, the margins often meeting those of the opposing petiole; no stipules.

FLOWERS: June-July; perfect. Flowers terminal or in the upper leaf axils in short racemes of 2-5 flowers; pedicels 2.5 mm long, green, glabrous, recurved; the 2 bracts below the ovary glabrous and ciliate; calyx lobes 5, triangular, light green, 0.5 mm long, ciliate; corolla campanulate, 2.5-3 mm long, 3-3.3 mm wide, pink, glabrous outside, villous inside; lobes 5, rounded, 2.5-3 mm long, pink or white, erect or slightly spreading, glabrous; ·stamens 5, attached to the throat of the corolla, filaments 1.5 mm long, white; anthers yellow, about 1 mm long, not extending beyond the corolla; style 3 mm long, greenish, glabrous; stigma capitate.

FRUIT: August. Ovoid, 8-9 mm long, 6-8 mm wide, white, smooth, not lustrous, old calyx lobes persistent and often white; 2 seeds per fruit. Seeds white, elliptical, 4-5 mm long, 2-2.3 mm wide, 1.4-1.6 mm thick, ventral surface nearly flat, the dorsal side rounded, ends acute, smooth or minutely striate.

TWIGS: 0.8 mm diameter, yellow-brown, covered with fine, curled hairs, especially near the nodes; leaf scars indefinite, old leaf bases often remain; pith pale brown, or the stem hollow, one-half of stem. Buds yellow-brown, ovate, 2 mm long, 1 mm wide, acute, the outer scales slightly keeled and often spreading, curly pubescent.

TRUNK: Bark of the upper branches tight or broken and peeled laterally, that of the lower branches and trunk exfoliating into long, thin, narrow shreds, gray-brown to red-brown. Wood soft, greenish-white, with a dark brown pith.

HABITAT: Wooded hillsides or on open, rocky slopes, in either moist or dry soil.

RANGE: Alaska, south to California, east across Colorado and Nebraska to Virginia, north to Quebec, and west to British Columbia.

S. albus is a rhizomatous shrub but rarely produces colonies of any size. The plants usually occur singly or about 1 meter from the next nearest stem. Our plants are var. albus, and the range extends into Montana where it meets the range of var. laevigatus. The range above is for the species. Var. albus is a smaller plant with smaller fruits than var. laevigatus: 8-9 mm compared to 10-15 mm. It is also a much smaller plant than S. occidentalis, which grows in the same area. The fruits of S. occidentalis are about the same size as those of S. albus but occur in groups of 10-20 and are sessile or nearly so; those of S. albus are in groups of 1-5 and have a definite pedicel. Another obvious difference is in the flowers; the stamens and pistil of S. albus do not extend beyond the corolla, while those of S. occidentalis do. The only other Symphoricarpos of our area is S. orbiculatus, which has red fruits and is restricted to the southern part of the region.

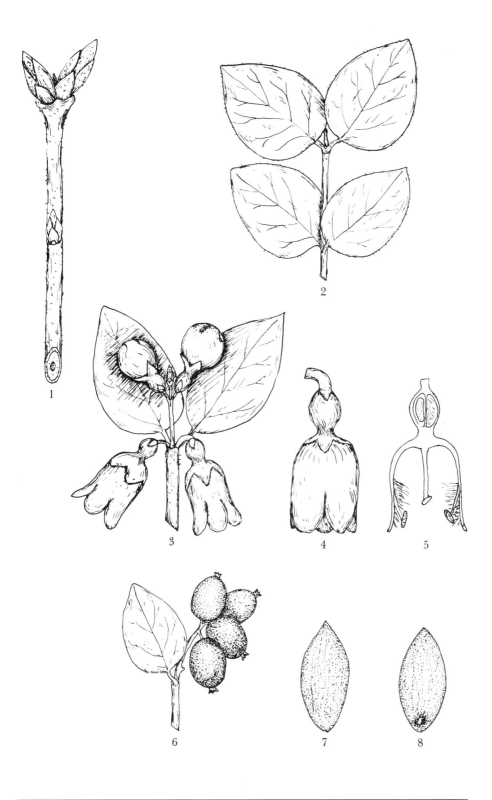

1. Winter twig
2. Leaves
3. Flowers
4. Flower
5. Diagram of flower
6. Fruit
7. Dorsal view of seed
8. Ventral view of seed

CAPRIFOLIACEAE
Symphoricarpos occidentalis Hook.

Wolfberry, western snowberry

A rhizomatous shrub to 1 m high, often in large colonies.

LEAVES: Opposite, simple. Broadly ovate, orbicular or elliptic, 3-6 cm long, 2-3.5 cm wide; margin entire or with coarse, rounded teeth, ciliate; tip acute or obtuse, often mucronate, the base broadly cuneate to slightly rounded; upper surface yellow-green or gray-green and glabrate; lower surface paler and pubescent; petiole 5-8 mm long, pubescent; no stipules.

FLOWERS: June to August; perfect. Terminal or axillary in 10-20 flowered, short spikes or racemes, the flowers sessile or nearly so; 2 bracts 3 mm long at the base of the raceme, and 2 smaller, ciliate bracts 2 mm long at the base of each flower. Hypanthium green, ovoid, 2.8-3.2 mm long, 1.9-2.1 mm wide, glabrous, glaucous; calyx above the hypanthium 1.8 mm long; calyx lobes 5, deltoid, 1-1.5 mm long, green, glaucous, ciliate; corolla campanulate, white or pink, tube 4.3 mm long, glabrous outside, densely hairy inside with long, white hairs; lobes 5, ovate, 3.8 mm long, acute or obtuse; stamens 5, inserted at the top of the corolla tube; filaments 4-4.5 mm long, usually curved, hairy at the base; anthers exserted, 2.5 mm long, yellow; disk 5-lobed, greenish, 0.8 mm diameter; style 6.5 mm long, white with long white hairs at the middle; stigmas capitate, greenish.

FRUIT: September-October. Fruits in globose or cylindric racemes up to 5 cm long; fruits globose, white, 6-9 mm diameter, sessile, occasionally a few brownish dots on the surface, calyx lobes persistent; 2 seeds. Seeds hard, oval to elliptic, flattened, 2.3-3 mm long, 1.5-2 mm wide, 1-1.4 mm thick, smooth but finely striate, pale yellow.

TWIGS: 0.8-1 mm diameter, flexible, gray-brown, pubescent at first, becoming glabrous; leaf scars half-round, bundles indistinct; pith pale brown, continuous, one-third of stem. Buds ovoid, 2.5 mm long, acute, the scales brown, slightly pubescent, the tips often spreading.

TRUNK: Bark gray-brown, peeling into wide, often rigid sheets attached at one edge and curling outward, appears shaggy. Wood hard, white, often brown around the small pith.

HABITAT: Dry, rocky hillsides, ravine banks, sandy flats, open woods, roadsides, pastures, and prairies. Occasionally in moist soil.

RANGE: British Columbia, east to Ontario, south to Michigan, Illinois, and northern Missouri, southwest to Oklahoma and New Mexico, and north to Montana. Introduced further east.

When the fruits are not present this species is often confused with the red-fruited *S. orbiculatus*. It differs from that species in having somewhat glaucous leaves; larger flowers with the style and stamens exserted; thick, rigid flakes of curled bark; and a whitened, salmon-color inner bark. *S. orbiculatus* has yellow-green leaves; small flowers with the style and stamens enclosed; the bark with papery flakes or thin, shredded, papery strips; and a red-brown inner bark. The range of the two species overlaps in north-central Kansas and east-central Nebraska.

The small thickets scattered through a pasture or an open woods furnish good cover and nesting sites for small birds and mammals. A few birds eat the fruits during early winter. It is a good soil binder for the prevention of erosion and also holds the snow during winter. It is quite common on the open-woods flood plain of the Arkansas River in western Kansas, as well as all through the northern prairie states.

470

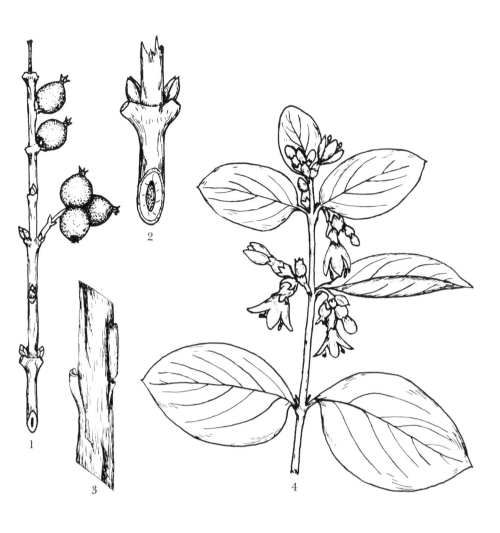

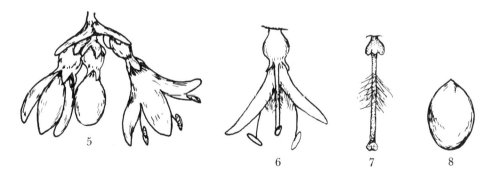

1. Winter twig with fruit
2. Detail of twig
3. Peeling bark of trunk
4. Leaves with flower clusters

5. Flowers
6. Section of flower
7. Pistil
8. Seed

CAPRIFOLIACEAE
Symphoricarpos orbiculatus
Moench

Coralberry, buckbrush, Indian currant

A rhizomatous shrub to 1 m high, usually about 60 cm, forming dense colonies.

LEAVES: Opposite, simple. Ovate to oval, 3-5 cm long, 1.5-3.5 cm wide; margin entire, ciliate, rarely coarsely toothed; tip rounded to obtuse, often mucronate, the base rounded to truncate; upper surface yellow-green, dull, glabrous or glabrate; lower surface paler and pubescent; petiole 1-3 mm long, pubescent; no stipules.

FLOWERS: Early July; perfect. Flowers in very short spikes, appearing clustered in the leaf axils, the short peduncle sharply bent downward. Flowers sessile, 1 large and 2 small bracts at the base of the hypanthium; hypanthium ovoid, 1.7-1.9 mm long, green, glabrous; calyx above the hypanthium short, the 5 lobes acute, 1 mm long, green; corolla narrowly campanulate, 3 mm long, greenish-white or purplish, glabrous outside and villous inside; lobes 5, obtuse, 1.3 mm long; stamens 5, alternate with the corolla lobes and inserted on the upper part of the corolla tube; filaments white, 1 mm long, pubescent at the base; anthers 1 mm long, yellow; style 2.5 mm long, greenish, enlarged at the base, pubescent with long white hairs at the middle; stigma capitate, indistinctly 5-lobed, green.

FRUIT: October. Clustered on the short peduncle; fruit sessile, globose to obovoid, 4.5-6 mm diameter, red, smooth; calyx persistent; 2 seeds in each fruit. Seeds hard, 2.7-3 mm long, 1.8-2 mm wide, 1.2-1.3 mm thick, ovate to elliptic, flattened on one side, white, smooth.

TWIGS: 0.8-1 mm diameter, brown, flexible; young twigs with curved white hairs, becoming glabrous; leaf scars small, wide crescent-shaped, usually roughened and indistinct; 1 bundle scar; pith brownish, continuous, one-fifth of stem. Buds gray-brown, ovoid, acute, 0.5 mm long.

TRUNK: Bark brown, peeling into small, short flakes which are easily rubbed off or shredded into long, thin strips. Wood soft, nearly white, with a small pith.

HABITAT: Rocky woodlands, thickets, roadsides, fence rows, pastures, prairies, upper edge of stream banks, rich flood plains, sandy or rocky hillsides and flats, usually in dry soil.

RANGE: Pennsylvania, west to Minnesota, southwest to Nebraska, south to Texas, east to Florida, and northwest to Ohio. Introduced in New England.

This is the only red-fruited *Symphoricarpos,* and it appears only in the southeastern part of our area so should not be confused with other shrubs.

It is considered a weedy plant in agricultural regions, tending to take over a pasture and encroaching on the prairies. It forms thickets by sending out runners as long as 4 meters. The plants are usually about 60 cm high, but in wooded areas may reach 1 meter. The young shoots and runners are often eaten by cattle or sheep and this tends to keep the plant under control.

It is a good cover plant for birds and small mammals, but should not be planted unless it can be controlled. The fruits are persistent through early winter but apparently are not eaten to any extent by the birds. Often some of the smaller birds build their nests in the dense thickets.

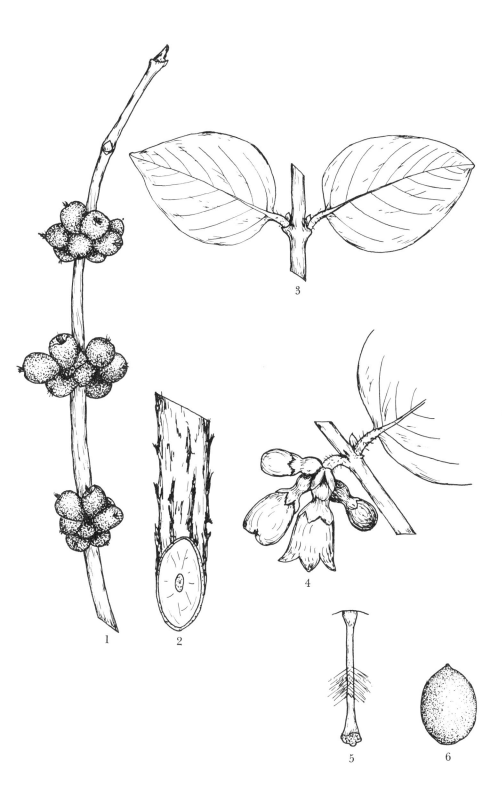

1. Winter twig with fruit
2. Bark of older stem
3. Leaves

4. Flower cluster
5. Pistil
6. Seed

CAPRIFOLIACEAE
Linnaea borealis L. ssp. longiflora (Torr.) Hult.

Linnaea americana Forbes

Twin flower

A repent, suffrutescent plant up to 60 cm long, hardly a shrub, but with the main stem woody.

LEAVES: Opposite, evergreen, simple; veins impressed above and the laterals obscure beneath. Leaves rotund, obovate or elliptic, 1.5-2 cm long, 8-14 mm wide, tip rounded to obtuse, base broadly cuneate and decurrent on the petiole; margin entire or with 1-4 crenate teeth toward the tip; upper surface dark green, with a few scattered, stiff hairs; lower side pale and glabrous; petiole 3-5 mm long, trough-shaped, stiffly ciliate, the hairs slightly retrorse, the petiole base broad, about halfway around the stem; no stipules.

FLOWERS: June-July; perfect. Peduncles erect, 5-7 cm long, slender, terminal on a short leafy branch, pubescent with short, retrorsely curled hairs and longer, straight, glandular hairs; 2 flowers on each peduncle, pedicels 14-16 mm long, short pubescent and glandular pubescent, 2 oblong ciliate bracts about 3 mm long at the base of the pedicels, and 1-2 small bracts just below the flower; pedicels curved at the apex so the 2 flowers are at right angles to the peduncle; sepals 5, lance-linear, 2.5-3 mm long, green, ciliate, and glandular; corolla tube campanulate, 8-11 mm long, pink, glabrous outside and long hairy inside; the 5 blunt lobes 3-4 mm long and somewhat spreading; stamens 4, paired, the upper 2 filaments 8-10 mm long, white, attached about midway on the corolla tube, the lower 2 filaments 5-6 mm long and attached near the base of the tube; anthers pale yellow; ovary globose or ovoid, occasionally with a prolonged neck, green, short pubescent and glandular pubescent, 1.2-1.5 mm long, subtended by 2 ovate bracts 1-1.5 mm long, pubescent and densely glandular stipitate, and 2 small, stiffly ciliate, lance-ovate bracts; style slender, 11-13 mm long, exserted; stigma capitate.

FRUIT: Early August. Fruit ovoid, 2.5 mm long, 1.25 mm wide, red-brown, with a few short, glandular hairs and nearly surrounded by the glutinous bracts; 1 seed. Seed ovoid or ellipsoid, 1.5 mm long, 0.8-1 mm diameter, apiculate, the surface smooth with a crease on one side, pale yellow-brown.

TWIGS: 0.7-0.9 mm diameter, red-brown, flexible, semiglossy, and with semirigid, curled, retrorse pubescence; leaves evergreen and the leaf scars on old stems indefinite; pith white, porous, or the stem hollow, three-fourths of twig. Buds naked, consisting of a series of opposite, scale-like leaves 2-2.5 mm long, 1 mm wide without petioles, green, ciliate, often with minute red spots on the outer scale leaves; buds often alternate, that is, an axillary bud on one side of the stem, and the next bud above will be on the opposite side of the stem.

TRUNK: Bark red-brown, smooth, and with retrorse, curled pubescence. Wood soft, greenish-white, mostly of porous pith, hardly wood.

HABITAT: Moist hillsides, usually in the woods or on shaded, moist flood plains; often growing with moss and sphagnum.

RANGE: Alaska, east across northern Canada to Newfoundland, south to West Virginia, westerly to South Dakota, south into Colorado, west to northern California, and north to British Columbia. The Eurasian form is also found in Alaska.

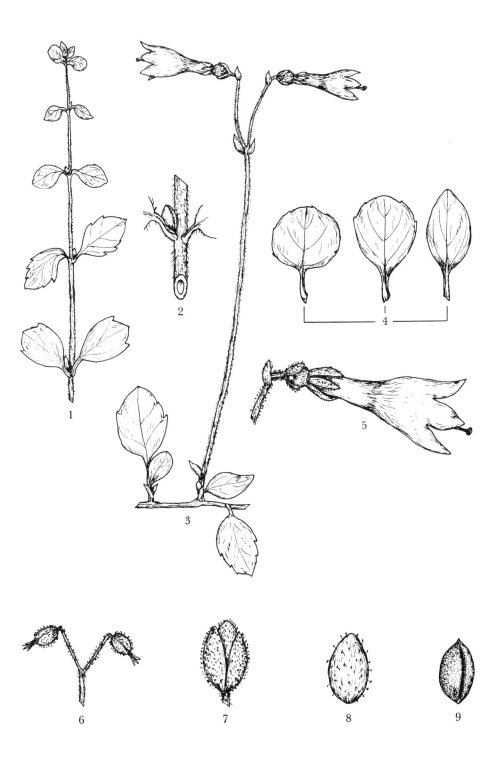

1. Winter twig, usually horizontal
2. Detail of twig
3. Leaves and flowering stalk
4. Leaf variation
5. One flower
6. Fruit
7. Fruit with bracts
8. Fruit
9. Seed

CAPRIFOLIACEAE
Viburnum edule (Michx.) Raf.

V. opulus L. var. *edule* Michx.;
V. pauciflorum Pylaie ex T. & G.;
V. eradiatum (Oakes) House

Mooseberry, squashberry

A shrub to 1 m high, erect or somewhat decumbent.

LEAVES: Opposite, simple, usually 3-lobed but often without lobes, the main veins of the lobes originating 3-8 mm above the petiole. Blade of the lobed leaves broadly ovate, 5-10 cm long, 5-9 cm wide; unlobed leaves broadly elliptic and definitely pinnately veined; tip acute or abruptly short acuminate, base cuneate, truncate or subcordate; margin ciliate, coarsely toothed, 1-2.5 teeth per cm, mucronate, the sinuses broadly rounded; upper surface dark green, glabrous; lower surface paler, pubescent and glandular along the veins, densely so in the vein angles; soft, somewhat leathery; often a large gland on the blade near the petiole; petiole 2-2.5 cm long, grooved above, sparsely pubescent; no stipules.

FLOWERS: Early June; all perfect. Cymes glandular, 2-4 cm across, loosely flowered, 15-25 flowers; cymes on short, lateral branches with 2 leaves; peduncles 1-2 cm long, branches 5-12 mm long; pedicels 1-1.5 mm long, a linear bract 2-5 mm long at the base of the branches and pedicels, green with reddish or brown margin; hypanthium cylindric to narrowly campanulate, 1.5 mm long, green, glabrous; lobes 5, short, 0.25 mm long, 0.75 mm wide, acute, pinkish tip; corolla 5-6 mm across, white, campanulate, 1.5 mm deep, 2 mm across, lobes spreading, oblong, 1.5-2 mm long, rounded; stamens 5, filaments 1 mm long, white, inserted midway on the corolla tube between the lobes; anthers yellow; style short, stout, 1 mm long; stigma capitate. Corolla and stamens fall as a unit.

FRUIT: August. Single or in clusters of 2-10. Drupes ellipsoid, 8-11 mm long, 6-9 mm wide, bright red, smooth, glossy, translucent, soft, acrid, juicy, with 1 seed. Stone obovate, flat, 7.3 mm long, 5.2 mm wide, 2 mm thick, yellow-brown or pinkish, apex often retuse, surface dull, muriculate.

TWIGS: 1.3-1.6 mm diameter, yellow-brown, 3 ridges extending downward from the leaf scars, glabrous, the lenticels inconspicuous; 1 year old twigs gray-brown; leaf scars deeply U-shaped or V-shaped; 3 bundle scars; pith white, continuous, about one-third of stem. Buds purplish-red, scales in pairs, the outer 2 completely connate, middle pair papery and partly connate, the inner scales purplish, distinct; terminal bud ellipsoid, keeled on 2 sides, 6-8 mm long, 2-2.5 mm wide, glabrous; laterals obovoid, rounded at the apex, keeled on the sides, stalked, often yellowish at the base.

TRUNK: Bark gray-brown, thin, slightly split but not furrowed, the pustule-like lenticels evident, inner bark green. Wood fine-grained, soft, nearly white, with a large, white pith.

HABITAT: Moist, wooded hillsides, usually in heavy shade; rock ledges along creek banks, beneath trees or large bushes.

RANGE: Alaska, east through northern Quebec to Newfoundland, south to Pennsylvania, west across Wisconsin and South Dakota to Oregon, and north to British Columbia. Also in Colorado.

V. edule is normally inconspicuous, but the bright red, autumn foliage causes it to stand out quite sharply. It is not a dense shrub, often with one main stem and only 2-3 branches. Only about 40 percent of the flowers produce mature fruit.

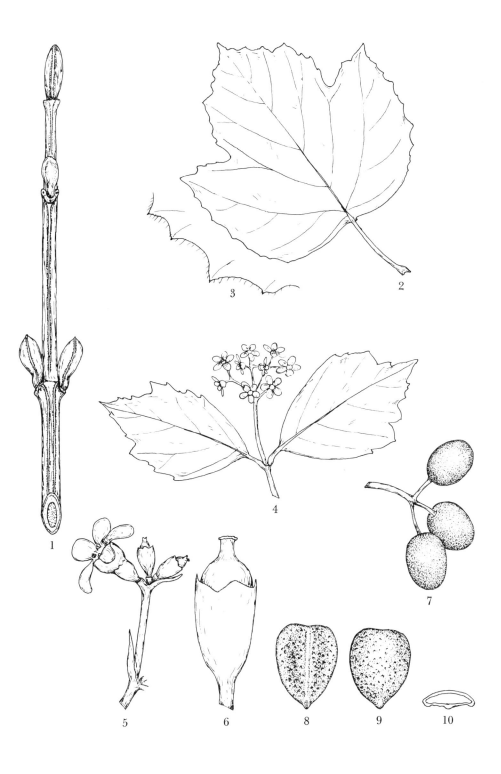

1. Winter twig
2. Leaf
3. Leaf margin
4. Flowering cyme
5. Flower
6. Calyx and pistil
7. Fruits
8. Ventral view of seed
9. Dorsal view of seed
10. Transverse section of seed, dorsal side up

477

CAPRIFOLIACEAE
Viburnum lentago L.

Sheepberry, nannyberry, wild raisin

A shrub to 5 m high, with the branches on the upper one-third; often in small open colonies.

LEAVES: Opposite, simple. Ovate or narrowly ovate, 6-8 cm long, 2.5-3.5 cm wide; tip abruptly acuminate, occasionally acute, base cuneate or rounded; margin sharply and finely serrate, 6-10 teeth per cm, spreading or incurved, callous at the tip, sinuses rounded; upper surface dark green, glabrous, the lower surface only slightly paler, sparsely and finely stellate-peltate pubescent, with red stellate hairs on the midrib; thin, coriaceous; petiole 1-1.5 cm long, grooved above, winged at least part way, wing often with irregular margin, ciliate, a few red stellate hairs and a few white hairs; no stipules.

FLOWERS: June; all perfect. Cymes sessile, 3-5 rays with red scurfy hairs, round-topped, many flowers; a small bract 0.5 mm long at the base of the branches; pedicels 1-2 mm long, glabrous or with scattered red scurf; hypanthium obconic, 1.5-2 mm long, 1 mm wide, pale green, glabrous; lobes 5, triangular, 0.5-1 mm long, glabrous, greenish-white or pinkish; corolla gamepetalous, lobes divided nearly to the base, 2.5-3 mm long, 2 mm wide, oval to ovate, spreading; stamens 5, inserted on the base of the corolla between the lobes; filaments 3-4 mm long, white; anthers lemon-yellow; pistil stout, broad based, 1-1.5 mm long, nearly 1 mm wide, green; stigma capitate, irregular.

FRUIT: September. Drooping or erect clusters of 15-30 fruits. Drupes ellipsoid, 10-13 mm long, 9-11 mm wide, somewhat flattened, dark blue to nearly black with a waxy bloom, glossy when the bloom is removed, smooth, 1-seeded. Stone broadly oval, often nearly circular, 7-9 mm long, 7-8.5 mm wide, 1.4-2 mm thick, sides slightly rounded, surface muriculate.

TWIGS: 2.8-3 mm diameter, yellow to reddish-brown in the first year, gray-green in the second year; pruinose, smooth, glabrous, rigid, the lenticels nearly circular and pale orange; leaf scars narrow, angular; 3 bundle scars; pith white, continuous, rather solid, one-third of stem. Terminal flower bud 16-23 mm long, 5-6 mm thick at the bulge, lanceolate with the bulge one-third of the way up from the base, tip long tapered, acute, often curved; terminal leaf bud 10-12 mm long, 3.5 mm wide, tapered from the base, tip acute or obtuse; lateral buds 3-9 mm long, 2-2.5 mm wide, one of the opposing pair usually much larger than the other; scales valvate, purple-brown, scurfy.

TRUNK: Bark of branches gray, smooth except for the raised lenticels; bark of the trunk grayish red-brown, with thin, somewhat squarish plates. Wood hard, fine-grained, nearly white, with no line of sapwood separation.

HABITAT: Open woods, hillsides, stream banks, margin of a woods, along roadways; dry or moist conditions, in sandy, rocky, rich or clay soils.

RANGE: Southern Manitoba, east to Quebec and New England, south to New Jersey and in the mountains to Georgia; Pennsylvania, west to Illinois and South Dakota, north to North Dakota; one collection record from Missouri and one, possibly escaped, from Colorado.

One of the most distinctive characteristics of *V. lentago* is the abruptly acuminate leaf tip. Specimens of *V. rufidulum* with sparse, red tomentum on the petiole have often been misidentified as *V. lentago*. However, *V. lentago* is more northern and the ranges of the two species do not overlap in our area. It is also confused with the more southern *V. prunifolium,* which has shorter and definitely pinkish winter buds.

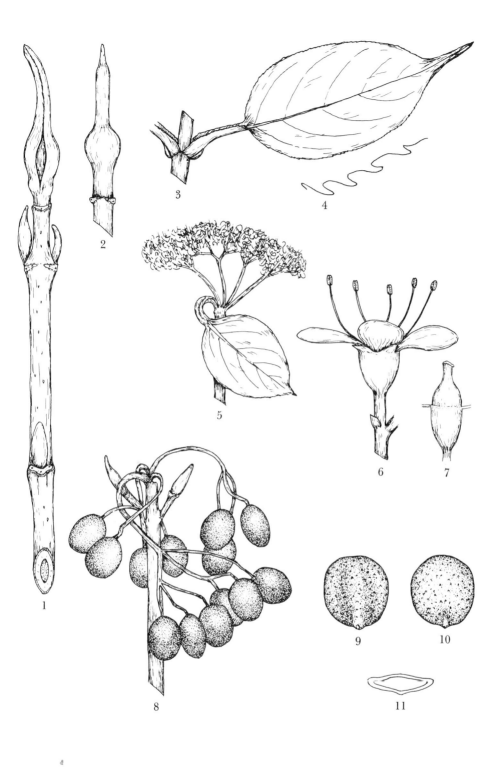

1. Winter twig
2. Edge view of end bud
3. Leaf
4. Leaf margin
5. Inflorescence
6. Flower

7. Pistil
8. Fruit cluster
9. Ventral view of seed
10. Dorsal view of seed
11. Transverse section of seed, dorsal side up

CAPRIFOLIACEAE
Viburnum opulus L. var. *americanum* Ait.

V. trilobum Marsh.; *V. opulus* ssp. *trilobum* R. T. Clausen; not *V. americanum* Mill.

Highbush cranberry

A shrub to 4 m high, the branches high; often in colonies.

LEAVES: Opposite, simple, 3-lobed, palmately veined, the veins prominent below, lobe sinuses rounded. More or less rotund in outline, 5-12 cm long and about the same width; lobe tips acuminate, often divergently falcate; leaf base broadly cuneate, truncate or rounded; margin coarsely toothed, 1-3 teeth per cm, the lobes occasionally entire, stiffly ciliate; upper surface dark green, a few stiff hairs near the margin and a few scattered sessile and short stipitate glands on the veins; main veins impressed near the petiole; lower surface paler, glabrous or stiffly pubescent, a few glands on the veins and tufts of hairs in the vein axils; petiole 12-25 mm long, 2 or more circular glands, often stalked, at the apex, glabrous or pubescent; stipules petiolar, linear, glandular, 4-6 mm long, often slightly enlarged at the tip.

FLOWERS: Mid-June; perfect or neutral. Cymes terminal, 8-13 cm across, nearly flat-topped, the branches glabrous or finely pubescent, often glandular; a linear, early deciduous bract at the base of each branch, the lower bracts up to 17 mm long; pedicels 2-3 mm long; outer flowers neutral with expanded petals, 2-2.5 cm across; hypanthium of perfect flowers narrowly campanulate, 1.5 mm long, 1 mm wide, enlarged slightly to form a shallow calyx with 5 hardly discernible lobes on the fertile flowers and definitely triangular on the neutral flowers; corolla tube campanulate, 1.5 mm long, 1.5 mm wide, white; petals 5, ovate with rounded tip, 1-1.2 mm long, 1.5 mm wide, white, recurved; tube with some pubescence at the base inside; stamens 5, alternate with the petals, filaments tapered, 4 mm long, white, long exserted; anthers yellow; petals and stamens fall as a unit; stylopodium stout, long dome-shaped, 0.75-1 mm long, 0.5-0.7 mm wide, whitish; stigmas 2-3, sessile on the stylopodium, globular, yellowish.

FRUIT: Late August-September. Clusters 5-8 cm across, pendent, 10-35 drupes; drupes bright red, ellipsoid to globose, 8-12 mm long, 7-10 mm wide, smooth, glossy, firm, 1-seeded; stones broadly obovate, 8.6 mm long, 7.7 mm wide, 2.7 mm thick, slightly concave on the dorsal side and rounded on the ventral, a small tip at the base, yellow-brown to pinkish, surface muriculate.

TWIGS: 2.4-2.7 mm diameter, rigid, tan-brown, glabrous, smooth, slightly ridged, lenticels not noticeable; leaf scars shallow U-shaped, dark brown; 3 bundle scars; pith white, continuous, slightly 6-angled, three-fourths of stem. Buds ovate, 5-8 mm long, 2-3 mm wide, purplish-red, stalked, rounded tip, glabrous; scales valvate, the outer ones totally fused, the inner scales green, fused; bud gummy inside.

TRUNK: Bark light gray-brown, tight but with irregular, longitudinal cracks. Wood light, soft, close-grained, nearly white, with no distinguishable sapwood.

HABITAT: Wooded lakeshores and low, wet woods; moist, springy, rocky, brushy hillsides.

RANGE: British Columbia across southern Canada to Newfoundland, south to Pennsylvania, west across northern Illinois to South Dakota, Wyoming, Idaho, and Washington.

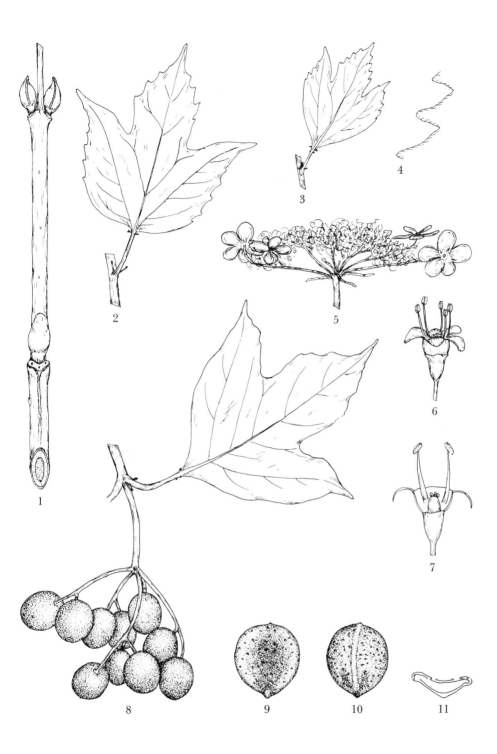

1. Winter twig
2. Leaf
3. Leaf variation
4. Leaf margin
5. Inflorescence
6. Flower
7. Cutaway of flower
8. Fruit cluster
9. Dorsal view of seed
10. Ventral view of seed
11. Transverse section of seed, dorsal side up

CAPRIFOLIACEAE
Viburnum prunifolium L.

Black haw, sweet haw

Open shrub to 2 m high, the branches near the top.

LEAVES: Opposite, simple. Oval to elliptic, 4-8 cm long, 2.5-4.5 cm wide; acute to abruptly short acuminate, the base cuneate; margin with incurved teeth, 7-9 per cm; dark yellow-green and glabrous above, paler and glabrate below, often a few rusty, stellate hairs at first; petiole 5-12 mm long, with or without a wing, glabrous, broadly grooved above and often reddish; no stipules.

FLOWERS: April-May; perfect. Cymes rounded, sessile or on a very short stalk, usually with 4 main rays, all branches glabrous. Hypanthium cylindric, 2-2.3 mm long, green, glabrous; calyx lobes 5, obtuse, 1 mm long, green, glabrous; corolla rotate, the 5 lobes white, ovate, 3 mm long; stamens 5, inserted at the base of the corolla and alternate with the petals, filaments 4.5-5 mm long, white, the anthers yellow; the exposed rounded top of the ovary glabrous, green; style white, 1-1.2 mm long, broadened at the base; stigma capitate, hardly wider than the style. At maturity, the petals and stamens fall as a unit.

FRUIT: September. Pedicels 3-6 mm long, red, glabrous; fruit compressed oval, 12-13 mm long, 8-9 mm wide, 6-7 mm thick, bluish-black with a slight bloom, the skin smooth and tough, the flesh mealy, sweet, and edible; calyx and style persistent; 1 seed. Seeds oval, flat, 8-11 mm long, 6-8 mm wide, 2-2.8 mm thick, dark brown, muricate, the surfaces somewhat rounded.

TWIGS: 1.4-1.8 mm diameter, light brown at first, becoming gray-brown, glabrate but with some reddish scurf; leaf scars crescent-shaped with large ends and middle; 3 bundle scars; pith white, continuous, one-third of stem. Flower buds 12-14 mm long, 4-5 mm thick at the widest part, abruptly bulged below the middle, tapered above to an obtuse apex and below to a broad base, the scales valvate, pinkish-gray, covered with a silvery scurf; lateral buds 4-7 mm long, or shorter on the lower part of the stem.

TRUNK: Bark dark gray, appears checkered with shallow furrows and flat, squarish ridges. Wood hard, heavy, brownish, with a wide white sapwood.

HABITAT: Low woods along streams, flood plains, at the base of a bluff or on the drier, upper slopes of ravines.

RANGE: Florida to Texas, north to Kansas and Iowa, east to Connecticut, and south to Georgia.

This species and *V. rufidulum* are the only Viburnums in the southern part of our area and their ranges do not overlap with others in any part of the central prairie states. In the winter condition, the two species can be separated by the definite gray-pink buds of *V. prunifolium* and the dark red-brown buds of *V. rufidulum*. During the summer, the dark red-rufescent tomentum on the petioles, the broadly winged petiole and the lustrous, coriaceous leaves of *V. rufidulum* should distinguish it from *V. prunifolium*, which is without these characters. Occasionally there appears to be some intergradation between the two, or perhaps a wide variation in each, and individual specimens may be difficult to determine.

The fruits of both species are sweet and quite palatable but are seldom used in the home, partly because of the comparative scarcity and partly from the dryness and the small amount of flesh surrounding the rather large seed.

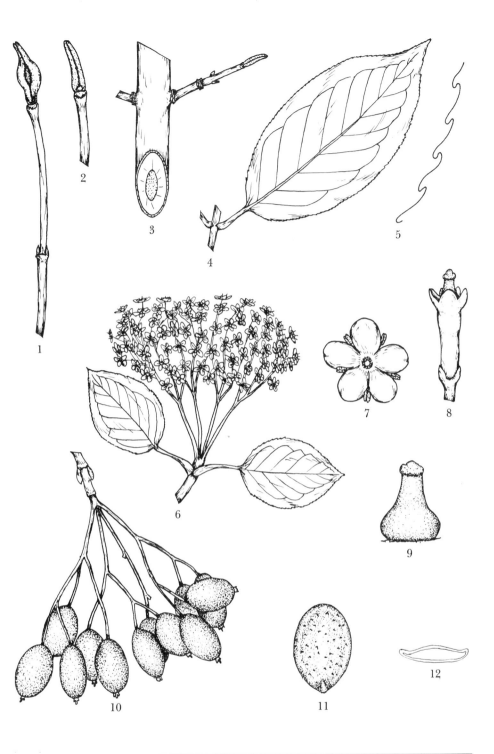

1. Winter twig with flower bud
2. Twig with leaf bud
3. Spur branch
4. Leaf
5. Leaf margin
6. Inflorescence
7. Flower

8. Calyx and pistil
9. Pistil
10. Fruit cluster
11. Seed
12. Transverse section of seed, ventral side up

CAPRIFOLIACEAE
Viburnum rafinesquianum Schult.

V. affine Bush var. *hypomalacum* Blake;
V. pubescens Torr., not (Ait.) Pursh;
incl. var. *affine* (Bush) House

Downy arrowwood

A shrub to 2 m high with low branches, dense in the open and rather scraggly when in the woods.

LEAVES: Opposite, simple, a lateral vein or main branch directly to each tooth. Ovate to broadly elliptic, 4-7 cm long, 3-5 cm wide, tip acute or somewhat acuminate, base mainly rounded or subcordate, occasionally broadly cuneate; margin coarsely toothed, 1-2 teeth per cm, ciliate, usually not toothed near the base; upper surface dark yellow-green, glabrous to sparsely appressed pubescent, red glandular near the base; lower surface slightly paler, variable from nearly glabrous with pilose veins to densely pilose over the whole surface, some of the hairs stellate; soft, leathery; petiole 1-6 mm long, shortest on the leaves just below the inflorescence, sharply grooved above, pubescent; stipules linear, 5-10 mm long, green, pubescent, often glandular on the margin and tip, ciliate, persistent, petiolar and often longer than the petiole, occasionally with 4 stipules on each petiole.

FLOWERS: Mid-June; perfect. Terminal cymes, rounded, 3-5 cm across, peduncle 2-3 cm long, branches and pedicels glandular puberulent; pedicels up to 5 mm long; a linear glandular bract at the base of the branches and pedicels, the basal ones early deciduous; flowers 6-8 mm across; hypanthium green, cylindric, 2.5-2.7 mm long, 1.6-1.7 mm wide, laterally flattened, finely glandular puberulent; calyx lobes 5, broadly triangular, 0.75 mm long, 1 mm wide, acute, green or reddish; corolla rotate, the tube flat, saucer-shaped, 4 mm across; petals 5, rounded, 2.5-3 mm long, 2.5-3.5 mm wide, white, spreading; stamens 5, attached to the corolla tube; filaments 5-6 mm long, erect or spreading, white; anthers pale yellow; stylopodium bell-shaped, 1.5 mm long, 1.4 mm wide, yellowish; stigma lobes 3-4, greenish.

FRUIT: Late August. Cymes 3-6 cm across, 2-15 fruits, erect; drupes ellipsoid or nearly globose, 8-10 mm long, 6-9 mm wide, smooth, glossy, dark purple to nearly black, not glaucous, sepals and style persistent. Stone oval, flattened, 7 mm long, 5 mm wide, 2.5 mm thick, dark red-brown, the ventral surface ridged along the margins and on the raphe, the dorsal side rounded, both surfaces muriculate.

TWIGS: 1.3-1.5 mm diameter, rigid, brown to gray-brown, pruinose, the lenticels circular, yellowish; leaf scars crescent-shaped with broad ends; 3 bundle scars; pith white, continuous, over one-half of stem. Terminal buds ovoid, acute, 5 mm long, 2.3-2.5 mm wide, not flattened, red-brown, glabrous, the scales paired, ciliate, lightly keeled, a small lateral bud on each side of the main bud; lateral buds on the lower stem similar, smaller, 2-2.5 mm long.

TRUNK: Bark gray, often with a purplish cast, tight, smooth except for the lenticels. Wood hard, fine-grained, light brown, with a white pith.

HABITAT: Variable; dry, wooded ridges to springy hillsides, or in open areas on rocky hillsides; usually in rocky, clay soil.

RANGE: Manitoba, east to Quebec, south to North Carolina and Georgia, west to Arkansas and Oklahoma, north through Missouri to eastern South Dakota and North Dakota.

In our area var. *rafinesquianum* with pilose leaves and long stipules cannot be differentiated from var. *affine,* which is without those characters. Both conditions may exist in the same colony, or even a recombination of the characters may appear on the same plant.

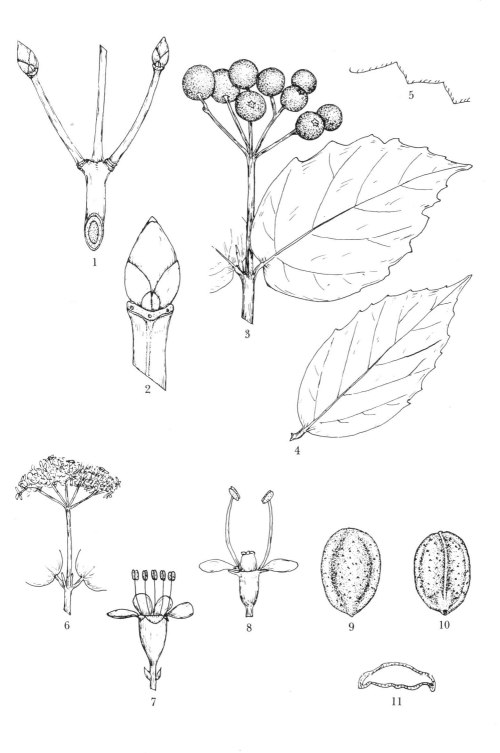

1. Winter twig
2. Detail of twig
3. Leaf and fruits
4. Leaf variation
5. Leaf margin
6. Inflorescence

7. Flower
8. Diagrammatic cutaway of flower
9. Dorsal view of seed
10. Ventral view of seed
11. Transverse section of seed, dorsal side up

CAPRIFOLIACEAE
Viburnum rufidulum Raf.

V. rufotomentosum Small; *V. rufidulum*
(var.) *margarettae* Ashe; *V. prunifolium*
L. var. *ferrugineum* T. & G.

Southern black haw, wild raisin,
rusty nannyberry

A shrub to 3 m high, the branches on young plants usually coarse.

LEAVES: Opposite, simple. Ovate, obovate, oval or broadly elliptic, 4-8 cm long, 3-6 cm wide, the small leaves often obcordate; margin finely serrate, 4-7 teeth per cm, tip acute, rounded or indented; base cuneate, subtruncate or rounded; upper surface dark green, lustrous, a few rusty hairs on the midrib; lower surface paler, rusty tomentose when young, scattered hairs on mature leaves, the midrib often retains the tomentum; thick, coriaceous or thin and firm; petiole 5-12 mm long, winged, densely or lightly red-brown, stellate pubescent; no stipules.

FLOWERS: Late April; perfect. A flat-topped or rounded cyme 6-8 cm across, 3-4 rays, each 2-3 cm long, glabrous or with a few scattered hairs; cymes sessile, 120-130 flowers, each with a small bract at the base. Hypanthium cylindric, 1.7-2 mm long, green, glabrous; calyx lobes 5, distinct, 0.5 mm long, acute; corolla rotate, white; petals 5, ovate to oval, 3.5-4 mm long; stamens 5, attached to the corolla between the lobes; filaments white, 3.5-4.5 mm long; anthers pale yellow; style stout, 1.8-2 mm long; stigma 3-4 lobes, greenish, no larger than the style.

FRUIT: October. Drooping cymes; rays and pedicels glabrate; pedicels 3-5 mm long, red or purple. Drupe ellipsoid, flattened, 13-15 mm long, 9-11 mm wide, 7-7.5 mm thick, reddish when immature, finally dark blue to black, glaucous, a minute tip at the apex; flesh sweet and edible; 1 seed. Stone oval, flat, 9-11 mm long, 6.5-7 mm wide, 2.5-3 mm thick, dark brown; a definite tip at the base, the raphe keel short; the ventral surface with 2 rounded ridges and concave between; the dorsal side slightly rounded; surface muriculate.

TWIGS: 1.8-2.2 mm diameter, fairly rigid, gray-brown, glaucous, densely tomentose with red-brown hairs at first, becoming nearly glabrous; leaf scars wide crescent-shaped, raised; 3 bundle scars; pith white, continuous, one-third of stem. Flower buds 9-12 mm long, 5-8 mm wide, obtuse, swollen abruptly just below the middle; 1 pair of valvate scales, red-brown, tomentose; leaf buds 5-7 mm long, 2-2.5 mm wide, rounded ends, red-brown and tomentose.

TRUNK: Bark dark gray or nearly black, shallow fissures and short, blocky ridges. Wood hard, red-brown, with a wide white sapwood.

HABITAT: Woods, roadsides, ravines; rocky dry soils or rich moist woodland soils; more common on rocky, well-drained hillsides.

RANGE: Virginia, west to Ohio and Kansas, south to Texas, east to Florida, and north to North Carolina.

The shape and texture of the leaves of *V. rufidulum* are quite variable and often cause a misidentification. The young leaves are often thin and hardly coriaceous; and should they have an acute tip, as they often do, they are mistaken for *V. prunifolium*. However, they usually have a rufescent tomentum that should distinguish the species. Occasionally a young plant will have coarse branches, 3-4 mm diameter, but the branches on old plants are always smaller.

In some areas the plant becomes a small tree, depending partly on the amount and density of the overgrowth. The crown is usually high and irregular, but in the open the branches may come close to the ground. The plants are mostly single, but occasionally may be found in small colonies.

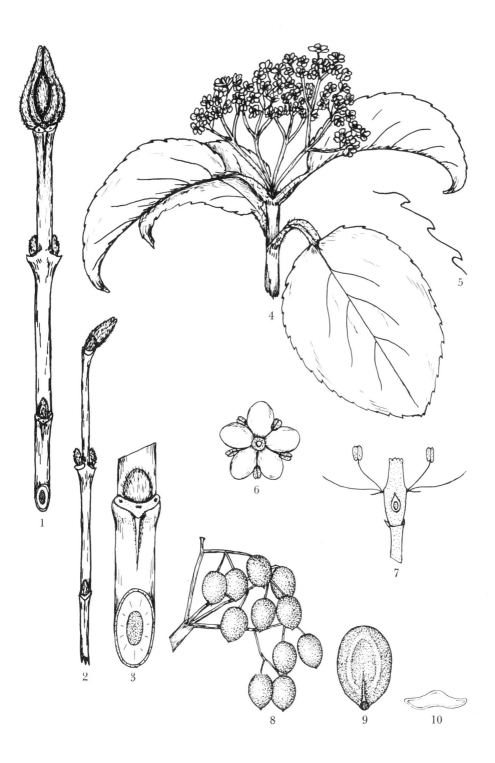

1. Winter twig with flower bud
2. Twig with leaf bud
3. Detail of twig
4. Leaves and inflorescence
5. Leaf margin

6. Flower
7. Diagram of flower
8. Fruit cluster
9. Ventral view of seed
10. Transverse section of seed, dorsal side up

CAPRIFOLIACEAE
Sambucus canadensis L.

Incl. var. *submollis* Rehd.

Elderberry

A shrub to 2 m high, forming colonies, the branches near the top.

LEAVES: Opposite, pinnately compound, 15-20 cm long, 5-7 leaflets. Leaflets elliptic to ovate, 6-12 cm long, 3-5 cm wide, the terminal leaflet widest near the middle or toward the outer end; margin sharply serrate with about 4 teeth per cm; tip acuminate, base cuneate to slightly rounded; upper surface dark green, semilustrous, glabrous; lower surface pale, slightly pubescent, especially along the midrib; petiole 5-8 cm long; petiolules 3-6 mm long; petiole, rachis, and petiolules grooved above. Occasionally the lower leaflets are 3-parted.

FLOWERS: May-June; perfect. Terminal cymes 8-18 cm wide, rounded or flat-topped, usually 5 rays, the ray branches glabrous, green fading to white toward the flowers; flowers sessile or with a white pedicel 1-4 mm long, usually a sessile flower at the base of the upper cyme branches. Hypanthium white, ovoid, 1-1.2 mm long, glabrous; calyx lobes 5, white, triangular, 0.8-1 mm long, the margin often irregular; corolla radiate, petals 5, white, 2.5-3 mm long, orbicular to broadly elliptic, the margin usually revolute; stamens 5, attached to the base of the corolla between the petals; filaments white, 1.8-2 mm long; anthers pale yellow, 1 mm long; top of the ovary 1-1.3 mm wide, white; style obsolete; stigmas 3, white, 0.6-0.7 mm long.

FRUIT: August-September. Rounded or flat-topped clusters, usually drooping when ripe. Pedicels thin, 3-4 mm long, glabrous. Fruit berry-like, globose, 4-6 mm diameter, deep purple, smooth, glossy; 3-4 seeds. Seeds obovate, 2.5-3 mm long, 1.6-2 mm wide, 1 mm thick, yellow, rough, usually with 2 flattened surfaces and one rounded surface.

TWIGS: 3-4 mm diameter, coarse, rigid, glabrous, light yellow-brown or gray-brown; the lenticels prominent; leaf scars crescent-shaped to shield-shaped, wide; 5 bundle scars; pith white, continuous, three-fourths of stem. Buds 1.5-3 mm long, ovoid, red-brown, glabrous.

TRUNK: Bark yellow-brown, not fissured; lenticels large and raised, causing a rough bark. Wood white, soft, with a large pith.

HABITAT: Usually in moist, rich soils; along the margin of a woods, stream banks, fence rows, and railroad banks.

RANGE: Nova Scotia, west to Quebec and Manitoba, south to Kansas and Texas, east to Florida, and north to New England. Also in Mexico and the West Indies.

This is the only elderberry in the southern part of our area and is not easily mistaken for any other plant. The large clusters of white, sweet-scented flowers and of dark purple fruits are outstanding characteristics. The only other elderberry of the area is the red-fruited berry of the northern states.

The individual plants of elderberry are short-lived but new shoots are produced from the roots and the colony is perpetuated. The branches are coarse, mostly pith and are easily broken. This was formerly one of the main sources of pith for sectioning plant material and for other scientific purposes.

The fruit is edible but not palatable when fresh. However, it is excellent when made into pies, jams, puddings, and beverages. Birds and mammals are common visitors to the plants when the fruit is ripe.

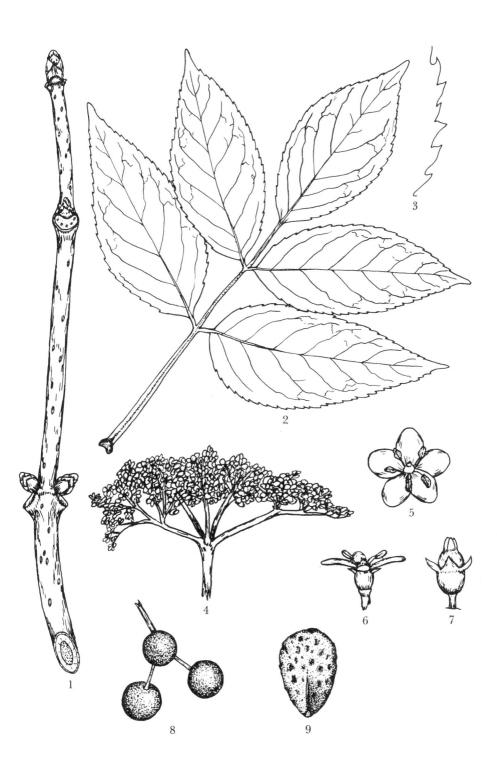

1. Winter twig
2. Leaf
3. Leaflet margin
4. Flowering cyme
5. Face view of flower
6. Side view of flower
7. Pistil, the petals and stamens removed
8. Fruit
9. Seed

CAPRIFOLIACEAE
Sambucus racemosa L. ssp. *pubens*
(Michx.) Koehne

S. pubens Michx.; *S. pubens* var. *leuco-carpa* T. & G.; *S. racemosa* L.
f. *leucocarpa* House; *S. pubens*
f. *leucocarpa* (T. & G.) Fern.

Red elderberry

A coarse shrub to 1.5 m high, not forming colonies.

LEAVES: Opposite, pinnately compound, 8-18 cm long, 5-7 leaflets. Leaflets lance-ovate, oblong or elliptic, 5-10 cm long, 1.5-3 cm wide; tip acuminate, base of terminal leaflet cuneate, the laterals oblique with the basal side rounded; sharply serrate, 3-6 teeth per cm, often callous-tipped; upper surface yellow-green with a few hairs, lower surface paler and pubescent, the hairs coarse; petiole 3-5 cm long, grooved above, glabrous or with short, coarse hairs; terminal petiolule 6-20 mm long, the laterals sessile or with petiolule to 4 mm; stipules linear, 1-2 mm long, or knob-like and about 2 mm across.

FLOWERS: May-June; perfect. Paniculate cyme terminal on new growth, 4-5 cm long, 3-4 cm wide, pyramidal with a definite axis, glabrous or coarsely pubescent. Hypanthium cylindric, 1-1.5 mm long, green, glabrous; calyx lobes 5, triangular, about 0.5 mm long, reflexed at anthesis, united at the base to a narrow rim, greenish-white; corolla radiate, petals 5, oblong, 3 mm long, 1.5 mm wide, reflexed, end rounded and often minutely toothed; stamens 5, alternate with the petals and inserted on the corolla base; filaments 1.5 mm long, white; anthers 1-1.5 mm long, the filament often divided into 2 short branches and the anther locules separated; top of ovary low dome-shaped, 1-1.5 mm wide, green, glabrous; style obsolete; stigmas 3, short cylindric, whitish.

FRUIT: August. Conical clusters terminating a branch. The berry-like fruit globular, 5-7 mm diameter, bright red, semiglossy, the old floral parts persistent; 2-4 seeds per fruit. Seeds ovate-oblong, flattened on the sides and the dorsal side rounded, 2.4-2.7 mm long, 1.8 mm wide, 1 mm thick, yellow, muricate.

TWIGS: 4-6 mm diameter, gray-green or brownish, short pubescent or glabrous; lenticels prominent; leaf scars large, semicircular to cordate, often somewhat glossy; 5 bundle scars; pith yellow-brown, continuous, firm, most of stem. Buds ovoid or obovoid, 9-12 mm long, 5-6 mm wide, short-stalked, nearly glabrous, the lower scales short and brownish-purple, deciduous; inner scales purple with a white, scarious, ciliate margin.

TRUNK: Bark tight, greenish-brown or gray-brown, smooth, except for the raised lenticels. Wood is only a narrow ring of soft, greenish-white material around the brown pith.

HABITAT: Rich or rocky soil in moist open woods or on creek banks, often in rock crevices in shaded areas.

RANGE: Alberta, east across southern Canada to Newfoundland, south to West Virginia, and in the mountains of Georgia and Tennessee, west to Ohio, Iowa, and South Dakota, and northwest to Montana.

Our plants are var. *pubens* with pubescence on the leaves, whereas var. *microbotrys* (Rydb.) Kearney & Peebles is glabrous or with a few hairs, and is more western. Some of the plants of our area are sparsely pubescent and might be placed in var. *microbotrys*. All of the specimens examined from Colorado were var. *microbotrys* and almost totally glabrous. One of the distinguishing characteristics of the winter twig in our variety is the large, definitely purple, winter buds. These stand out sharply, especially if the ground is covered with snow. The brown pith would distinguish this plant in purely vegetative form from the southern *S. canadensis*.

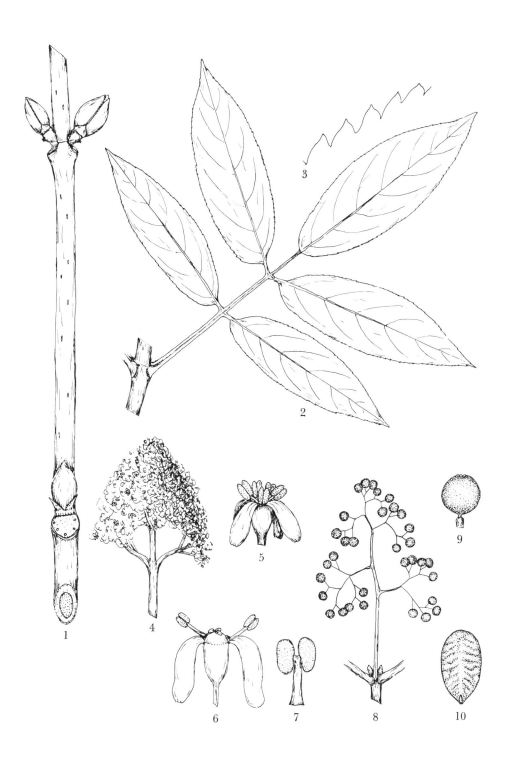

1. Winter twig
2. Leaf
3. Leaflet margin
4. Inflorescence
5. Flower

6. Diagrammatic cutaway of flower
7. Stamen
8. Fruit cluster
9. Fruit
10. Seed

COMPOSITAE
Gutierrezia sarothrae (Pursh) Britt. & Rusby

G. juncea Greene; *G. diversifolia* Greene; *G. divaricata* (Nutt.) T. & G.; *G. euthamiae* T. & G.; *G. ioensis* Lunell; *G. linearis* Rydb.; *G. longifolia* Greene; *Solidago sarothrae* Pursh; *Xanthocephalum sarothrae* (Pursh) Shinners

Broom snakeweed

A dense suffrutescent shrub to 40 cm high with branches close to the ground; overwintering buds often 20 cm above ground.

LEAVES: Alternate, simple. Linear, 1-3.5 cm long, 1-1.5 mm wide, slightly trough-shaped; entire or with a few scabrous projections; base broad, sessile and semiclasping; dark yellow-green, punctate and slightly glutinous on both sides, the surface with scattered, short, stiff, barb-like projections.

FLOWERS: August - September, staminate or pistillate. Inflorescence a bracted, paniculate cyme; involucres narrow, 3-4 mm long, 1.3-1.4 mm wide, with 5 rows of imbricated bracts, 4 bracts per row; bracts 1.2-3.5 mm long, linear, acute, with a green midrib and scarious margins. Ray flowers 4-5, pistillate, fertile, tubular for 2 mm and expanding into a ray 2.5 mm long, 1 mm wide, tip obtuse; rays sometimes absent; ovary cylindric to obconic, 0.8 mm long, 0.5 mm wide, densely pubescent; pappus of several chaff-like, thin scales 0.6 mm long; style yellowish, 2 mm long; stigmas 2, linear, flattened, 0.8 mm long, yellow, hairy on the outer side. Disk flowers 2-3, yellow, tubular, 2-2.3 mm long, 0.8 mm thick, with 5 acute, recurved lobes 0.8 mm long; pappus of several thin scales 1-1.1 mm long; stamens 5, filaments distinct, 0.5 mm long, yellow; anthers 1.4 mm long, yellow, apiculate, connate around the style, which may be rudimentary.

FRUIT: October - November. Achenes cylindric, 1.7-1.9 mm long, 0.5-0.6 mm wide, brown, the 10 low ridges densely pubescent; pappus of white chaff, scale-like, 0.5-0.8 mm long; 1-3 achenes in each involucre.

TWIGS: 0.8-1 mm diameter, green to brown, striate, flexible, coarsely pubescent and scurfy; leaf scars narrow, extending nearly halfway around the stem; 1-3 bundle scars; pith green, continuous, one-third of stem. Buds naked, no scales, only a cluster of minute, green leaves.

TRUNK: Bark brown, split into long thin plates with some shredding. Wood white, soft.

HABITAT: High plains, usually where the sod is broken; dry, overgrazed pastures, roadcuts and embankments; rocky, gravelly or sandy soils, chalk bluffs, and clay hills.

RANGE: In the plains from Washington, east to Manitoba and Minnesota, south to Texas, and west to California. Mexico.

G. sarothrae has one main trunk with many branches within 5 cm of the ground. These branch and rebranch until the crown is a dense structure of fine twigs. It flowers and fruits late in the season, and a freeze does not appear to harm the immature achenes. The leaves may remain on the plant until late winter.

This plant is just within the limits of a woody plant, but in part of its range is definitely woody, with winter buds some distance above the ground. It is one of the first plants to take over a barren area and will tolerate an overgrazed situation. Cattle browse sparingly on it during the winter months. Deer, antelope, and jack rabbits may also feed on the foliage and young stems. Quail and kangaroo rats are fond of the achenes, which form an important part of their diet.

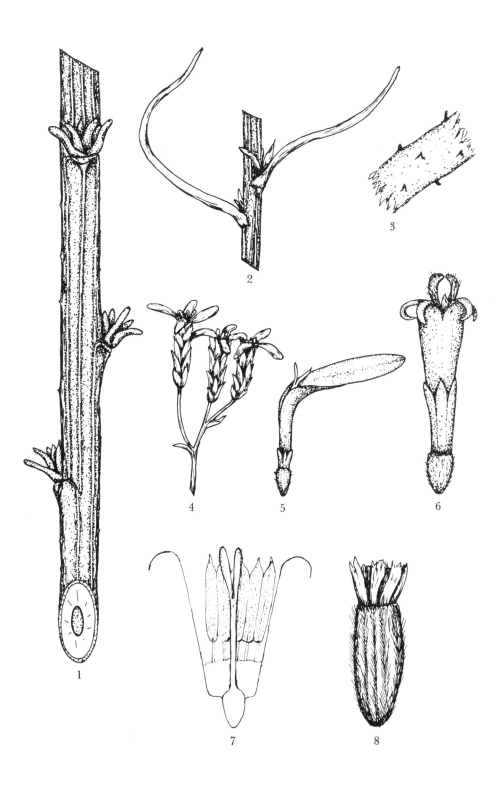

1. Winter twig
2. Leaves
3. Portion of leaf
4. Involucres with flowers
5. Ray flower
6. Disk flower
7. Cutaway of disk flower
8. Achene

COMPOSITAE
Chrysothamnus nauseosus (Pall.) Britt. ssp. graveolens (Nutt.) Piper

Chrysocoma graveolens Nutt.; Bigelowia graveolens Gray; Aster graveolens Kuntze

Rabbitbush

A dense shrub with short trunk and ascending branches to 2 m high.

LEAVES: Alternate, simple. Linear, 3-8 cm long, 1.5-2.2 mm wide, somewhat trough-shaped, lateral veins obscure; margin entire; tip acute, base broad, semiclasping, sessile; dark green and glabrate to tomentulose on both sides; stipules absent.

FLOWERS: Early September; perfect. Flowers in bracted, corymbose heads, terminal on new growth; peduncles green, scurfy-dotted. Involucres 8-9 mm long, 2-2.3 mm wide, mostly 5-flowered; the bracts in 4 rows of 4-5 bracts each, imbricated, lanceolate to linear, 2-7.5 mm long, 0.5-1 mm wide, glabrous with scarious, ciliate margins, yellow-green, strongly convex with an acute tip, the margins revolute near the tip, giving an acuminate appearance; corolla tubular, becoming wider and usually bent at the middle, 8.5-9.5 mm long, 1 mm wide at the top; lobes 5, acute, erect or slightly spreading, 0.8-0.9 mm long, thick margin; pappus 6-7 mm long, white, subplumose, attached at the top of the ovary; stamens 5, filaments distinct, yellow, 5.5 mm long, slender, attached near the base of the corolla tube; anthers yellow, 3 mm long, connate around the style, acuminate tip; ovary narrowly conical, 1.8-2.4 mm long, with slightly appressed white pubescence; style 9-9.5 mm long, yellow; stigmas 2, linear, 3 mm long, pubescent, exserted.

FRUIT: November. Achenes cylindric, 5-7 mm long, 0.8-1 mm wide, base acute, brown, 4-6 small ridges, sparsely pubescent. Pappus of buff, subplumose straight hairs, 5-7 mm long.

TWIGS: 1.4-1.6 mm diameter, rigid, yellow-green, covered with dense, tightly appressed tomentum with the appearance of bark; leaf scars narrowly crescent-shaped; 3 bundle scars; pith white, continuous, one-half of stem. Buds ovoid,

0.6-0.7 mm long, often concealed by old leaf bases.

TRUNK: Bark gray-brown, with shallow fissures, the flat ridges anastomosed, fibrous and slightly shredded. Wood hard, light brown, with a narrow, light sapwood.

HABITAT: Dry plains, roadside embankments and steep ravine walls; sandy or clay soil, usually in areas of broken sod.

RANGE: From North Dakota and Montana south to Texas, west to Arizona and Utah.

In our area ssp. graveolens is the more common form of C. nauseosus and has been described above. The following are the main characteristics of ssp. nauseosus: outer phyllaries tomentose, leaves densely white tomentose, leaves 2-5 cm long, 0.5-1.5 mm wide, tomentum of the stem nearly white, plants about 1 meter high, uncommon in our range. These are the only two subspecies reported for our area, but other subspecies and varieties are common to the west and southwest.

The bushes of C. nauseosus ssp. graveolens are quite dense, and occasionally are of sufficient number to form a heavy cover over the ground, killing or replacing the grasses. Jack rabbits, deer, and antelope browse on the foliage, which at certain seasons forms a large part of their diet. Smaller mammals, especially the ground squirrel, feed on the fruits and use the shrub as a cover for their dens.

● Chrysothamnus nauseosus nauseosus
■ Chrysothamnus nauseosus graveolens

494

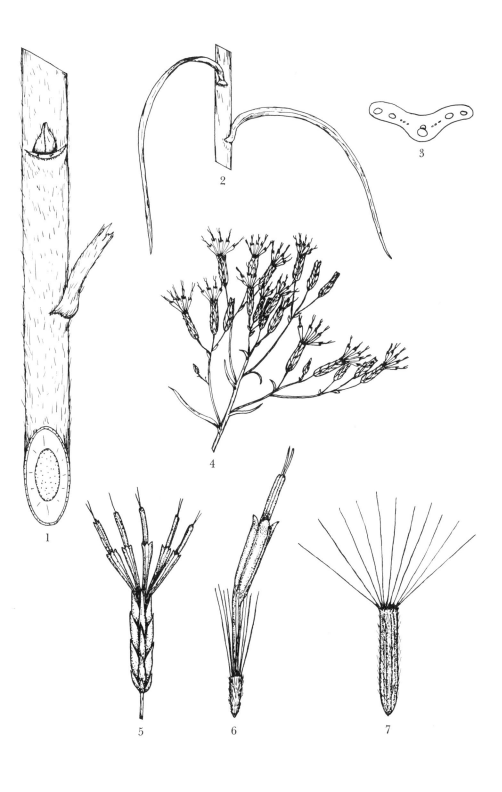

1. Winter twig
2. Leaves
3. Transverse section of leaf
4. Inflorescence
5. Involucre with flowers
6. One flower
7. Achene

COMPOSITAE
Chrysothamnus parryi (A. Gray) Greene ssp. howardii (Parry) H. & C.

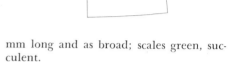

Bigelovia howardii Parry; Chrysothamnus howardii Greene; Linosyris howardii Parry

Chrysothamnus, rabbitbush

A low, spreading shrub to 40 cm high with numerous branches near the ground.

LEAVES: Alternate, simple, somewhat V-shaped in section, linear, 3-5 cm long, 1-2 mm wide, margin entire; acute at the tip; sessile, slightly narrower just above the broad base; surfaces gray-green, densely tomentose on both sides, the midrib hardly visible; stipules absent.

FLOWERS: August; perfect. Inflorescence paniculate, 4-6 cm long, terminating a leafy stem, often overtopped by the upper leaves. Involucres 8-9 mm long, 2-3 mm diameter with 5 rows of narrow, imbricated phyllaries, the short outer phyllaries greenish, pubescent, ciliate; inner phyllaries yellowish, with a hyaline margin; 4-6 flowers per involucre; corolla yellow, cylindric, 5-6 mm long, 1 mm wide at the top with 5 spreading, triangular, acute lobes 1-1.5 mm long; pappus of white, barbed hairs about 7 mm long; stamens 5, filaments pale yellow, slender, distinct, 7-8 mm long; anthers 3 mm long, thin, flattish, acute, more or less united, forming a tube around the style; ovary obconic, 2.5-3 mm long, 0.7 mm wide, greenish, ribbed, densely pubescent; style yellowish, 8 mm long, the 2 stigmas linear, 2 mm long.

FRUIT: September. Involucres spreading to release the achenes. Achene obconic, 5-6 mm long, tapering from a pointed base to 0.5 mm diameter at the top, brown, ribbed, densely pubescent; pappus tawny, barbellate, spreading, 6-8 mm long.

TWIGS: 1 mm diameter, greenish-white with a dense, compact tomentum that appears to be bark; the bark itself green, soft, and succulent; leaves break off at the narrow portion just above the broadened base, and many dried leaves persistent; pith white, continuous, one-half of stem. Old inflorescence persistent over winter. Buds concealed beneath the gibbous leaf bases, broadly ovoid, 0.7-1 mm long and as broad; scales green, succulent.

TRUNK: Bark brown, with low, anastomosing ridges, somewhat loose and scaly. Wood hard, fine-grained, rays obvious, brown with a wide, nearly white sapwood.

HABITAT: Dry, semiarid areas, usually rocky or clay soil; fairly common in the eroded areas along the west edge of the Nebraska panhandle.

RANGE: Wyoming, Colorado, and western South Dakota and Nebraska.

Chrysothamnus is more common to the west and southwest of our area, most of them with the center of population in the Great Basin Region. Our plants, then, are along the extreme eastern edge of the range. They are important browse plants for larger mammals as well as jack rabbits and are large enough to give cover and protection for these animals.

C. parryi has several subspecies or varieties but ssp. howardii, as far as is known, is the only one extending this far east. It is a small plant, rarely over 40 cm high, and has several erect stems from the base. The dense, white tomentum on the leaves gives the plant a gray appearance even from some distance.

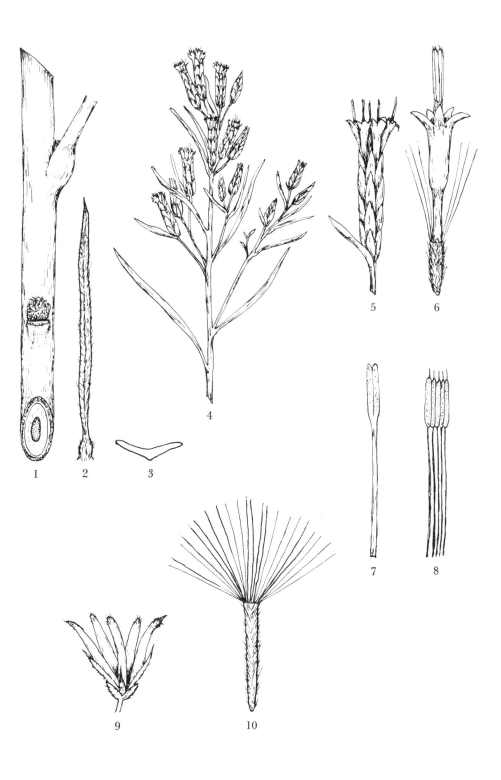

1. Winter twig
2. Leaf
3. Transverse section of leaf
4. Inflorescence
5. Involucre with flowers
6. One flower
7. Pistil
8. Stamens
9. Involucre at seed dispersal time
10. Achene

497

COMPOSITAE
Chrysothamnus pulchellus
(A. Gray) Greene

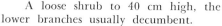

C. pulchellus var. *baileyi* (Woot. &
Standl.) H. & C.; *C. baileyi* Woot. &
Standl.; *Bigelovia pulchellus* A. Gray;
Linosyris pulchellus A. Gray

Rabbitbush

A loose shrub to 40 cm high, the
lower branches usually decumbent.

LEAVES: Alternate, simple. Linear,
1.5-2.5 cm long, 2-2.5 mm wide, ciliolate-
barbed; tip mucronate; base broad, somewhat clasping, sessile; both surfaces green,
granular, the lateral veins obscured; glabrous, thick; stipules absent.

FLOWERS: July-August; perfect.
Corymbose heads terminal on new
growth. Involucres 1.3-1.5 cm long, 4
mm thick, the bracts in 5 rows of 5-6
per row; bracts ovate to obovate, 7-13
mm long, 2.5-3.5 mm wide, keeled, short
acuminate, white with a green tip, thin
margins, slightly spreading at the tip;
5 flowers in each involucre. Ovary obconic, 2-2.5 mm long, slightly ribbed,
greenish, glabrous; pappus white, straight,
barbellate, 1.5-1.6 cm long; corolla tubular, 12-12.5 mm long, 1.5 mm wide at the
top, white, lobes 5, acute, 1 mm long,
erect; stamens 5, the filaments adnate to
the corolla for 3 mm, free for 6-7 mm,
flattened, white at base, yellow and
abruptly narrowed below the anther; anthers 4.5 mm long, yellow, connate
around the style; style 8 mm long, white;
stigmas 2, linear, 6 mm long, flattened,
appressed or slightly divergent, exserted,
pubescent on the outer side.

FRUIT: November. Achene obconic,
5.8-6.1 mm long, 1.2-1.3 mm wide, light
brown, glabrous or a few hairs at the
summit, 4-angled. Pappus dense, straight,
barbellate, 13-14 mm long, buffy, spreading.

TWIGS: 0.5-0.8 mm diameter, rigid,
glabrous, yellowish or green when young,
becoming gray-brown; bark splitting on
branches of the season. Buds hidden under old, persistent leaf bases; pith white,
continuous, one-tenth of stem.

TRUNK: Bark gray to gray-brown,
exfoliated into irregular tight flakes.
Wood hard, brown, with a narrow, light
sapwood.

HABITAT: Dry, sandy soil; often
in loose sand on the tops or sides of low
sandhills.

RANGE: Texas, New Mexico, Arizona, Colorado, Utah, Kansas.

C. pulchellus is a low, rounded,
open shrub to 40 cm high. The short
trunks are often twisted, decumbent, and
quite stocky. Occasionally the sand is
blown from around the base and the
upper portion of the roots shows above
ground. This does not appear to disturb
further growth of the plant.

As the name "rabbitbush" might im-
ply, rabbits eat the foliage and the young
fruiting heads and use the plant as protection from the sun. Mice and kangaroo
rats eat the mature achenes or store the
fruits in their burrows for future use.

C. pulchellus can be distinguished
from the other species of *Chrysothamnus*
in our area by the short leaves with
barbellate margin and by the large involucre, over 1 cm long.

498

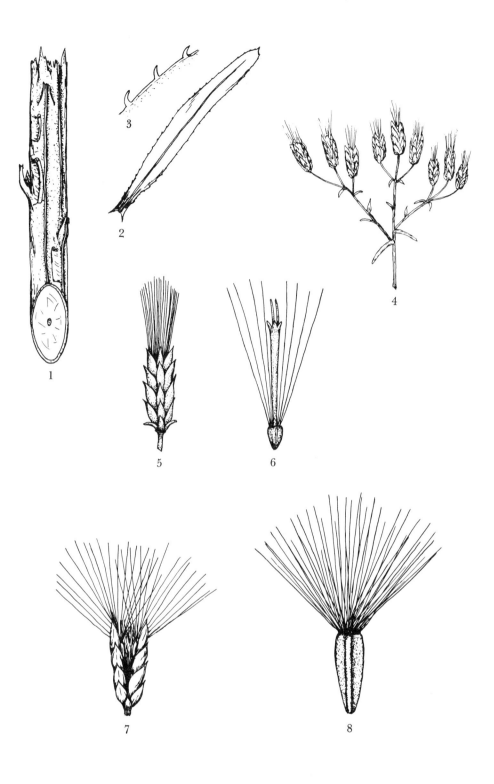

1. Winter twig
2. Leaf
3. Leaf margin
4. Inflorescence

5. Involucre
6. Flower
7. Involucre at seed dispersal time
8. Achene

COMPOSITAE
Chrysothamnus viscidiflorus
(Hook.) Nutt.

C. viscidiflorus var. *tortifolius* Greene;
C. tortifolius Greene; *Bigelowi douglasii*
Gray var. *tortifolia* Gray

Twist-leaf rabbitbush

A small, rounded shrub to 1 m high.
LEAVES: Alternate, simple. Linear,
1-4 cm long, 2-3 mm wide, usually
twisted, 1 main vein with 2 indistinct
veins from the base, tip definitely pointed,
base broad, sessile; margin with small
barbs pointing forward; yellow-green;
glabrous or with minute hairs below; no
stipules.
FLOWERS: Late August; perfect.
Inflorescence a terminal corymb with
leaves reduced to bracts, flat-topped, 5-6
cm high, 5-6 cm across; involucres 6-8
mm long with 5 rows of 3-4 phyllaries,
appressed, 2-6 mm long, keeled, thin
margin, yellow; involucre 5-flowered; pappus of many stiff hairs, antrorsely barbed,
white; corolla tubular, yellow, 4 mm
long, with 5 oblong, acute spreading
lobes 1 mm long; 5 stamens, filaments 3
mm long, free, attached to the corolla
about one-third of the way up from the
base at a small bulge on the tube; anthers 1.5 mm long, united around the
style, apiculate, yellow-brown; ovary obovoid, 1 mm long, 0.7 mm wide, densely
appressed pubescent; style 5-5.5 mm long,
white glabrous, divided near the tip, the
two stigmas cylindric, 0.7-1 mm long,
acute, pale yellow.
FRUIT: October. Involucral bracts
spreading to release the achenes. Achenes
slender, wedge-shaped, 3.5-4 mm long,
somewhat flattened, pale brown, 5-ribbed,
appressed pubescent. Pappus pale yellow-
brown, 5 mm long, minutely and antrorsely barbed.
TWIGS: 1-1.3 mm diameter, light
yellow, smooth and semiglossy; young
twigs greenish brown and puberulent;
leaf scars crescent-shaped, raised; 3 bundle scars; pith white, continuous, one-
third of stem. Buds broadly conical, 2-2.5
mm long, brown, somewhat resinous, the
2 outer scales short and often spreading
at the tip.
TRUNK: Bark light brown with
thin, fibrous shreds. Wood medium hard,
fine-grained, pale brown, with a narrow,
yellow sapwood.
HABITAT: Dry open prairies and
broad valleys, occasionally on open woods
hillsides in the foothills. Sandy or red
gypsum soil.
RANGE: Southern British Columbia, to Montana, south to Western Nebraska and New Mexico, and west to
California.

The only known location of *C. viscidiflorus* in our area is in Sioux County,
Nebraska, the most northwesterly county.
Only a few plants are growing in this
small spot. The group was located and
pointed out to the author by L. C. Anderson, who is doing extensive research
on the genus.
One of the most interesting and outstanding characteristics of the plant is the
twisted, glabrous leaves. Our plants are
small, no more than 1 meter high, and
several of them have been browsed and
broken or stunted in some manner. Farther west, the plant is a rather dense,
somewhat globular shrub varying from 50
to 150 cm high and often as wide.

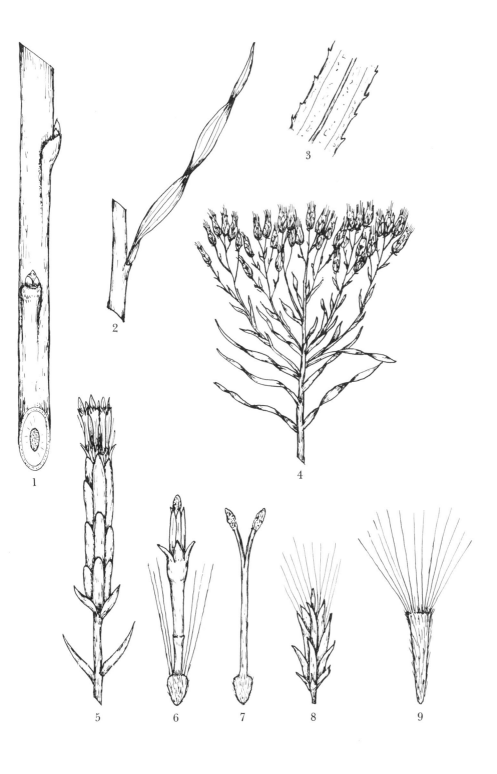

1. Winter twig
2. Leaf
3. Underside of leaf
4. Inflorescence
5. Involucre with flowers
6. One flower
7. Pistil
8. Involucre at seed dispersal time
9. Achene

COMPOSITAE
Baccharis salicina T. & G.

Willow baccharis, groundsel tree

A dense shrub to 1.5 m high, stems erect, clustered.

LEAVES: Alternate, simple. Elliptic to narrowly obovate, 3.5-4.5 cm long, 6-13 mm wide; coarsely 2-3 toothed or entire; tip acute to obtuse, base acuminate; gray-green on both sides, glabrous or with scurfy scales, granular; leaves sessile or on a petiole 1-2 mm long; 3 main veins, the lateral veins often obscure.

FLOWERS: July-August; dioecious. Staminate flowers in terminal panicles 8-15 cm long, 5-8 cm wide, flowers somewhat crowded. Involucre 5 mm long, 3.5 mm wide, 30-40 flowers; bracts imbricate, 2.5-2.6 mm long, 1-1.4 mm wide, greenish with irregular margin; the outer bracts broader and slightly shorter than the inner ones; corolla tubular, white, 3-4 mm long; lobes 5, ovate, acute, 1.5 mm long, strongly recurved; stamens 5, filaments white, adnate to the corolla tube for 2 mm, free for 1.5 mm; anthers 1.5 mm long, yellow, exserted, connate around the vestigial pistil, the pore terminal; pappus of a few kinky hairs 2.5-3 mm long. Pistillate flowers in open terminal panicles 7-16 cm long, 5-8 cm wide; involucre 6-6.3 mm long, 3 mm wide, 25-35 flowers; bracts imbricate, 1.5-6 mm long, 1-1.3 mm wide, greenish, ciliate, outer bracts the smallest; corolla narrowly tubular, 3.5-4 mm long; lobes 5, minute, appressed to the style; ovary cylindric, 0.8-1 mm long, glabrous, ribbed, whitish; style 4 mm long, exserted; stigma lobes 2, linear, divergent, green to reddish; pappus a single ring of straight hairs, 6-8 mm long.

FRUIT: September. Bracts of involucre spreading to release the 15-20 achenes. Achenes cylindric, 1.2-1.8 mm long, 0.5-0.7 mm wide, 10-ridged, stramineous, a small flange at the truncate apex, the base rounded, with a minute tip; pappus a single ring of buffy hairs at the summit, 8.6-11 mm long.

TWIGS: 1-2.5 mm diameter, flexible, erect, glabrous, scurfy, pale brown, ridged; leaf scars narrow, crescent-shaped or broadly V-shaped, enlarged at the center and ends, raised; bundles indistinct; pith white, continuous, one-half of stem. Buds ovoid to globose, 1.2-2 mm long, often with a short stalk, the scales small and scurfy.

TRUNK: Bark light brown, exfoliating into flakes with loose edges. Wood medium hard, white.

HABITAT: Open areas near water, in sandy soil of pastures, lakeshores, sandy flood plains or along stream banks, often in subsaline soil.

RANGE: Texas, New Mexico, Colorado, Kansas.

This is a common plant around lakes and stream flood plains in southwestern Kansas, occasionally at the edge of the water in wet sandy soil. Although it is quite common in low pastures the cattle do not browse heavily on it. In these low areas near water the erect branches furnish good nesting sites for such birds as the red-winged blackbird.

Another species, *B. wrightii* A. Gray, is also found in southwestern Kansas and it is often listed as suffrutescent. In our area it dies to the ground each year, forming a gnarled crown at or just beneath the ground surface. It is a dry prairie plant, 20-30 cm high, and grows on sandy or gravel hillsides.

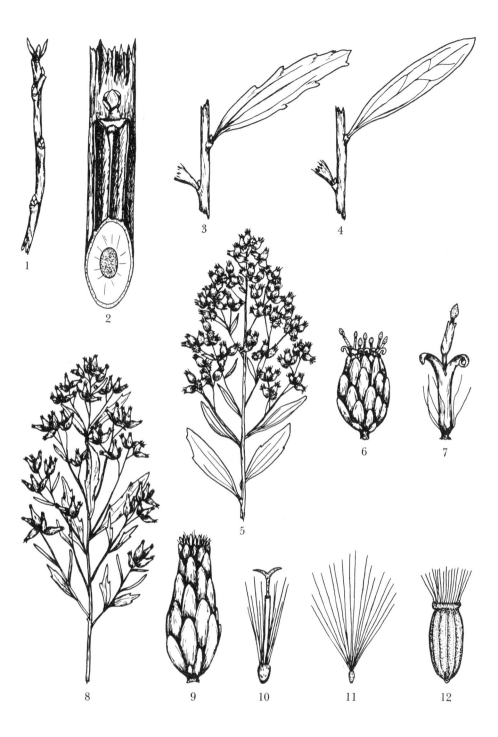

1. Winter twig
2. Detail of twig
3. Typical leaf
4. Leaf variation
5. Staminate inflorescence
6. Staminate involucre
7. Staminate flower

8. Pistillate inflorescence
9. Pistillate involucre
10. Pistillate flower
11. Achene
12. Detail of achene (pappus shortened in drawing)

COMPOSITAE
Hymenoxys scaposa (DC.) Parker

Actinea scaposa (DC.) Kuntze; *Tetraneuris scaposa* (DC.) Greene; *T. linearis* (Nutt.) Greene; *T. fastigiata* Greene; *T. stenophylla* Rydb.

Bitterweed

A suffrutescent shrub to 25 cm high, branched near the ground; new shoots often 20 cm above ground.

LEAVES: Alternate, simple, tardily deciduous, usually clustered near the end of a stem. Narrow, linear or slightly wider toward the apex, 3-6 cm long, 1.5-4 mm wide; margin entire, rolled inward forming a trough; tip acute; the blade tapered toward the base but without a definite petiole, then abruptly expanding to a clasping base 3-4 times as wide as the blade; the broadened base densely, long white ciliate; blade gray-green or yellow-green on both sides, glandular punctate, glabrate or pubescent.

FLOWERS: June-August. Flower heads on terminal, solitary, pubescent scapes 10-20 cm long; involucre 5 mm high, 8 mm wide; bracts 20-30, ovate, 4.5-5 mm long, 2 mm wide, densely pubescent on the outside and margin; ray flowers 10-15, pistillate, fertile; corolla tubular for ·1.5 mm, yellow, hairy, the ligule 7-10 mm long, 4-6 mm wide, obovate, truncate, 3-lobed, pubescent on the outer side, yellow, often with greenish veins; ovary narrowly obconic, 2-2.5 mm long, densely hairy; style 2 mm long, yellow; stigma 2-lobed, flattened, divergent; pappus of 5-6 membranaceous, awn-tipped, obovate scales 2 mm long, 1 mm wide. Disk flowers 25-30, staminate; corolla tubular, 3-4 mm long, abruptly narrowed at the base, yellow, pubescent; lobes 5, papillose; stamens 5, filaments yellow, 0.8-1 mm long, partly connate; anthers 1.8 mm long; pappus as in the ray flowers.

FRUIT: August-September. Involucral bracts spreading; achene obconic, pale tan, 2.5-2.9 mm long, 0.9-1 mm wide, white pubescent, ridged; scale pappus 1.5-2 mm long, including the subulate tip.

TWIGS: 2.5-3 mm diameter, gray-brown, covered by the old leaf bases, often with long silky hairs remaining on the leaf base; pith white, continuous, one-sixth of stem. Buds naked, winter as a minute, green tip on the end of the stem.

TRUNK: Bark gray, flaky, soft and not fibrous; bark of branches rough with old leaf bases. Definitely ligneous, but the wood is soft, brown, with a lighter sapwood.

HABITAT: High, dry hills, rocky soils, pastures, roadsides, wherever the sod is broken but not cultivated, preferably in limestone soils.

RANGE: Utah, to South Dakota, south to Texas, New Mexico, and Colorado.

Hymenoxys scaposa is one of those plants that normally would not be termed a woody plant and a great many of them, perhaps, are not. However, many of our plants reach a height of 25 cm, not including the flowering scape, have secondary thickening of the stem, and the naked wintering buds are produced as high as 20 cm above the ground. These plants have 20-50 ultimate branches with the leaves of the current season densely imbricated around the upper 2-3 cm of each branch. The lower 5-20 cm of the branches and main stem is barren or with the persistent, brown, dried leaf bases of the past few years. The flowering scape is terminal on the new growth and the base of it is surrounded by the leaves of the current season.

This plant hardly fits either the keys or the description of any variety given in recent manuals. Therefore, in this study it has not been given a varietal name.

1. Winter twig
2. Leaves
3. Transverse section of leaf
4. Base of leaf
5. Flowering branches
6. Ray flower
7. Disk flower
8. Achene

COMPOSITAE
Artemisia cana Pursh

Artemisia viscidula (Osterh.) Rydb.;
A. columbiensis Nutt.

Wild sagebrush

A densely branched shrub to 1.2 m high, often branched at ground level.

LEAVES: Alternate, simple, a few leaves remain all winter. Linear to lance-linear, 4-9 cm long, 2-7 mm wide, reduced to bracts in the inflorescence; acute at the tip and tapered to a broadened, sessile base; margin entire; both surfaces densely pubescent, the hairs mostly malpighiaceous; no stipules.

FLOWERS: September; perfect. Inflorescence a panicle 15-30 cm long, terminal on a leafy branch; involucres with 3-10 flowers and subtended by a leaf-like bract; involucre 4-6 mm long, 3-4 mm wide, 8-10 bracts, the outer ones short, ovate, acute, the dorsal side convex and tomentose; middle bracts ciliate; the inner ones narrowly ovate to oblong, nearly flat, glabrous at the base, green in the center, with a wide scarious margin. Flowers yellow, tubular, 4 mm long, not including the protruding style; no calyx; corolla tube 2.5 mm long, abruptly enlarged below the middle, glandular, the 5 small, acute lobes erect, spreading or recurved; stamens 5, alternate with the corolla lobes and inserted near the base of the tube; filament delicate, anthers 1.5 mm long, caudate, longer than the filaments; ovary obconic, 1-1.5 mm long, style 2.5 mm long with 2 divergent, papillate, stigmatic limbs 1 mm long.

FRUIT: October. Involucral bracts spreading to release the achenes; receptacle naked; achene obovoid, 2.5 mm long and 1 mm wide at the truncate outer end; angular with 5-6 heavy ribs, light brown, thin-walled. Seeds obovoid, dark brown, striate, lightly ribbed.

TWIGS: 1-2.1 mm diameter, rigid, white, with a dense, tight coating of malpighiaceous hairs, yellow-brown beneath the hairs; leaf scars indistinct, obscured by leaf base remains; pith white, continuous, one-half of stem. Buds naked, the embryonic leaves densely pubescent.

TRUNK: Bark tan to gray-brown, thin, exfoliating into long, thin, fibrous strips, or into flakes on the younger branches. Wood hard, fine-grained, pale brown, with a wide, yellow sapwood; the rays dark brown and hollow, giving a mottled, porous appearance to the wood.

HABITAT: Valleys, low hillsides, and flat prairies, mainly in sandy soil with more moisture than is customary for the other woody sages of our area.

RANGE: British Columbia, east to Saskatchewan, south through western North Dakota to northwestern Nebraska, Colorado, and New Mexico, north to Utah, and west to northern California.

Artemisia cana is usually found on the flat areas bordering the prairie creeks in dry regions. It is also common on the eroded side of hills but always near the bottom. *A. tridentata* may be higher up on the same hillside, while *A. filifolia* grows chiefly in the dry, sandy areas, either on hills or flats.

A. cana can be identified, generally, in our area, by a combination of 3 characters. First, it is a woody plant; second, the crushed leaves have a definite smell of sage; third, the leaves are somewhat tapered toward both ends and usually 4-5 mm wide at the middle. *A. tridentata* has obovate, truncate, 3-toothed leaves and *A. filifolia* has filiform leaves.

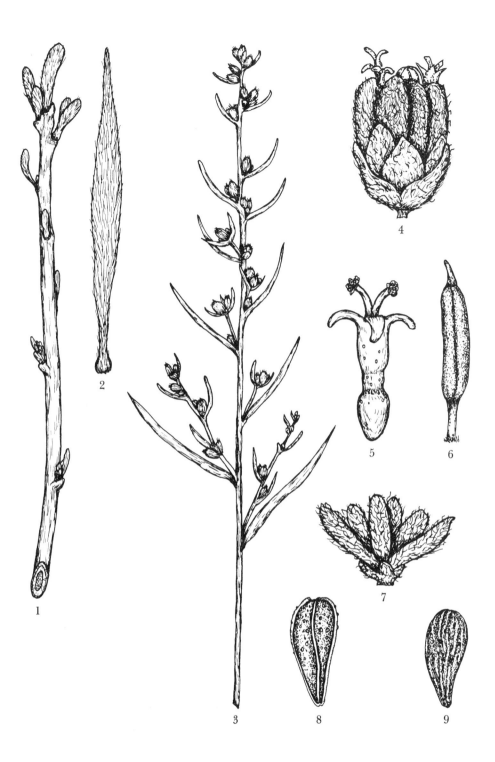

1. Winter twig
2. Leaf
3. Inflorescence
4. Involucre with flowers
5. One flower

6. Stamen
7. Involucre at time of seed dispersal
8. Achene
9. Seed

COMPOSITAE
Artemisia filifolia Torr.

A. plattensis Nutt.

Sandhill sage

A low, dense shrub to 1 m high with branches spreading from ground level.

LEAVES: Alternate, simple, usually clustered. Filiform, 3.3-5.3 cm long; margin entire or with long filiform lobes; gray-green on all sides, appressed pubescent, with 4 lines of white, tomentulose hairs; sessile; many leaves remain on the plant all winter; no stipules.

FLOWERS: Early August. Terminal leafy panicles 15-20 cm long, 6-10 cm wide, with numerous densely flowered branches, the pubescence mainly of malpighiaceous hairs; peduncles 0.6-1 mm long, sharply reflexed. Involucre 1.8 mm long and as wide, the bracts imbricate, 0.9-1.5 mm long, broadly oval, cupped, densely woolly and ciliate; outer flowers 2-4, pistillate, the corolla tubular, narrow and membranaceous, 0.6-0.7 mm long, glandular and minutely lobed; ovary obconic, 0.7-0.8 mm long, ridged on 2 sides, glabrous; style exserted, 0.5 mm long, glabrous, the 2 stigmas linear, divergent, papillose, 0.4 mm long. Inner flowers 3-5, corolla funnel-form, 1.5 mm long, 0.6 mm wide at the top, yellow, puberulent, with 5 acute lobes; stamens 5, anthers nearly sessile, 0.7 mm long, connate and apiculate; pistil vestigial, with erose, stubby stigma lobes.

FRUIT: September-October. Scales of the involucre white woolly, not spreading; achene brown, obovoid, 0.7-1 mm long, 0.4-0.5 mm wide, faintly 4-ribbed, glabrous, no pappus, the floral parts often persistent. Achene wall thin, nearly transparent, the seed brown and smooth. Of the many plants checked in the field, only about 3 percent of the ovules had matured properly.

TWIGS: 1.3 mm diameter, flexible, canescent, the hairs mostly malpighiaceous, becoming glabrate and brown or gray-brown; leaf bases persistent for some time and the leaf scars irregular or indefinite; pith white, continuous, one-third of stem. Buds 1-1.3 mm long, globular, gray-pubescent, the scales broadly ovate, strongly cupped, glabrous inside.

TRUNK: Bark gray-brown, exfoliating into long, thin shreds, or with shallow furrows and broad, flat, loose ridges. Wood hard, brown.

HABITAT: Sandhills, sandy flood plains, pastures, roadsides.

RANGE: South Dakota and Wyoming, southwest to Arizona and northern Mexico, east to western Texas, north through western Oklahoma and Kansas to western Nebraska.

This is the only woody sage of our area that has filiform leaves. It has a more southern range than the other two woody sages and overlaps their range only in western Nebraska and southwestern South Dakota. In local areas it has completely taken over the low sandy hills and replaced the grass. Cattle seldom browse on it. It is a good soil binder and furnishes cover for small mammals and birds. Wood rats commonly build their dens in the large shrubs.

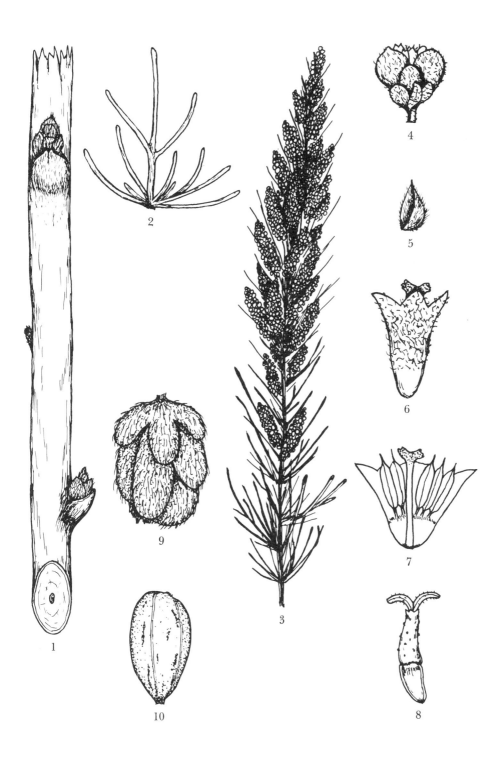

1. Winter twig
2. Cluster of leaves
3. Inflorescence
4. Involucre
5. Involucral bract
6. Staminate flower
7. Cutaway of flower
8. Pistillate flower
9. Involucre at time of seed dispersal
10. Achene

COMPOSITAE
Artemisia tridentata Nutt.

A. vaseyana Rydb., *A. tridentata* var.
angustifolia Gray, *A. angusta* Rydb.

Western sagebrush

A shrub to 1.5 m high, branched at ground level, the branches often growing together, giving the appearance of a trunk 15-18 cm diameter. In our area usually only 20-35 cm high.

LEAVES: Alternate, simple, often overwintering but becoming dry. Wedge-shaped, with the large end out, 10-30 mm long, 3-8 mm wide at the tip, 3 rounded teeth on the truncate end; upper leaves, especially in the inflorescence, entire; tapered to a broad, sessile base; both surfaces densely appressed pubescent, hairs mostly malpighiaceous; the surface glandular beneath the hairs; no stipules.

FLOWERS: September; perfect. Inflorescence a terminal, narrow panicle 5-15 cm long, involucres subtended by leaf-like bracts reduced in size toward the tip of the inflorescence; involucres 2-4 mm high, 1-4 flowers and 8-10 bracts, the outer bracts ovate, short, the dorsal side convex, green and densely pubescent; inner bracts oblong, with a green center and scarious margin, glabrous at the base, flat. Flowers tubular, enlarged about the middle, pale yellow, 2.5-3.5 mm long, glandular, no pappus; lobes 5, triangular, erect or recurved, often tinged with brown; stamens 5, filaments white, inserted near the base of the corolla tube; anthers caudate, yellow, longer than the filament; ovary obconic, 1 mm long, slightly ridged, whitish; style stout, about the same length as the corolla tube, divided into 2, divergent, flattened stigmatic limbs.

FRUIT: October. Involucre spreading to release the achenes, receptacle naked. Achene obovoid, pale brown, flattened, angular, 5-6 low ribs, thin wall, glandular, pubescent, 2-2.5 mm long, 1 mm wide, outer end rounded. Seeds brown, 2 mm long, 1 mm wide, flattened obovate, with low, longitudinal ridges.

TWIGS: 1-1.5 mm diameter, rigid, green, often brown toward the base of the current season growth, densely pubescent; leaf scars concealed by old leaf bases; pith white at first, later brown, continuous, one-half of stem; buds ovoid, about 1 mm long, blunt, densely pubescent.

TRUNK: Bark gray-brown, shredded into long, flat, thin strips, the rays appear as minute elliptical open slits, sometimes corky-filled. Wood medium hard, fine-grained, light brown, with a narrow, yellow sapwood, often indistinguishable on old trunks; rays in the wood dark brown or open, giving a mottled appearance to the wood.

HABITAT: Dry, sandy, or rocky hillsides, usually in the open, seldom on flat plains or in alkaline soil.

RANGE: British Columbia, south to Baja California, east to New Mexico, north to western Nebraska and North Dakota, west to Alberta. Absent or rare in the Black Hills.

This is an easy sage to identify, for nothing else around has small, wedge-shaped leaves with 3 teeth at the apex. Farther west it becomes a dense shrub up to 1.5 meters high and equally as broad, the branches often decumbent at the base.

The foliage is browsed heavily by antelope and deer, the plants often showing the effect. The sage hen eats both the foliage and the flowering or fruiting involucres, as well as utilizing the cover furnished by the plant.

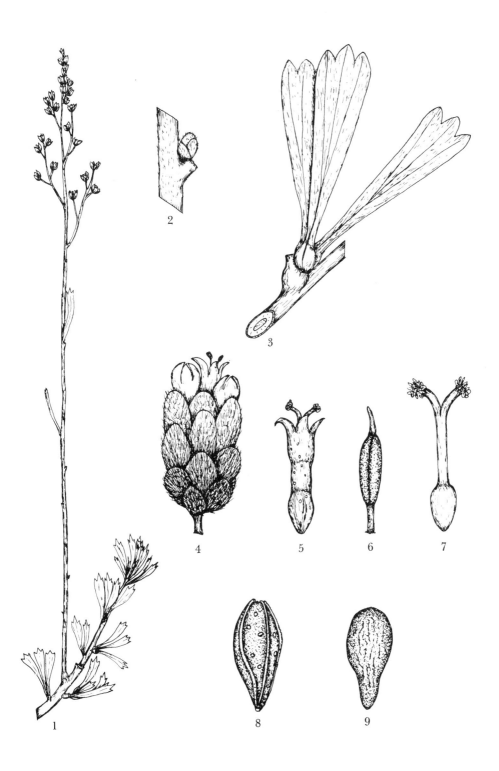

1. Winter twig with old inflorescence
2. Detail of bud
3. Leaves
4. Involucre
5. One flower

6. Stamen
7. Pistil
8. Achene
9. Seed

SPECIES NOT INCLUDED

Acer ginnala Maxim. O. A. Stevens (1963) states that it is planted and can be expected to establish itself. Not known to do so as yet.

Acer platanoides L. Cultivated, rarely escaping.

Acer pseudoplatanus L. Cultivated, rarely escaping.

Acer rubrum L. Often reported for Kansas but is more eastern and southern, and no specimens were found for our area.

Aesculus hippocastanum L. Cultivated, not escaping.

Amelanchier canadensis (L.) Medic. A more eastern species. Our *A. arborea* has been called *A. canadensis* forma *nuda* in some manuals.

Amelanchier interior Nielson. Listed for North Dakota by Fernald (1950), no specimens seen.

Amelanchier laevis Wieg. Listed for Kansas by Rydberg (1932) and Barkley (1968), but is a more eastern species. No specimens seen for our area.

Amelanchier spicata (Lam.) Koch. Listed for Kansas by Barkley (1968). A more eastern species according to Gleason (1963). The name has been misapplied to *A. arborea*.

Aristolochia macrophylla Lam. Listed for Kansas by Rydberg (1932). A more eastern species.

Artemisia bigelovi Gray. Listed for South Dakota, but was misidentified and the name changed later.

Arundinaria gigantea (Walt.) Chapm. Introduced and established in only one known location in Kansas, the colony having been there for years.

Baccharis neglecta Britt. Listed for Kansas by Rydberg (1932). All specimens examined were *B. salicina*.

Berberis aquifolium Pursh. Listed for Nebraska as *Mahonia aquifolium* (Pursh) Nutt. by Rydberg (1932), who also erroneously states that it is synonymous with *B. repens*.

Berberis thunbergii DC. Cultivated, no escaped specimens were seen.

Berberis vulgaris L. Formerly planted and often escaping or persisting, now almost entirely eradicated. No escaped specimens found.

Betula alba L. A European species, planted but not escaping in our area.

Betula caerulea Blanch. Listed for South Dakota by Rydberg (1932). According to Fernald (1950), a hybrid between *B. populifolia* Marsh. and *B. caerulea-grandis* Blanch., neither of which occurs in our area.

Betula cordifolia Regel. Listed for North Dakota by Rydberg (1932). Fernald (1950) places it as a variety of *B. papyrifera* Marsh. No specimens seen.

Betula pendula Roth. Cultivated, not escaping.

Betula sandbergii Britt. A hybrid between *B. papyrifera* Marsh. and *B. glandulosa* Michx. var. *glandulifera* Regel. Occasionally found in North Dakota.

Broussonetia papyrifera (L.) Vent. Introduced and escaping in only one known area.

Bumelia lycioides (L.) Pers. Listed for Kansas by Rydberg (1932). A more southern species.

Carpinus caroliniana Walt. Listed for Kansas by Winter (1936), for Nebraska by Pool (1951) and for South Dakota by Winter and Van Bruggen (1959), but no specimens could be located.

Carya glabra (Mill.) Sweet. Listed for Kansas by Gleason (1963) and Barkley (1968). A more eastern and southeastern species.

Carya ovalis (Wang.) Sarg. Listed for Kansas by Gates (1940) and Barkley (1968), but the specimens examined were *C. texana*. Little (1953) lists it as a synonym of *C. glabra*.

Ceanothus sanguineus Pursh. Listed for

South Dakota by Rydberg (1932), Gleason (1963), and Fernald (1950), but no specimens were found for our area. Winter and Van Bruggen (1959) do not include it.

Cornus rugosa Lam. Listed for North Dakota by Rydberg (1932) as *Svida rugosa* (Lam.) Rydb., but is a more eastern species.

Crataegus douglasii Lindl. Listed for the Dakotas by Hitchcock et al. (1955-1969), but no specimens were found.

Crataegus erythropoda Ashe. Listed for South Dakota by McIntosh (1931), but is a more southwestern species.

Crataegus rotundifolia Moench. O. A. Stevens (1963) groups all the *Crataegus* of North Dakota as this species.

Crataegus succulenta Link var. *neofluvialis* (Ashe) Palmer. Listed for Kansas by Gates (1940) as *C. neofluviatilis* Ashe and as synonymous with *C. succulenta* Link.

Eriogonum corymbosum Benth. Listed for Kansas by Rydberg (1932), but no specimens were seen.

Euonymus americanus L. One specimen was found labeled "Coffey Co. (Kansas), near Leroy, Sept. 1929." Placed here as excluded because of a lack of information on the label, and a search of the area has failed to produce it.

Haplopappus armerioides (Nutt.) Gray. Suffrutescent, could have been included but was omitted because of a lack of fresh specimens of all stages. Specimens were found from Nebraska, South and North Dakota.

Juglans cinerea L. Listed for Nebraska by Pool (1951), for Kansas and North Dakota by Rydberg (1932) and Gleason (1963), for North Dakota by Fernald (1950), but no native specimens could be found.

Koelreuteria paniculata Laxm. Listed by Barkley (1968) as escaping in one location, no specimens seen.

Lonicera hirsuta Eat. var. *interior* Gl. Listed for Nebraska by Fernald (1950), but no specimens were seen.

Morus nigra L. Introduced from Italy and is not hardy in our area. Reports are probably based on the black fruited *M. alba*.

Periploca graeca L. Cultivated, rarely escaping. Listed for Kansas by Gates (1940), Fernald (1950), and Gleason (1963). One specimen seen, but the information was insufficient to include it.

Philadelphus coronarius L. Cultivated, rarely escaping.

Populus canadensis Moench. Listed for Kansas as an escapee, but no specimen was located.

Populus candicans Ait. Cultivated. According to Gleason (1963), a plant of questionable origin and seldom escaping.

Populus grandidentata Michx. Listed for North Dakota by Gleason (1963), no specimen could be located.

Populus nigra L. var. *italica* (Muenchh.) Koehne. Listed by Barkley (1968) as seldom escaping, no specimen found.

Prunus nigra Ait. Listed for South Dakota by Gleason (1963), but no specimen was located. One specimen from North Dakota appears to be this species, but until further collections can be made, must be placed as excluded.

Pyrus angustifolia Ait. Listed for Kansas by Gleason (1963) and Rydberg (1932). A more eastern species, and no specimen was found.

Pyrus coronaria L., incl. var. *dasycalyx* (Rehd.) Fern. Listed for Kansas by Rydberg (1932), Gleason (1963), Fernald (1950), and Barkley (1968). Specimens examined were *P. ioensis*.

Quercus bicolor Willd. Listed for Kansas by Barkley (1968) and Rydberg (1932), and for Nebraska by Fernald (1950) and Pool (1951). Winter (1936) states that it is questionable. Several specimens so labeled were examined, and they were either *Q. muehlenbergii* Engelm. or there was a question about the data.

Quercus coccinea Muenchh. Listed for Nebraska by Rydberg (1932) and Pool (1951). Winter (1936) states this is questionable. A more eastern species and the specimens examined were *Q. velutina* Lam.

Quercus ellipsoidalis E. J. Hill. Listed for Nebraska by Winter (1936). The specimens examined were *Q. velutina* Lam.

Quercus michauxii Nutt. Listed for Kansas by Barkley (1968) as a doubtful species for Kansas. The specimens examined were *Q. muehlenbergii* Engelm. var. *alexanderi* (Britt.) Trel.

Quercus triloba Michx. Listed for Nebraska by Winter (1936), but no specimen seen. Kansas specimens labeled *Q. falcata* Michx. var. *triloba* (Michx.) Nutt. were examined and found to be *Q. velutina* Lam.

Rhamnus caroliniana Walt. Listed for

Kansas by Rydberg (1932). Cultivated in our area, but no escaped specimens were found.

Rhus trilobata Nutt. Appears to be a species of the Rocky Mountains, but additional work needs to be done on the *aromatica-trilobata* complex.

Rhus typhina L. Cultivated, seldom escaping. No records for our area.

Ribes aureum Pursh. Listed for Nebraska by Winter (1936) and for South Dakota by Rydberg (1932). All specimens examined from our area are here placed under *R. odoratum,* although some of the more western specimens tend toward *R. aureum,* but the distinguishing characteristics were not constant.

Ribes grossularia L. Cultivated, seldom escaping but often persisting.

Ribes hudsonianum Richards. Listed for South Dakota. One specimen was located but was too incomplete for positive identification.

Ribes vulgare Lam. Cultivated, seldom escaping.

Robinia hispida L. Cultivated, rarely escaping in our area.

Rosa eglanteria L. Cultivated, seldom escaping.

Rosa micrantha Sm. Cultivated, rarely escaping.

Rosa palustris Marsh. Listed for Kansas by Barkley (1968). Specimens examined were *R. suffulta* Greene, *R. carolina* L., or *R. foliolosa* Nutt.

Rosa pimpinellifolia L. Listed by Gates (1940) as escaping in Kansas; no specimens were located.

Rosa rubiginosa L. Synonymous with *R. eglanteria* L.

Rosa rudiuscula Greene. Listed for Kansas by Gates (1940). Steyermark (1963) states that the original description was based on a specimen intermediate between *R. carolina* L. and *R. arkansana* Porter var. *suffulta* (Greene) Cockerell, and suggests the name be abandoned.

Rosa spinosissima L. Cultivated. Listed by Barkley (1968) as escaping in Kansas.

Rubus hispidus L. Listed for Kansas by Rydberg (1932), but is a more eastern species.

Rubus idaeus L. var. *canadensis* Richards. Listed for South Dakota by Fernald (1950), but no specimen was located.

Rubus mollior Bailey. Listed for Kansas by Fernald (1950). The species is somewhat questionable and is prob-

ably synonymous with *R. ostryifolius* Rydb.

Rubus odoratus L. Listed as sometimes escaping in Kansas by Barkley (1968), and by Rydberg (1932) as *Rubacer odoratum* (L.) Rydb. as escaping in Kansas.

Salix babylonica L. Cultivated, rarely escaping.

Salix drummondiana Barratt in Hook. var. *subcoerulea* (Piper) Ball. Listed by Winter and Van Bruggen (1959) for South Dakota, but Froiland (1962) does not mention it. No specimens were found.

Salix fragilis L. Cultivated, occasionally escaping.

Salix geyeriana Anderss. Listed for South Dakota by Rydberg (1932). A more western species, and no specimens were located.

Salix glauca L. Listed for South Dakota by Rydberg (1932). A more northwestern species, and no specimens were found for our area.

Salix gracilis Anderss. var. *tectoris* Fern. Listed for Nebraska by Steyermark (1963), but no specimens so labeled were seen. If this variety is synonymous with *S. petiolaris,* then it is present.

Salix lasiandra Benth. var. *caudata* (Nutt.) Sudw. (*Salix pentandra* L. var. *caudata* Nutt.). Cultivated, rarely found as an escapee, known from only two locations, one in the Black Hills and one in Kansas.

Salix prinoides Pursh. Listed for South Dakota by Rydberg (1932). Possibly a variety of *S. discolor.*

Salix purpurea L. Cultivated, escapes rarely.

Sambucus microbotrys Rydb. Listed for South Dakota by Rydberg (1932). It is synonymous with *S. racemosa* L. ssp. *pubens* (Michx.) House var. *microbotrys* (Rydb.) Kearney and Peebles and is a species of the western mountains.

Smilax glauca Walt. Listed for Nebraska by Rydberg (1932) and for Kansas by Barkley (1968). Specimens examined were either *S. hispida* Muhl. or *S. bona-nox* L. *S. glauca* is a southeastern species and does not enter our area.

Smilax pseudo-china L. Listed for Kansas by Gates (1940), but is a more southeastern species.

Smilax rotundifolia L. Listed for Kansas by Rydberg (1932) and Barkley

(1968). The specimens examined were all *S. hispida* Muhl., probably misidentified because of the flattened spines, which is also a characteristic of *S. hispida* in our area.

Sorbaria sorbifolia (L.) A. Br. Introduced and escaping in eastern South Dakota, according to Winter and Van Bruggen (1959).

Sorbus aucuparia L. Listed by Barkley (1968) as escaping in Kansas, but no specimens were seen.

Spiraea densiflora Nutt. Listed for South Dakota by Rydberg (1932) and McIntosh (1931). A northwestern species, and those examined were *S. betulifolia* Pall. var. *lucida* (Dougl.) C. L. Hitchc.

Spiraea japonica L. f. Listed by Barkley (1968) as escaping in Kansas. No specimens were located.

Spiraea latifolia (Ait.) Borkh. Listed for Nebraska by Rydberg (1932). A more eastern species, and no specimens were found for our area.

Spiraea salicifolia L. Listed as an escapee in Kansas by Rydberg (1932), but no specimens were seen.

Spiraea tomentosa L. Listed for Kansas by Rydberg (1932). An eastern species, and no specimens were located.

Toxicodendron toxicarium (Salisb.) Gillis. Cited by Gillis (Rhodora 73: 408), for Kansas, but the specimen was not located.

Vaccinium myrtilloides Michx. Synonymous with *Cyanococcus canadense* Kalm, and listed for Kansas by Rydberg (1932). It is a northern and eastern species, and apparently has not been located in our area.

Vitis rotundifolia Michx. Listed by Rydberg (1932) as *Muscadina rotundifolia* (Michx.) Small for Kansas. It is a southern species, and not found in our area.

Yucca baccata Torr. Listed for Kansas by Rydberg (1932), but it is a southwestern species and does not occur in this area.

Yucca filamentosa L. Cultivated, may escape or persist, but most of our specimens so labelled are probably *Y. smalliana* Fern.

Ziziphus jujuba Mill. Introduced and listed as well established in a Research Reserve in southeastern Kansas.

SELECTED REFERENCES

Barkley, T. M. 1968. *A Manual of the Flowering Plants of Kansas.* The Kansas State University Endowment Association, Manhattan, Kansas.

Benson, Lyman, and Robert A. Darrow. 1954. *The Trees and Shrubs of the Southwestern Deserts.* The University of New Mexico Press, Albuquerque.

Correll, Donovan Stewart, and Marshall Conring Johnston. 1970. *Manual of the Vascular Plants of Texas.* Texas Research Foundation, Renner, Texas.

Fernald, Merritt Lyndon. 1950. *Gray's Manual of Botany,* 8th edition. American Book Company, New York.

Froiland, Sven G. 1962. *The Genus Salix (Willows) in the Black Hills of South Dakota.* U. S. Department of Agriculture, Forest Service, Washington, D. C.

Gates, Frank C. 1940. *Annotated List of the Plants of Kansas.* Kansas Academy of Sciences. Printed by the Kansas State Printing Plant, Topeka.

Gillis, William T. 1971. "The Systematics and Ecology of Poison Ivy and the Poison Oaks (Toxicodendron, Anacardiaceae)." *Rhodora* 73: 72-159, 161-237, 370-443, 465-540.

Gleason, Henry A., and Arthur Cronquist. 1963. *Manual of Vascular Plants of Northeastern United States and Adjacent Canada.* D. Van Nostrand Co., Inc., New York.

Harrington, H. D. 1954. *Manual of the Plants of Colorado.* Sage Books, Denver.

Hitchcock, C. Leo, Arthur Cronquist, Marion Ownbey, and J. W. Thompson. 1955-1969. *Vascular Plants of the Pacific Northwest.* University of Washington Press, Seattle.

Kearney, Thomas H., and Robert H. Peebles. 1960. *Arizona Flora,* 2nd edition. University of California Press, Berkeley.

Little, Elbert L., Jr. 1953. *Check List of Native and Naturalized Trees of the United States.* United States Forest Service, Washington, D. C.

McIntosh, Arthur C. 1931. "A Botanical Survey of the Black Hills of South Dakota." *Black Hills Engineer,* South Dakota School of Mines, Rapid City.

Pool, Raymond J. 1951. *Handbook of Nebraska Trees.* University of Nebraska. Lincoln.

Rosendahl, Carl Otto. 1955. *Trees and Shrubs of the Upper Midwest.* University of Minnesota Press, Minneapolis.

Rydberg, Per Axel. 1932. *Flora of the Prairies and Plains of Central North America.* New York Botanical Garden, New York.

Stevens, Orin Alva. 1963. *Handbook of North Dakota Plants.* North Dakota Institute for Regional Studies, Fargo.

Steyermark, Julian A. 1963. *Flora of Missouri.* Iowa State University Press, Ames.

Van Haverbeke, David F. 1968. *A Population Analysis of Juniperus in the Missouri River Basin.* University of Nebraska, Lincoln.

Waterfall, U. T. 1969. *Keys to the Flora of Oklahoma,* 4th edition. Oklahoma State University, Stillwater.

Winter, John Mack. 1936. *An Analysis of the Flowering Plants of Nebraska.* University of Nebraska, Lincoln.

Winter, John Mack, Clara K. Winter, and Theodore Van Bruggen. 1959. *A Checklist of the Vascular Plants of South Dakota.* University of South Dakota, Vermillion. (Mimeographed.)

GLOSSARY

Abaxial. Dorsal side; away from the axis of the organ; the basal side of a pine-cone scale.

Abortive. Not developed.

Accrescent. Enlarged in age from the usual size.

Achene. A dry, indehiscent, one-locular fruit.

Acicular. Needle-like, slender and rounded in section.

Acuminate. Gradually tapered to a point.

Acute. Tapered sharply to a point.

Adaxial. Ventral side; toward the axis of the organ; the apical side of a pine-cone scale.

Adnate. United with a different kind of organ.

Aggregate fruit. A fleshy fruit formed from one flower, the several pistils becoming fleshy and on a common receptacle, as in the blackberry.

Alternate. At different levels, such as buds on some stems; located between organs, as opposed to in front of the organ.

Ament. Catkin.

Anastomose. To connect by cross ridges or veins.

Anthesis. The time of flowering or maturation of floral organs.

Antrorse. Directed toward the apex.

Apical. Toward the apex.

Apiculate. With a short, definite point.

Appressed. Lying close to the surface of another structure.

Attenuate. Tapered to a long slender point, usually used to describe a leaf base.

Axis. The central, elongated part to which organs are attached.

Barbed. Rigid points, directed either forward (antrorse) or backward (retrorse).

Barbellate. Minutely barbed.

Berry. A pulpy fruit containing several seeds.

Bipinnate. Twice pinnate.

Biserrate. Double toothed.

Blade. The wide portion of a leaf.

Bloom. A whitish, powdery covering on some fruits.

Bole. The unbranched trunk of a tree.

Bract. A small, leaf-like structure, usually subtending an organ.

Bristle. A stiff hair, but not hard enough to be a spine.

Caducous. Early deciduous.

Caespitose. Low and mat- or tuft-forming.

Callous. Having a hardened texture, obviously different from the rest of the organ.

Campanulate. Bell-shaped.

Canescent. Gray pubescent.

Capitate. Like a ball or disk of greater diameter than the supporting structure.

Capsule. A dry, dehiscent fruit with more than one carpel.

Castaneous. Chestnut-brown color.

Catkin. A spike-like inflorescence with scaly bracts, usually flexuous and pendulous.

Cauline. Pertaining to the stem.

Cilia. Marginal hairs.

Claw. The abruptly narrowed, basal portion of some petals.

Coaetaneous. Expanding at the same time.

Compressed. Somewhat flattened from the sides.

Connate. Joined; two like structures fused.

Cordate. Heart-shaped.

Coriaceous. Leathery.

Corymb. A broad inflorescence with flat or convex top; a modified raceme, flowering from the outside inward.

Cucullate. With a hood or an incurved apex.

Cuneate. Wedge-shaped, with the point downward.

Cyme. A broad, flat-topped inflorescence, flowering from the center outward.

Decumbent. The main stem reclining above but not on the ground, the branches ascending.

Decurrent. Extending downward, as one structure extending downward onto another.

Decussate. Having opposite leaves, but the pairs alternating at right angles to those above or below.

Dehiscent. Opening in some manner.

Deltoid. The shape of an equilateral triangle.

Dentate. With pointed teeth extending forward.

Depressed. Somewhat flattened from the apex.

Dichotomous. Branching in pairs.

Dioecious. With the staminate and pistillate flowers on separate plants.

Disk. A fleshy development of the receptacle, nectaries, or staminodes around the pistil.

Distinct. Separate; not joined.

Dorsal. Relating to the back, or away from the axis; abaxial.

Downy. With soft, short hairs.

Drupe. A fleshy fruit containing a hard stone.

Ellipsoid. Solid with an elliptical outline.

Elliptical. Widest at the middle and rounded equally toward both ends, the width about one-half the length.

Emarginate. With a shallow notch at the apex.

Entire. Without marginal teeth or lobes.

Erose. With irregular or jagged margin.

Exfoliating. Peeling in thin layers.

Exserted. Projecting beyond, as stamens extending beyond a tubular corolla.

Falcate. Flat, tapered and curved.

Filiform. Thread-like.

Fimbriate. Fringed.

Flabellate. Fan-shaped.

Floccose. Having soft hairs in patches over the surface.

Floricane. The flowering, second-year development of a primocane, as in *Rubus*.

Foliaceous. Leaf-like in appearance.

Fruticose. Shrubby, the branches woody.

Fulvous. Light red-brown; tawny.

Fusiform. Widest at the middle and tapered toward both ends.

Gibbous. Swollen on one side.

Glabrate. Becoming glabrous, used to indicate a few scattered hairs.

Glabrous. Without hairs.

Glandular punctate. With minute glands hardly raised above the surface.

Glaucous. Whitened with a bloom, often waxy.

Hilum. The scar of attachment on a seed.

Hirsute. Pubescent with short, stiff hairs.

Hyaline. Transparent or translucent.

Hypanthium. An enlarged structure under the calyx; often used to include such things as the undeveloped fruit and calyx.

Imbricate. Overlapping.

Incised. Sharply and irregularly cut.

Included. Not extended beyond the enclosing organ.

Indehiscent. Remaining permanently closed; fruits which do not split open.

Inferior ovary. An ovary below the attachment of the perianth and stamens.

Inserted. Attached to another structure.

Introrse. Facing inward.

Involucre. The group of bracts surrounding a flower or head of flowers, as in the composites and hickories.

Keel. A central, dorsal ridge.

Laciniate. Irregularly torn or shredded.

Lanceolate. Long and narrow, with a broad base, and evenly tapered to the top.

Lateral. Pertaining to the sides.

Leaflet. A leaf-like division of a compound leaf.

Lenticel. Corky spots on bark, corresponding to a stoma.

Ligneous. Woody.

Linear. Long and narrow, with nearly parallel sides.

Lobe. An extended section of an organ, as the lobe of a leaf.

Locule. A cavity, usually of the ovary or anther.

Malpighiaceous hairs. Straight hairs attached at the middle, with both ends free.

Marcescent. Withered but remaining attached.

Midrib. The central rib or vein of a leaf.

Monoecious. With stamens and pistils in separate flowers but on the same plant.

Mucronate. Tipped with a small, abrupt point, an extension of the midrib.

Multiple fruit. A fleshy fruit formed from several flowers on a common axis as in the mulberry.

Muricate. Rough, with short, firm, sharp or rounded points.

Nut. A hard, 1-seeded, indehiscent fruit; not to be confused with the term acorn, which includes the involucre.

Obconical. Conical, with the apical end large.

Oblanceolate. Lanceolate, with the wide end outward.

Oblate. A sphere flattened at the ends like a tangerine.

Oblique. With unequal, slanting sides.

Oblong. Somewhat rectangular, the sides nearly parallel.

Obovate. Egg-shaped in outline, with the large end out, used for a plane surface.

Obovoid. An egg-shaped solid, with the large end out.

Oval. Broadly elliptical.

Ovate. Egg-shaped in outline, with the small end out.

Ovoid. An egg-shaped solid, with the small end out.

Palmate. Radiate from a common center.

Panicle. A branched and rebranched inflorescence, each flower with a pedicel.

Papilionaceous flower. A zygomorphic flower, with a posterior standard petal, two lateral petals, and two keel petals.

Pedicel. The stalk of a single flower.

Peduncle. A leafless axis bearing several flowers.

Peltate. With the petiole or stalk (of a scale) fastened near the center.

Pendent. Hanging or drooping; pendulous.

Perfect. Having both stamens and pistil in one flower.

Perfoliate. With a base extending around the stem.

Perianth. The calyx and corolla, that which encloses the essential floral parts.

Persistent. Remaining over a long period.

Petal. A division of the corolla, whether distinct or connate.

Petiole. The stalk of a leaf; the portion below the lowest leaflets of a compound leaf.

Petiolule. The stalk of a leaflet.

Phyllary. A bract of the involucre in the composites.

Pilose. Having soft, long, straight hairs.

Pinna. One division of a bipinnately compound leaf.

Pistil. The ovule bearing organ of a flower.

Pistillate. Bearing pistils.

Pith. The spongy center of a stem.

Plicate. Folded lengthwise.

Polygamous. Having unisexual and bisexual flowers on the same plant.

Porrect. Directed outward at nearly right angles.

Precocious. Developing early, such as the flower appearing before the leaves.

Prickle. A small, slender, sharp outgrowth from the bark.

Primocane. A vegetative, first-year cane, as in the blackberry.

Procumbent. Trailing on the ground but not rooting at the nodes.

Prostrate. Flat on the ground.

Pruinose. Having a bloom, a gray waxy covering.

Puberulent. With very short, fine hairs.

Pubescent. Covered with hairs, a general term.

Pulvinus. A swelling at the base of a petiole.

Punctate. With minute holes, as if pricked by a pin.

Pyriform. Pear-shaped.

Quadrate. Somewhat square.

Raceme. An inflorescence consisting of a single axis and pedicelled flowers.

Racemose. Resembling a raceme.

Rachis. The axis of an inflorescence; the axis of a pinnately compound leaf above the lowest leaflets, corresponding to a midrib.

Radiate. Diverging from a central point.

Raphe. The ridge or adnate funicle on the ventral side of a seed.

Ray. A branch of an umbel; a medullary ray in woody stems; the outer, strapshaped corolla of some composites.

Receptacle. The expanded end of a stem which bears the organs of the flower or the fruits.

Recurved. Curved downward.

Reflexed. Abruptly bent downward.

Reniform. Kidney-shaped.

Repent. Creeping and rooting at the nodes.

Reticulate. With outstanding veins or anastomosing lines.

Retrorse. Directed toward the base.

Retuse. With a small notch at the apex.

Revolute. With the margin rolled toward the underside.

Rhizome. A horizontal, underground stem rooting at the nodes, with shoots ascending above the ground.

Rhombic. Diamond-shaped.

Rib. The primary vein of a leaf; midrib.

Rostrate. Having a beak.

Rosulate. Rosette-like.

Rudimentary. Undeveloped.

Rufous. Red-brown.

Rugose. A wrinkled appearance.

Sagittate. Arrowhead-shaped, with basal lobes turned outward or backward.

Samara. A winged, indehiscent fruit.

Scabrous. Covered with harsh hairs that feel rough when rubbed.

Scalariform. Ladder-like.

Scarious. Thin, dry, and membranaceous.

Scurfy. Covered with minute flakes.

Sepal. A division of a calyx; calyx lobe.

Sericeous. Covered with long, fine, appressed, silky hairs.

Serrate. With sharp, forward-pointing teeth.

Sessile. Without a stalk.

Sinuate. Having a wavy margin, too shallow to be lobed.

Sinus. The angle between two lobes or teeth.

Spatulate. Narrow, wider at the outer end.

Spike. An inflorescence with sessile flowers on an elongated axis.

Stamen. The pollen-bearing organ of a flower.

Staminode. A sterile structure corresponding to a stamen.

Stellate. Star-shaped; used to describe a radiately branched hair.

Stigma. The portion of the pistil which receives the pollen.

Stipel. Structure at the base of a petiolule.

Stipule. A small structure at the base of a leaf.

Stramineous. Straw-colored.

Striate. With longitudinal streaks or fine ridges.

Strict. Definitely upright.

Strigose. With appressed, straight, stiff hairs.

Strobile. A reproductive organ with imbricated bracts or scales, as a pine cone.

Style. The portion of a pistil between stigma and ovary.

Stylopodium. A disk- or dome-like structure at the base of the style.

Subulate. Awl-shaped.

Succulent. Fleshy and juicy.

Suffruticose. Woody only near the base; suffrutescent.

Suture. A junction, usually a line of opening.

Tendril. A slender, twining structure from a leaf or stem.

Tepal. A sepal or petal when the two are not readily differentiated.

Terete. Circular in section.

Thorn. A coarse, long spine.

Thyrse. An ovoid, compact panicle, as in the lilac or grape.

Tomentose. Densely covered with long matted hairs.

Trailing. Prostrate without rooting.

Trifoliolate. Having 3 leaflets; not to be confused with trifoliate, which means three leaves.

Truncate. As if cut off transversely, or squared off.

Tuberculate. Bearing small, usually rounded, processes.

Tumid. Swollen.

Umbel. An inflorescence with the peduncles or pedicels coming from one level.

Umbo. A protuberance on the exposed surface (apophysis) of a scale on the pine cone.

Undulate. With the margin wavy up and down.

Valvate. Having valves; equal structures, such as bud scales, meeting along the edges but not overlapping.

Veins. Loosely used for the fibrovascular tissue of a leaf.

Velutinous. Velvety.

Ventral. The inner surface of an organ; adaxial.

Vestigial. Rudimentary and not functional.

Villous. With long, soft hairs but not interwoven; villose.

Viscid. Sticky.

Whorl. With 3 or more leaves at the same level on the stem.

Wing. An extension of the margin.

Woolly. Covered with long matted hairs.

INDEX

The italicized words are synonyms.